Basic and Clinical Science Course

Fundamentals and Principles of Ophthalmology

Section 2

2001–2002

AF838625

THE FOUNDATION
OF THE AMERICAN ACADEMY
OF OPHTHALMOLOGY

LEO
LIFELONG
EDUCATION FOR THE
OPHTHALMOLOGIST®

LEO

The Basic and Clinical Science Course is one component of the Lifelong Education for the Ophthalmologist (LEO) framework, which assists members in planning their continuing medical education. LEO includes an array of clinical education products that members may select to form individualized, self-directed learning plans for updating their clinical knowledge. Active members or fellows who use LEO components may accumulate sufficient CME credits to earn the LEO Award. Contact the Academy's Clinical Education Division for further information on LEO.

The American Academy of Ophthalmology is accredited by the Accreditation Council for Continuing Medical Education to provide continuing medical education for physicians.

The American Academy of Ophthalmology designates this CME activity for up to 40 hours in category 1 credit toward the AMA Physician's Recognition Award. Each physician should claim only those hours of credit that he/she has actually spent in the educational activity.

The Academy provides this material for educational purposes only. It is not intended to represent the only or best method or procedure in every case, nor to replace a physician's own judgment or give specific advice for case management. Including all indications, contraindications, side effects, and alternative agents for each drug or treatment is beyond the scope of this material. All information and recommendations should be verified, prior to use, with current information included in the manufacturers' package inserts or other independent sources, and considered in light of the patient's condition and history. Reference to certain drugs, instruments, and other products in this publication is made for illustrative purposes only and is not intended to constitute an endorsement of such. Some material may include information on applications that are not considered community standard, that reflect indications not included in approved FDA labeling, or that are approved for use only in restricted research settings. The FDA has stated that it is the responsibility of the physician to determine the FDA status of each drug or device he or she wishes to use, and to use them with appropriate patient consent in compliance with applicable law. The Academy specifically disclaims any and all liability for injury or other damages of any kind, from negligence or otherwise, for any and all claims that may arise from the use of any recommendations or other information contained herein.

The authors state the following financial relationships:
Prof. Bron: consultant, Allergan; Novartis/CIBA; and Eli Lilley

The other authors state that they have no significant financial interest or other relationship with the manufacturer of any commercial product discussed in the chapters that they contributed to this publication or with the manufacturer of any competing commercial product.

Copyright © 2001
The Foundation of the American Academy of Ophthalmology
All rights reserved
Printed in the United States of America

Basic and Clinical Science Course

Thomas J. Liesegang, MD, Jacksonville, Florida
Senior Secretary for Clinical Education

Thomas A. Deutsch, MD, Chicago, Illinois
Secretary for Instruction

M. Gilbert Grand, MD, St. Louis, Missouri
BCSC Course Chair

Section 2

Faculty Responsible for This Edition

Gerhard W. Cibis, MD, *Chair,* Kansas City, Missouri

Ata A. Abdel-Latif, PhD, Augusta, Georgia

Anthony J. Bron, BSc, FRCS, FRCOphth, Oxford, England

K.V. Chalam, MD, Jacksonville, Florida

Brenda J. Tripathi, PhD, Columbia, South Carolina

Ramesh C. Tripathi, MD, PhD, Columbia, South Carolina

Janey Wiggs, MD, PhD, Boston, Massachusetts

Aazy A. Aaby, MD, Portland, Oregon
Practicing Ophthalmologists Advisory Committee for Education

Recent Past Faculty

Robert E. Anderson, MD, PhD
H. Dwight Cavanagh, MD, PhD
Emily Y. Chew, MD
Richard B. Einaugler, MD
Frederick L. Ferris III, MD
Gerald A. Fishman, MD
Mitchell H. Friedlaender, MD
Edward K. Isbey, Jr, MD
Henry J. Kaplan, MD
Randy H. Kardon, MD, PhD
Sidney Lerman, MD
Richard A. Lewis, MD
Joel S. Mindel, MD
Kenneth H. Musson, MD
Paul F. Palmberg, MD, PhD
James M. Richard, MD
Morton E. Smith, MD
Elise Torczynski, MD
F.J.G.M. van Kuijk, MD, PhD
David L. Verlee, MD
Stephen R. Waltman, MD
Martin Wand, MD
Thomas A. Weingeist, MD, PhD
Richard G. Weleber, MD
Norman F. Woodlief, MD

In addition, the Academy gratefully acknowledges the contributions of numerous past faculty and advisory committee members who have played an important role in the development of previous editions of the Basic and Clinical Science Course.

American Academy of Ophthalmology Staff

Kathryn A. Hecht, EdD
Vice President, Clinical Education

Hal Straus
Director, Publications Department

Margaret Denny
Managing Editor

Jack Daniel
Medical Editor

Maxine Garrett
Administrative Coordinator

The Foundation of the American Academy of Ophthalmology
655 Beach Street
Box 7424
San Francisco, CA 94120-7424

CONTENTS

GENERAL INTRODUCTION

The Basic and Clinical Science Course (BCSC) is designed to meet the needs of residents and practitioners for a comprehensive yet concise curriculum of the field of ophthalmology. The BCSC has developed from its original brief outline format, which relied heavily on outside readings, to a more convenient and educationally useful self-contained text. The Academy regularly updates and revises the course, with the goals of integrating the basic science and clinical practice of ophthalmology and of keeping ophthalmologists current with new developments in the various subspecialties.

The BCSC incorporates the effort and expertise of more than 70 ophthalmologists, organized into 12 section faculties, working with Academy editorial staff. In addition, the course continues to benefit from many lasting contributions made by the faculties of previous editions. Members of the Academy's Practicing Ophthalmologists Advisory Committee for Education serve on each faculty and, as a group, review every volume before and after major revisions.

Organization of the Course

The 12 sections of the Basic and Clinical Science Course are numbered as follows to reflect a logical order of study, proceeding from fundamental subjects to anatomic subdivisions:

1. Update on General Medicine
2. Fundamentals and Principles of Ophthalmology
3. Optics, Refraction, and Contact Lenses
4. Ophthalmic Pathology and Intraocular Tumors
5. Neuro-Ophthalmology
6. Pediatric Ophthalmology and Strabismus
7. Orbit, Eyelids, and Lacrimal System
8. External Disease and Cornea
9. Intraocular Inflammation and Uveitis
10. Glaucoma
11. Lens and Cataract
12. Retina and Vitreous

In addition, a comprehensive Master Index allows the reader to easily locate subjects throughout the entire series.

References

Readers who wish to explore specific topics in greater detail may consult the journal references cited within each chapter and the Basic Texts listed at the back of the book. These references are intended to be selective rather than exhaustive, chosen by the BCSC faculty as being important, current, and readily available to residents and practitioners.

Related Academy educational materials are also listed in the appropriate sections. They include books, audiovisual materials, self-assessment programs, clinical modules, and interactive programs.

Study Questions and CME Credit

Each volume of the BCSC is designed as an independent study activity for ophthalmology residents and practitioners. The learning objectives for this volume are stated on the facing page. The text, illustrations, and references provide the information necessary to achieve the objectives; while the study questions allow readers to test their understanding of the material and their mastery of the objectives. Further, physicians who wish to claim CME credit for this educational activity must complete the study questions and submit the answers, together with the signed Credit Reporting Form and Section Evaluation (these forms are located at the end of the book). Requests for CME credit must be submitted within 3 years of the date of purchase.

Conclusion

The Basic and Clinical Science Course has expanded greatly over the years, with the addition of much new text and numerous illustrations. Recent editions have sought to place a greater emphasis on clinical applicability, while maintaining a solid foundation in basic science. As with any educational program, it reflects the experience of its authors. As its faculties change and as medicine progresses, new viewpoints are always emerging on controversial subjects and techniques. Not all alternate approaches can be included in this series; as with any educational endeavor, the learner should seek additional sources, including such carefully balanced opinions as the Academy's Preferred Practice Patterns.

The BCSC faculty and staff are continuously striving to improve the educational usefulness of the course; you, the reader, can contribute to this ongoing process. If you have any suggestions or questions about the series, please do not hesitate to contact the faculty or the managing editor.

The authors, editors, and reviewers hope that your study of the BCSC will be of lasting value and that each section will serve as a practical resource for quality patient care.

OBJECTIVES FOR BCSC SECTION 2

Upon completion of BCSC Section 2, *Fundamentals and Principles of Ophthalmology,* the reader should be able to:

- □ Identify the bones making up the orbital walls and the orbital foramina
- □ Identify the origin and pathways of cranial nerves I–VII
- □ Identify the origin and insertions of the extraocular muscles and use CT and MRI studies to point out the extraocular muscles, optic nerve, and lacrimal gland in axial and coronal views of the orbit
- □ Describe the distribution of the arterial and venous circulations of the orbit and optic nerve
- □ Summarize the structural-functional relationships of the outflow pathways for aqueous humor of the eye
- □ Delineate the events of early embryogenesis that are important for the subsequent development of the eye and orbit
- □ Identify the roles of growth factors, homeobox genes, and neural crest cells in the genesis of the eye
- □ Describe the sequence of events in the differentiation of the ocular tissues during embryonic and fetal development of the eye
- □ Recognize and characterize congenital anomalies of the eye that arise as a result of genetic factors or environmental effects during development
- □ Review the stages in the development of the eye and the correlation between congenital ocular disorders and the timing of an insult to the embryo
- □ Appreciate how the human genome is organized and the role of genetic mutations in health and disease
- □ Explain how DNA can be manipulated in the laboratory to map and to clone genes, identify genes from surrounding DNA, and create transgenic and knockout animals
- □ Demonstrate how appropriate diagnosis and management of genetic diseases can lead to better patient care
- □ Assess the role of the ophthalmologist in the provision of genetic counseling
- □ Identify the biochemical composition of the various parts of the eye and its secretions
- □ Review new concepts regarding the interaction between membrane proteins and G proteins and how this affects ocular functions, such as rhodopsin with transducin in the conversion of "light-stimulus" to "electric-signal"
- □ Discuss the biochemical derangements in diabetes and how they lead to its ocular complications, such as diabetic retinopathy and cataract formation
- □ List the varied functions of the retinal pigment epithelium such as phagocytosis and vitamin A metabolism and their relationship to retinal diseases
- □ Summarize the role of free radicals and antioxidants
- □ Recognize the features of the eye that facilitate or impede drug delivery
- □ Cite the basic principles underlying the use of autonomic therapeutic agents in a variety of ocular conditions

- □ List the indications, contraindications, mechanisms of action, and side effects of various drugs in the management of glaucoma
- □ Describe the mechanisms of action of antibiotics, antivirals, and antifungal medications: their indications, dosages, and side effects
- □ Discuss the anesthetic agents used in ophthalmology, their dosages and adverse effects
- □ Recognize therapeutic drugs on the horizon and in the process of being introduced into clinical practice in the immediate future

PART 1

ANATOMY

The classic reference books listed below are especially useful in the study of ophthalmic anatomy. Rather than being cited repeatedly in the text, these works are listed here for the reader's general reference. Other references are interspersed throughout the text where relevant.

Beard C, Quickert MH. *Anatomy of the Orbit: A Dissection Manual.* 3rd ed. Birmingham, AL: Aesculapius Publishing Co; 1988.

Duke-Elder S, ed. *System of Ophthalmology.* Vol II. *The Anatomy of the Visual System.* St Louis: Mosby; 1976.

Dutton JJ. *Atlas of Clinical and Surgical Orbital Anatomy.* Philadelphia: Saunders; 1994.

Fine BS, Yanoff M. *Ocular Histology: A Text and Atlas.* 2nd ed. Hagerstown, MD: Harper & Row; 1979.

Hogan MJ, Alvarado JA, Weddell JE. *Histology of the Human Eye.* Philadelphia: Saunders; 1971.

Mausolf FA. *The Anatomy of the Ocular Adnexa.* Springfield, IL: Charles C Thomas; 1975.

Miller NR, Newman NJ, eds. *Walsh and Hoyt's Clinical Neuro-Ophthalmology.* 5th ed. Baltimore: Williams & Wilkins; 1997.

Reeh MJ, Wobig JL, Wirtschafter JD. *Ophthalmic Anatomy.* San Francisco: American Academy of Ophthalmology; 1981.

Snell RS, Lemp MA. *Clinical Anatomy of the Eye.* Boston: Blackwell Scientific Publications; 1989.

Tasman W, Jaeger EA, eds. *Duane's Clinical Ophthalmology.* Philadelphia: Lippincott; 1994.

Zide BM, Jelks GW. *Surgical Anatomy of the Orbit.* New York: Raven Press; 1985.

CHAPTER I

Orbit and Ocular Adnexa

Orbital Anatomy

Periorbital Sinuses

The eyes lie within two bony orbits, located on either side of the root of the nose. Each orbit is pear-shaped, with the optic nerve representing the stem. The medial walls of the orbits are almost parallel. They border the nasal cavity anteriorly and the ethmoidal air cells and the sphenoid sinus posteriorly. In the adult, the lateral wall of each orbit forms an angle of approximately 45° with the medial plane. The lateral walls border the middle cranial, temporal, and pterygopalatine fossae. Superior to the orbit are the anterior cranial fossa and the frontal and supraorbital sinus. The maxillary sinus and the palatine air cells are located inferiorly.

The *periorbital sinuses* offer a route for the spread of infection. Mucoceles occasionally arise from the sinuses and may confuse the clinician in the differential diagnosis of orbital tumors. The locations of the paranasal air sinuses and their relation to anatomical features of the skull are shown in Figures I-1 and I-2. Figure I-1 also shows the distribution of pain originating from sinusitis.

BCSC Section 7, *Orbit, Eyelids, and Lacrimal System.*

Doxanas MT, Anderson RL. *Clinical Orbital Anatomy.* Baltimore: Williams & Wilkins; 1984:232.

Orbital Volume

The volume of each adult orbit is slightly less than 30 cc. The orbital entrance averages about 35 mm in height and 45 mm in width. The maximum width is about 1 cm (behind the anterior orbital margin). In adults, the depth of the orbit varies from 40 to 45 mm from the orbital entrance to the orbital apex. Both race and sex affect each of these measurements.

Bony Orbit

Seven bones make up the bony orbit (Fig I-3):

- Frontal
- Zygomatic
- Maxillary
- Ethmoidal
- Sphenoid
- Lacrimal
- Palatine

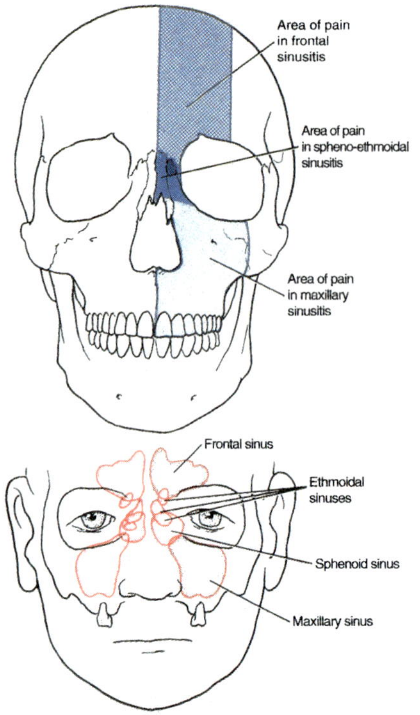

FIG I-1—**Top**, Bones of the face, showing regions where pain is experienced in sinusitis. **Bottom**, Positions of paranasal sinuses relative to the face. (Reproduced with permission from Snell RS, Lemp MA. *Clinical Anatomy of the Eye.* Boston: Blackwell; 1989.)

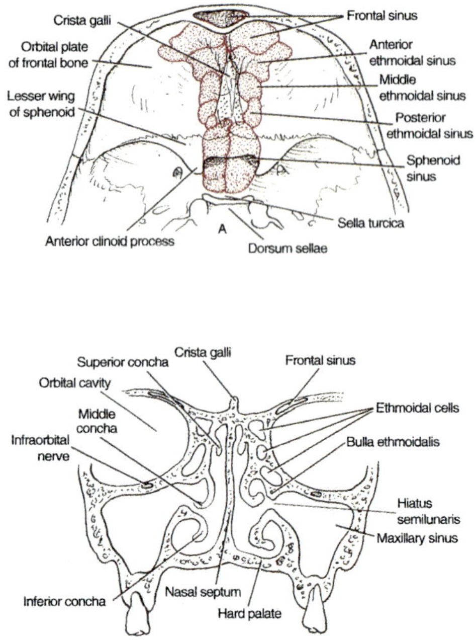

FIG I-2—**Top**, Position of the paranasal sinuses relative to the anterior cranial fossa, in axial view. **Bottom**, Coronal section through the nasal cavity, showing the ethmoidal and maxillary sinuses. (Reproduced with permission from Snell RS, Lemp MA. *Clinical Anatomy of the Eye.* Boston: Blackwell; 1989.)

Orbital Margin

The *orbital margin* forms a quadrilateral spiral (Fig I-4), whose superior margin is formed by the frontal bone, which is interrupted medially by the *supraorbital notch.* The medial margin is formed above by the frontal bone and below by the *posterior lacrimal crest* of the lacrimal bone and the *anterior lacrimal crest* of the maxillary bone. The inferior margin derives from the maxillary and zygomatic bones. Laterally, the zygomatic and frontal bones complete the rim.

Orbital Roof

The *orbital roof* is formed from both the orbital plate of the frontal bone and the lesser wing of the sphenoid bone (Fig I-5). The fossa for the lacrimal gland, lying anterolaterally behind the zygomatic process of the frontal bone, resides within the orbital roof. Medially, *the fovea trochlearis,* located on the frontal bone approximately 4

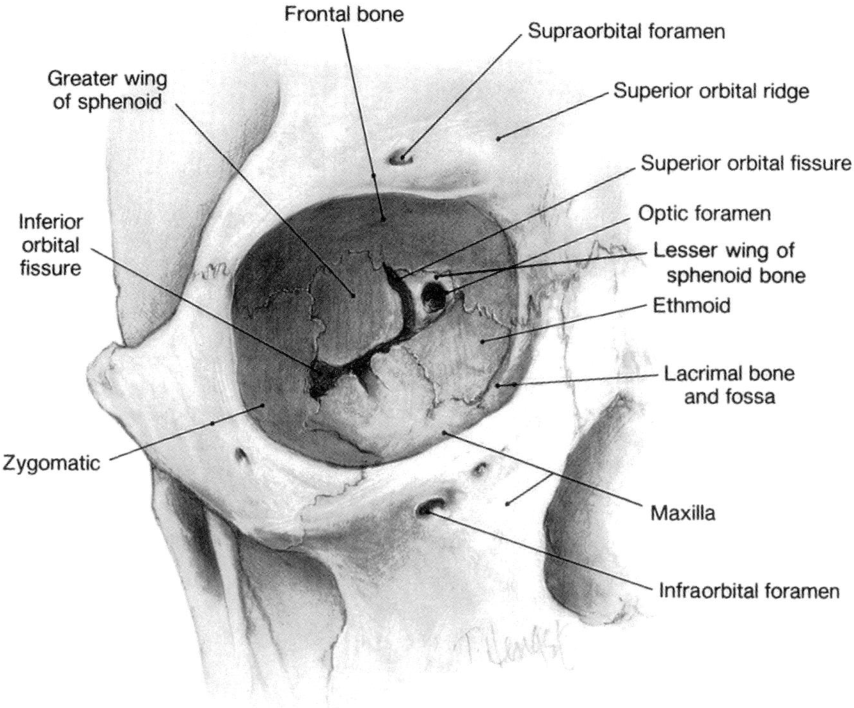

FIG I-3—Frontal view of bony right orbit. (Reproduced with permission from Doxanas MT, Anderson RL. *Clinical Orbital Anatomy.* Baltimore: Williams & Wilkins; 1984.)

mm from the orbital margin, forms the pulley of the superior oblique muscle where the trochlea, a curved plate of hyaline cartilage, is attached.

Helveston EM, Merriam WW, Ellis FD, et al. The trochlea. A study of the anatomy and physiology. *Ophthalmology.* 1982;89:124–133.

Medial Orbital Wall

The *medial wall of the orbit* is formed from four bones (Fig I-6):

- Frontal process of the maxilla
- Lacrimal bone
- Orbital plate of the ethmoid
- Lesser wing of the sphenoid

The ethmoidal bone makes up the largest portion of the medial wall. The lacrimal fossa is formed by the frontal process of the maxillary and the lacrimal bone. Below, the lacrimal fossa is continuous with the bony nasolacrimal canal, which extends into the inferior meatus of the nose. The paper-thin structure of the medial wall is reflected in its name, *lamina papyracea.*

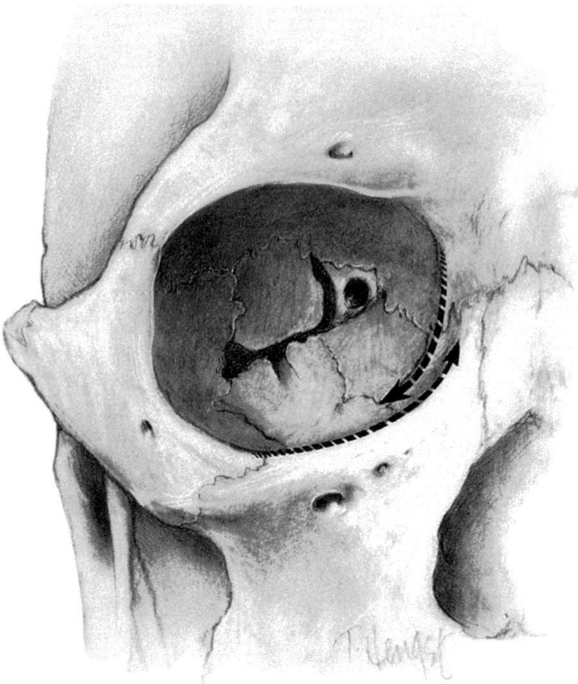

FIG I-4—Right orbital margin. Note the relationship between the anterior lacrimal crest of the maxillary bone and posterior lacrimal crest of the lacrimal bone. (Reproduced with permission from Doxanas MT, Anderson RL. *Clinical Orbital Anatomy.* Baltimore: Williams & Wilkins; 1984.)

Orbital Floor

The *floor of the orbit,* which is the roof of the maxillary antrum, or sinus, is composed of three bones (Fig I-7):

- Maxillary
- Palatine
- Orbital plate of the zygomatic

The *infraorbital groove* traverses the floor and descends anteriorly into a canal. It exits as the *infraorbital foramen* below the orbital margin of the maxillary bone. Arising from the floor of the orbit just lateral to the opening of the nasolacrimal canal is the *inferior oblique muscle,* the only extraocular muscle that does not originate from the orbital apex. The floor of the orbit slopes downward approximately 20° from posterior to anterior.

Blunt trauma to the soft tissues of the orbit may cause dehiscence of the fragile bony floor. Clinical features of such a blowout fracture may include diplopia, enophthalmos, hypoesthesia in the distribution of the infraorbital nerve, entrapment of orbital tissues, a positive forced duction test, and radiographic evidence of a fluid

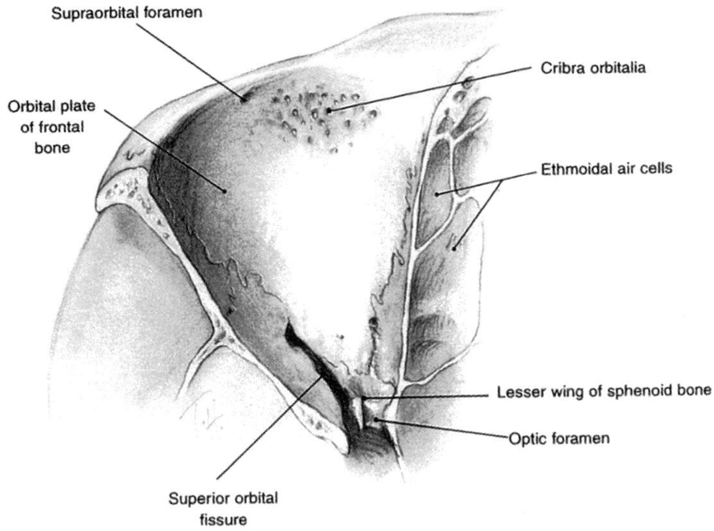

FIG I-5—View from below looking up into orbital roof (superior orbital wall). (Reproduced with permission from Doxanas MT, Anderson RL. *Clinical Orbital Anatomy.* Baltimore: Williams & Wilkins; 1984.)

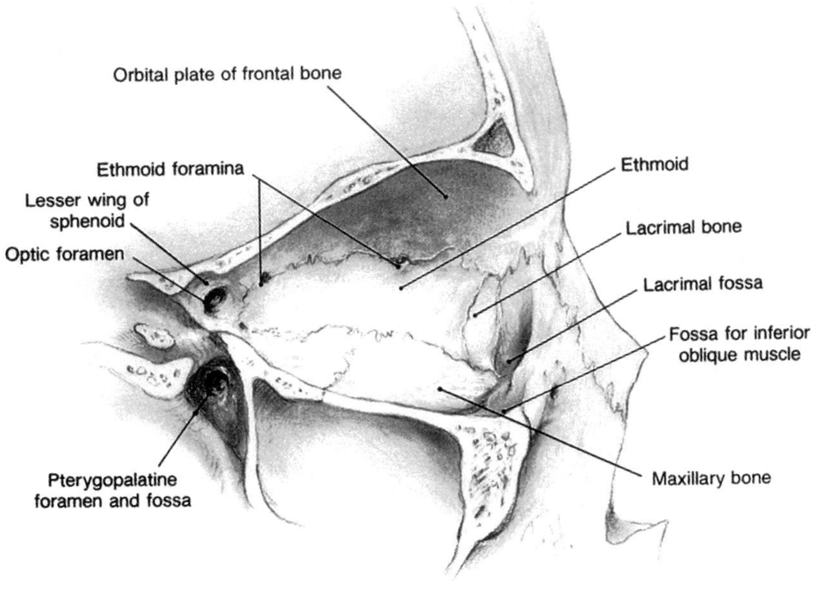

FIG I-6—Right medial orbital wall as viewed from lateral side. (Reproduced with permission from Doxanas MT, Anderson RL. *Clinical Orbital Anatomy.* Baltimore: Williams & Wilkins; 1984.)

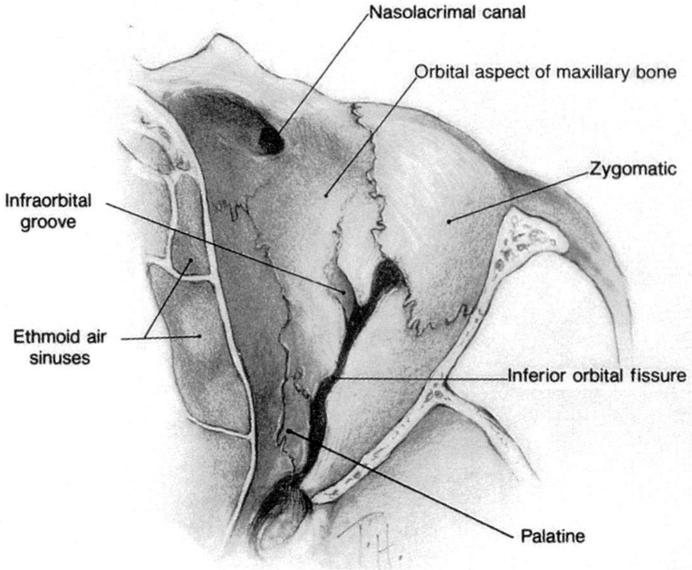

FIG I-7—Right orbital floor and inferior orbital fissure. (Reproduced with permission from Doxanas MT, Anderson RL. *Clinical Orbital Anatomy.* Baltimore: Williams & Wilkins; 1984.)

level or cloudiness in the maxillary sinus. (For more information on such fractures, see chapter VI in BCSC Section 7, *Orbit, Eyelids, and Lacrimal System.*)

Lateral Orbital Wall

The *lateral wall of the orbit* is the thickest and strongest of the orbital walls and is formed from two bones (Fig I-8):

- Zygomatic
- Greater wing of the sphenoid

The lateral orbital tubercle *(Whitnall's tubercle)*, a small elevation of the orbital margin of the zygomatic bone, lies approximately 11 mm below the frontozygomatic suture. This important landmark is the site of attachment for the following:

- Check ligament of the lateral rectus muscle
- Suspensory ligament of the eyeball
- Lateral palpebral ligament
- Aponeurosis of the levator muscle

Orbital Foramina, Ducts, Canals, and Fissures

Foramina The *optic foramen* leads from the middle cranial fossa to the apex of the orbit. It is directed forward, laterally, and somewhat downward and conducts the optic nerve, the ophthalmic artery, and sympathetic fibers from the carotid plexus

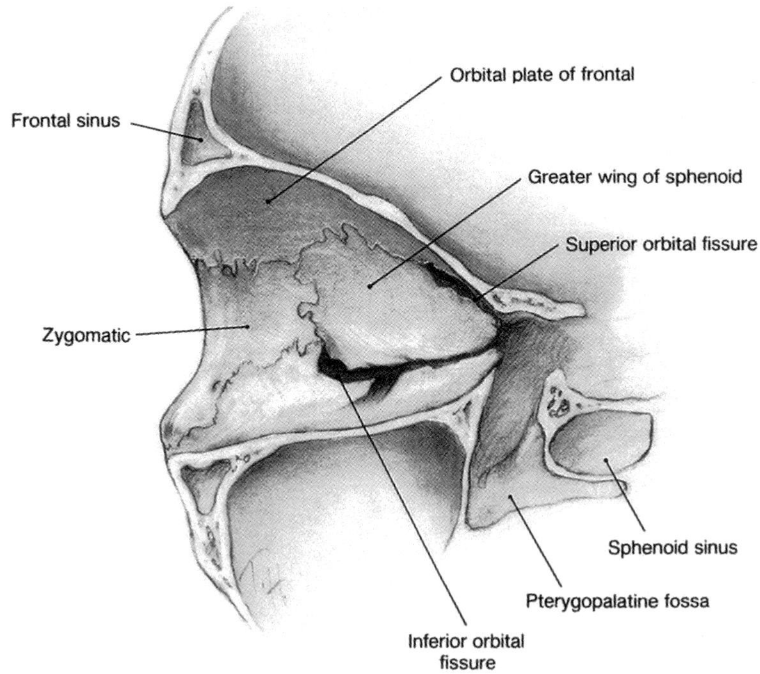

FIG I-8—Right lateral orbital wall as viewed from medial side. (Reproduced with permission from Doxanas MT, Anderson RL. *Clinical Orbital Anatomy.* Baltimore: Williams & Wilkins; 1984.)

(Figs I-9, I-10). The optic foramen passes through the lesser wing of the sphenoid bone. The *supraorbital foramen,* or notch, is located at the medial third of the superior margin of the orbit. It transmits blood vessels and the supraorbital nerve, which is a branch of the ophthalmic division (V_1) of cranial nerve V (trigeminal). The *anterior ethmoidal foramen* is located at the frontoethmoidal suture and transmits the anterior ethmoidal vessels and nerve. The *posterior ethmoidal foramen* lies at the junction of the roof and the medial wall of the orbit and transmits the posterior ethmoidal vessels and nerve through the frontal bone. The *zygomatic foramen* lies in the lateral aspect of the zygomatic bone and contains zygomaticofacial and zygomaticotemporal branches of the zygomatic nerve and the zygomatic artery.

Nasolacrimal duct The *nasolacrimal duct* travels inferiorly from the lacrimal fossa into the *inferior meatus* of the nose.

Infraorbital canal The *infraorbital canal* continues anteriorly from the infraorbital groove and exits 4 mm below the inferior orbital margin, where it transmits the infraorbital nerve, which is a branch of V_2 (the maxillary division of cranial nerve V).

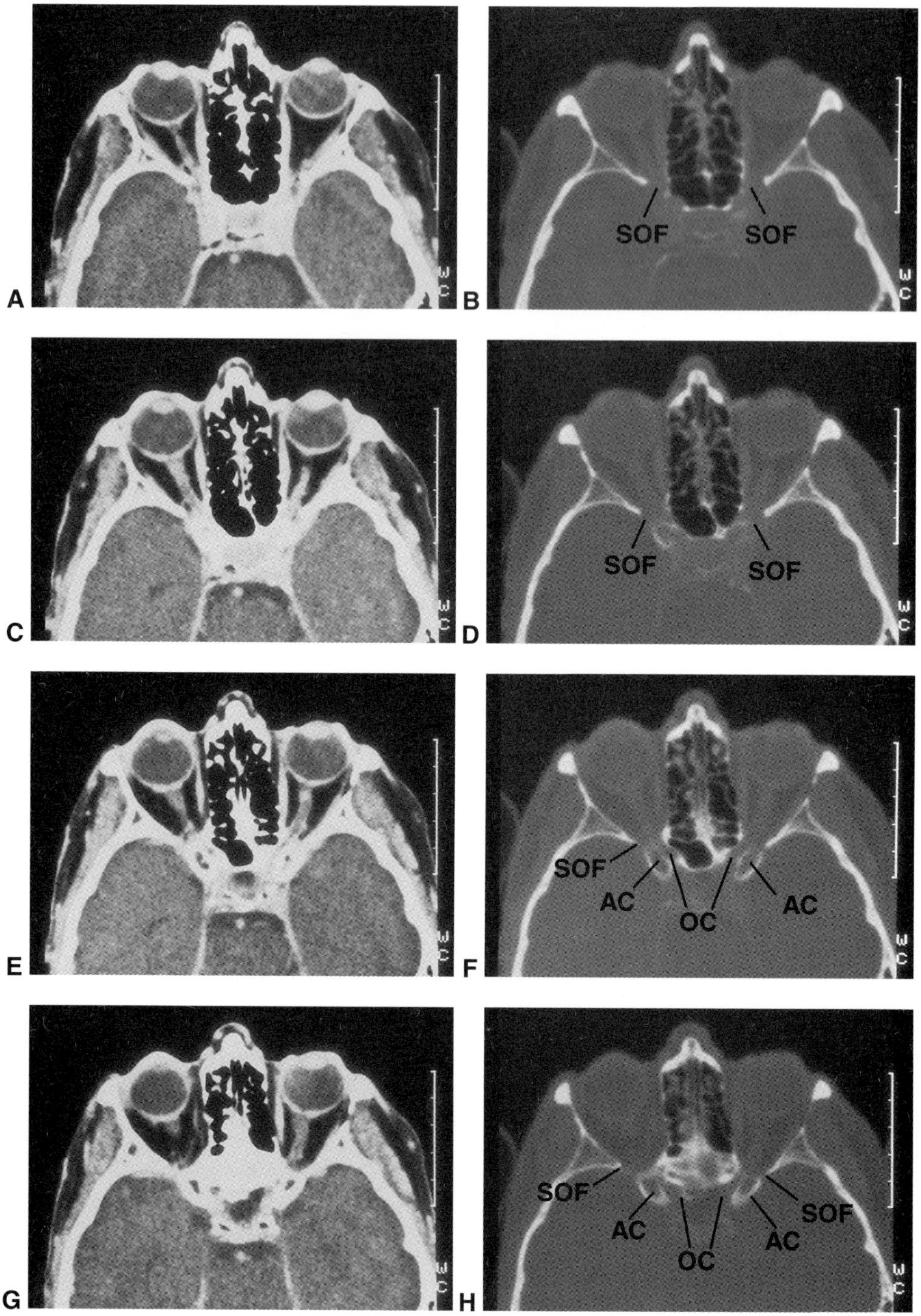

FIG I-9—Series of axial computed tomography scans. Each compares tissue and corresponding bone-window density through the optic canal *(OC)* and superior orbital fissure *(SOF)*. The superior orbital fissure passes above and below the plane of the optic canal and is commonly mistaken for the optic canal. The optic canal lies in the same plane as the anterior clinoid processes *(AC)* and may be cut obliquely in scans so that the entire canal length does not always appear in one section. Four different planes of section are shown in this series: **A–B**, Plane one is below the canal; **C–D**, Plane two is just under the canal; **E–F**, Plane three is at the canal; **G–H**, Plane four is just at the top of and above the canal.

Fissures The *superior orbital fissure* is located between the greater and the lesser wings of the sphenoid bone and lies below and lateral to the optic foramen. It is approximately 22 mm long and is spanned by the common tendinous ring of the rectus muscles *(annulus of Zinn)*. Above the ring, the superior orbital fissure transmits:

- Lacrimal nerve of V_1
- Frontal nerve of V_1
- Cranial nerve IV

Within the ring or between the two heads of the rectus muscle are the following:

- Superior and inferior divisions of cranial nerve III (oculomotor)
- Nasociliary branch of cranial nerve V_1
- Sympathetic roots of the ciliary ganglion
- Cranial nerve VI (abducens)
- Superior ophthalmic vein

Occasionally, below the ring:

- Inferior ophthalmic vein

The *inferior orbital fissure* lies just below the superior fissure between the lateral wall and the floor of the orbit, giving access to the pterygopalatine and inferotemporal fossae. Hence, it is close to the foramen rotundum and pterygoid canal. It transmits the infraorbital and zygomatic branches of cranial nerve V_2, an orbital nerve from the pterygopalatine ganglion, and the inferior ophthalmic vein. The inferior orbital vein connects with the pterygoid plexus before the vein enters the cavernous sinus.

Cranial Nerves

Six of the twelve cranial nerves (CN II–VII) directly innervate the eye and periocular tissues. Because certain tumors affecting cranial nerve I (olfactory) can give rise to important ophthalmic signs and symptoms, familiarity with the anatomy of this nerve is also important for the ophthalmologist. (Chapter IV discusses the central and peripheral connections of cranial nerves I–VII.)

Ciliary Ganglion

The *ciliary ganglion* is located approximately 1 cm in front of the annulus of Zinn, on the *lateral* side of the ophthalmic artery between the optic nerve and the lateral rectus muscle (Figs I-11, I-12). It receives three roots:

- A long *sensory root* arises from the nasociliary branch of cranial nerve V_1. It is 10–12 mm long and contains sensory fibers from the cornea, the iris, and the ciliary body.
- A short *motor root* arises from the inferior division of cranial nerve III, which also supplies the inferior oblique muscle. The fibers of the motor root synapse in the ganglion, and the postganglionic fibers carry *parasympathetic axons* to supply the iris sphincter.
- The *sympathetic root* comes from the plexus around the internal carotid artery. It enters the orbit through the superior orbital fissure within the tendinous ring, *passes through the ciliary ganglion without synapse,* and innervates ocular blood vessels and possibly the dilator muscle.

A

B

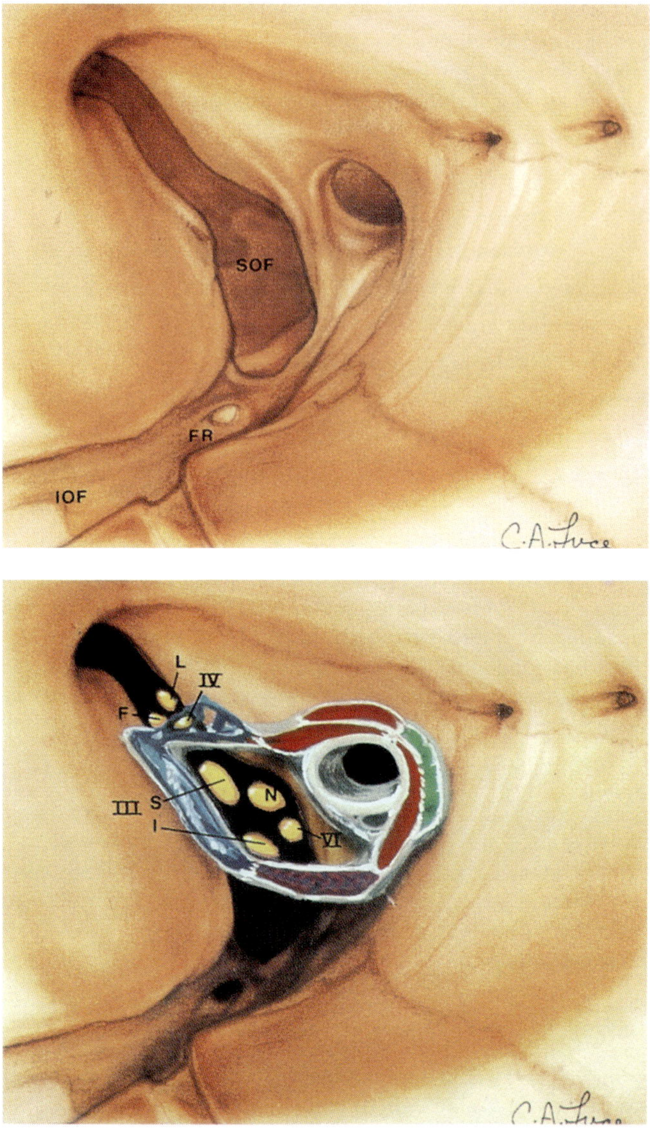

FIG I-10—**A**, The superior orbital fissure *(SOF)* widens medially where it lies below the level of the optic foramen. Its total length is 22 mm. Note the foramen rotundum *(FR)* just inferior to the confluence between the SOF and inferior orbital fissure *(IOF)*. **B**, The common tendinous ring, or annulus of Zinn, divides the SOF. The extraocular muscles arise from this common ring. The portion of the annulus that is formed by the origin of the lateral rectus muscle *(blue)* divides the SOF into two compartments. The area encircled by the annulus is termed the oculomotor foramen, which opens into the middle cranial fossa and transmits cranial nerve III *(III)*, superior *(S)* and inferior *(I)* divisions; cranial nerve VI *(VI)*; the nasociliary *(N)* branch of cranial nerve V_1; ophthalmic veins; the orbital branch of the middle meningeal artery (occasionally); and sympathetic nerve fibers. Above the annulus, note cranial nerve IV *(IV)* and the frontal *(F)* and lacrimal *(L)* branches of cranial nerve V_1 and cranial nerve IV. It is important to realize that the frontal and lacrimal branches of the ophthalmic division of cranial nerve V and cranial nerve IV *enter the orbit outside the muscle cone.* (**A** and **B** reproduced with permission from Zide BM, Jelks GW. *Surgical Anatomy of the Orbit.* New York: Raven; 1985.)

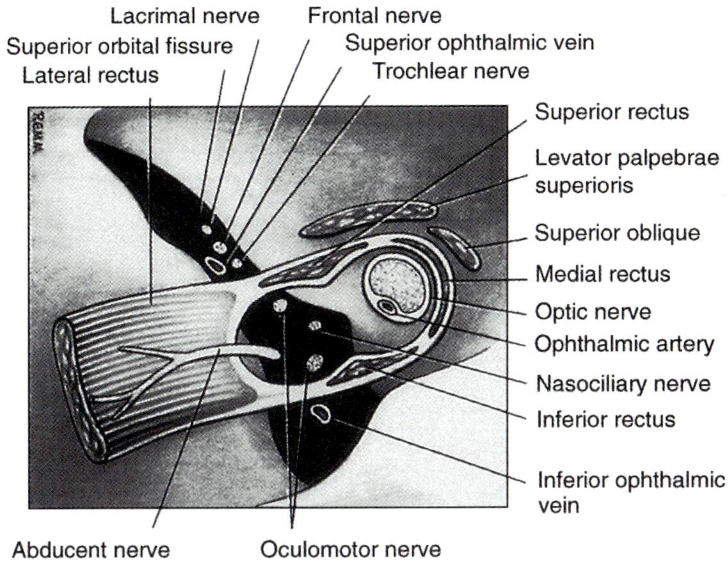

C

FIG I-10 (cont.)—**C**, Structures passing through the superior orbital fissure. (From Bron AJ, Tripathi RC, Tripathi BJ. *Wolff's Anatomy of the Eye and Orbit.* 8th ed. London, New York: Chapman & Hall; 1997.)

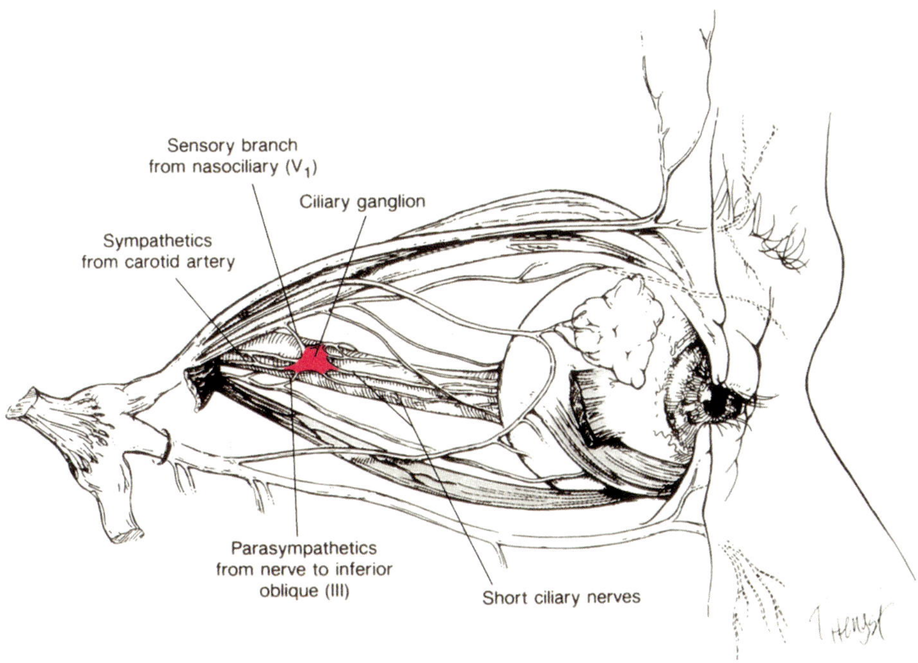

FIG I-11—Contributions to the ciliary ganglion. (Reproduced with permission from Doxanas MT, Anderson RL. *Clinical Orbital Anatomy.* Baltimore: Williams & Wilkins; 1984.)

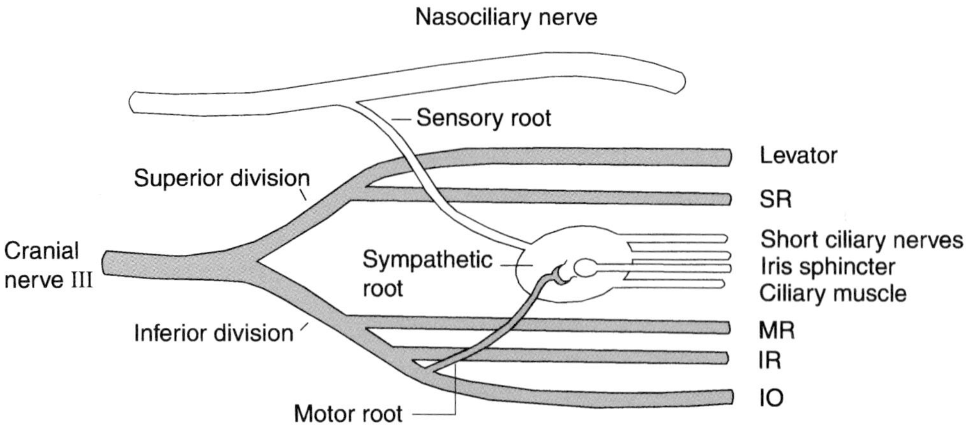

FIG I-12—Cranial nerve III and ciliary ganglion. *SR,* superior rectus; *MR,* medial rectus; *IR,* inferior rectus; *IO,* inferior oblique. (Illustration by Sylvia Barker.)

Branches of the Ciliary Ganglion

Only the parasympathetic fibers synapse in the ciliary ganglion. The sympathetic fibers are postganglionic from the superior cervical ganglion and pass through it without synapse. Sensory fibers from cell bodies in the trigeminal ganglion carry sensation from the eye, orbit, and face. Together, the nonsynapsing sympathetic fibers, the sensory fibers, and the myelinated, fast-conducting postganglionic parasympathetic fibers form the short ciliary nerves.

Short Ciliary Nerves

Two groups, totaling six to ten short ciliary nerves, arise from the ciliary ganglion. They travel on both sides of the optic nerve and, together with the long ciliary nerves, pierce the sclera around the optic nerve. They pass anteriorly between the choroid and the sclera into the ciliary muscle, where they form a plexus that supplies the cornea, the ciliary body, and the iris.

Extraocular Muscles

Knowing the location, origin, and insertion of the extraocular muscles is important to understanding the action of these muscles on the globe. These features are shown in Figures I-13 through I-16.

Extraocular Muscle Insertions

The four rectus muscles insert anteriorly on the globe along the *spiral of Tillaux* (Fig I-17). The medial rectus muscle inserts nearest to the limbus, and the superior rectus muscle inserts farthest from the limbus (Table I-1). The relationship between the muscle insertions and the ora serrata is clinically important. A misdirected suture passed through the insertion of the superior rectus muscle could perforate the retina.

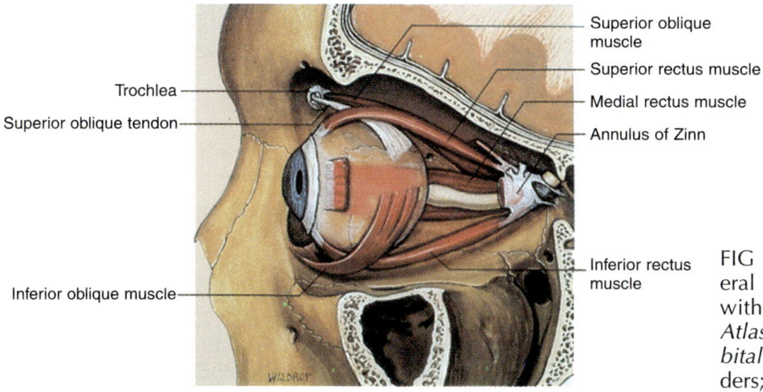

FIG I-13—Extraocular muscles, lateral composite view. (Reproduced with permission from Dutton JJ. *Atlas of Clinical and Surgical Orbital Anatomy.* Philadelphia: Saunders; 1994.)

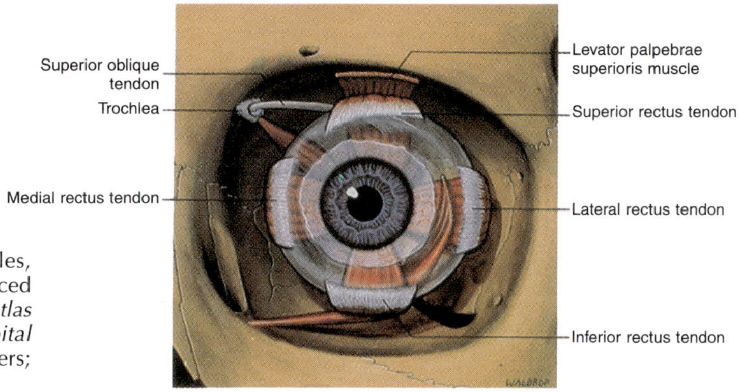

FIG I-14—Extraocular muscles, frontal composite view. (Reproduced with permission from Dutton JJ. *Atlas of Clinical and Surgical Orbital Anatomy.* Philadelphia: Saunders; 1994.)

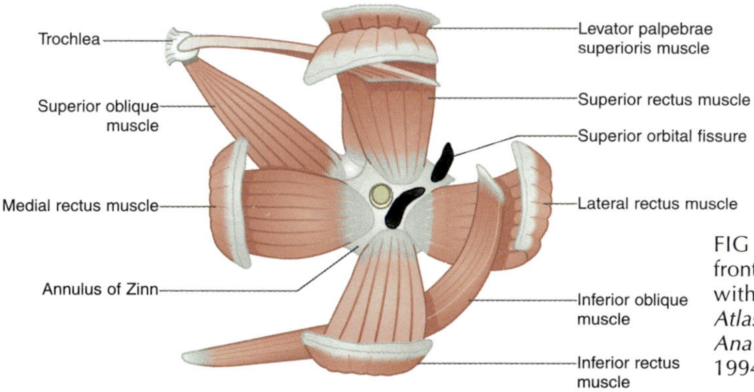

FIG I-15—Extraocular muscles, frontal view, left eye. (Reproduced with permission from Dutton JJ. *Atlas of Clinical and Surgical Orbital Anatomy.* Philadelphia: Saunders; 1994.)

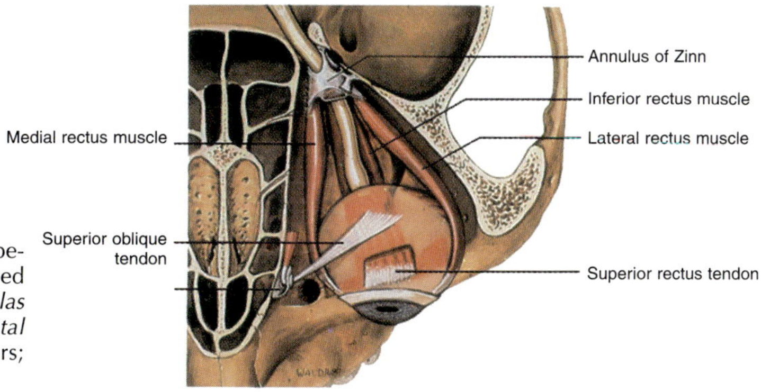

FIG I-16—Extraocular muscles, superior composite view. (Reproduced with permission from Dutton JJ. *Atlas of Clinical and Surgical Orbital Anatomy.* Philadelphia: Saunders; 1994.)

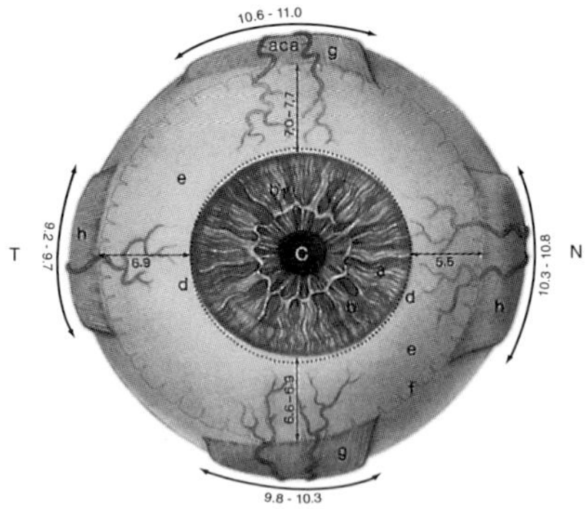

FIG I-17— Anterior view of the right globe. *N,* nasal; *T,* temporal; *a,* cornea (anterior to iris); *b,* iris; *c,* pupil; *d,* corneal margin; *e,* sclera; *f,* ora serrata (*dotted line*; note that this is located more posteriorly on the nasal side); *g,* superior and inferior recti; *h,* horizontal recti; *aca,* anterior ciliary arteries. The distance from muscle insertion to limbus increases as you follow the insertions around from medial, to inferior, to lateral, and to superior rectus. (From Hogan MJ, Alvarado JA, Weddell JE. *Histology of the Human Eye.* Philadelphia: Saunders; 1971.)

TABLE I-1

COMPARISON OF EXTRAOCULAR MUSCLES

MUSCLE	ORIGIN	INSERTION	BLOOD SUPPLY	SIZE
Medial Rectus	Annulus of Zinn	Medially, in horizontal meridian 5.5 mm from limbus	Inferior muscular branch of ophthalmic artery	40.8 mm long; tendon:3.7 mm long, 10.3 mm wide
Inferior	Annulus of Zinn at orbital apex	Inferiorly, in vertical meridian 6.5 mm from limbus	Inferior muscular branch of ophthalmic artery and infraorbital artery	40 mm long; tendon: 5.5 mm long, 9.8 mm wide
Lateral Rectus	Annulus of Zinn spanning the superior orbital fissure	Laterally, in horizontal meridian 6.9 mm from limbus	Lacrimal artery	40.6 mm long; tendon: 8 mm long, 9.2 mm wide
Superior Rectus	Annulus of Zinn at orbital apex	Superiorly, in vertical meridian 7.7 mm from limbus	Superior muscular branch of ophthalmic artery	41.8 mm long; tendon: 5.8 mm long, 10.6 mm wide
Superior Oblique	Medial to optic foramen, between annulus of Zinn and periorbita	To trochlea, through pulley, at orbital rim, then hooking back under superior rectus, inserting posterior to center of rotation	Superior muscular branch of ophthalmic artery	40 mm long; tendon: 20 mm long, 10.8 mm wide
Inferior Oblique	From a depression on orbital floor near orbital rim (maxilla)	Posterior inferior temporal quadrant at level of macula; posterior to center of rotation	Inferior branch of ophthalmic artery and infraorbital artery	37 mm long; no tendon: 9.6 mm wide at insertion

Extraocular Muscle Distribution in the Orbit

Figures I-15 and I-16 show the arrangement of the extraocular muscles within the orbit. Note the relationship between the oblique extraocular muscles and the superior, medial, and inferior rectus muscles.

The location of the extraocular muscles within the orbit and their relationship to surrounding nerves and bone are illustrated in coronal, cross-sectional views (Figs I-18, I-19) and correlated with coronal MRI of the orbit (Fig I-20). Longitudinal, axial views are shown in Figures I-21 and I-22.

Extraocular Muscle Origins

The annulus of Zinn consists of superior and inferior orbital tendons. The lateral and medial rectus muscles originate from the upper tendon; the inferior rectus arises from the lower tendon.

The superior oblique muscle originates from the periosteum of the body of the sphenoid bone, above and medial to the optic foramen. The inferior oblique muscle originates *anteriorly,* from a shallow depression in the orbital plate of the maxillary bone at the anteromedial corner of the orbital floor near the lacrimal fossa. The inferior oblique muscle passes from its origin posteriorly, laterally, and superiorly to insert into the globe.

Blood Supply to the Extraocular Muscles

The inferior and superior muscular branches of the ophthalmic artery, lacrimal artery, and infraorbital artery supply the extraocular muscles. The lateral rectus muscle is supplied by a *single* vessel derived from the lacrimal artery; the other rectus

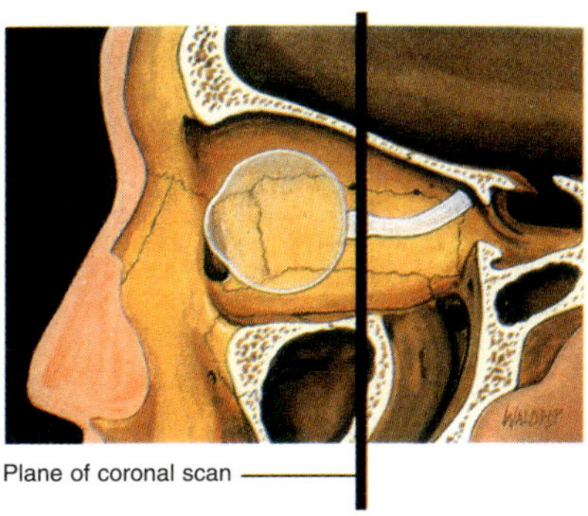

FIG I-18—Location of the plane of section shown in Figure I-19. (Reproduced with permission from Dutton JJ. *Atlas of Clinical and Surgical Orbital Anatomy.* Philadelphia: Saunders; 1994.)

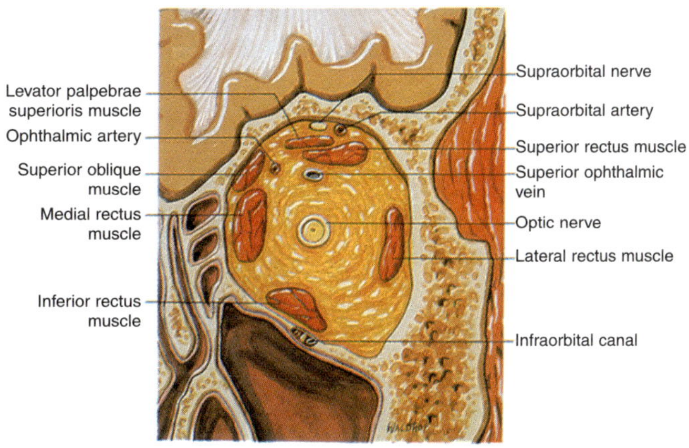

FIG I-19—Coronal section through the central orbit just posterior to the globe. (Reproduced with permission from Dutton JJ. *Atlas of Clinical and Surgical Orbital Anatomy.* Philadelphia: Saunders; 1994.)

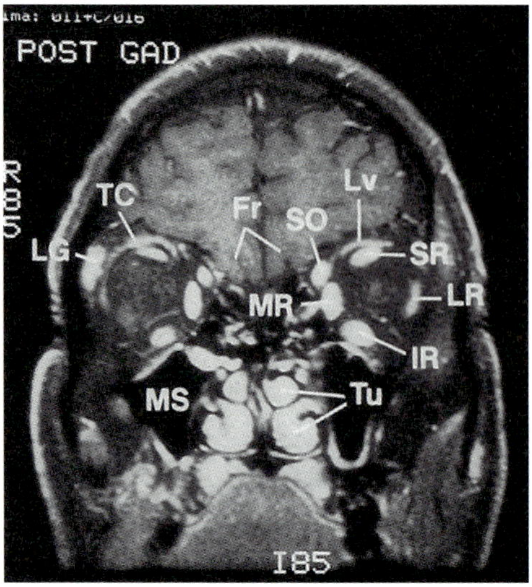

FIG I-20—Coronal magnetic resonance image with fat suppression and gadolinium enhancement through midorbit in a patient with Graves disease. All of the muscles are enlarged and show normal enhancement. *SR,* superior rectus; *Lv,* levator; *SO,* superior oblique; *MR,* medial rectus; *IR,* inferior rectus; *LR,* lateral rectus; *TC,* Tenon's capsule; *LG,* lacrimal gland; *Tu,* turbinates; *MS,* maxillary sinus; *Fr,* frontal lobe of brain. This is the same plane of section shown in Figures I-18 and I-19.

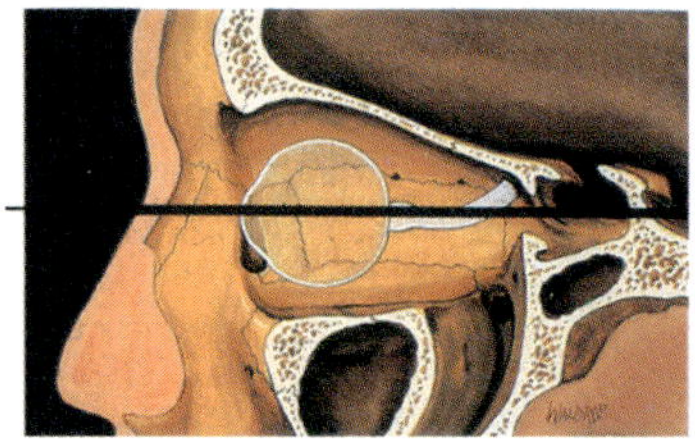

FIG I-21—Location of the plane of section shown in Figure I-22. (Reproduced with permission from Dutton JJ. *Atlas of Clinical and Surgical Orbital Anatomy.* Philadelphia: Saunders; 1994.)

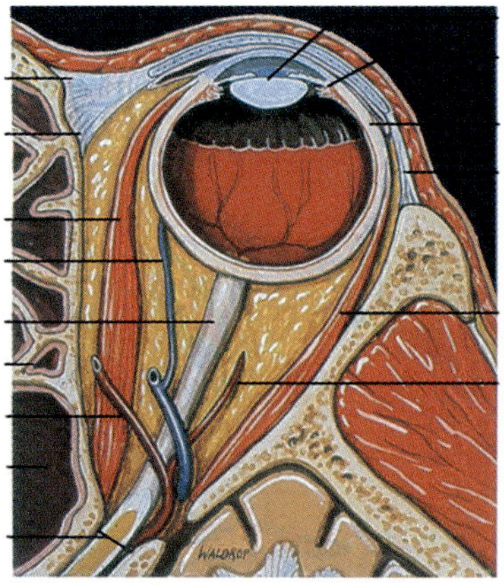

FIG I-22—Axial section through the midorbit at the level of the optic nerve. The third portion of the ophthalmic artery crosses the nerve in the posterior orbit. (Reproduced with permission from Dutton JJ. *Atlas of Clinical and Surgical Orbital Anatomy.* Philadelphia: Saunders; 1994.)

muscles receive *two* anterior ciliary arteries that communicate with the major arterial circle of the ciliary body via perforating scleral vessels. The vascular supply and drainage of orbital structures is discussed later under the heading of Vascular Supply and Drainage of the Orbit.

Hayreh SS, Scott WE. Fluorescein iris angiography. I, Normal pattern. II, Disturbances in iris circulation following strabismus operation on the various recti. *Arch Ophthalmol.* 1978;96:1383–1400.

Extraocular Muscle Structure

Extraocular muscle, unlike skeletal muscle, is poorly fasciculated and has a high ratio of nerve axons to muscle fibers: 1:5 for extraocular muscle compared to 1:100 for skeletal muscle. Fiber diameter is uniform in infancy, but in the adult, the inner (global) fibers are larger (10–40 μm) than the outer (orbital) fibers (5–15 μm). Fibers run mainly from one end of the muscle belly to the other.

A connective tissue *epimysium* surrounds the entire muscle, a *perimysium* demarcates the fascicles, and an *endomysium* surrounds each fiber. The fiber membrane is the *sarcolemma,* and its sarcoplasm contains parallel, striated *myofibrils,* 1–2 μm across.

Each repeating unit is a *sarcomere.* The myofibrils are composed of interdigitating myofilaments whose overlap gives rise to the banding pattern. The thick myofilaments are composed of *myosin,* the thin myofilaments of *actin,* attached to the electron-dense Z disks (responsible for the Z bands). The thin myofilaments occupy the light-appearing I band; the thick myofilaments (with some thin ones) occupy the A band. In the zone of thick filaments, a lucent H band is split by the M band.

Within the thick filaments, the myosin contains light and heavy *meromyosin* molecules in staggered array. The latter contain binding sites for actin and myosin *adenosine triphosphatase (ATPase).* Thin filaments contain two double-helical molecules, the *tropomyosin* lying within the *actin* grooves and shielding the myosin-binding sites. The protein *troponin* is distributed along the tropomyosin at intervals. Contraction of twitch-type muscle occurs when a *propagated nerve action potential* arrives at the motor end plate and acetylcholine packages are released across the synaptic cleft. Binding with postjunctional receptors causes depolarization, with a leak of sodium and potassium ions across the membrane, which generates an *endplate potential.* This leads to a rise of calcium within the fiber. Binding of calcium to troponin causes a conformational change that exposes the myosin-binding sites along the actin molecule. Interaction between the myosin and actin causes the filaments to slide over one another, which is the basis of contraction. The energy is released by the action of myosin ATPase, located in the globular head of the myosin molecule, where the actin-binding site is also housed.

Muscle fiber type The fibers of extraocular muscle are a mixture of the *twitch* type, resembling skeletal muscle and innervated by single *motor end plates (en plaque),* and the *tonic* type, innervated multiply with diffuse grapelike endings *(en grappe).* Twitch-type fibers have a regular myofibril arrangement, or *Fibrillenstruktur.* Tonic muscle fibers do not exhibit a propagated action potential; instead, there is a passive electrotonic spread of depolarization across the fiber surface from multiple sites. Tonic fibers are more primitive, resembling the tonic fibers of amphibians and birds, with slow contractile and unusual pharmacologic properties (Table I-2).

TABLE I-2

EXTRAOCULAR MUSCLES

	FIBRILLENSTRUKTUR	FELDERSTRUKTUR
	"fibril" (fast, twitch)	"field" (slow, tonic)
myofibrils	well defined	poorly defined
sarcoplasm	abundant	sparse
sarcomere	well developed	poorly developed
T-system	regular	absent, or aberrant
Z-line	straight	zigzag course
M-line	well marked	absent
nuclei	located peripherally	located centrally or eccentrically
innervation	thick; heavily myelinated	thin
neuromuscular junction	en plaque (single)	en grappe (grapelike)
synaptic vesicles	agranular	granular/agranular
acetylcholine	twitch contraction	tonic contraction

Fiber classification Fibers of differing color, width, contractile properties, and histochemical profiles are distributed throughout each extraocular muscle. Those in the region closest to the globe are called *global,* and those closest to the orbit are called *orbital.* Twitch and tonic fibers may be distinguished histochemically on the basis of staining for myosin ATPase and for oxidative and glycolytic enzymes. A current classification for extraocular muscle is as follows:

Orbital zone. Type 1, singly innervated fast-twitch fibers, rich in mitochondria and possessing both oxidative and glycolytic properties, account for 80% of this zone.

Type 2, multiply innervated tonic fibers make up 20%. Their tonic properties are located toward the tips of the fibers, with a twitch capability toward the fiber center.

Global zone. Type 3 fibers (30%) are red, singly innervated, fast-twitch, fatigue-resistant fibers, rich in mitochondria but with both oxidative and glycolytic properties.

Type 4 fibers (about 25%) are singly innervated fast-twitch fibers, differing from type 3.

Type 5 fibers (about 30%) are pale, singly innervated, fast-twitch fibers with low fatigue resistance.

Type 6 fibers (about 10%) are multiply innervated tonic fibers containing a mixture of slow myosins, evenly distributed along the fiber.

Spencer RF, Porter JD. Structural organization of the extraocular muscles. In: Buttner-Ennever JA, ed. *Neuroanatomy of the Oculomotor System.* Amsterdam: Elsevier; 1988.

Porter JD, Baker RS, Ragusa RJ, et al. Extraocular muscles: basic and clinical aspects of structure and function. *Surv Ophthalmol.* 1995;39(6):451–484.

A classification of extraocular muscles based on their expression of different myosins is as follows: Many orbital zone fibers contain *neonatal* myosins coexpressed with "fast" extraocular myosin, distinct from that found in skeletal muscle. The proportion of neonatal myosin declines with age, but unlike the case with striated muscle, it doesn't disappear. The orbital zone also contains *slow-twitch* and tonic fibers. Global fibers mainly contain specific fast extraocular myosin within twitch-type fibers. There are also slow-twitch fibers, which are unusual in that they are multiply innervated.

Sensory endings Extraocular muscle contains various sensory endings, including palisade endings, muscle spindles, and Golgi tendon organs. Probably the most important of these in human extraocular muscle is the *palisade ending,* or *myotendinous cylinder.*

The palisade endings are found at the myotendinous junction and consist of encapsulated nerve endings, which are cupped to receive a *single* extrafusal muscle fiber tip. They are associated with global, multiply innervated fibers. It is assumed that they are activated by compression of the nerve terminals during muscle contraction.

Ruskell GL. The fine structure of innervated myotendinous cylinders in extraocular muscles of rhesus monkeys. *J Neurocytol.* 1978;7(6):693–708.

The *Golgi tendon organ* is made up of a capsule containing 15–20 extrafusal muscle fibers that enter through a tight collar. They terminate in braided collagen fibrils that interweave with large-diameter, group 1b afferent axons. In skeletal muscle, the organ is partly located in muscle and partly in tendon; in extraocular muscle, the organ is completely located in tendon. Contraction of the muscle belly compresses the axons and signals increased muscle tension.

Muscle spindles signal changes in length of muscles and rate of change. They are composed of a capsule enclosing intrafusal fibers arranged parallel to the muscle mass. Grouped nuclei are found in *nuclear bag spindles,* and linearly arranged nuclei are found in *nuclear chain spindles.* Contraction of the main muscle mass shortens the spindles and reduces the signal. *Gamma efferents* to the intrafusal fibers control the sensitivity of length detection. Human rectus muscles contain 22–71 spindles per muscle, mainly in their proximal or distal parts. However, the importance of these spindles to extraocular muscle proprioception is uncertain. Some species, including certain primates, have *no* spindles. In human extraocular muscle, the spindles may be functionally redundant and perhaps play a relatively small sensory role.

Pharmacologic and toxicologic considerations The muscle relaxant *succinylcholine* has different effects on skeletal and extraocular muscles. In skeletal muscle, it is a *depolarizing blocker* of neuromuscular transmission; in extraocular muscle, it selectively activates multiply innervated global and possibly orbital fibers. Because this action causes a slow muscular contraction, succinylcholine may cause a rise in the intraocular pressure during induction for general anesthesia. Therefore, this drug is contraindicated in patients with penetrating ocular injuries.

Local anesthetics of the aminoacyl type (eg, lidocaine) are myotoxic and may cause extraocular muscle palsy when used for local anesthesia around the orbit (peribulbar or retrobulbar injection). On injection into muscle, these drugs induce a calcium overload that disrupts the sarcolemma and may cause persistent muscle palsy following ocular surgery. Toxicity results from direct injection of muscles and, in the monkey, chiefly affects global, pale, singly innervated muscles.

Botulinum toxin acts by blocking the calcium-dependent release of acetylcholine at the neuromuscular junction. The *A serotype* is used clinically to induce transient weakening of extraocular muscles in the management of strabismus, relaxation of the levator muscle to induce a protective ptosis, and relaxation of facial and other muscles recruited in dystonic muscle disorders.

Paralysis is due to secondary denervation atrophy of the injected muscles. Recovery is due to motor neuron sprouting and functional reinnervation. Although the effect of botulinum toxin is therefore reversible, it can cause permanent damage to extraocular muscles. In the monkey, there is a specific loss of orbital, singly innervated muscles. Experimentally, the effects are more profound in young monkeys, perhaps because the muscles are still developing.

Spencer RF, Porter JD. Structural organization of the extraocular muscles. In: Buttner-Ennever JA, ed. *Neuroanatomy of the Oculomotor System.* Amsterdam: Elsevier; 1988.

Eyelids

The palpebral fissure is the exposed zone between the upper and lower eyelids. Normally, the adult fissure is 27–30 mm long and 8–11 mm wide. The upper eyelid, more mobile than the lower, can be raised 15 mm by the action of the levator muscle alone. If the frontalis muscle of the brow is used, the palpebral fissure can be widened an additional 2 mm. The levator muscle is innervated by cranial nerve III, and the orbicularis oculi muscle is its antagonist. Important changes in the fissure occur in thyroid disease, myasthenia gravis, congenital ptosis, levator disinsertion, Horner syndrome, facial palsy, and third nerve palsy (Fig I-23).

Anatomy

The upper eyelid can be divided into eight anatomical segments from the dermal surface inward (Figs I-24 through I-27):

Skin The eyelid skin, the thinnest in the body, contains fine hairs, sebaceous glands, and sweat glands. A superior eyelid fold is present near the upper border of the tarsus, where the levator aponeurosis establishes its first insertional attachments, except in persons of Asian descent. The aponeurosis forms its firmest attachments on the anterior aspect of the tarsus about 3 mm superior to the eyelid margin.

Margin The eyelid margin contains several important landmarks. A small opening, the *punctum* of the canaliculus, presents medially at the summit of each lacrimal papilla. The upper punctum, normally hidden by slight internal rotation, is located more medially. The lower is usually apposed to the globe and is not normally visible without eversion.

Along the entire length of the free margin of the eyelid is the delicate *gray line* (or *intermarginal sulcus*), corresponding histologically to the most superficial portion of the orbicularis muscle, the *muscle of Riolan,* and to the avascular plane of the lid. The eyelashes (or *cilia*) arise anterior to this line; behind it are the openings of the *tarsal* (or *meibomian*) glands just anterior to the mucocutaneous junction.

Wulc AE, Dryden RM, Khatchaturian T. Where is the gray line? *Arch Ophthalmol.* 1987;105:1092–1098.

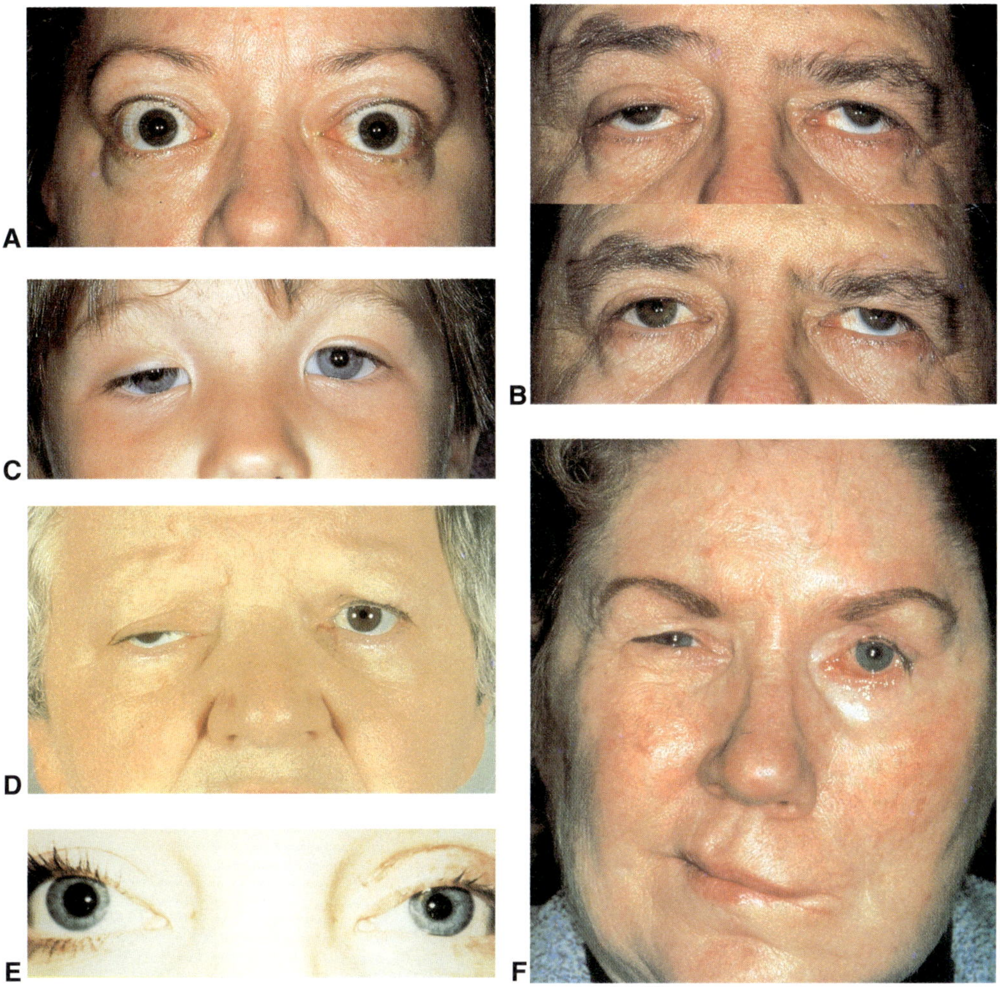

FIG I-23—Changes in the eyelid fissure. **A**, Graves disease stare. **B**, Myasthenia gravis (*top,* right ptosis before Tensilon; *bottom,* ptosis after Tensilon). **C**, Congenital ptosis of right eye. **D**, Levator disinsertion. **E**, Horner syndrome or oculosympathetic denervation of left eye. **F**, Left seventh (facial) nerve palsy. (Photographs courtesy of Jeffrey Nerad, MD.)

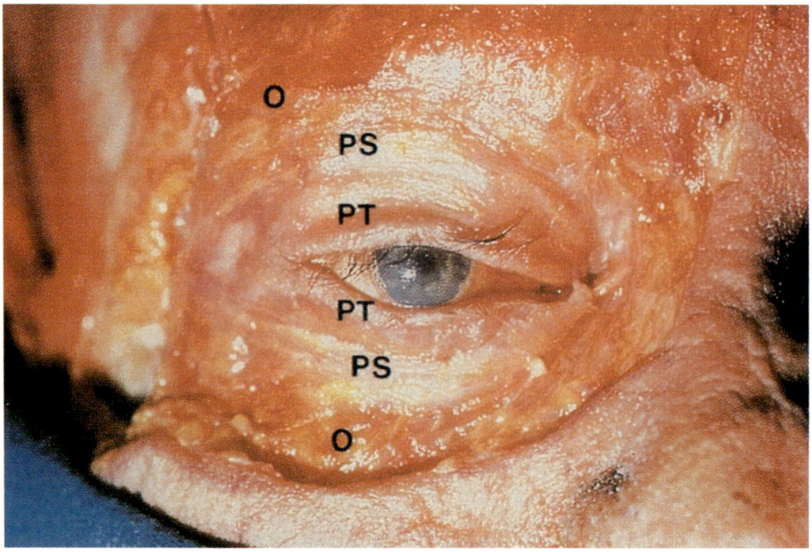

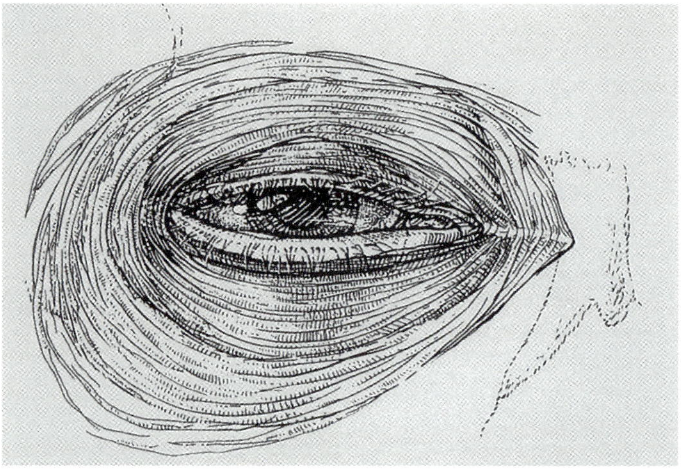

FIG I-24—The periorbital skin has been removed in the top photograph, exposing orbicularis oculi muscle that is innervated by cranial nerve VII. This muscle acts as an antagonist to the levator palpebrae superioris muscle innervated by cranial nerve III. The orbicularis muscle is divided into the palpebral and orbital *(O)* portions. The palpebral portion is further subdivided into pretarsal *(PT)* and preseptal *(PS)* portions. The bottom diagram depicts the arrangement of muscle fibers of the orbicularis muscle. (Reproduced with permission from Zide BM, Jelks GW. *Surgical Anatomy of the Orbit.* New York: Raven; 1985.)

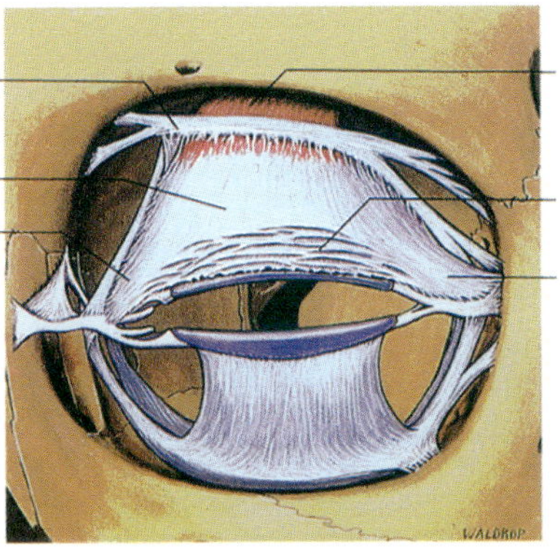

A

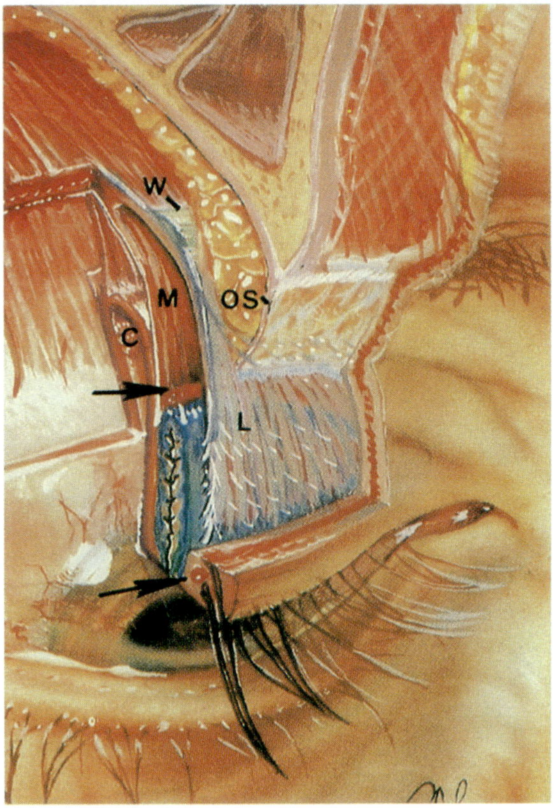

B

FIG I-25—**A**, The upper and lower tarsal plates and their attachments to the levator aponeurosis and to Whitnall's ligament. (Reproduced with permission from Dutton JJ. *Atlas of Clinical and Surgical Orbital Anatomy.* Philadelphia: Saunders; 1994.) **B**, The three-dimensional organization of the upper eyelid. For convenience, the upper eyelid may be divided into anterior and posterior lamellae. The *anterior lamella* consists of the skin and orbicularis muscle and its associated fascial and vascular structures. Note the marginal artery *(lower arrow)* approximately 3.0–3.5 mm above the eyelid margin. The *posterior lamella* consists of the levator aponeurosis *(L)*, tarsus *(blue)*, Müller's muscle *(M)*, and conjunctival lining *(C)*. At a variable height above the superior edge of the tarsus, the orbital septum *(OS)* forms the anterior border of the preaponeurotic fat space. The peripheral arterial arcade is situated *(upper arrow)* at the level of the superior edge of the tarsus, posterior to the levator aponeurosis within the so-called *pretarsal space*. The levator muscle usually becomes aponeurotic at the equator of the globe in the superior orbit. The aponeurosis courses anteriorly to insert onto the lower two thirds of the anterior tarsal plate. The levator muscle provides origin to Müller's muscle *(M)*, the nonstriated, sympathetically innervated elevator of the upper eyelid, which inserts into the superior edge of the tarsus and into the conjunctiva of the superior fornix. The superior transverse ligament of Whitnall *(W)* is noted as a fascial condensation along the upper aspect of the levator muscle. This ligament attaches to the trochlear fascia medially and the fascia of the orbital lobe of the lacrimal gland laterally. (Reproduced with permission from Zide BM, Jelks GW. *Surgical Anatomy of the Orbit.* New York: Raven; 1985.)

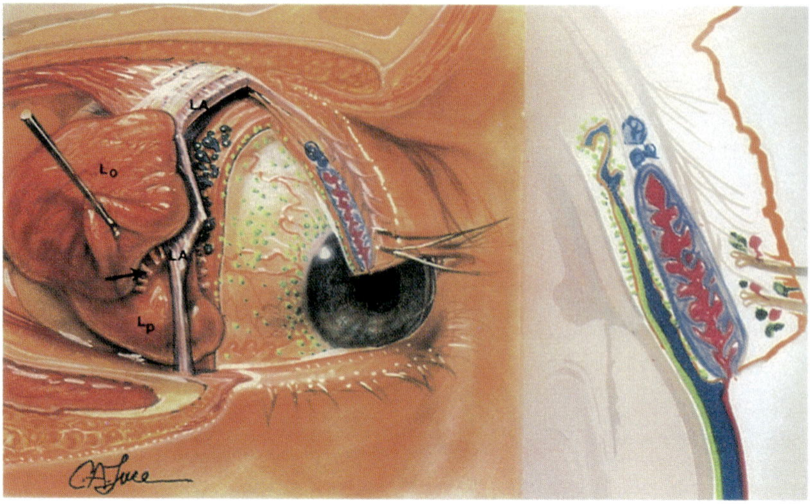

FIG I-26—The lacrimal secretory system: (1) the conjunctival and tarsal mucin-secreting goblet cells *(green)* produce a mucoprotein layer covering the epithelial surface of the cornea and conjunctiva; (2) the accessory lacrimal exocrine glands of Krause and Wolfring are present in the subconjunctival tissues *(blue)* and contribute to the aqueous layer of the precorneal tear film; and (3) oil-producing meibomian glands and palpebral glands of Zeis and Moll *(pink)*. The orbital lobe of the lacrimal gland *(Lo)* and the palpebral lobe of the lacrimal gland *(Lp)* are separated by the lateral horn of the levator palpebrae superioris *(LA)*. The tear ducts *(arrow)* from the orbital portion traverse the palpebral portion. (Reproduced with permission from Zide BM, Jelks GW. *Surgical Anatomy of the Orbit.* New York: Raven; 1985.)

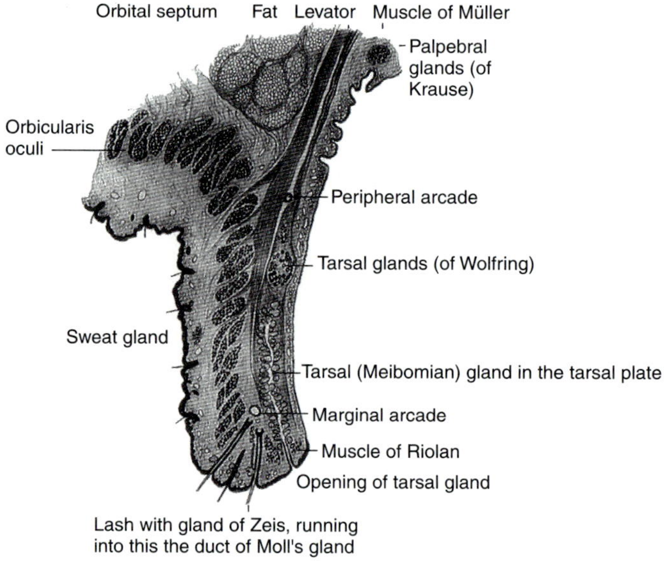

FIG I-27—Cross section of upper eyelid. Note position of cilia, tarsal gland orifices, and mucocutaneous border. (From Bron AJ, Tripathi RC, Tripathi BJ. *Wolff's Anatomy of the Eye and Orbit.* 8th ed. London, New York: Chapman & Hall; 1997.)

TABLE I-3

GLANDS OF THE EYE AND ADNEXA

GLANDS	LOCATION	SECRETION	CONTENT
Lacrimal	orbital gland	exocrine	aqueous
	palpebral gland	exocrine	aqueous
Accessory lacrimal	plica, caruncle	exocrine	aqueous
Krause	eyelid	exocrine	aqueous
Wolfring	eyelid	exocrine	aqueous
Meibomian	tarsus	holocrine	oily
Zeis	follicles of cilia	holocrine	oily
	eyelid, caruncle	holocrine	oily
Moll	eyelid	eccrine	sweat
Goblet cell	conjunctiva	holocrine	mucous
	plica, caruncle	holocrine	mucous

The eyelashes are arranged in two or three irregular rows along the anterior dermal edge of the eyelid margin. They are usually longer and more numerous on the upper than the lower eyelid. The margins contain the *glands of Zeis* (modified sebaceous glands associated with the cilia) and the *glands of Moll,* which are apocrine sweat glands of skin (Table I-3).

Subcutaneous connective tissue The loose connective tissue of the eyelid contains no fat. Blood or other fluids can accumulate beneath the skin and result in rapid and dramatic swelling of the lids.

Orbicularis oculi muscle The orbicularis muscle is arranged in several concentric bands around the palpebral fissure and can be subdivided into orbital and palpebral parts. The muscle fibers are short and connected by *myomyous junctions.* They have the smallest diameter of all the facial muscles. Innervation is by the facial nerve, and end plates are arranged in clusters over the entire length of the muscle. This arrangement may influence the action of botulinum A toxin used in the treatment of blepharospasm.

Lander T, Wirtschafter JD, McLoon LK. Orbicularis oculi muscle fibers are relatively short and heterogeneous in length. *Invest Ophthalmol Vis Sci.* 1996;37(9):1732–1739.

The *orbital part* inserts in a complex way into the medial canthal tendon and into other portions of the orbital rim and the corrugator supercilii muscle. The orbital part acts like a sphincter and functions solely as a voluntary muscle.

The *palpebral part* of the orbicularis functions both voluntarily and involuntarily in spontaneous and reflex blinking. The preseptal and pretarsal portions unite along the superior palpebral furrow. The pretarsal muscle is firmly adherent to the tarsus.

Orbicularis fibers extend to the eyelid margin, where there is a small bundle of striated muscle fibers called the *muscle of Riolan.* Disinsertion of the lower eyelid retractors from the tarsus may result in laxity of the lower eyelid, followed by spastic entropion.

Orbital septum A thin sheet of connective tissue called the *orbital septum* encircles the orbit as an extension of the periosteum of the roof and the floor of the orbit (Fig I-28). It also attaches to the anterior surface of the levator muscle. Posterior to the orbital septum is the orbital fat. In both the upper and lower eyelids, the orbital septum attaches to the aponeurosis. The orbital septum thus provides a barrier to anterior or posterior extravasation of blood or the spread of inflammation. The intermuscular orbital septa can be identified in coronal MRI studies with fat suppression and gadolinium enhancement.

Superiorly, the septum is attached firmly to the periosteum of the superior half of the orbital margin. It passes medially in front of the trochlea and continues along the medial margin of the orbit, along the margin of the frontal process of the maxillary bone, and on to the inferior margin of the orbit. Here, the septum also delimits the lateral spread of edema, inflammation, or blood trapped anterior to it and appears clinically as a dramatic barrier to these processes.

Levator muscle The *levator palpebrae superioris* muscle originates from a short tendon that blends with the superior rectus and the superior oblique muscles at the apex of the orbit. The body of the levator muscle overlies the superior rectus as it travels anteriorly toward the eyelid. *Whitnall's ligament* results from a condensation of tissue surrounding the superior rectus and levator muscles. Near Whitnall's ligament, the levator muscle changes direction from horizontal to more vertical, and it divides anteriorly into the aponeurosis and posteriorly into the superior tarsal (Müller's) muscle.

The aponeurosis inserts into the anterior surface of the tarsus and by medial and lateral horns into the canthal tendons. The fibrous elements of the aponeurosis pass through the orbicularis muscle and insert subcutaneously to produce the superior eyelid fold. The aponeurosis also inserts into the trochlea of the superior oblique muscle and into the fibrous tissue bridging the supraorbital notch. Attachments also exist with the conjunctiva of the upper fornix and with the orbital septum.

The levator muscle and tendon are 50–55 mm long. The muscle, which elevates the upper eyelid, is 40 mm long and is innervated by the superior division of cranial nerve III.

Müller's muscle *Müller's muscle* is a smooth (nonstriated), sympathetically innervated muscle that originates from the undersurface of the levator muscle in the upper eyelid. A similar smooth muscle arises from the capsulopalpebral head of the inferior rectus in the lower eyelid. Müller's muscle attaches to the upper border of the upper tarsus and to the conjunctiva of the upper fornix. The capsulopalpebral muscle, which is much weaker than Müller's muscle, attaches to the lower border of the lower tarsus. Actuation of Müller's muscle gives rise to a "stare" in thyroid ophthalmopathy. Sympathetic denervation causes ptosis in Horner syndrome.

Tarsus The *tarsal plates* consist of dense connective tissue, not cartilage. They are attached to the orbital margin by the medial and lateral palpebral ligaments (Fig I-29). The length (29 mm) and thickness (1 mm) of the upper and lower tarsal plates are similar. The upper tarsus is almost three times as wide vertically (11 mm) as the lower tarsus (4 mm).

A

B

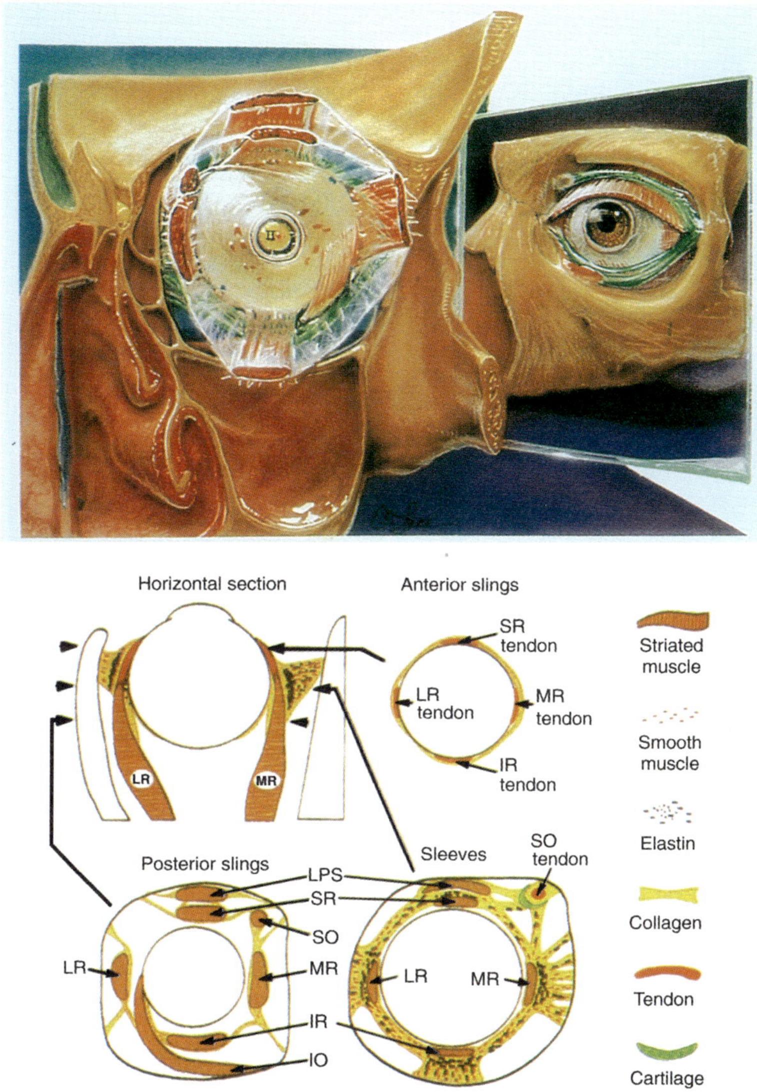

FIG I-28—**A**, Orbital septum. More posteriorly, the intervening intramuscular septa are thinner and may become discontinuous, thus opening the muscle cone. Tenon's capsule continues posteriorly, adherent to the globe to eventually encircle the optic nerve, which penetrates it. More anteriorly, as the intermuscular septa thicken, the connective mass between the inferior oblique and inferior rectus muscles forms the inferior suspensory ligament of Lockwood *(green)*. Similarly, a thickening or condensation of fascia (ie, Whitnall's ligament) may be noted superiorly overlying or surrounding the levator muscle. (Reproduced with permission from Zide BM, Jelks GW. *Surgical Anatomy of the Orbit.* New York: Raven; 1985.) **B**, Diagram of the orbital connective tissues. *IR,* inferior rectus; *LPS,* levator palpebrae superioris; LR, lateral rectus; *MR,* medial rectus; *SO,* superior oblique; *SR,* superior rectus. (From Demer JL, Miller JM, Poukens V, et al. Evidence for fibromuscular pulleys of the recti extraocular muscles. *Invest Ophthalmol Vis Sci.* 1995;36:1125. ©Association for Research in Vision and Ophthalmology.)

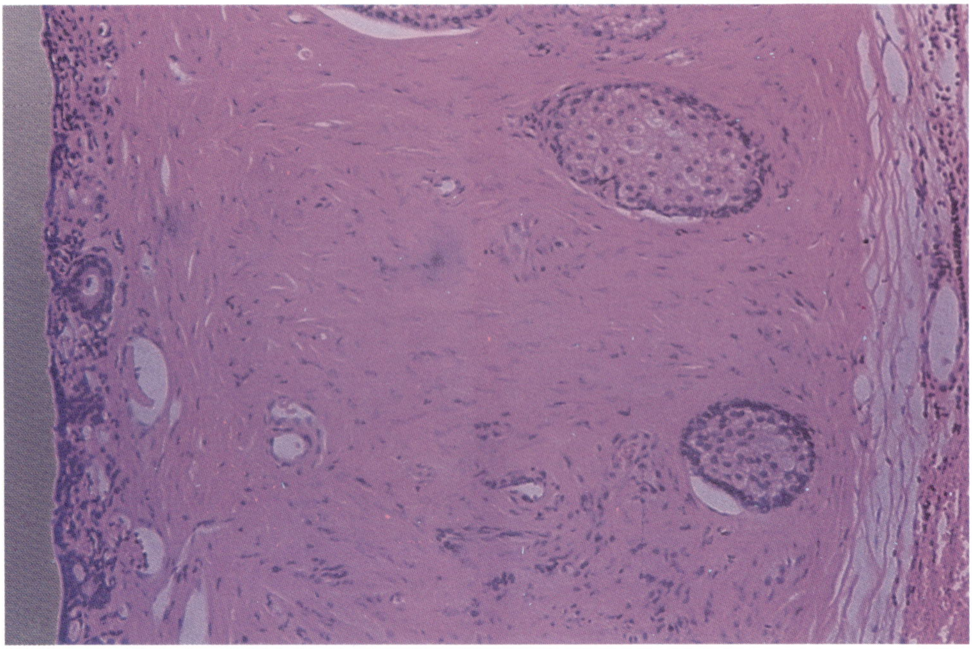

FIG I-29—Palpebral conjunctiva and meibomian glands within the tarsus of the eyelid (H&E ×32). (Photograph courtesy of Thomas A. Weingeist, PhD, MD.)

The *meibomian glands* (Fig I-30) are modified holocrine sebaceous glands that are oriented vertically in parallel rows through the tarsus. Their distribution and number within the eyelid can be observed by transillumination (Fig I-31). A single row of 30–40 meibomian orifices is present in the upper eyelid, but there are only 20–30 in the lower lid. Oil from these orifices forms a reservoir on the skin of the lid margin and is spread onto the tear film with each blink.

The hair bulbs of the cilia are located anterior to the tarsus and the meibomian gland orifices. Misdirection in the orientation of the eyelashes (trichiasis) or aberrant growth through the orifices of meibomian glands (distichiasis) may occur as either a congenital or an acquired defect and occasionally is hereditary.

Conjunctiva The *palpebral conjunctiva* is a transparent vascularized membrane covered by a nonkeratinized epithelium that lines the inner surface of the eyelids. Continuous with the conjunctival fornices (cul-de-sacs), it merges with the bulbar conjunctiva before terminating at the limbus.

Vascular Supply of Eyelids

The blood supply of the eyelids is derived from the facial system, which arises from the external carotid artery, and the orbital system, which originates from the internal carotid artery along branches of the ophthalmic artery (Fig I-32). The superficial and deep plexuses of arteries provide a vast blood supply to the upper and lower eyelids. The facial artery becomes the angular artery as it passes upward, forward, and lateral to the nose, where it serves as an important landmark in dacryocystorhinostomy surgery.

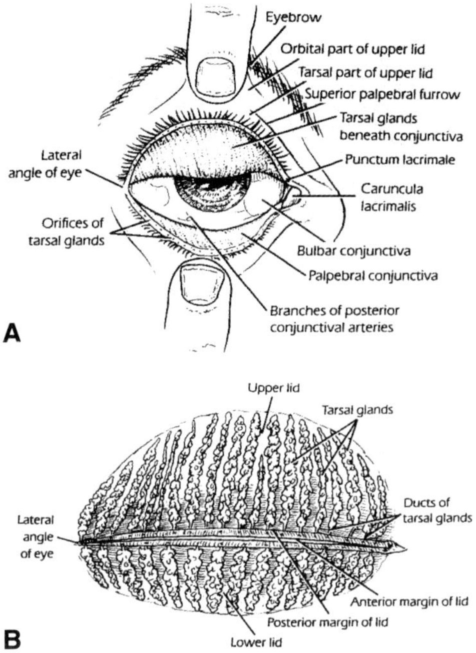

FIG I-30—**A**, Complete eversion of the upper eyelid of the right eye, made possible by the stiffness of the superior tarsal plate; the lower eyelid is pulled downward. Note the orifices of the tarsal (meibomian) glands and the lacrimal puncta; note also the branches of the posterior conjunctival arteries. **B**, Posterior view of the eyelids with the palpebral fissure nearly closed. Note the tarsal glands with their short ducts and orifices. The palpebral conjunctiva has been removed to show the tarsal glands in situ. (Reproduced with permission from Snell RS, Lemp MA. *Clinical Anatomy of the Eye.* Boston: Blackwell; 1989.)

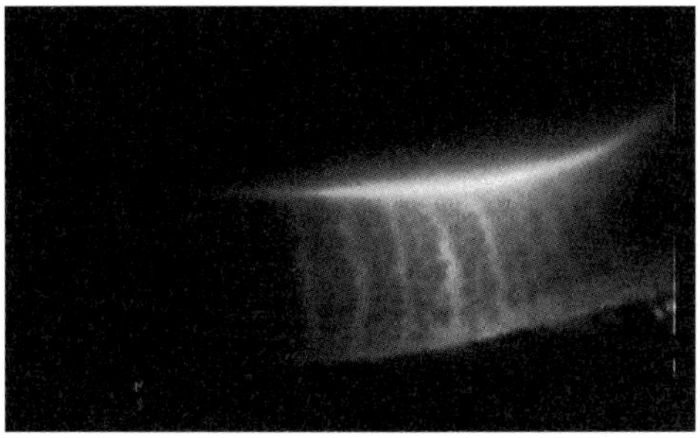

FIG I-31—Distribution of the meibomian glands in the lower eyelid as revealed by infrared transillumination of the eyelid. The glands appear as dark gray linear structures. (Photograph courtesy of William Mathers, MD.)

The *marginal arterial arcade* is located 3 mm from the free border of the eyelid, just above the ciliary follicles. It is either between the tarsal plate and the orbicularis or within the tarsus. A smaller peripheral arcade runs along the upper margin of the tarsal plate within Müller's muscle.

The venous drainage of the eyelids can also be divided into two portions:

- A superficial, or pretarsal, system that drains into the internal and external jugular veins
- A deep, or post-tarsal, system that eventually flows into the cavernous sinus

Lymphatics of the Eyelids

Lymphatic vessels are found in the lids and conjunctiva but neither lymphatic vessels nor nodes are present in the orbit. Lymphatic drainage from the eyelids parallels the course of the veins. Two groups of lymphatics exist:

- A medial group that drains into the submandibular lymph nodes
- A lateral group that drains into the superficial preauricular lymph nodes

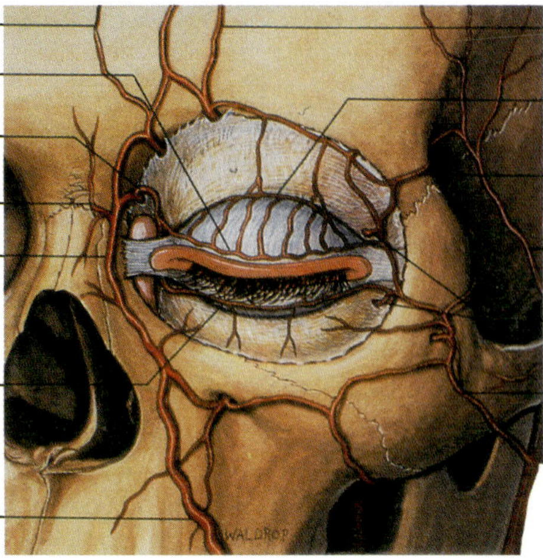

FIG I-32—Periorbital and eyelid arteries, frontal view. (Reproduced with permission from Dutton JJ. *Atlas of Clinical and Surgical Orbital Anatomy.* Philadelphia: Saunders; 1994.)

Accessory Eyelid Structures

Caruncle The *caruncle* is a small, fleshy, ovoid structure attached to the inferomedial side of the plica semilunaris. As a piece of modified skin, it contains sebaceous glands and fine, colorless hairs. The surface is covered by nonkeratinized, stratified squamous epithelium.

Plica semilunaris The *plica semilunaris* is a narrow, highly vascular, crescent-shaped fold of the conjunctiva located lateral to and partly under the caruncle. Its lateral border is free and separated from the bulbar conjunctiva, which it resembles histologically. The epithelium of the plica is rich in goblet cells. Its stroma contains fat and some nonstriated muscle. This vestigial structure is analogous to the nictitating membrane, or third eyelid, of dogs and other animals.

Lacrimal Gland and Excretory System

Lacrimal Gland

The main *lacrimal gland* is located in a shallow depression within the orbital part of the frontal bone. The gland is separated from the orbit by fibroadipose tissue and divided in two parts by a lateral expansion of the levator aponeurosis (Fig I-33). The smaller, palpebral part can be seen in the superolateral conjunctival fornix when the upper eyelid is everted. An isthmus of glandular tissue occasionally exists between the palpebral lobe and the main orbital gland.

A variable number of thin-walled excretory ducts, blood vessels, lymphatics, and nerves pass from the orbital into the *palpebral lacrimal gland* (Fig I-34). The

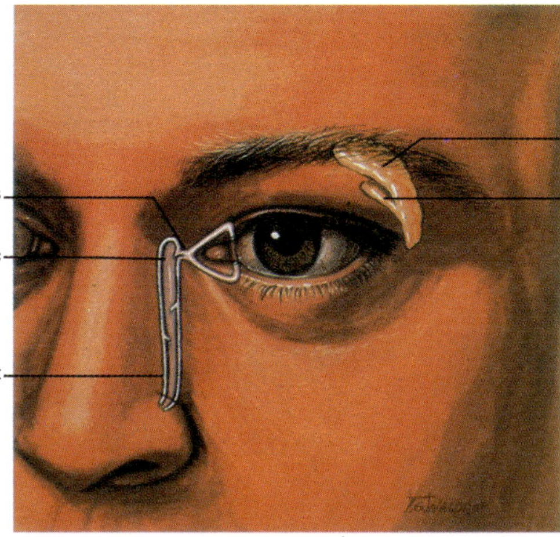

FIG I-33—Lacrimal system. (Reproduced with permission from Dutton JJ. *Atlas of Clinical and Surgical Orbital Anatomy.* Philadelphia: Saunders; 1994.)

ducts continue downward, and about 12 empty into the conjunctival fornix approximately 5 mm above the superior margin of the upper tarsus. Because the lacrimal excretory ducts pass through the palpebral portion of the gland, biopsy of the lacrimal gland is usually performed on the main part to avoid sacrificing the ducts.

The lacrimal glands are exocrine glands, and they produce a serous secretion. The body of each gland contains two cell types (Fig I-35):

- Acinar cells, which line the lumen of the gland
- Myoepithelial cells, which surround the parenchyma and are covered by a basement membrane

The lacrimal artery, a branch of the ophthalmic artery, supplies the gland. The lacrimal gland receives secretomotor cholinergic, vasoactive intestinal polypeptide (VIP)-ergic, and sympathetic nerve fibers in addition to a sensory innervation via the lacrimal nerve (CN V_1). Cyclic adenosine monophosphate is the second messenger for VIP and β-adrenergic stimulation of the gland; cholinergic stimulation acts through an inositol 1,4,5-triphosphate–activated protein kinase C. The gland also contains α_1-adrenergic receptors. The gland's neuroanatomy governs both reflex and psychogenic stimulation and is extremely complex. See BCSC Section 5, *Neuro-Ophthalmology.*

Accessory Glands

The accessory lacrimal *glands of Krause and Wolfring* are located at the proximal lid borders or in the fornices and are cytologically identical to the main lacrimal gland, receiving a similar innervation. They account for about 10% of the total lacrimal secretory mass.

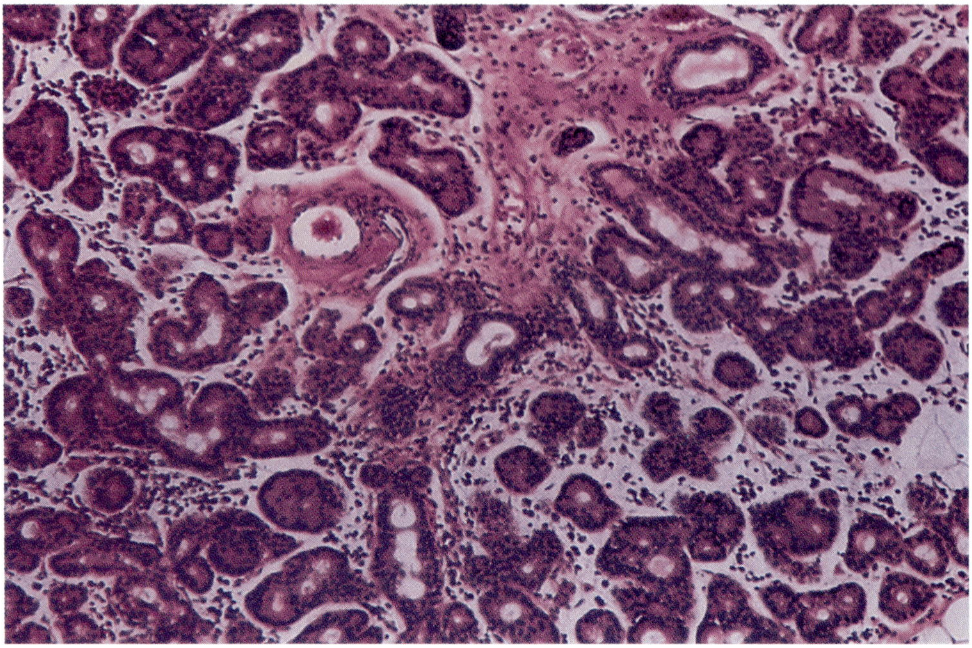

FIG I-34—Orbital lacrimal gland and ducts (H&E ×32). (Photograph courtesy of Thomas A. Weingeist, PhD, MD.)

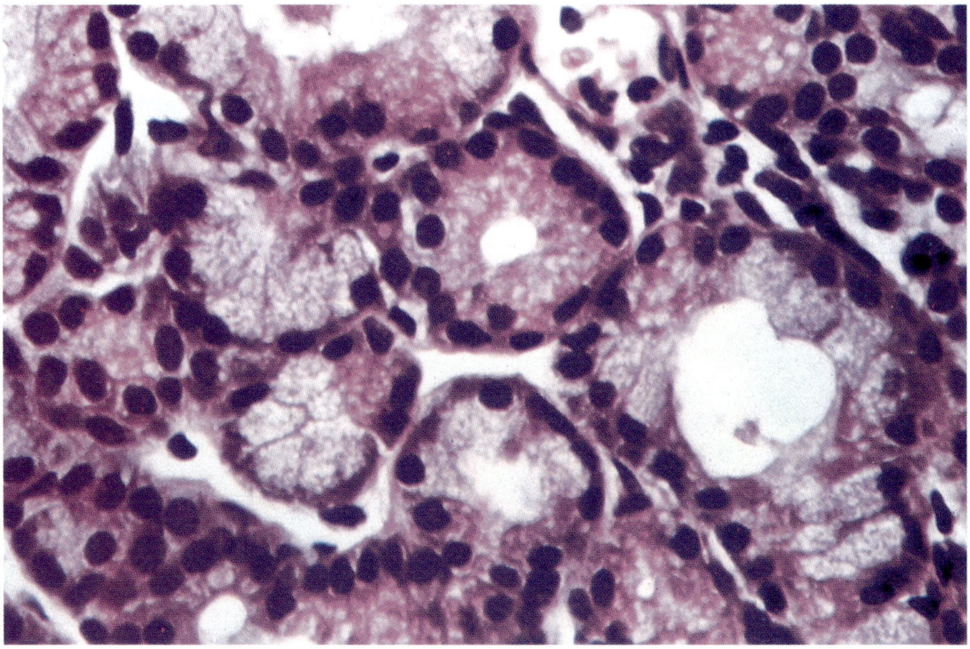

FIG I-35—Higher magnification of lacrimal gland lobules. Note that the acinar cells forming the lobules are surrounded by myoepithelial cells that contain flattened nuclei (H&E ×64). (Photograph courtesy of Thomas A. Weingeist, PhD, MD.)

Lacrimal Excretory System

The lacrimal drainage system includes the upper and lower puncta, canaliculi, the tear sac, and the nasolacrimal duct (Figs I-36, I-37). The *lacrimal papillae* are located at the extreme nasal border of the lids at their junction with the inner canthus. The puncta are directed posteriorly into the tear lake at the inner canthus. Each tiny opening, or *lacrimal punctum,* is 0.3 mm in diameter. The *inferior punctum* is 6.5 mm from the medial canthus; the *superior punctum* is 6.0 mm from it. These openings lead to the *lacrimal canaliculi,* the *lacrimal sac,* and finally the *nasolacrimal duct* to the nose. In 90% of subjects, the canaliculi join to form a common canaliculus. In about 30% of full-term neonates, the outlet of the nasolacrimal duct is closed and may remain so for up to 6 months. Occasionally, probing may be necessary to achieve patency. BCSC Section 7, *Orbit, Eyelids, and Lacrimal System,* discusses these issues in detail.

The lacrimal puncta and the canaliculi are lined with stratified squamous nonkeratinized epithelium that merges with the epithelium of the eyelid margins. Near the lacrimal sac, the epithelium changes to two layers: a superficial columnar layer and a deep, flattened cell layer. Goblet cells and occasional cilia are present. In the canaliculi, the substantia propria consists of collagenous connective tissue and elastic fibers. The wall of the sac resembles adenoid tissue and has a rich venous plexus and many elastic fibers. The angular vein and artery, important surgical landmarks in dacryocystorhinostomy, can be seen 3–4 mm medial to the inner canthus.

Conjunctiva

The *conjunctiva* can be divided into three geographic zones: palpebral, fornical, and bulbar (Fig I-38). The *palpebral part* begins at the mucocutaneous junction of the eyelid and covers its inner surface. This part adheres firmly to the tarsus. The tissue becomes redundant and freely movable in the fornices *(fornical conjunctiva),* where it becomes enmeshed with fibrous elements of the levator aponeurosis and Müller's muscle in the upper eyelid. In the lower eyelid, fibrous expansions of the inferior rectus muscle sheath fuse with the inferior tarsal muscle, the equivalent of Müller's muscle. The conjunctiva is reflected at the cul-de-sac and attaches to the globe. The delicate *bulbar conjunctiva* is freely movable but fuses with Tenon's capsule and inserts into the limbus.

Anterior ciliary arteries supply blood to the bulbar conjunctiva. The tarsal conjunctiva is supplied by branches of the marginal arcades of the lids. The proximal arcade, running along the upper border of the lid, sends branches proximally to supply the fornical and then the bulbar conjunctiva as the *posterior conjunctival arteries.* The limbal blood supply derives from the ciliary arteries through the anterior conjunctival arteries. The vascular watershed between the anterior and posterior territories lies about 3 or 4 mm from the limbus. Its innervation is derived from the ophthalmic division of cranial nerve V. The conjunctiva is a mucous membrane consisting of a nonkeratinizing squamous epithelium with numerous goblet cells and a thin, richly vascularized substantia propria containing lymphatic vessels, plasma cells, macrophages, and mast cells. A lymphoid layer extends from the bulbar conjunctiva to the subtarsal folds of the lids. In places, specialized aggregations of *conjunctiva-associated lymphoid tissue (CALT)* correspond to *mucosa-associated lymphoid tissue (MALT)* elsewhere and comprise collections of T and B lymphocytes underlying a modified epithelium. These regions are concerned with antigen processing.

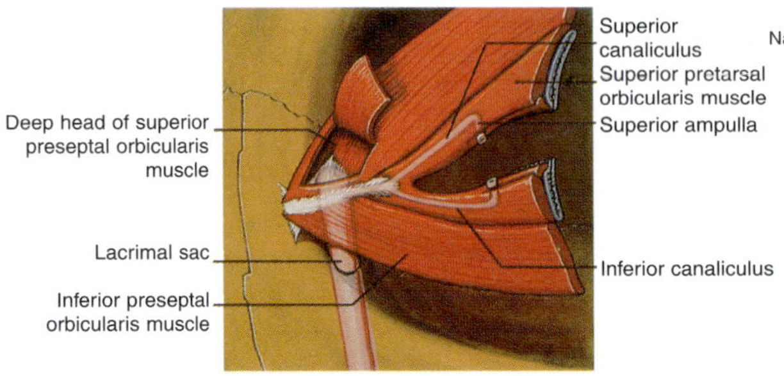

FIG I-36—Lacrimal drainage system and the orbicularis muscle. (Reproduced with permission from Dutton JJ. *Atlas of Clinical and Surgical Orbital Anatomy.* Philadelphia: Saunders; 1994.)

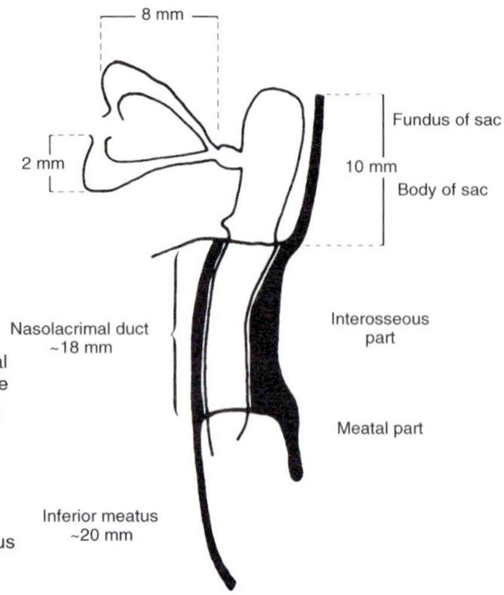

FIG I-37—Lacrimal excretory system. (Illustration by Thomas A. Weingeist, PhD, MD.)

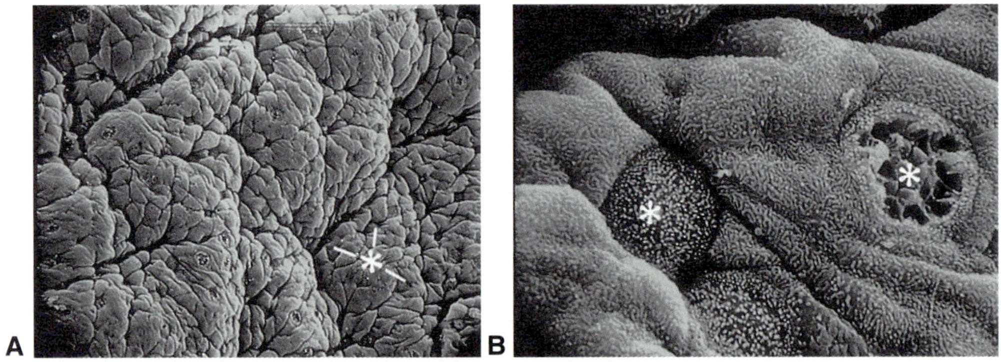

FIG I-38—Scanning electron micrographs of bulbar conjunctiva. **A**, Low magnification. **B**, Higher magnification showing nonkeratinized epithelium with interspersed goblet cells *(*)* at different stages of secretion. Before secreting, the goblet cell surface has many microvilli, which are lost as mucin accumulates within the cell in preparation for secretion.

The *conjunctival epithelium* varies from two to five cells in thickness. The basal cells are cuboidal and evolve into flattened polyhedral cells as they reach the surface. The goblet cells (unicellular mucous glands) are concentrated in the inferior and medial portion of the conjunctiva, especially in the region of the caruncle and plica semilunaris. They are sparsely distributed throughout the remainder of the conjunctiva and are absent in the limbal region.

Knop N, Knop E. Conjunctiva-associated lymphoid tissue in the human eye. *Invest Ophthalmol Vis Sci.* 2000;41:1270–1279.

Tenon's Capsule

Tenon's capsule is composed entirely of compactly arranged collagen fibers and a few fibroblasts. Anteriorly, it fuses with the conjunctiva slightly posterior to the corneoscleral junction. Posteriorly, it is perforated by the optic nerve sheath and by the posterior ciliary vessels and nerves and becomes more attenuated (see Fig I-28A). The vortex veins pass through the capsule near the equator of the globe. Tenon's capsule and the intermuscular fibrous membranes surrounding the four rectus muscles fuse to form a type of fibrous sling or support. The position of the recti relative to the orbit is further stabilized during eye movements by *connective tissue sleeves* or *pulleys* through which the recti pass close to the globe. These pulleys contain collagen, elastin, and smooth muscle and are most developed around the horizontal recti (see Fig I-28B). They probably influence the direction of pull of the extraocular muscles during eye movements. Additional fibromuscular septa extend from the pulleys and Tenon's capsule to the orbital walls.

Clark RA, Rosenbaum AL, Demer JL. Magnetic resonance imaging after surgical transposition defines the anteroposterior location of the rectus muscle pulleys. *J AAPOS.* 1999;3(1):9–14.

Demer JL, Miller JM, Poukens V, et al. Evidence for fibromuscular pulleys of the recti extraocular muscles. *Invest Ophthalmol Vis Sci.* 1995;36(6):1125–1136.

Demer JL, Poukens V, Miller JM, et al. Innervation of extraocular pulley smooth muscle in monkeys and humans. *Invest Ophthalmol Vis Sci.* 1997;38(9):1774–1785.

Koornneef L. *Spatial Aspects of Orbital Musculo-fibrous Tissue in Man.* Amsterdam: Swets & Zeitlinger; 1977.

The *suspensory ligament of Lockwood* is a fusion of the sheath of the inferior rectus muscle, the inferior tarsal muscle, and the check ligaments of the medial and lateral rectus muscles. It supports the globe and attaches to the medial and lateral retinacula. The retinacula are formed from thickened periosteum attached to the zygomatic bone in the lateral canthal region and to the lacrimal bone in the medial canthal area.

Vascular Supply and Drainage of the Orbit (Figs I-39 through I-44)

Anterior and Posterior Ciliary Arteries

Approximately 20 short *posterior ciliary arteries* and 10 short *posterior ciliary nerves* enter the globe in a ring around the optic nerve (see Figs I-39 through I-41). Usually, two long ciliary arteries and nerves enter the sclera on either side of the optic nerve close to the horizontal meridian. The course of these vessels can usually be followed

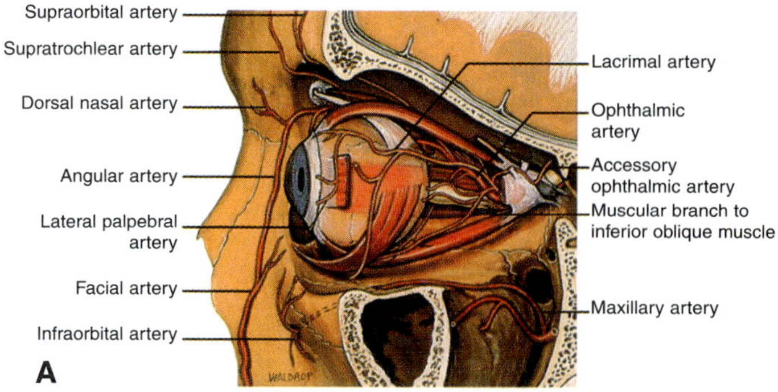

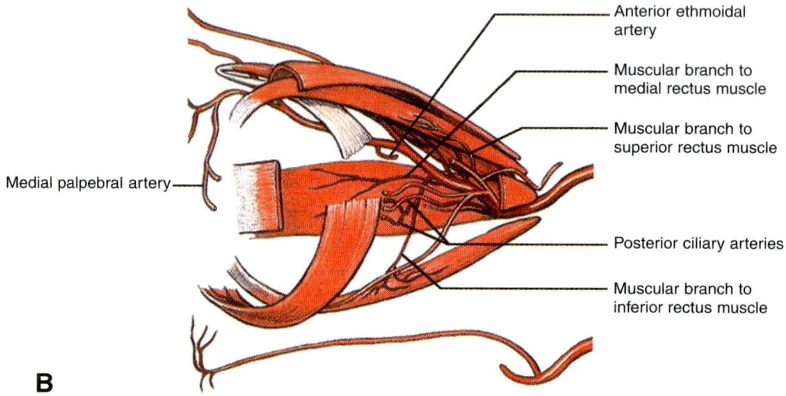

FIG I-39—Orbital arteries. **A**, Lateral view with extraocular muscles, composite view. **B**, Central dissection. (Reproduced with permission from Dutton JJ. *Atlas of Clinical and Surgical Orbital Anatomy.* Philadelphia: Saunders; 1994.)

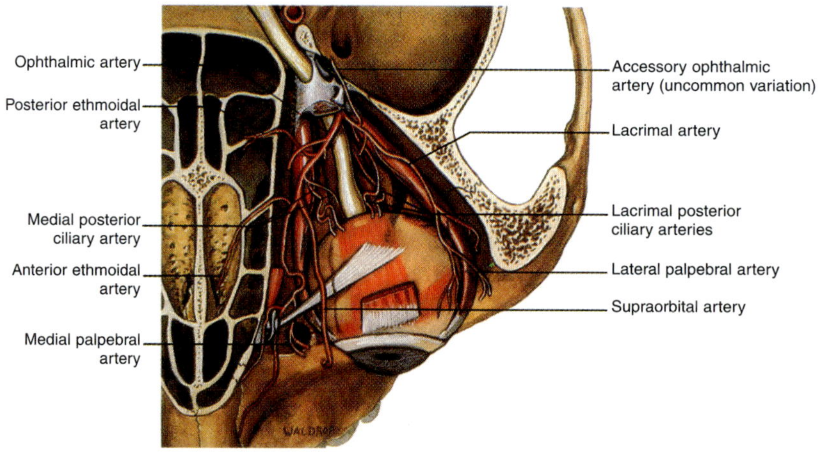

FIG I-40—Orbital arteries, superior composite view. (Reproduced with permission from Dutton JJ. *Atlas of Clinical and Surgical Orbital Anatomy.* Philadelphia: Saunders; 1994.)

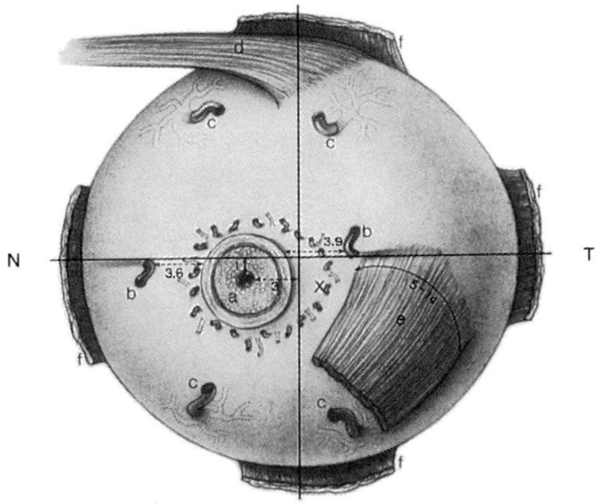

FIG I-41—Posterior view of the right globe. *N,* nasal; *T,* temporal; *a,* optic nerve; *b,* long posterior ciliary arteries and nerves; *c,* vortex veins; *d,* superior oblique muscle; *e,* inferior oblique muscle; *f,* rectus muscles; *X,* approximate position of the macula. (From Hogan MJ, Alvarado JA, Weddell JE. *Histology of the Human Eye.* Philadelphia: Saunders; 1971.)

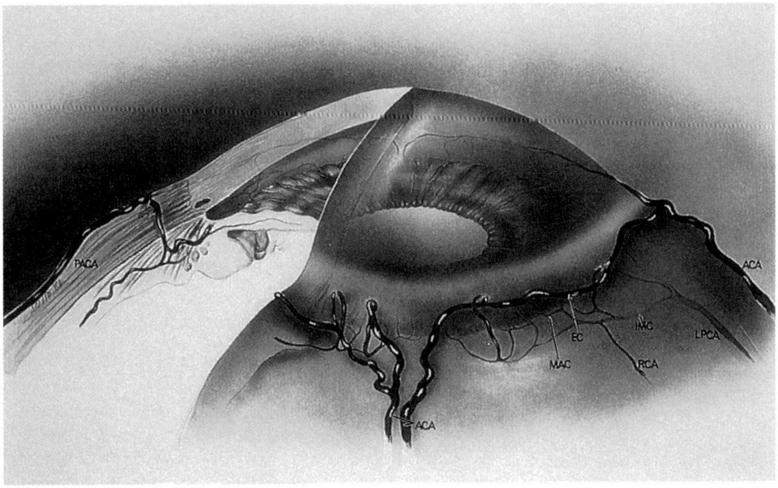

FIG I-42—Three-dimensional representation of the multilevel collateral circulation in the primate anterior uvea in both surface and cutaway views. *ACA,* anterior ciliary artery; *LPCA,* long posterior ciliary artery; *PACA,* posterior perforating anterior ciliary artery; *RCA,* recurrent ciliary artery; *EC,* episcleral circle; *IMC,* intramuscular circle; *MAC,* major arterial circle. To the left, in cross section, perforating branches of the anterior ciliary artery are shown as they pass through the sclera to supply the intramuscular circle and major arterial circle. (Reproduced with permission from Morrison JC, van Buskirk EM. Anterior collateral circulation in the primate eye. *Ophthalmology.* 1983;90:707.)

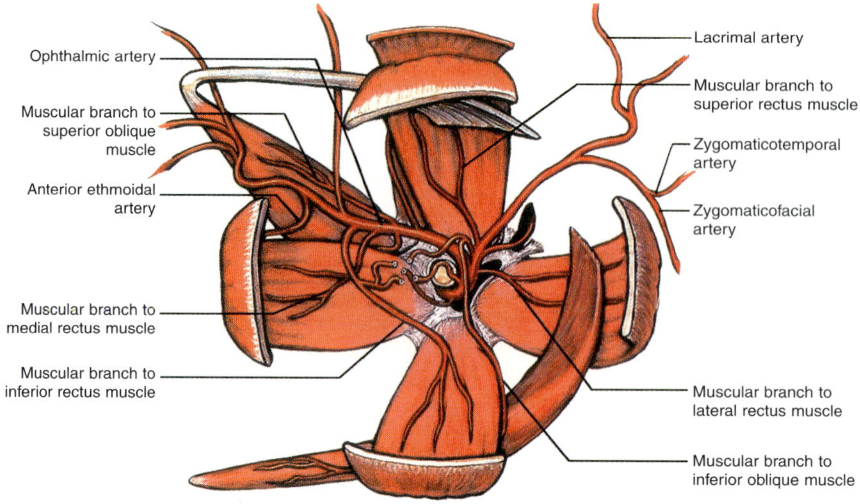

FIG I-43—Orbital arteries, frontal view with extraocular muscles. (Reproduced with permission from Dutton JJ. *Atlas of Clinical and Surgical Orbital Anatomy.* Philadelphia: Saunders; 1994.)

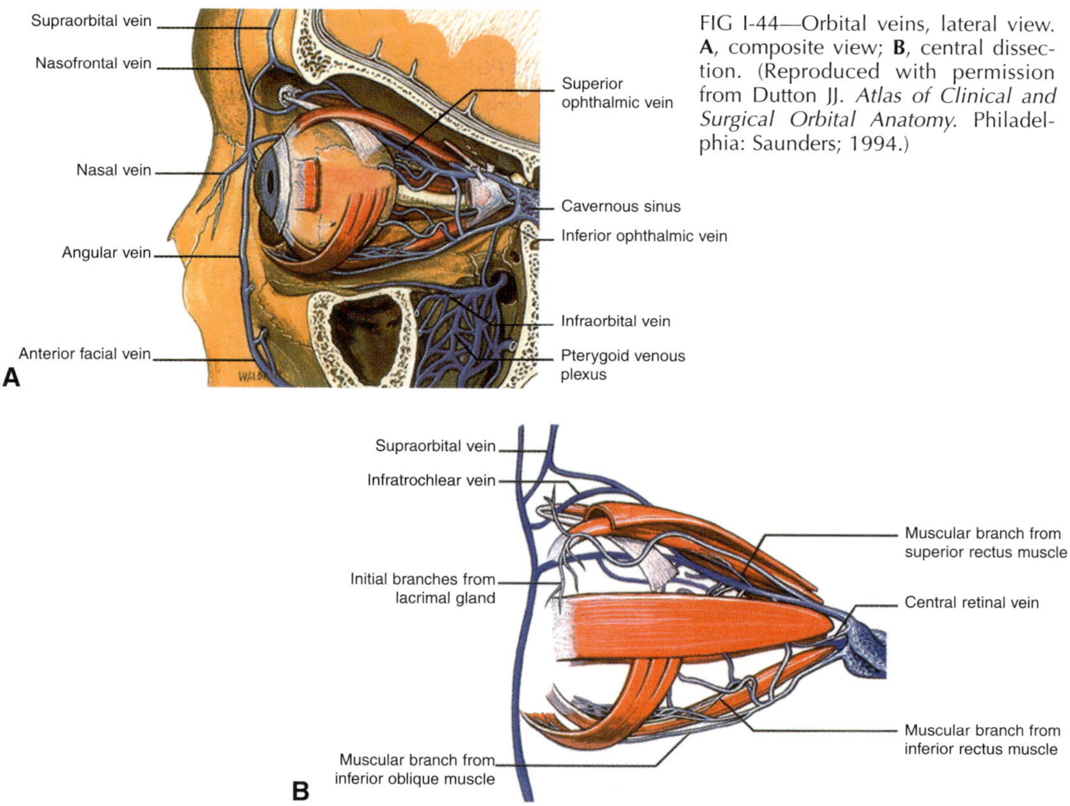

FIG I-44—Orbital veins, lateral view. **A**, composite view; **B**, central dissection. (Reproduced with permission from Dutton JJ. *Atlas of Clinical and Surgical Orbital Anatomy.* Philadelphia: Saunders; 1994.)

for a short distance in the suprachoroidal space. The posterior ciliary vessels originate from the ophthalmic artery and supply the whole uveal tract, the cilioretinal arteries, the sclera, the margin of the cornea, and the adjacent conjunctiva. Occlusion of the posterior ciliary vessels (as in giant cell arteritis) may have profound consequences for the eye, such as anterior ischemic optic neuropathy.

Hayreh SS, Baines JA. Occlusion of the posterior ciliary artery. I, Effects on choroidal circulation. II, Chorio-retinal lesions. III, Effects on the optic nerve head. *Br J Ophthalmol.* 1972;56:719–764.

The *anterior ciliary arteries* also arise from the ophthalmic artery and usually supply (in pairs) the superior, medial, and inferior rectus muscles (see Fig I-43). A single *anterior ciliary vessel* enters the lateral rectus muscle from the lacrimal artery. The anterior and posterior ciliary vessels usually anastomose with the long posterior ciliary vessels via anastomoses that perforate the sclera anterior to the rectus muscle insertions. Within the eye, the posterior ciliary vessel forms the *intramuscular circle of the iris,* from which branches supply the *major arterial circle* (which is usually discontinuous). This circle lies within the apex of the ciliary muscle, which it supplies together with the iris. The iris vessels have a radial arrangement that is visible upon slit-lamp examination in lightly pigmented blue irises. The radial arrangement is clearly distinguishable from the irregular new iris vessels formed in *rubeosis iridis.* Anterior segment ischemia often results in segmental areas of sphincter palsy in the pupil and in iris atrophy.

Hayreh SS, Scott WE. Fluorescein iris angiography. I, Normal pattern. II, Disturbances in iris circulation following strabismus operation on the various recti. *Arch Ophthalmol.* 1978;96:1383–1400.

Vortex Veins

The *vortex veins* drain the venous system of the choroid, ciliary body, and iris. Each eye contains four to seven (or more) veins. One or more veins are usually located in each quadrant and exit 14–25 mm from the limbus between the rectus muscles. The ampullae of the vortex veins are 8–9 mm from the ora serrata and are visible by indirect ophthalmoscopy. A circle connecting these ampullae corresponds roughly to the equator and divides the central or posterior fundus from the peripheral portion.

Rutnin U. Fundus appearance in normal eyes. I, The choroid. *Am J Ophthalmol.* 1967; 64(5):821–839.

Rutnin U, Schepens CL. Fundus appearance in normal eyes. II, The standard peripheral fundus and developmental variations. *Am J Ophthalmol.* 1967;64(5):840–852.

CHAPTER II

Topographic Features of the Globe

The adult human eye averages 24 mm in diameter. The normal anteroposterior diameter varies between 21 and 26 mm. This measurement is characteristically smaller in hyperopia and larger in myopia and buphthalmos. The anteroposterior diameter is approximately 16 mm at birth but reaches about 23 mm by 3 years of age. The eye reaches maximum size before puberty. The transverse vertical diameter is less variable.

Important surface features of the globe, such as the vortex veins, the posterior ciliary arteries and nerves, and the optic nerve and its surrounding meningeal sheaths, are illustrated in chapter I.

Cornea

The cornea occupies the center of the anterior pole of the globe. Since the sclera and conjunctiva overlap the cornea anteriorly, slightly more above and below than medially and laterally, the cornea appears *elliptical* when viewed from the front. In the adult, it measures about 12 mm in the horizontal meridian and about 11 mm in the vertical. From behind, viewed at its posterior landmark (Schwalbe's line—the termination of Descemet's membrane), the circumference of the cornea appears circular. The limbus, which borders the cornea and the sclera, is gray and translucent. The extraocular muscles insert at various distances posterior to the limbus, as shown in Figure I-14 in chapter I. Also see Table I-1.

Sclera

In contrast to the transparent cornea, the sclera is opaque and white. It is thinnest at the insertions of the rectus muscles (0.3 mm) and increases to about 1 mm thick posteriorly. The sclera becomes thin and sievelike at the lamina cribrosa, where the axons of the ganglion cells exit to form the optic nerve.

The insertions of the extraocular oblique muscles can be seen from the posterior aspect of the eyeball, as shown in Figure I-41. The inferior oblique muscle has little (if any) tendon. Its medial border inserts near the fovea, and its lateral border inserts more anteriorly. The superior oblique muscle inserts via its tendon mostly posteriorly at the anatomical equator of the eye and temporal to the vertical meridian. The muscle passes anteriorly on its way to the trochlea (see Fig I-15).

CHAPTER III

The Eye

Precorneal Tear Film

The exposed surfaces of the cornea and globe are covered by the tear film, which is composed of three layers:

- A *superficial oily layer* produced predominantly by the meibomian glands
- A *middle aqueous layer* produced by the main and accessory lacrimal glands
- A *deep mucin layer* derived from the conjunctival goblet cells; the surface cells of the cornea and conjunctiva also express a mucinous glycocalyx

Maintenance of the precorneal tear film is vital for normal corneal function. In addition to lubricating the surface of the cornea and conjunctiva, tears produce a smooth optical surface, provide oxygen and other nutrients, and contain immunoglobulins, lysozyme, and lactoferrin. Aberrations in the tear film result from a variety of diseases (eg, dry eye) that profoundly affect the integrity of the surface.

Cornea (Fig III-1)

Characteristics of the Central and Peripheral Cornea

The cornea and aqueous humor together form a positive lens of about 43 D in air and constitute the main refractive element of the eye. The central third of the cornea is nearly spherical and measures about 4 mm in diameter in the normal eye. Because the posterior surface of the cornea is more curved than the anterior surface, the central cornea is thinner (0.5 mm) than the peripheral cornea (0.7 mm). The cornea becomes flatter in the periphery, but the rate of flattening is not symmetrical. Flattening is more extensive nasally and superiorly than temporally and inferiorly. This topography is important in contact lens fitting. BCSC Section 8, *External Disease and Cornea,* discusses the cornea in detail.

Epithelium and Basal Lamina

The anterior surface of the cornea is derived from surface ectoderm and is covered by a nonkeratinized, stratified squamous epithelium whose basal columnar layer is attached to a basal lamina by *hemidesmosomes* (Fig III-2). The basal cells have a width of 12 µm and a density of about 6000 cells/mm^2. The hemidesmosomes contain bullous pemphigoid antigens and $\alpha_6\beta_4$ integrin, which attaches to the *anchoring filaments* (laminin type 5) of the basal lamina. Attachment is further extended into Bowman's layer by *anchoring fibrils* (type VII collagen), which terminate in an *anchoring plaque*. The occasional recurrence of corneal erosion following a traumatic corneal abrasion may be due to improper formation of hemidesmosomes after an epithelial abrasion.

Overlying the basal cell layer are two or three layers of polygonal "wing" cells. The superficial corneal epithelial cells are extremely thin (30 µm) and are attached

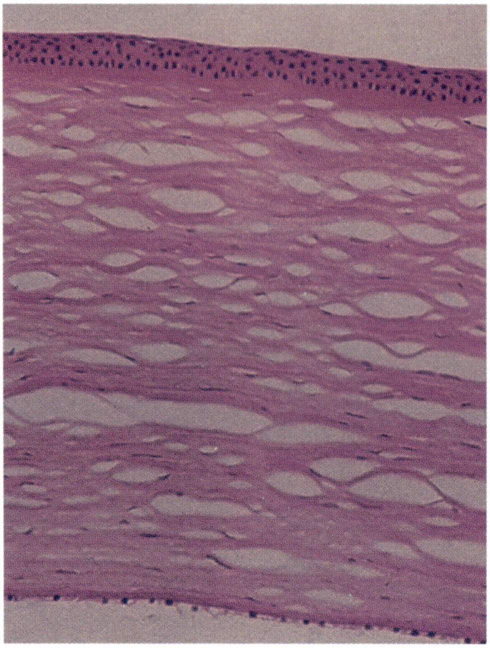

FIG III-1—Cornea. The empty spaces in the stroma are artifactitious (H&E ×32). (Photograph courtesy of Thomas A. Weingeist, PhD, MD.)

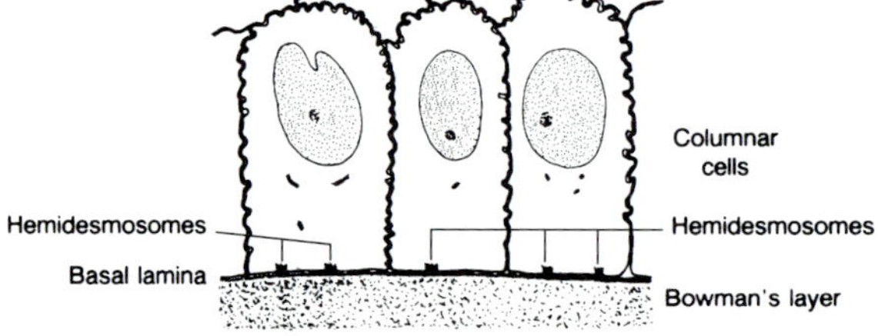

FIG III-2—The corneal epithelium and Bowman's layer, showing hemidesmosomes along the basal lamina. (Illustration by Thomas A. Weingeist, PhD, MD.)

to one another by *occluding zonules*. These zonules confer the properties of a semipermeable membrane to the epithelium. Microplicae and microvilli make their apical surfaces highly irregular; however, the precorneal tear film renders the surfaces optically smooth. Although the deeper epithelial cells are firmly attached to one another by *desmosomes*, they migrate continuously from the basal region toward the tear film, into which they are shed; they also migrate centripetally from their *stem cell source* at the limbus. Division of the slow-cycling stem cells gives rise to a progeny of daughter cells *(transient amplifying cells)*, whose division serves to maintain the corneal epithelium. Diffuse damage to the limbal stem cells (eg, by chemical burns, trachoma) leads to chronic epithelial surface defects.

Fine BS, Yanoff M. *Ocular Histology: A Text and Atlas.* 2nd ed. Hagerstown, MD: Harper & Row; 1979:163–168.

Gipson IK, Spurr-Michaud SJ, Tisdale AS: Anchoring fibrils form a complex network in human and rabbit cornea. *Invest Ophthalmol Vis Sci.* 1987;28:212–220.

Nonepithelial Cells

Nonepithelial cells may appear within the corneal epithelial layer. Wandering histiocytes, macrophages, lymphocytes, and pigmented melanocytes are frequent components of the peripheral cornea. Antigen-presenting *Langerhans' cells* are found peripherally and move centrally with age or in response to keratitis.

Bowman's Layer

Beneath the basal lamina is *Bowman's membrane*, or *Bowman's layer*, which is a tough layer consisting of randomly dispersed collagen fibrils. It is 8–14 µm thick, and its posterior border merges with the corneal stroma. Unlike Descemet's membrane, it is not restored after injury but is replaced by scar tissue.

Stroma

The *stroma* constitutes about 90% of the total corneal thickness in humans. It is composed of collagen-producing keratocytes, ground substance, and collagen lamellae. The collagen fibrils form obliquely oriented lamellae in the anterior third of the stroma (with some interlacing) and parallel lamellae in the posterior two thirds. The corneal collagen fibrils extend across the entire diameter of the cornea, finally winding circumferentially around the limbus. They are remarkably uniform in size and separation; this regularity determines the transparency of the cornea. Separation of the collagen fibrils by edema fluid leads to stromal clouding. The macroperiodicity of the fibrils (640 Å) is typical of collagen. The stroma's collagen types are I, III, V, and VI. Type VII forms the anchoring fibril of the epithelium.

The ground substance of the cornea consists of proteoglycans that run along and between the collagen fibrils. Their glycosaminoglycan components (eg, keratan sulfate) are highly charged and account for the swelling property of the stroma. The keratocytes lie between the corneal lamellae and synthesize both collagen and proteoglycans. Ultrastructurally, they resemble fibrocytes.

The cornea has about 2.4 million keratocytes, which occupy about 5% of the stromal volume; the density is higher anteriorly (1058 cells/mm^2) than posteriorly (771 cells/mm^2). Keratocytes are highly active cells, rich in mitochondria, rough endoplasmic reticulum, and Golgi apparatus. They have attachment structures, communicate by gap junctions, and have unusual fenestrations in their plasma membranes. Their flat profile and even distribution in the coronal plane ensure a minimum disturbance of light transmission. Studies with vital dyes suggest that there may be at least three different types of keratocyte.

Müller LJ, Pels L, Vrensen GF. Novel aspects of the ultrastructural organization of human corneal keratocytes. *Invest Ophthalmol Vis Sci.* 1995;36(13):2557–2567.

Mustonen RK, McDonald MB, Srivannaboon S, et al. Normal human corneal cell populations evaluated by in vivo scanning slit confocal microscopy. *Cornea.* 1998; 17(5):485–492.

Descemet's Membrane (Fig III-3)

The basal lamina of the corneal endothelium, *Descemet's membrane,* is periodic acid–Schiff (PAS)-positive (as is the lens capsule) and of unusual thickness. At birth, it is 3–4 µm thick; thickness increases throughout life to an adult level of 10–12 µm. It is composed of an *anterior banded zone* that develops in utero and a *posterior nonbanded zone* that is laid down by the corneal endothelium throughout life (Fig III-4). These zones provide a historical record of the synthetic function of the endothelium. Like other basal laminae, Descemet's membrane is rich in type IV collagen.

Peripheral excrescences of Descemet's membrane, known as *Hassall-Henle warts,* are common, especially among elderly persons. Central excrescences *(cornea guttata)* also appear with increasing age.

Murphy C, Alvarado J, Juster R. Prenatal and postnatal growth of the human Descemet's membrane. *Invest Ophthalmol Vis Sci.* 1984;25:1402–1415.

Endothelium

The *corneal endothelium* is composed of a single layer, mostly of hexagonal cells derived from the *neural crest* (Fig III-5). The corneal endothelium is therefore of neuroectodermal origin. About 500,000 cells are present, with a density of about 3000 cells/mm^2.

Johnston MC, Noden DM, Hazelton RD, et al. Origins of avian ocular and periocular tissues. *Exp Eye Res.* 1979;29:27–43.

The size, shape, and morphology of the endothelial cells can be observed by specular microscopy at the slit lamp. The apical surfaces of these cells face the anterior chamber; their basal surfaces abut Descemet's membrane. Typically, young endothelial cells have a large nucleus and abundant mitochondria. The active transport of ions by these cells leads to the transfer of water from the corneal stroma and the maintenance of stromal deturgescence and transparency. Mitosis of the endothelium is rare in humans, and the overall number of endothelial cells decreases with age.

Adjacent endothelial cells interdigitate in a complex way and form a variety of adherent junctions, but desmosomes are never seen between normal cells. In cross section, pinocytotic vesicles and a *terminal web* (a meshwork of fine fibrils that increases the density of the cytoplasm) can be seen toward the apical surface of the

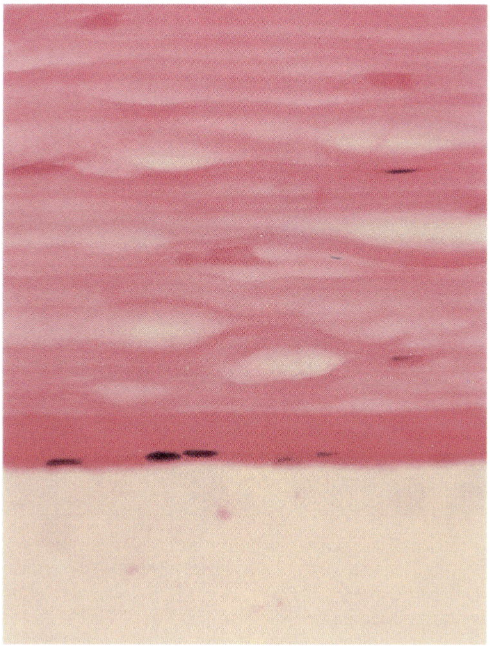

See Figure III-4 below for diagram of this portion

FIG III-3—Posterior cornea. Note the appearance of Descemet's membrane and the corneal endothelium (H&E ×64). (Photograph courtesy of Thomas A. Weingeist, PhD, MD.)

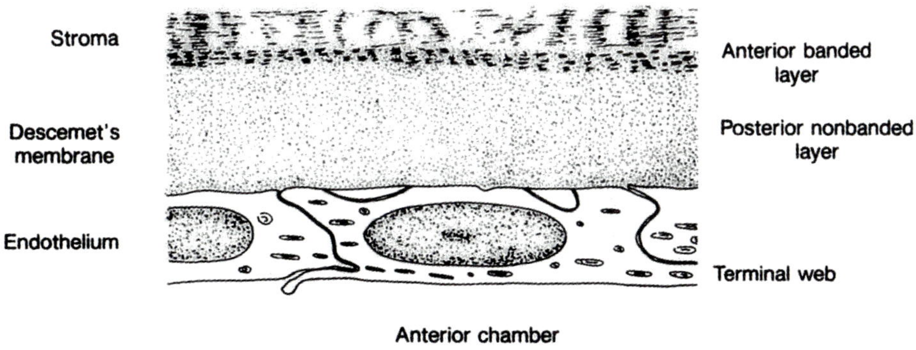

FIG III-4—Corneal endothelium and Descemet's membrane. (Illustration by Thomas A. Weingeist, PhD, MD.)

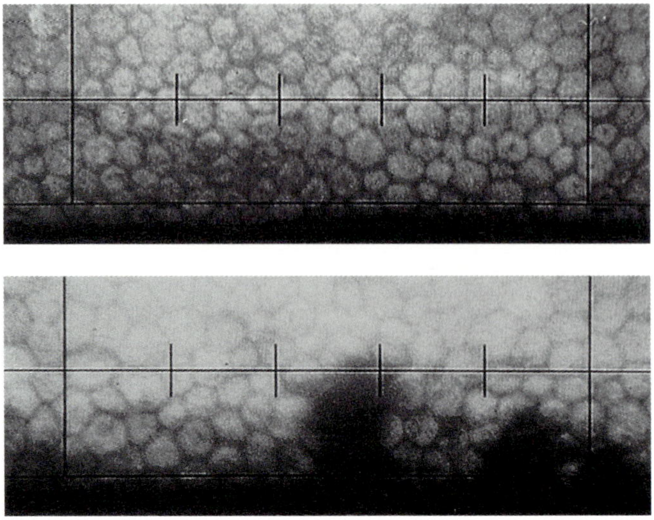

A

B

FIG III-5—Specular micrographs of the corneal endothelium. **A**, Normal patient. **B**, Patient with Fuchs endothelial dystrophy. Both are taken at the same magnification. Bottom micrograph shows larger, more irregular cells (polymegethism); the three dark areas toward the bottom are cornea guttata. (Photographs courtesy of David Palay, MD, and David Litoff, MD.)

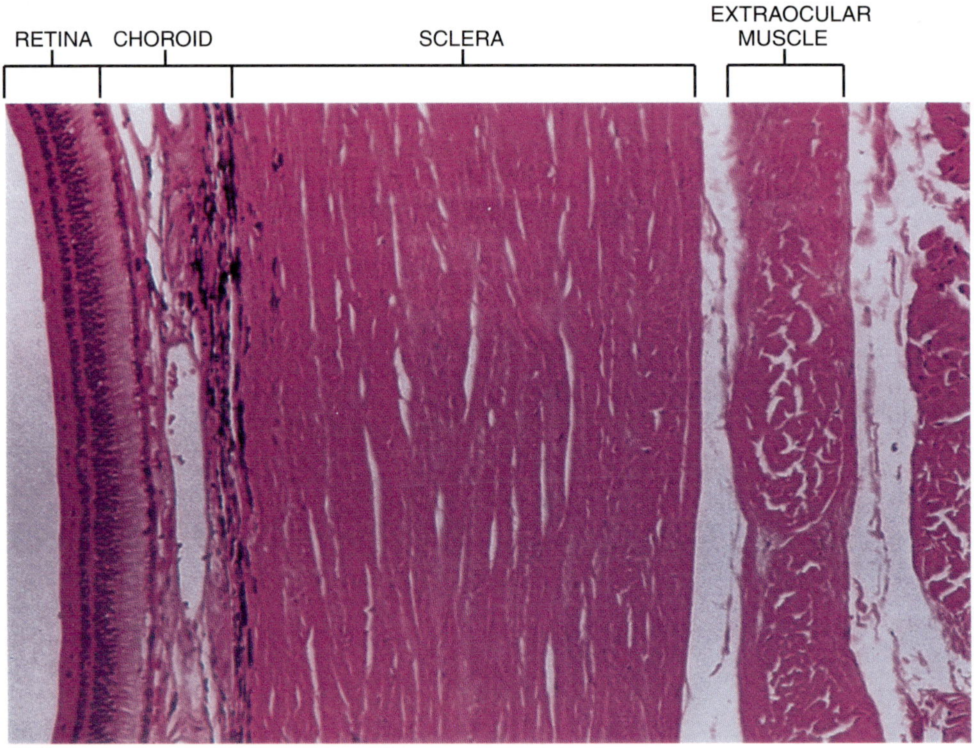

FIG III-6—Retina, choroid, sclera, and extraocular muscle (H&E ×8). (Photograph courtesy of Thomas A. Weingeist, PhD, MD.)

cells. *Junctional complexes* are present at the overlapping apicolateral boundaries of contiguous cells. They form a significant but lesser barrier to ion and water flow than the tight junctions of the epithelium.

Iwamoto T, Smelser GK. Electron microscopy of the human corneal endothelium with reference to transport mechanisms. *Invest Ophthalmol.* 1965;4:270–284.

Endothelial cell dysfunction and loss—through surgical injury, inflammation, or inherited disease (eg, Fuchs endothelial dystrophy)—may cause endothelial decompensation, stromal edema, and visual failure. In humans, endothelial mitosis is limited, and destruction of cells causes cell density to decrease and residual cells to spread and enlarge.

Sclera (Fig III-6)

The *sclera* covers the posterior four fifths of the surface of the globe, with an anterior opening for the cornea and a posterior opening for the optic nerve. The tendons of the rectus muscles insert into the superficial scleral collagen. Tenon's capsule invests the sclera and rectus muscles anteriorly, and both are overlain by the bulbar conjunctiva. The capsule and conjunctiva fuse near the limbus.

The sclera is thinnest (0.3 mm) behind the insertions of the rectus muscles and thickest (1.0 mm) at the posterior pole around the optic nerve head. It measures 0.4–0.5 mm at the equator and is 0.6 mm thick anterior to the muscle insertions. Because of the thinness of the sclera, strabismus and retinal detachment surgery require careful placement of sutures. Scleral rupture following blunt trauma can occur at a number of sites: in a circumferential arc parallel to the corneal limbus opposite the site of impact, at the insertion of the rectus muscles, or at the equator of the globe. The most common site is the superonasal quadrant near the limbus.

The sclera, like the cornea, is essentially avascular except for the superficial vessels of the episclera and the intrascleral vascular plexus located just posterior to the limbus. A number of channels, or *emissaria,* penetrate the sclera for the passage of arteries, veins, and nerves. Extraocular extension of malignant melanoma of the choroid often occurs by way of the emissaria.

Branches of the ciliary nerves that supply the cornea sometimes leave the sclera to form loops posterior to the nasal and temporal limbus. These *nerve loops* are often pigmented and have been mistaken for uveal tissue or malignant melanoma.

Anteriorly, the *episclera* consists of a dense vascular connective tissue that merges deeply with the superficial sclera and superficially with Tenon's capsule and the conjunctiva. The *scleral stroma* is composed of bundles of collagen, fibroblasts, and a moderate amount of ground substance. Collagen fibrils of the sclera vary in size and shape and have been shown to taper at their ends, indicating that they are not continuous fibers as in the cornea. In general, the outer scleral collagen fibers have a larger diameter (1600 Å) than the inner collagen fibers (1000 Å). The inner layer of the sclera *(lamina fusca)* blends imperceptibly with the suprachoroidal and supraciliary lamellae of the uveal tract. The collagen fibers in this portion of the sclera branch and intermingle with the outer ciliary body and choroid. The bundles of collagen fibers contain electron-dense bodies, fibroblasts, and melanocytes. The opaque, porcelain-white appearance of the sclera contrasts markedly with the transparency of the cornea and is primarily due to two things: the greater variation in fibril separation and diameter, and the greater degree of fibril interweave in the sclera.

Limbus

The transition zone between the peripheral cornea and the anterior sclera, known as the *limbus,* is defined differently by anatomists, pathologists, and clinicians. Although not a distinct anatomical structure, it is important for two reasons: its relationship to the chamber angle and its use as a surgical landmark. The following structures are included in the limbus:

- Conjunctiva and limbal palisades
- Tenon's capsule
- Episclera
- Corneoscleral stroma
- Aqueous outflow apparatus

The transition from opaque sclera to clear cornea occurs gradually over 1.0–1.5 mm and is difficult to define histologically. The corneoscleral junction begins centrally in a plane connecting the end of Bowman's layer and the termination of Descemet's membrane. Internally, its posterior limit is *Schwalbe's line,* the anterior tip of the scleral spur. Pathologists consider the posterior limit of the limbus to be formed by another plane perpendicular to the surface of the eye, about 1.5 mm posterior to the termination of Bowman's layer in the horizontal meridian and 2.0 mm posterior in the vertical meridian, where there is greater scleral overlap (Fig III-7).

The surgical limbus can be divided conceptually into two equal zones: (1) an anterior bluish gray zone overlying clear cornea and extending from Bowman's layer to Schwalbe's line and (2) a posterior white zone overlying the trabecular meshwork and extending from Schwalbe's line to the scleral spur or iris root. Familiarity with these landmarks is essential to the surgeon performing cataract extraction or a glaucoma-filtering procedure.

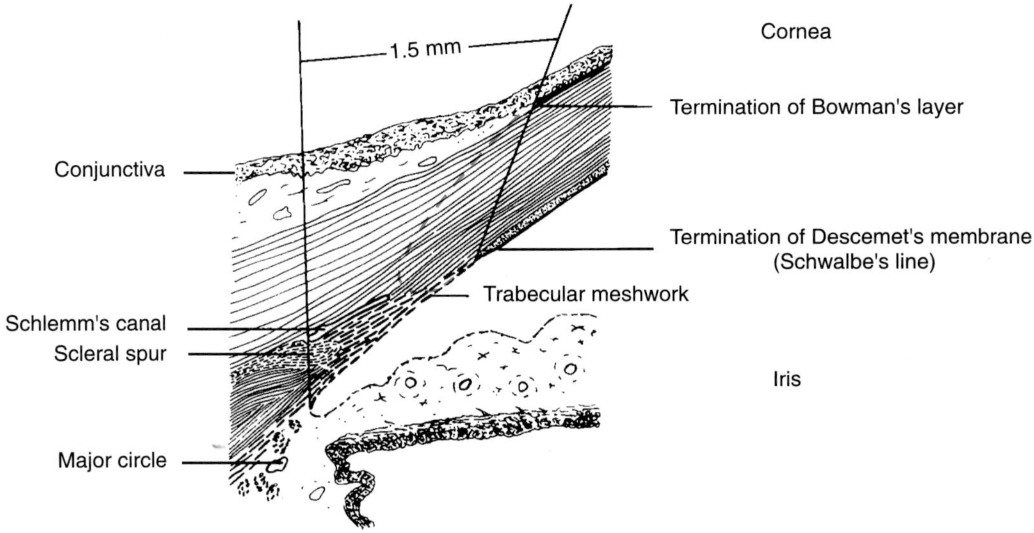

FIG III-7—Anterior chamber angle and limbus, depicting concept of limbus. *Solid lines* represent the limbus as seen by pathologists; the *green dotted line* represents the limbus as seen by anatomists. (Illustration by Thomas A. Weingeist, PhD, MD.)

Jaffe NS, Jaffe MS, Jaffe GF. *Cataract Surgery and Its Complications.* 5th ed. St Louis: Mosby; 1990:46–50.

Anterior Chamber

The *anterior chamber* is bordered anteriorly by the cornea and posteriorly by the iris diaphragm and the pupil. The *anterior chamber angle,* which lies at the junction of the cornea and the iris, consists of the following structures (Fig III-8):

- Schwalbe's line
- Schlemm's canal and the trabecular meshwork
- Scleral spur
- Anterior border of the ciliary body (where its longitudinal fibers insert into the scleral spur)
- Iris

The depth of the anterior chamber varies. It is deeper in aphakia, pseudophakia, and myopia and shallower in hyperopia. In the normal adult emmetropic eye, the anterior chamber is about 3 mm deep at its center and reaches its narrowest point slightly central to the angle recess. The volume of the anterior chamber is about 200 μl in the emmetrope.

The anterior chamber is filled with aqueous humor, which is produced by the ciliary epithelium in the posterior chamber. The fluid passes through the pupil aperture and drains chiefly by the *conventional pathway* through the trabecular meshwork into Schlemm's canal and partly by the nonconventional *uveoscleral drainage pathway,* across the ciliary body into the supraciliary space. The uveoscleral pathway accounts for less than 10% of aqueous outflow. BCSC Section 10, *Glaucoma,* discusses the anterior chamber and aqueous humor in detail.

High-resolution ultrasound biomicroscopy provides detailed two-dimensional views of the anterior segment of the eye and is performed in vivo (Fig III-9). This allows the clinician to view the relationship of the structures in the anterior segment under different pathologic conditions.

Pavlin CJ, Sherar MD, Foster FS. Subsurface ultrasound microscopic imaging of the intact eye. *Ophthalmology.* 1990;97:244–250.

The *internal scleral sulcus* accommodates the *canal of Schlemm* externally and *trabecular meshwork* internally. Schwalbe's line, the periphery of Descemet's membrane, forms the anterior margin of the sulcus; the *scleral spur* is its posterior landmark. The scleral spur receives the insertion of the longitudinal ciliary muscle, and contraction opens up the trabecular spaces. Contractile cells are found within the scleral spur, as are structures resembling mechanoreceptors, which receive a sensory innervation.

Myofibroblast-like *scleral spur cells* with contractile properties are disposed circumferentially within the scleral spur. They are connected by elastic tissue to the trabecular meshwork; experimentally, stimulation with vasoactive intestinal polypeptide (VIP) or calcitonin gene–related peptide (CGRP) causes an increase in outflow facility. Individual scleral spur cells are innervated by unmyelinated axons, the terminals of which contact the cell membranes of the spur cells without an intervening basal lamina. The nerve fibers in this region are immunoreactive for neuropeptide Y, substance P, CGRP, VIP, and nitrous oxide and therefore are mediated by sympathetic, sensory, and pterygopalatine nerve pathways. There are no cholinergic fibers.

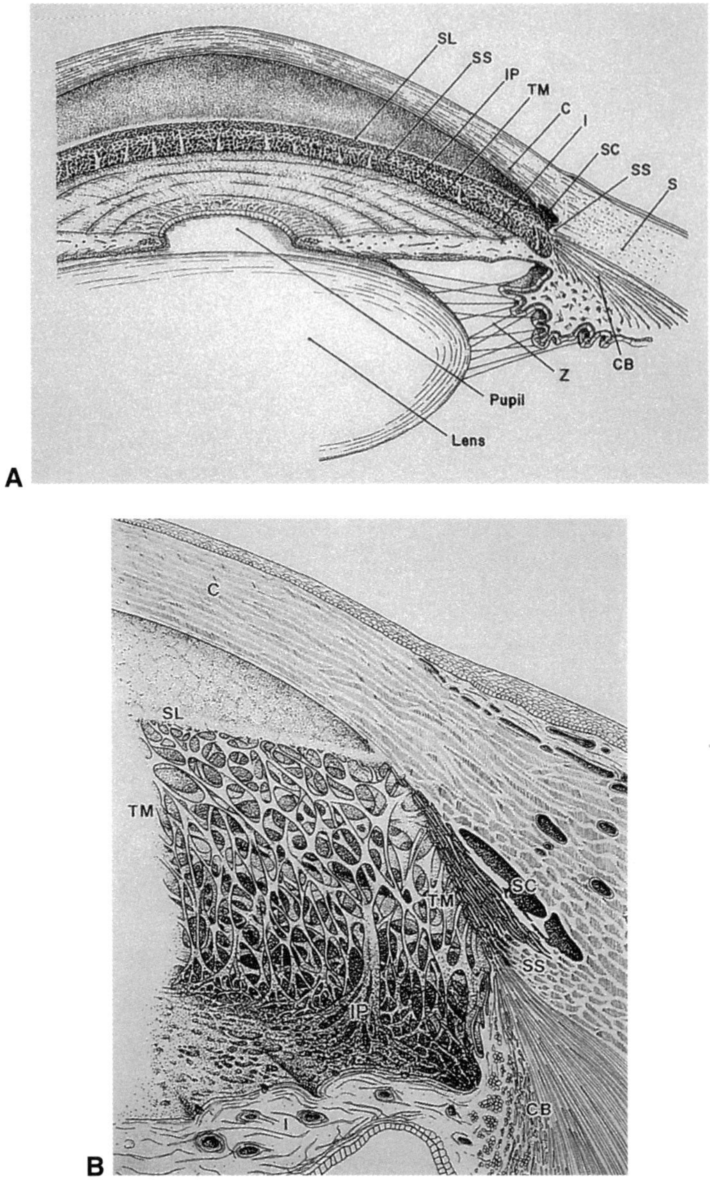

FIG III-8—Semidiagrammatic representation of the structures of the angle of the anterior chamber and ciliary body. **A**, Composite gonioscopic and cross-sectional view of the anterior segment of the eye. **B**, Enlarged view. Note the superimposed trabecular sheets with intratrabecular spaces through which aqueous humor percolates to reach Schlemm's canal. *SL,* Schwalbe's line; *SS,* scleral spur; *IP,* iris process; *TM,* trabecular meshwork; *C,* cornea; *I,* iris; *SC,* Schlemm's canal; *S,* sclera; *CB,* ciliary body; *Z,* zonular fibers.

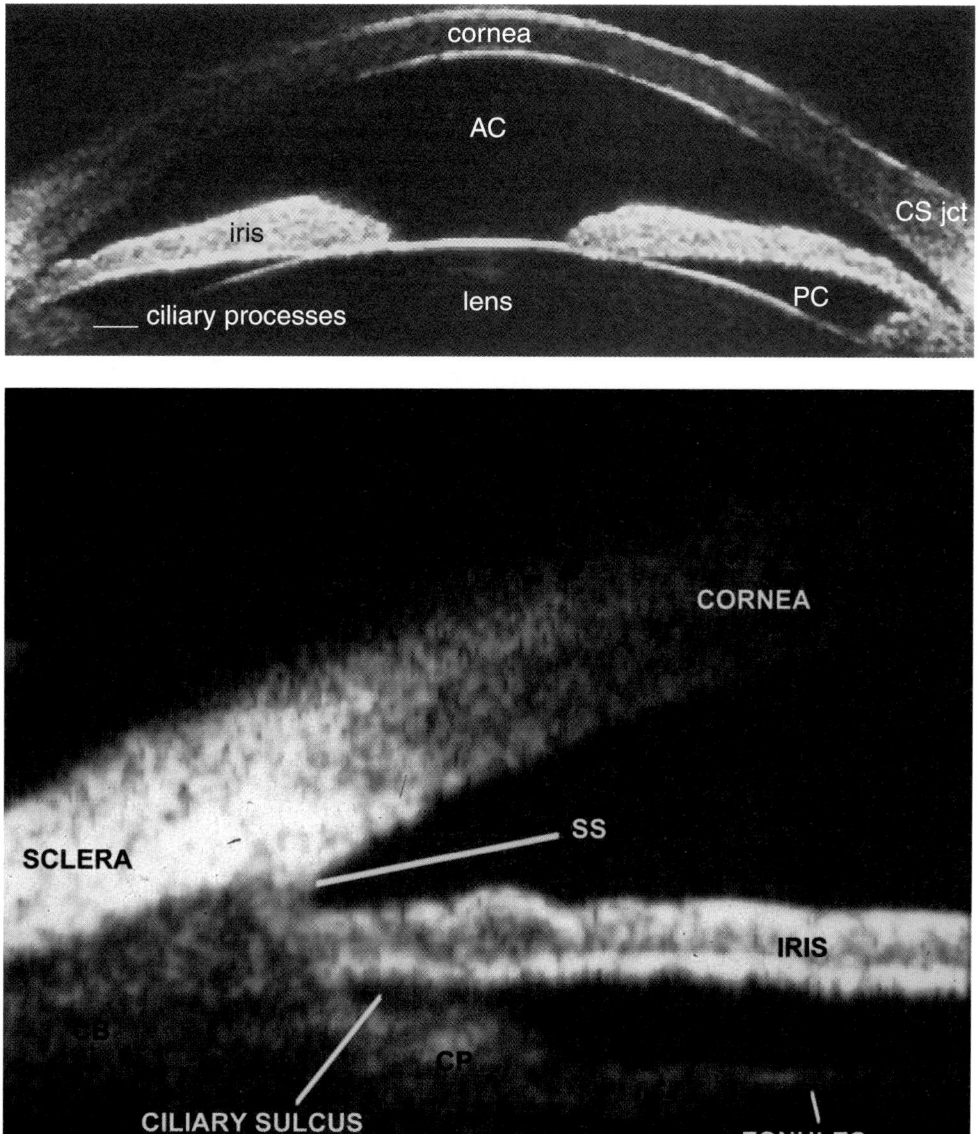

FIG III-9—**A**, Ultrasound biomicroscopic composite image of the anterior segment, including the anterior chamber *(AC)*. The iris is slightly convex, indicating mild pupillary block. The corneoscleral junction *(CS jct)*, ciliary processes, and posterior chamber *(PC)* region are clearly imaged. The angle is narrow but open. Iris–lens contact is small. (Photograph courtesy of Charles Pavlin, MD.) **B**, High-resolution ultrasound image of the anterior segment. Note the location of the ciliary sulcus. *CB,* ciliary body; *CP,* ciliary process; *SS,* scleral spur. (Photograph courtesy of Dr. K. Nischal, MD.)

Tamm ER, Koch TA, Mayer B, et al. Innervation of myofibroblast-like scleral spur cells in human and monkey eyes. *Invest Ophthalmol Vis Sci.* 1995;36:1633–1644.

Myelinated nerve fibers passing forward from the ciliary region to the inner aspect of the scleral spur give branches both to the meshwork and to *club-shaped endings* in the scleral spur. These endings have the morphologic features of mechanoreceptors found elsewhere in the body such as in the carotid. The endings are incompletely covered by a Schwann-cell sheath and make contact with extracellular matrix materials such as elastin. Various functions have been proposed for these endings, including (1) proprioception to the ciliary muscle, which inserts into the scleral spur, signaling contraction of the scleral spur cells, and (2) baroreception in response to changes in intraocular pressure.

Trabecular Meshwork

The relationship of the trabecular meshwork (see Fig III-7; Fig III-8) and Schlemm's canal to other structures is complex because the outflow apparatus is composed of tissue derived from cornea, sclera, iris, and ciliary body (see Fig III-9; Fig III-10).

The *trabecular meshwork* is a circular spongework of connective tissue lined by *trabeculocytes.* These cells have contractile properties, which may influence outflow resistance. They also have phagocytic properties. The meshwork is roughly triangular in cross section, with the apex at Schwalbe's line and the base formed by the scleral spur and the ciliary body. Some trabecular tissue passes posterior to the spur. The trabecular meshwork can be divided into three parts:

- Inner uveal portion
- Outer corneoscleral meshwork
- Pericanalicular tissue, which is directly adjacent to Schlemm's canal

The uveal and corneoscleral meshwork may be divided by an imaginary line drawn from Schwalbe's line to the scleral spur. The uveal meshwork lies internal and the corneoscleral meshwork lies external to this line.

Uveal Trabecular Meshwork

The *uveal meshwork* is composed of cordlike trabeculae with fewer elastic fibers than in the corneoscleral meshwork. The trabeculocytes usually contain pigment granules, and the trabecular apertures are less circular and larger than those of the corneoscleral meshwork.

Corneoscleral Meshwork

The *corneoscleral meshwork* consists of a series of thin, flat, perforated connective tissue sheets arranged in a laminar pattern. In both regions, each trabecular beam is covered by a monolayer of thin trabecular cells exhibiting multiple pinocytotic vesicles. The basal lamina of these cells forms the outer *cortex* of the trabecular beam; the inner *core* is of collagen and elastic fibers. There is an outer zone with circularly arranged collagen and an inner zone where the fibers are parallel to the axis of the beam or sheet. In the uveal sheets, the elastin is usually central in the core; in the corneoscleral meshwork, elastin is also found around the core. The presence of elastic tissue may impart an elastic recoil to the meshwork, promoting recovery after deformation, such as that induced by ciliary muscle contraction.

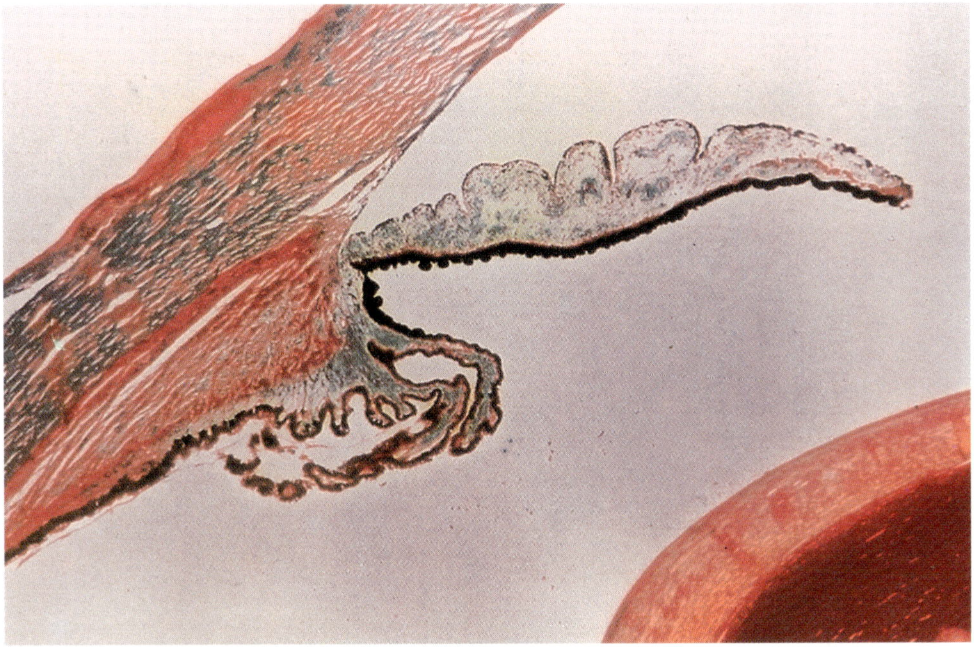

FIG III-10—Anterior chamber angle, ciliary body, and peripheral lens. Note the triangular shape of the ciliary body. The muscle fibers appear red in contrast with the connective tissue. The scleral spur is clearly delineated from the ciliary muscle in the region of the trabecular meshwork. The lens is artifactually displaced posteriorly. (Masson trichrome ×8). (Photograph courtesy of Thomas A. Weingeist, PhD, MD.)

Pericanalicular Connective Tissue

Pericanalicular connective tissue invests Schlemm's canal in its entire extent. On its trabecular aspect, between the outermost layers of the corneoscleral meshwork and the endothelial lining of Schlemm's canal, lies the *endothelial meshwork,* a multilayered collection of cells forming a loose network. Spaces exist between these cells, up to 10 μm in width, through which aqueous humor can percolate to reach the endothelial lining of Schlemm's canal. This region of the drainage system makes the greatest contribution to outflow resistance, partly because the pathway is narrow and tortuous and partly because of the resistance offered by extracellular proteoglycans and glycoproteins.

Schlemm's Canal

Schlemm's canal is a circular tube closely resembling a lymphatic vessel. It is formed by a continuous monolayer of nonfenestrated endothelium and a thin connective tissue wall. The basement membrane of the endothelium is poorly defined. The lateral walls of the endothelial cells are joined by tight junctions. Micropinocytotic vesicles are present at both the apical and the basal surfaces of the cells. Larger vesicles (so-called *giant vacuoles*) have been observed along the internal canal wall (Figs III-11,

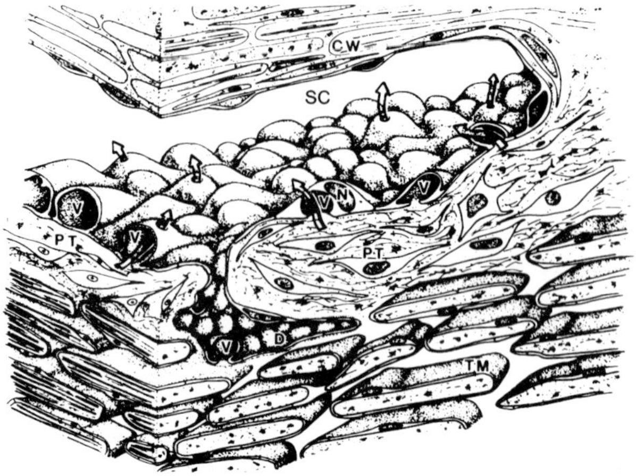

FIG III-11—The walls of Schlemm's canal *(SC)* and adjacent trabecular meshwork *(TM)*. The endothelial lining of the trabecular wall of Schlemm's canal is very irregular; normally, the cells show luminal bulges corresponding to cell nuclei *(N)* and macrovacuolar configuration *(V)*. The latter represents cellular invaginations from the basal aspect that eventually open on the apical aspect of the cell to form transcellular channels *(arrows)* through which aqueous humor flows down a pressure gradient. A diverticulum *(D)*—its endothelial lining continuous with that of the canal—is shown on the inner wall of Schlemm's canal next to macrovacuolar configurations. Such blind, tortuous diverticula course for a variable distance into the trabecular meshwork but remain separated from the open spaces of the meshwork by their continuous endothelial lining. The endothelial lining of the trabecular wall is supported by interrupted, irregular basement membrane and a zone of pericanalicular connective tissue *(PT)* of variable thickness. The cellular element predominates in this zone, and the fibrous elements, especially elastic fibers, are irregularly arranged in a netlike fashion. Here, the open spaces are narrower than those of the trabecular meshwork. The corneoscleral trabecular sheets show frequent branching, and the endothelial covering may be shared between adjacent sheets. The corneoscleral wall *(CW)* of Schlemm's canal is more compact than the trabecular wall, with a predominance of lamellar arrangement of collagen and elastic tissue.

III-12). These vacuoles are lined by a single membrane, and their size and number are increased by increasing intraocular pressure. They are thought to contribute to the pressure-dependent outflow of aqueous.

Collector Channels

About 25–30 *collector channels* arise from Schlemm's canal (Fig III-13) and drain into the deep and midscleral venous plexuses. Up to eight of these channels drain directly into the episcleral venous plexus as *aqueous veins,* which are visible in the conjunctiva by biomicroscopy.

Aging brings about a two- to three-fold thickening of trabecular sheets; the cortex thickens and the core thins. There is loss of endothelial cellularity, an increase in connective tissue (eg, in the endothelial meshwork), and an accumulation of debris in the meshwork and of glycosaminoglycans in the extracellular space. Such changes are exaggerated in chronic open-angle glaucoma.

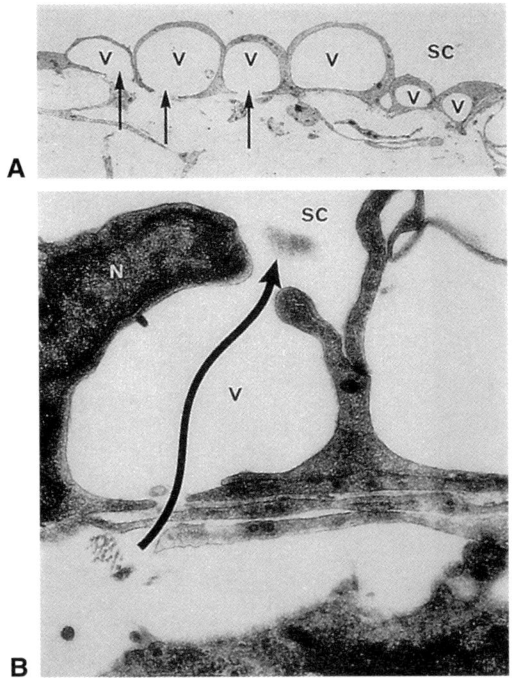

FIG III-12—**A**, Low-magnification electron micrograph of the endothelial lining of Schlemm's canal *(SC)* shows that the majority of the vacuolar configurations *(V)* at this level of section have direct communications *(arrows)* with the subendothelial extracellular spaces, which contain aqueous humor (×3970). **B**, Electron micrograph of a vacuolar structure that shows both basal and apical openings, thus constituting a vacuolar transcellular channel *(arrow)*. In this way, the fluid-containing extracellular space on the basal aspect of the cell is temporarily connected with a lumen of Schlemm's canal, allowing bulk outflow of aqueous humor. *N,* indented nucleus of the cell (×23,825).

Gonioscopy of the Chamber Angle—Clinical Anatomy

Much of the chamber angle anatomy may be observed by gonioscopy. This description refers to the superior angle as viewed through the lower mirror of a gonioprism. The gonioscopist should acquire the habit of looking first at the pupillary margin, then at the iris surface, and finally at the angle itself.

The surface of the iris is normally devoid of vessels. In lightly pigmented irides, radial vessels may be observed within the iris stroma. Undulations in the surface of the iris appear as a series of concentric rings as the examiner looks toward the chamber angle. The final iris roll is visible at the beginning of the angle recess, or *iris root,* the thinnest portion of the iris where it joins the ciliary body. Occasionally, iris processes extend from the surface of the iris into the trabecular area. Iris processes should not be confused with peripheral anterior synechiae, which are formed when the peripheral iris becomes adherent to the peripheral cornea. The visibility of the angle structures, the contour of the iris plane, and the depth of the chamber angle

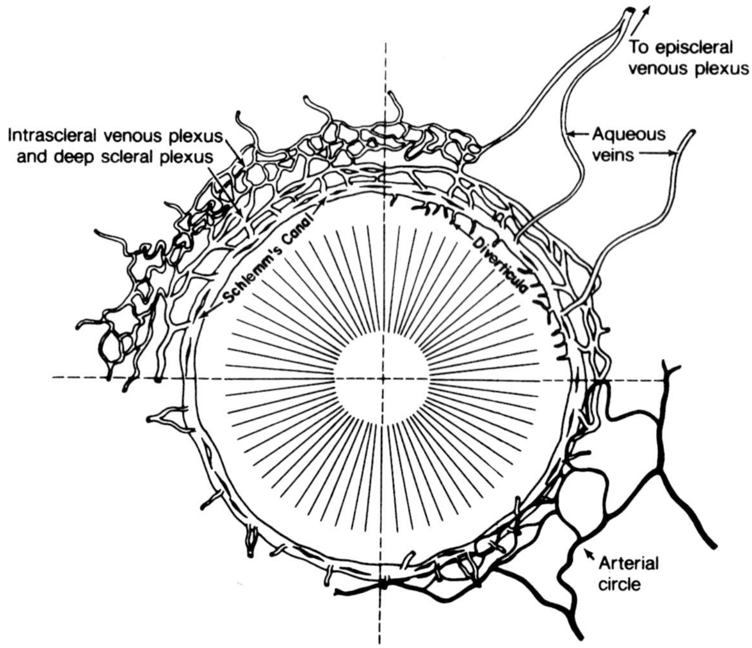

FIG III-13—Diagram of Schlemm's canal and relationships of the arteriolar and venous vascular supply. For clarity, the various systems have been limited to only parts of the circumference of the canal. Small, tortuous, blind diverticula (so-called *Sondermann's channels*) extend from the canal into the trabecular meshwork. Externally, the collector channels arising from Schlemm's canal anastomose to form the intrascleral and deep scleral venous plexuses. At irregular intervals around the circumference, aqueous veins arise from the intrascleral plexus and connect directly to the episcleral veins. The arteriolar supply closely approximates the canal, but no direct communication occurs between the two. (Reproduced with permission from Tripathi RC, Tripathi BJ. Functional anatomy of the anterior chamber angle. In: Jakobiec FA, ed. *Ocular Anatomy, Embryology, and Teratology.* Philadelphia: Harper & Row; 1982:276.)

should be noted since these observations are the basis for the classification of both primary and secondary glaucoma into open-angle or angle-closure glaucoma.

The landmarks of the drainage angle are as follows: Posteriorly, the longitudinal muscle fibers of the ciliary body insert into the scleral spur to form a dark, narrow *ciliary band* above the iris root. The *scleral spur* is the posterior landmark of the trabecular meshwork, visible as a white line of varying width. Most of the meshwork is accommodated between the scleral spur and Schwalbe's line, the peripheral margin of Descemet's membrane. During gonioscopy, obstruction of the episcleral veins may cause a reflux of blood into the collector channels and hence into Schlemm's canal, which becomes visible as a thin red stripe between the middle and the posterior third of the trabecular meshwork. This is a normal phenomenon. Otherwise, in some eyes, Schlemm's canal may appear as a faint gray line. Schwalbe's line can easily be identified by gonioscopy since the slit beam converges to a point in this location. It is sometimes hypertrophied. BCSC Section 10, *Glaucoma,* discusses gonioscopy in depth.

Certain pathologic features may be seen on gonioscopy. Dandruff-like deposits associated with pseudoexfoliation of the lens capsule may be visible, as may posterior synechiae caused by adhesions between the posterior surface of the iris and the anterior lens capsule. In certain conditions, such as pigment dispersion, some pigment is found over the trabecular meshwork, particularly inferiorly. It is also found in some normal angles, more in brown-eyed than in blue-eyed subjects. The melanin granules located in the trabecular meshwork are structurally identical to those found in the posterior pigmented layer of the iris. A fine network of new vessels may form pathologically on the iris surface *(rubeosis iridis)* and occlude the angle. The network is a response to posterior-segment ischemic factors released in diabetic retinopathy, central retinal vein occlusion, or severe carotid artery disease.

Uveal Tract

The *uveal tract* is the main vascular compartment of the eye. It consists of three parts:

- Iris
- Ciliary body (located in the anterior uvea)
- Choroid (located in the posterior uvea)

The uveal tract is firmly attached to the sclera at only three sites: the scleral spur, the exit points of the vortex veins, and the optic nerve. These attachments account for the characteristic anterior balloons formed in choroidal detachment.

Iris (Fig III-14)

The *iris* is the most anterior extension of the uveal tract. It is made up of blood vessels and connective tissue, in addition to the melanocytes and pigment cells that are responsible for its distinctive color. The mobility of the iris allows the pupil to change size. During mydriasis, the iris is thrown into a number of ridges and folds; during miosis, its anterior surface appears relatively smooth.

The *iris diaphragm* subdivides the anterior segment into the anterior and posterior chambers. Blunt trauma can cause the iris root (the thinnest portion of the iris) to disinsert from the ciliary body.

Stroma

The iris *stroma* is composed of pigmented cells *(melanocytes)* and nonpigmented cells, collagen fibrils, and a matrix containing hyaluronic acid. The aqueous humor flows freely through the loose stroma along the anterior border of the iris, which contains multiple crypts and crevices that vary in size, shape, and depth. This surface is covered by an interrupted layer of connective-tissue cells that merges with the ciliary body.

The overall structure of the iris stroma remains similar in irides of all colors. Differences in color are related to the amount of pigmentation in the *anterior border layer* and the deep stroma. The stroma of blue irides is lightly pigmented, and brown irides have a densely pigmented stroma that absorbs light.

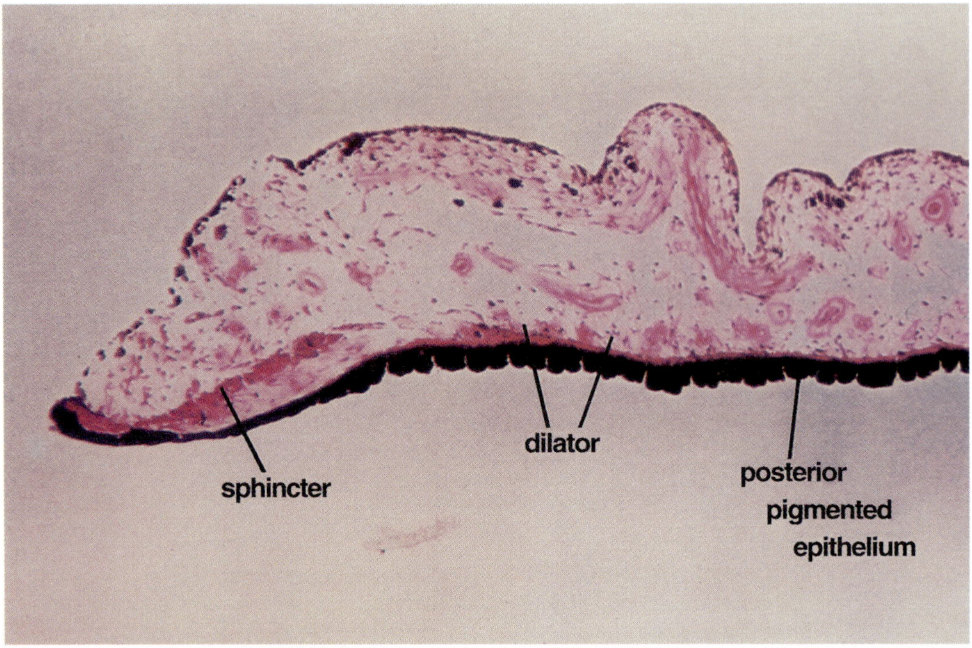

FIG III-14—Iris. Note the relationship between the sphincter and dilator muscles (H&E ×20). (Photograph courtesy of Thomas A. Weingeist, PhD, MD.)

Vessels and Nerves

Blood vessels form the bulk of the iris stroma. Most follow a radial course, arising from the major arterial circle and passing to the center of the pupil. In the region of the *collarette* (the thickest portion of the iris), anastomoses occur between the arterial and venous arcades to form the minor vascular circle of the iris, which is often incomplete. The *major arterial circle* is located at the apex of the ciliary body, not the iris. In humans, the anterior border layer is normally avascular. The diameter of the capillaries is relatively large. Their endothelium is nonfenestrated and is surrounded by a basement membrane, associated pericytes, and a zone of collagenous filaments. The intima has no internal elastic lamina.

Myelinated and nonmyelinated nerve fibers serve sensory, vasomotor, and muscular functions throughout the stroma.

Posterior Pigmented Layer

The posterior surface of the iris is densely pigmented and appears velvety smooth and uniform. It is continuous with the nonpigmented epithelium of the ciliary body and thence with the neurosensory portion of the retina. The polarity of its cells is maintained from embryogenesis. The basal surface of the pigmented layer borders the posterior chamber. The apical surface faces the stroma and adheres to the *anterior pigmented layer,* which gives rise to the *dilator muscle* (Fig III-15).

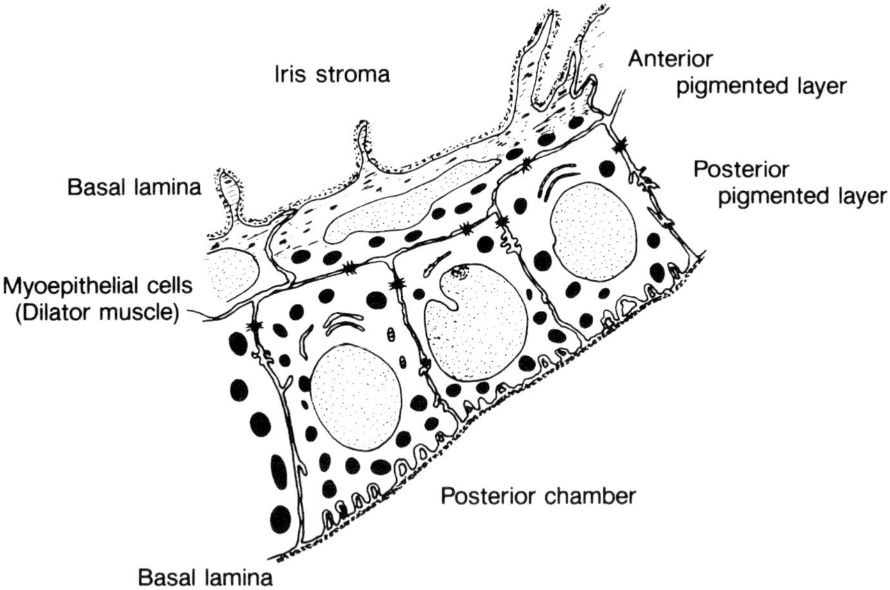

FIG III-15—Posterior layer of the iris. (Illustration by Thomas A. Weingeist, PhD, MD.)

The *posterior pigmented layer* of the iris curves around the pupillary margin and extends for a short distance onto the anterior border layer of the iris stroma as the *pupillary ruff.* In rubeosis iridis, the pigmented layer extends farther onto the anterior surface of the iris, a condition called *ectropion.* The term *ectropion uveae* is a misnomer, since all of these layers are derived from neuroectoderm.

Dilator Muscle

The *dilator muscle* is derived embryologically from the outer layer of the optic cup, which is neuroectoderm. It lies parallel and anterior to the posterior pigmented epithelium. The smooth muscle cells contain fine myofilaments and melanosomes. The myofibrils are confined mainly to the basal portion of the cells and extend anteriorly into the iris stroma. The melanosomes and the nucleus are in the apical region of each myoepithelial cell.

There is a dual sympathetic and parasympathetic innervation. The dilator muscle contracts in response to sympathetic α_1-adrenergic stimulation; cholinergic parasympathetic stimulation may have an inhibitory role.

The *first-order neuron* of the sympathetic chain begins in the ipsilateral posterolateral hypothalamus and passes through the brain stem to synapse in the intermediolateral gray matter of the spinal cord, chiefly at thoracic level 1. The *second-order preganglionic neuron* exits the spinal cord, passes over the pulmonary apex and through the stellate ganglion without synapsing, and synapses in the *superior cervical ganglion.* The *third-order postganglionic neuron* originates here, joins the internal carotid plexus, enters the cavernous sinus, and travels with the ophthalmic division of cranial nerve V to the orbit and then to the dilator muscle.

Sphincter Muscle

Like the dilator muscle, the *sphincter muscle* is derived from neuroectoderm. It is composed of a circular band of smooth muscle fibers and is located near the pupillary margin in the deep stroma, anterior to the pigment epithelium of the iris. Although a dual innervation has been demonstrated morphologically, the sphincter muscle receives its primary innervation from parasympathetic nerve fibers that originate in the cranial nerve III nucleus, and the sphincter muscle responds pharmacologically to muscarinic stimulation. The reciprocal sympathetic innervation to the sphincter appears to serve an inhibitory role, helping to relax the sphincter in darkness.

The fibers subserving the sphincter muscle leave the Edinger-Westphal subnucleus and follow the inferior division of cranial nerve III after it bifurcates in the cavernous sinus. The fibers continue in the branch supplying the inferior oblique muscle, exit, and synapse with postganglionic fibers in the ciliary ganglion. The *postganglionic fibers* travel with the *short ciliary nerves* to the iris sphincter. They are unusual in that they are myelinated, presumably reflecting a need for fast conduction.

Ciliary Body

The *ciliary body,* which is triangular in cross section, bridges the anterior and posterior segments (see Fig III-10). The apex of the ciliary body is directed posteriorly toward the ora serrata. Its base gives rise to the iris. The only attachment of the ciliary body to the sclera is at its base, via its longitudinal muscle fibers, where they insert into the scleral spur.

The ciliary body has two principal functions: aqueous humor formation and lens accommodation. It also plays a role in the trabecular and uveoscleral outflow of aqueous humor.

Ciliary Epithelium and Stroma

The ciliary body is 6–7 mm wide and consists of two parts: the pars plana and the pars plicata. The *pars plana* is a relatively avascular, smooth pigmented zone; it is 4 mm wide and extends from the ora serrata to the ciliary processes (Fig III-16). The safest posterior surgical approach to the vitreous cavity is through the pars plana, located 3–4 mm from the corneal limbus. The *pars plicata* is richly vascularized and consists of approximately 70 radial folds, or *ciliary processes.* The zonular fibers of the lens attach primarily in the valleys of the ciliary processes but also along the pars plana.

The *capillary plexus* of each ciliary process is supplied by arterioles as they pass anteriorly and posteriorly from the major arterial circle and is drained by one or two large venules located at the crest of each process. Sphincter tone within the arteriolar smooth muscle not only affects the capillary hydrostatic pressure gradient but also influences whether blood flows into the capillary plexus or directly to the draining choroidal vein, bypassing the plexus completely. Neuronal innervation of the vascular smooth muscle and humoral vasoactive substances may be important in determining regional blood flow, capillary surface area available for exchange of fluid, and hydrostatic capillary pressure. All of these affect the rate of aqueous humor formation.

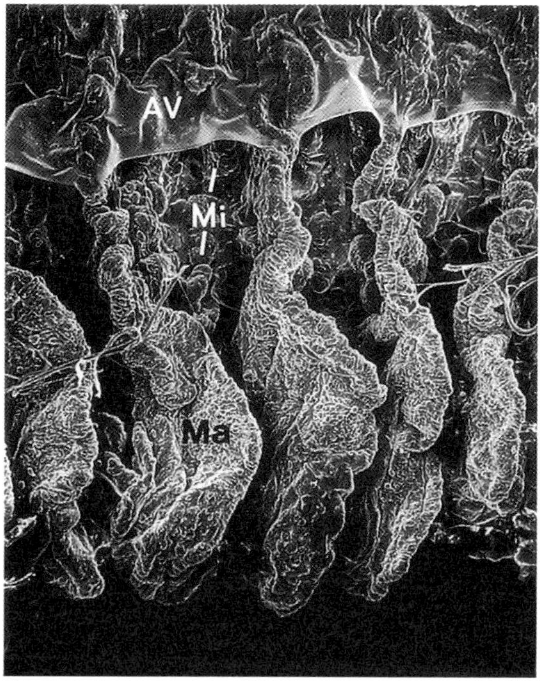

FIG III-16—Scanning electron micrograph of the human ciliary processes showing major processes *(Ma)* and minor processes *(Mi)*. Remnants of the anterior vitreous *(AV)* cover the pars plana (×60).

The ciliary body is lined by a double layer of epithelial cells, the nonpigmented and the pigmented epithelium. The inner *nonpigmented epithelium* is located between the aqueous humor of the posterior chamber and the outer pigmented epithelium. The apices of the nonpigmented and pigmented cell layers are fused by a complex system of junctions and cellular interdigitations. Along the lateral intercellular spaces, near the apical border of the nonpigmented epithelium, are tight junctions *(zonulae occludentes)* that maintain the blood–aqueous barrier. The basal surface of the nonpigmented epithelium, which borders the posterior chamber, is covered by the basal lamina, which is multilaminar in the valleys of the processes. The basal lamina of the pigmented epithelium is thick and more homogeneous than that of the nonpigmented epithelium and faces the iris stroma.

The *pigmented epithelium* is relatively uniform throughout the ciliary body (Fig III-17). Its cuboidal cells are characterized by multiple basal infoldings, a large nucleus, mitochondria, extensive endoplasmic reticulum, and many melanosomes. The nonpigmented epithelium tends to be cuboidal in the pars plana region but columnar in the pars plicata. It also has multiple basal infoldings, abundant mitochondria, and large nuclei. The endoplasmic reticulum and Golgi complex in these cells are important to aqueous humor formation. Occasional melanosomes are present, especially anteriorly, near the iris.

The *uveal portion of the ciliary body* consists of comparatively large fenestrated capillaries, collagen fibrils, and fibroblasts. The main arterial supply to the ciliary body comes from the long posterior and the anterior ciliary arteries, which join together to form a multilayered *arterial plexus* consisting of a superficial *episcleral plexus,* a deeper *intramuscular plexus,* and an incomplete *major arterial circle* often mistakenly attributed to the iris but actually located posterior to the anterior chamber angle recess, in the ciliary body. The major veins drain posteriorly through the vortex system, although some drainage also occurs through the intrascleral venous plexus and the episcleral veins into the limbal region.

Ciliary Muscle (Fig III-18)

Descriptions of the *ciliary muscle* suggest that it has three layers of fibers:

- Longitudinal
- Radial
- Circular

Most of the ciliary muscle is made up of an outer layer of longitudinal fibers that attach to the scleral spur. The radial muscle fibers arise in the midportion of the ciliary body, and the circular fibers are located in the innermost portion. Clinically, the three groups of muscle fibers function as a unit. Presbyopia is related to age-related changes in the lens (discussed below) rather than changes in the ciliary muscle. Even so, the muscle does change with age, with increasing amounts of connective tissue between the muscle bundles and a loss of elastic recoil after contraction.

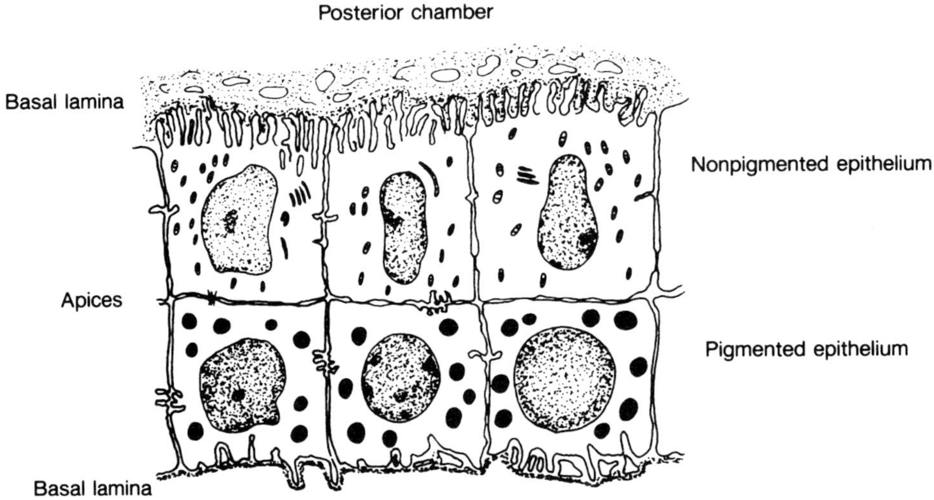

FIG III-17—Ciliary epithelium. (Illustration by Thomas A. Weingeist, PhD, MD.)

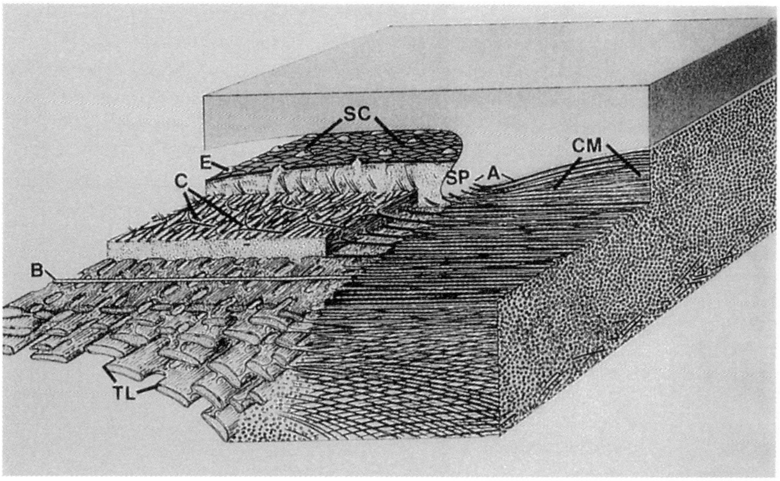

FIG III-18—Schematic drawing of the ciliary muscle–tendon attachments. Type A tendons *(A)* connect the anterior ciliary muscle tips to the sclera and scleral spur. Type B tendons *(B)* traverse through the entire trabecular meshwork and insert into the corneal stroma. Type C tendons *(C)* appear brushlike and fan out at 90° angles at their terminal ends. They become continuous with the cribriform meshwork of elastic fibers underlying Schlemm's canal that connect to the endothelial cells lining the canal. *CM,* ciliary muscle; *E,* endothelium of Schlemm's canal; *SC,* Schlemm's canal; *SP,* scleral spur; *TL,* trabecular lamellae. (Reproduced with permission from Rohen JW. The evolution of the primate eye in relation to the problem of glaucoma. In: Lütjen-Drecoll E, ed. *Basic Aspects of Glaucoma Research.* New York: Schattauer; 1982:21.)

The ciliary muscles behave like other smooth, nonstriated muscle fibers. Ultrastructural studies reveal that they contain multiple myofibrils with characteristic electron-dense attachment bodies, mitochondria, glycogen particles, and a prominent nucleus. The smooth muscle cells are surrounded by a basal lamina separated from the cell membrane by a 300 Å space. Bundles of fibers are surrounded by a thin fibroblastic sheath rather than by collagen. The muscle is rich in type VI collagen, which forms a sheath around the *anterior elastic tendons.* These tendons insert into the scleral spur and around the tips of the oblique and circular muscle fibers as they insert into the trabecular meshwork.

Lütjen-Drecoll E, Rittig M, Rauterburg J, et al. Immunomicroscopical study of type VI collagen in the trabecular meshwork of normal and glaucomatous eyes. *Exp Eye Res.* 1989;48:139–147.

Streeten BW. The ciliary body. In: Duane TD, Jaeger EA, eds: *Biomedical Foundations of Ophthalmology.* Philadelphia: Lippincott; 1995.

Both myelinated and nonmyelinated nerve fibers are observed throughout the ciliary muscle. Innervation is mainly derived from parasympathetic fibers of cranial nerve III via the short ciliary nerves. About 97% of these ciliary fibers are directed to the ciliary muscle, and about 3% to the iris sphincter. Sympathetic fibers have also been observed and may play a role in relaxing the muscle. Cholinergic drugs contract the ciliary muscle. Since some of the muscle fibers form tendinous attachments to the scleral spur, their contraction increases aqueous flow by opening up the spaces of the trabecular meshwork. Because these muscle–tendon attachments are not well formed in infants, miotics have only a minor influence on aqueous outflow in this age group.

Choroid

The *choroid*, the posterior portion of the uveal tract, nourishes the outer portion of the retina (Fig III-19). It averages 0.25 mm in thickness and consists of three layers of vessels:

- The choriocapillaris, the innermost layer
- A middle layer of small vessels
- An outer layer of large vessels

Perfusion of the choroid comes from both the long and the short posterior ciliary arteries and from the perforating anterior ciliary arteries (Fig III-20). Venous blood drains through the vortex system. Blood flow through the choroid is high compared to other tissues. As a result, the oxygen content of the choroidal venous blood is only 2%–3% less than that of the arterial blood.

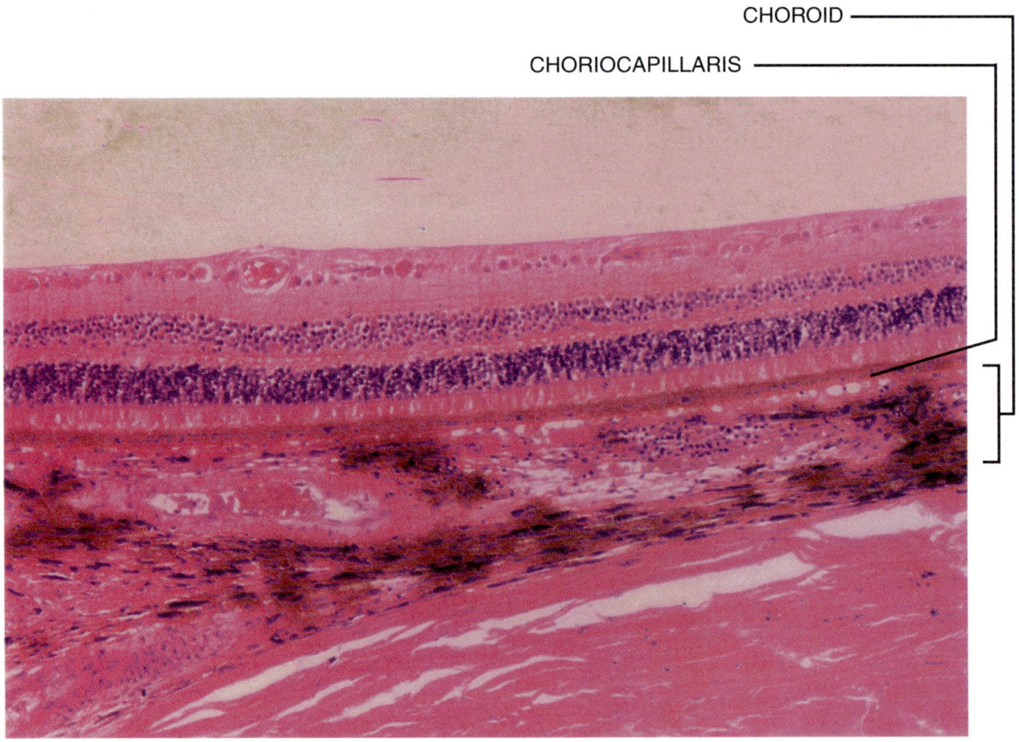

FIG III-19—Choroid. The choriocapillaris lies just below the retinal pigment epithelium. Beneath are a middle and outer vascular layer and multiple dendritic melanocytes (H&E ×32). (Photograph courtesy of Thomas A. Weingeist, PhD, MD.)

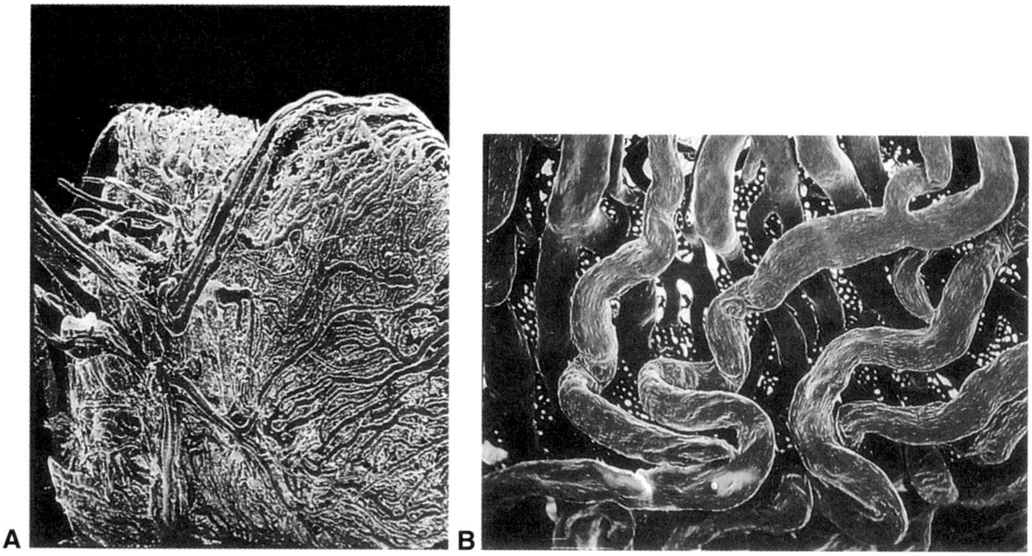

FIG III-20—Scanning electron micrograph of a vascular cast of the choroidal circulation as viewed from the scleral side of the posterior pole. The outer choroid consists of many interweaving branches of the short posterior ciliary arteries and vortex veins. Beneath these vessels, the choriocapillaris layer appears as a sheet or net of capillaries. **A**, Low magnification. **B**, Higher magnification. (Reproduced with permission from Kessel RG, Kardon RH. *Tissues and Organs: A Text-Atlas of Scanning Electron Microscopy.* New York: Freeman; 1979.)

Bruch's Membrane (Figs III-21, III-22)

Bruch's membrane is a PAS-positive lamina resulting from the fusion of the basal laminae of the retinal pigment epithelium (RPE) and the choriocapillaris of the choroid. It extends from the margin of the optic disc to the ora serrata, and ultrastructurally it has five elements:

- Basal lamina of the RPE
- Inner collagenous zone
- A thicker, porous band of elastic fibers
- Outer collagenous zone
- Basal lamina of the choriocapillaris

Bruch's membrane, therefore, consists of a series of connective tissue sheets that are highly permeable to small molecules such as fluorescein. Defects in Bruch's membrane develop spontaneously in myopia or pseudoxanthoma elasticum or result from trauma or inflammation. Subretinal neovascular membranes are important sequelae of these defects that can lead to disciform macular changes as part of age-related macular degeneration and ocular histoplasmosis syndrome.

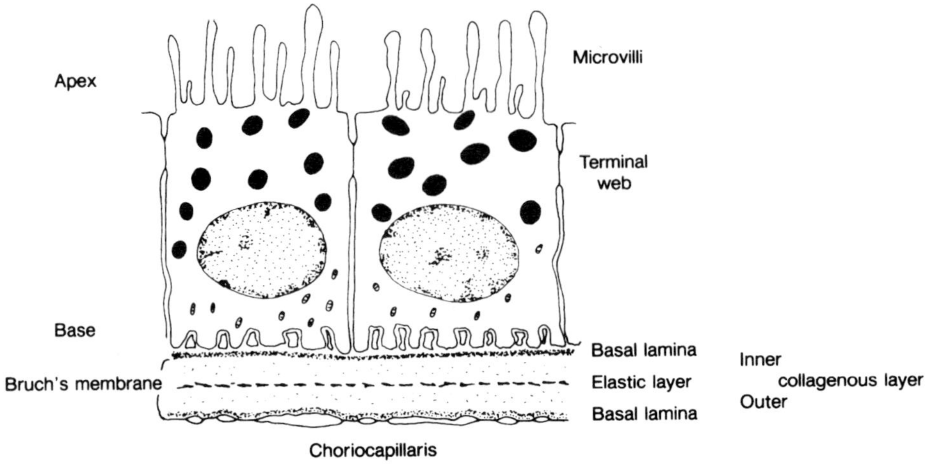

FIG III-21—Retinal pigment epithelium and Bruch's membrane. (Illustration by Thomas A. Weingeist, PhD, MD.)

Choriocapillaris

The *choriocapillaris* is a continuous layer of large capillaries (40–60 µm in diameter) lying in a single plane beneath the RPE (Fig III-23). The vessel walls are extremely thin and contain multiple fenestrations, especially on the surface facing the retina (Fig III-24). Pericytes are located along the outer wall.

Although in vitro and postmortem injection studies have indicated that the choriocapillaris is a continuous vascular system, ample clinical evidence suggests that, functionally, it acts like an end-arteriole system. In vivo fundus fluorescein angiography of the choriocapillaris reveals a lobular pattern, especially in the posterior pole. The lobule is supplied centrally by a precapillary arteriole that drains peripherally into the postcapillary venule. A lobular, mosaic pattern can be demonstrated in the early phases of fluorescein angiography, especially under experimental conditions if fluorescein is injected intra-arterially or with video fluorescein angiography with a fast sampling rate.

Studies using corrosion vascular casts and scanning electron microscopy have demonstrated that the angioarchitecture of the choriocapillaris is most dense at the macula and becomes less so toward the periphery (Fig III-25).

Fryczkowski AW, Sherman MD, Walker J. Observations on the lobular organization of the human choriocapillaris. *Int Ophthalmol.* 1991;15:109–120.

Hayreh SS. The choriocapillaris. *Albrecht von Graefe's Arch Klin Exp Ophthalmol.* 1974;192:165–179.

Torczynski E, Tso MO. The architecture of the choriocapillaris at the posterior pole. *Am J Ophthalmol.* 1976;81:428–440.

Yoneya S, Tso MO. Angioarchitecture of the human choroid. *Arch Ophthalmol.* 1987; 105:681–687.

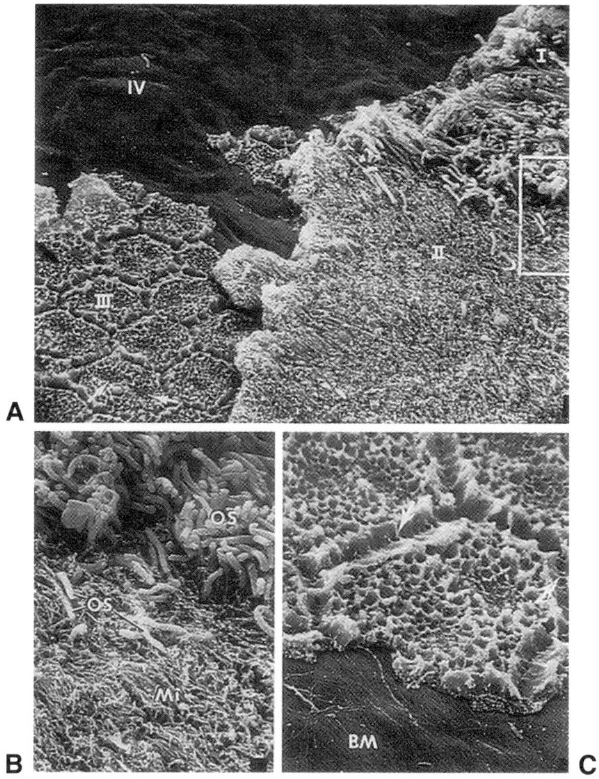

FIG III-22—Scanning electron micrograph of Bruch's membrane and pigmented epithelium–photoreceptor complex. **A**, View from the inner retinal side, with most of the retina removed, exposes the outer layers: *I*, the layer of the outer segments of photoreceptors; *II*, the microvillous layer of the apical end of the pigmented epithelium; *III*, the cytoplasm and polygonal cell borders of the pigmented epithelium; *IV*, Bruch's membrane. The *arrows* point to depressions in the cytoplasm where the cell nuclei were removed. The rectangle at right is magnified in the lower left photo. **B**, Relationship of the photoreceptor outer segments *(OS)* (inner segment torn away) to microvilli *(Mi)* of pigmented epithelium. **C**, Basal portion of pigmented epithelial cell in relation to Bruch's membrane *(BM)*. Adjacent epithelial cells are joined through junctional complexes *(arrows)*. The pigment granules have left small holes in the cytoplasm after tissue preparation, giving the cell a honeycomb appearance at this plane of section. (Reproduced with permission from Kessel RG, Kardon RH. *Tissues and Organs: A Text-Atlas of Scanning Electron Microscopy*. New York: Freeman; 1979.)

The middle and outer choroidal vessels are not fenestrated. The large vessels, typical of small arteries elsewhere, possess an internal elastic lamina and smooth muscle cells in the media. As a result, small molecules such as fluorescein, which diffuse across the endothelium of the choriocapillaris, do not leak through medium and large choroidal vessels. Abundant melanocytes as well as occasional macrophages, lymphocytes, mast cells, and plasma cells appear throughout the choroidal stroma. The intercellular space contains collagen fibers and nerve fibers. The degree of pigmentation observed ophthalmoscopically in the ocular fundus primarily de-

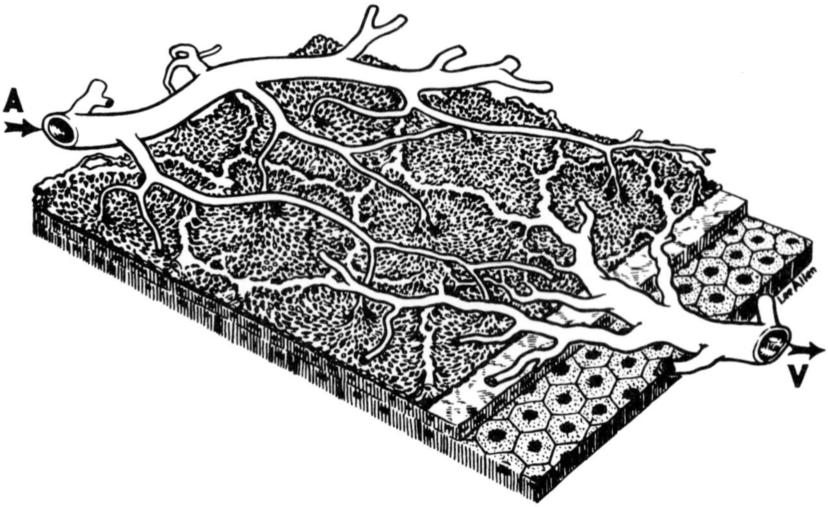

FIG III-23—Lobular pattern of choriocapillaris. Note that the retinal pigment epithelium is below. *A,* choroidal arteriole; *V,* choroidal venule. (Reproduced with permission from Hayreh SS. The choriocapillaris. *Albrecht von Graefe's Arch Klin Exp Ophthalmol.* 1974;192:165–179.)

pends on the number of pigmented melanocytes in the choroid. Melanosomes are absent from the RPE and choroid of albinos. In lightly pigmented eyes, pigmentation in the choroid is sparse compared with darkly pigmented eyes. The degree of pigmentation in the choroid must be considered when one is performing photocoagulation.

Lens

The *lens* is a biconvex structure located directly behind the posterior chamber and pupil (Fig III-26). It is the lesser of the two refractive elements in the dioptric system. The equatorial diameter is 6.5 mm at birth and increases in the first two to three decades of life, remaining in the region of 9–10 mm in diameter in late life. The anteroposterior width of the lens is about 3 mm at birth and increases after the second decade of life to about 6 mm at 80 years of age. This growth is accompanied by a shortening of the anterior radius of curvature of the lens, which would increase its optical power if it were not for a compensatory change in the refractive gradient across the lens substance.

In youth, accommodation for near vision is achieved by ciliary muscle contraction, which moves the ciliary muscle mass forward and inward. This relaxes zonular tension and allows the lens to assume a globular shape, causing a shortening of its anterior curvature. The increased lens thickness during accommodation is entirely due to a change in nuclear shape. With age, accommodative power is lost steadily and is completed by the age of 50 years; this loss is called *presbyopia.* Its causes include the increased size of the lens, altered mechanical relationships, and an increased stiffness of the lens nucleus secondary to changes in the crystalline proteins of the fiber cytoplasm.

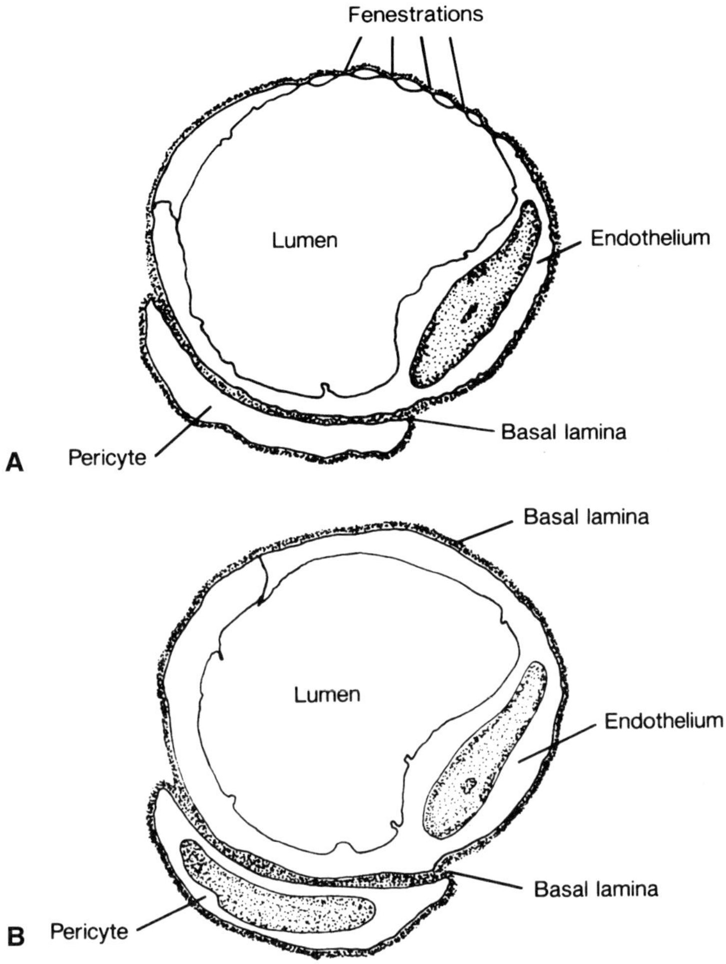

FIG III-24—**A**, Fenestrated choroidal capillary. **B**, Nonfenestrated retinal capillary. (Illustration by Thomas A. Weingeist, PhD, MD.)

The lens has certain unusual features. It lacks innervation and is avascular. After regression of the hyaloid vasculature during embryogenesis, the lens depends totally on the aqueous and vitreous for its nourishment. It is entirely enclosed from embryonic life by a basal lamina, the lens capsule. BCSC Section 11, *Lens and Cataract,* discusses the lens in depth.

Capsule

The lens is surrounded by a basal lamina, the *lens capsule,* which is a product of the lens epithelium (Fig III-27A). It is rich in type IV collagen and other matrix proteins.

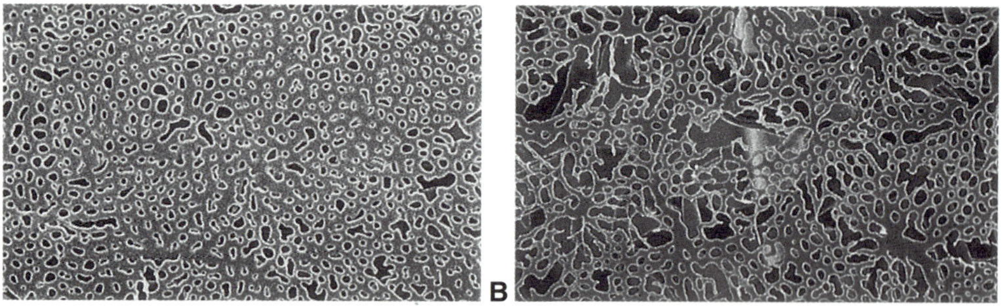

FIG III-25—**A**, Scanning electron micrograph of a vascular cast of the choriocapillaris at the macula in the posterior pole after tissue digestion, as seen from the retina side. The superficial retinal circulation was removed in this preparation. **B**, Similar view from a more anterior location, toward the ora serrata. The larger arteries and veins can be discerned underneath it. Anatomically, the choriocapillaris resembles a continuous sheet toward the macula, but the capillary plexus becomes much less dense in the periphery.

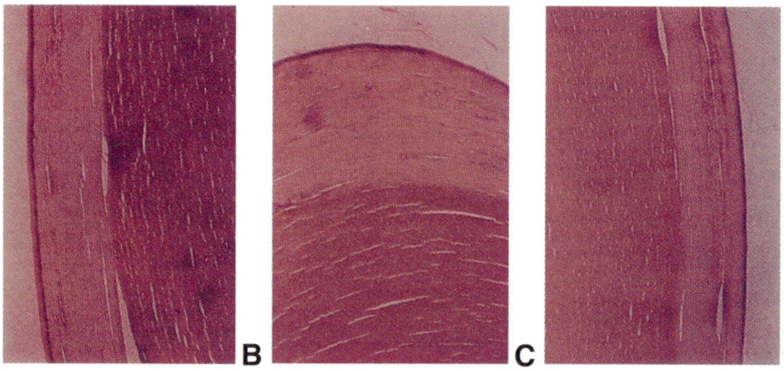

FIG III-26—**A**, Lens: anterior capsule, epithelium, and lens fibers. **B**, Equator of the lens. Note nuclei within the lens bow and the zonular fibers. **C**, Posterior lens capsule. Note absence of lens epithelium (H&E ×32). (Photographs courtesy of Thomas A. Weingeist, PhD, MD.)

Synthesis of the anterior lens capsule (which overlies the epithelium) proceeds throughout life, so that its thickness increases; while that of the posterior capsule remains relatively constant. Values of 15.5 μm for the anterior capsule and 2.8 μm for the posterior capsule have been cited for the adult lens.

Morphologically, the lens capsule consists of fine filaments arranged in lamellae, parallel to the surface. The anterior lens capsule contains a fibrogranular material, identified as laminin, which is absent from the posterior capsule at the ultrastructural level. The thinness of the posterior capsule creates a potential for rupture during extracapsular cataract extraction, particularly by phacoemulsification.

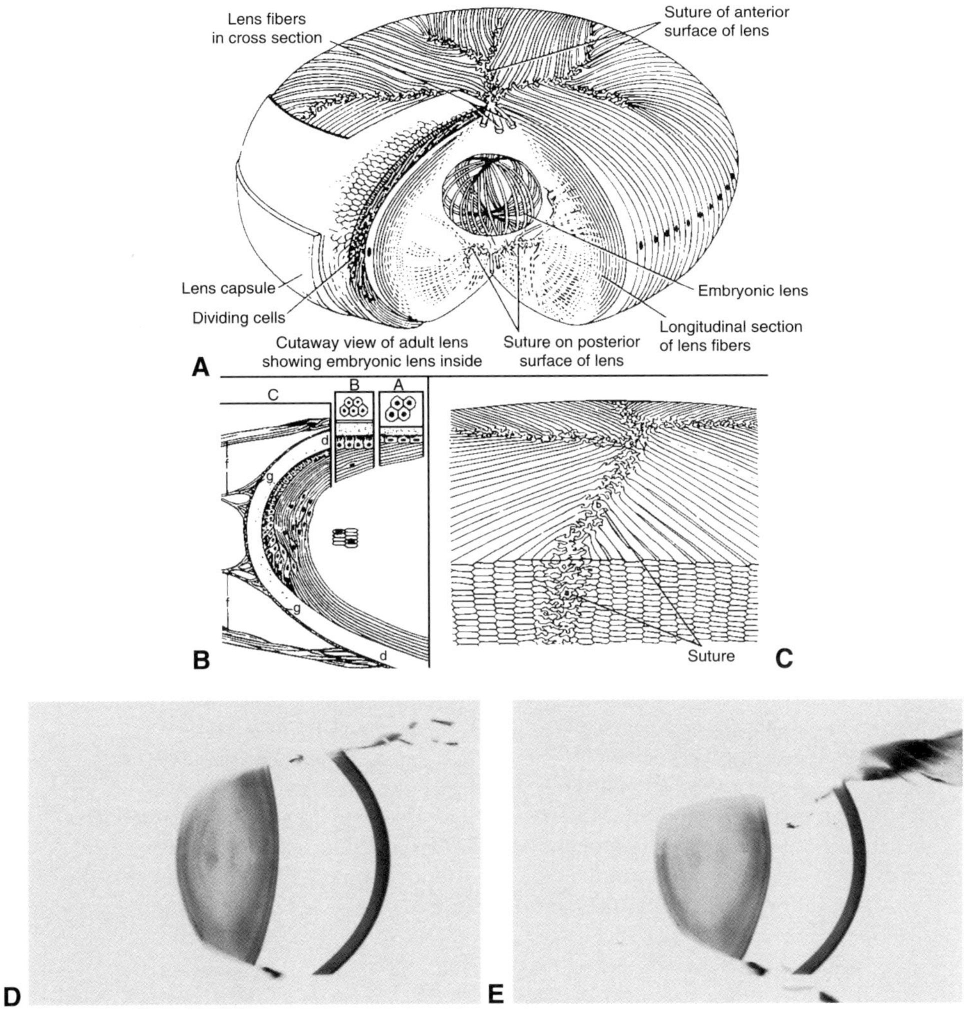

FIG III-27— Organization of the lens. At areas where lens cells converge and meet, sutures are formed. **A**, The embryonal nucleus has a Y-shaped suture at both the anterior and posterior poles. In the adult lens cortex, the organization of the sutures is more complex. At the equator, the lens epithelium can divide and the cells become highly elongated and ribbon-like, sending processes anteriorly and posteriorly. As new lens cells are formed, older cells come to lie in the deeper parts of the cortex. **B**, Diagram shows the difference in lens fibers at the anterior *(A)*, intermediate *(B)*, and equatorial zones *(C)* as shown in cross section and corresponding surface view. The lens capsule, or basement membrane of the lens epithelium *(d)*, is shown in relation to the zonular fibers *(f)* and their attachment to the lens *(g)*. **C**, Diagram shows a closer view of lens sutures. (**A–C** reproduced with permission from Kessel RG, Kardon RH. *Tissues and Organs: A Text-Atlas of Scanning Electron Microscopy.* New York: Freeman; 1979.) **D** and **E**, Optical sections of a young adult human lens (25-year-old female) demonstrated by Scheimpflug photography. The cornea is to the right. **D**, Lens in the nonaccommodated state. **E**, Lens during accommodation—note that the anterior radius of curvature is shortened in the latter case. (Scheimpflug photographs courtesy of Jane Koretz.)

Epithelium

The lens *epithelium* lies beneath the anterior and equatorial capsule but is absent under the posterior capsule. The basal aspects of the cells abut the lens capsule without specialized attachment sites. The apices of the cells face the interior of the lens, and the lateral borders interdigitate, with practically no intercellular space. Each cell contains a prominent nucleus but relatively few cytoplasmic organelles.

Regional differences in the lens epithelium are important. The *central zone* represents a stable population of cells whose numbers slowly decline with age. An *intermediate zone* of smaller cells shows occasional mitoses. Peripherally, there are meridional rows of cuboidal pre-equatorial cells that form the *germinative zone* of the lens. Here, cells undergo mitotic division, elongate anteriorly and posteriorly, and form the differentiated fiber cells of the lens. In the human lens, cell division continues throughout life and is responsible for the continued growth of the lens. Germinative cells left behind after phacoemulsification can give rise to posterior capsular opacification as a result of aberrant proliferation and cell migration. Visual loss can be relieved by incision of the capsule with an Nd:YAG laser.

Fibers

The lens has an outer cortex and an inner nucleus. The *nucleus* is the part of the fiber mass that is formed at birth, and the *cortex* forms as new fibers are added postnatally. In optical section with the slit lamp, lamellar *zones of discontinuity* are visible, differentiating the adult cortex into deep and superficial regions (Fig III-27B). The fiber cells are hexagonal in cross section, are spindle-shaped, and possess numerous interlocking finger-like projections (Fig III-28). Apart from the most superficial cortical fibers, the cytoplasm is homogeneous and contains few organelles. The high refractive index of the lens results from the high concentration of lens crystallins (α, β, and γ) in the fiber cytoplasm. The *lens sutures* are formed by the interdigitation of the anterior and posterior tips of the spindle-shaped fibers. In the fetal lens, this forms the anterior Y-shaped suture and the posterior inverted Y-shaped suture. As the lens ages, further branches are added to the sutures, each new set of branch points corresponding to the appearance of a fresh optical zone of discontinuity.

Zonule (Suspensory Ligament)

The lens is held in place by a system of zonular fibers that originate from the basal laminae of the nonpigmented epithelium of the pars plana and pars plicata of the ciliary body. These fibers chiefly attach to the lens capsule anterior and posterior to the equator. Each zonular fiber is made up of multiple filaments of fibrillin that merge with the equatorial lens capsule. In the Marfan syndrome, mutations in the fibrillin gene lead to weakening of the zonule and subluxation of the lens.

Streeten BW. Anatomy of the zonular apparatus. In: Duane TD, Jaeger EA, eds: *The Biomedical Foundations of Ophthalmology.* Philadelphia: Harper & Row; 1992.

When the eye is focused for distance, the zonule is under tension and the lens form is relatively flattened. During accommodation, contraction of the ciliary muscle moves the proximal attachment of the zonule forward and inward so that the lens becomes more globular and the eye adjusts for near vision.

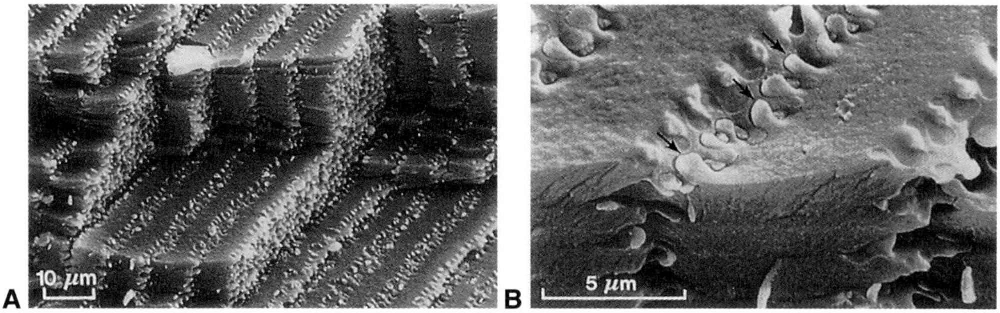

FIG III-28—**A** and **B**, Scanning electron micrographs of the relationship of lens fiber packing and interdigitation (*arrows* in **B**). (Reproduced with permission from Kessel RG, Kardon RH. *Tissues and Organs: A Text-Atlas of Scanning Electron Microscopy.* New York: Freeman; 1979.)

Retina

The *fundus oculi* is the part of the eye that is visible on ophthalmoscopy, including the retina and its vessels and the *optic nerve head* (or *optic disc*). The *macula,* 5–6 mm in diameter, lies between the temporal vascular arcades. At the macula's center lies the *fovea,* rich in cones and responsible for color vision and the highest visual acuity. In the far periphery, the *ora serrata* (the junction between the retina and the pars plana) can be seen by gonioscopy or indirect ophthalmoscopy. The red color of the fundus is due to the transmission of light reflected from the posterior sclera through the capillary bed of the choroid (Fig III-27C).

The *retina* is a thin, transparent structure that develops from the inner and outer layers of the optic cup. In cross section, from outer to inner retina, its layers are:

- RPE and its basal lamina
- Rod and cone inner and outer segments
- External limiting membrane
- Outer nuclear layer (nuclei of the photoreceptors)
- Outer plexiform layer
- Inner nuclear layer
- Inner plexiform layer
- Ganglion cell layer
- Nerve fiber layer (axons of the ganglion cells)
- Internal limiting membrane

The retina is also discussed in BCSC Section 12, *Retina and Vitreous.*

Retinal Pigment Epithelium

The structure of the outer pigmented epithelial layer is relatively simple compared with that of the overlying inner, or *neurosensory,* retina. The RPE consists of a monolayer of hexagonal cells that extends anteriorly from the optic disc to the ora serrata, where it merges with the pigmented epithelium of the ciliary body. Its structure is deceptively simple considering its many functions:

- Vitamin A metabolism
- Maintenance of the outer blood–retina barrier
- Phagocytosis of the photoreceptor outer segments
- Absorption of light (reduction of scatter)
- Heat exchange
- Formation of the basal lamina
- Production of the mucopolysaccharide matrix surrounding the outer segments
- Active transport of materials in and out of the RPE

Like other epithelial and endothelial cells, the RPE cells are polarized. The basal aspect is intricately folded and provides a large surface of attachment to the thin basal lamina that forms the inner layer of Bruch's membrane (see Figs III-21 and III-22). The apices have multiple villous processes that engage with the photoreceptor outer segments, embedded in a mucopolysaccharide matrix *(interphotoreceptor matrix)* containing chondroitin-6-sulfate, sialic acid, and hyaluronic acid. Separation of the RPE from the neurosensory retina is called *retinal detachment.*

Contiguous RPE cells are firmly attached by a series of lateral, intercellular *junctional complexes* (Fig III-29). The *zonulae occludentes* and *zonulae adherens* not only provide structural stability but also play an important role in maintenance of the outer blood–retina barrier. Zonulae occludentes consist of fused plasma membranes forming a circular band or belt between adjacent cells. A small intercellular space is present between zonulae adherens.

The retina and RPE show important regional differences (Fig III-30). The retina is thickest in the papillomacular bundle near the optic nerve (0.23 mm) and thinnest in the foveola (0.10 mm) and ora serrata (0.11 mm). RPE cells vary from 10 to 60 μm in diameter. RPE cells in the fovea are taller and thinner and contain more and larger melanosomes. These characteristics account in part for the decreased transmission of choroidal fluorescence observed during fundus fluorescein angiography. Cells in the periphery are broader, shorter, and less pigmented. The eye of a fetus or infant contains between 4 and 6 million RPE cells. Although the surface area of the eye increases appreciably with age, the increase in the number of RPE cells is relatively small. No mitotic figures are apparent within the RPE of the normal adult eye.

The cytoplasm of the RPE cells contains multiple round and ovoid pigment granules *(melanosomes).* These organelles develop in situ during formation of the optic cup and first appear as nonmelanized *premelanosomes.* Their development contrasts sharply with that of the pigment granules in uveal melanocytes, which are derived from the neural crest and later migrate into the uvea.

Lipofuscin granules probably arise from the disks of photoreceptor outer segments and represent residual bodies arising from phagosomal activity. This so-called *wear-and-tear pigment* is less electron-dense than the melanosomes, and its concentration increases gradually with age. Histologically, it is sudanophilic and exhibits a golden yellow autofluorescence.

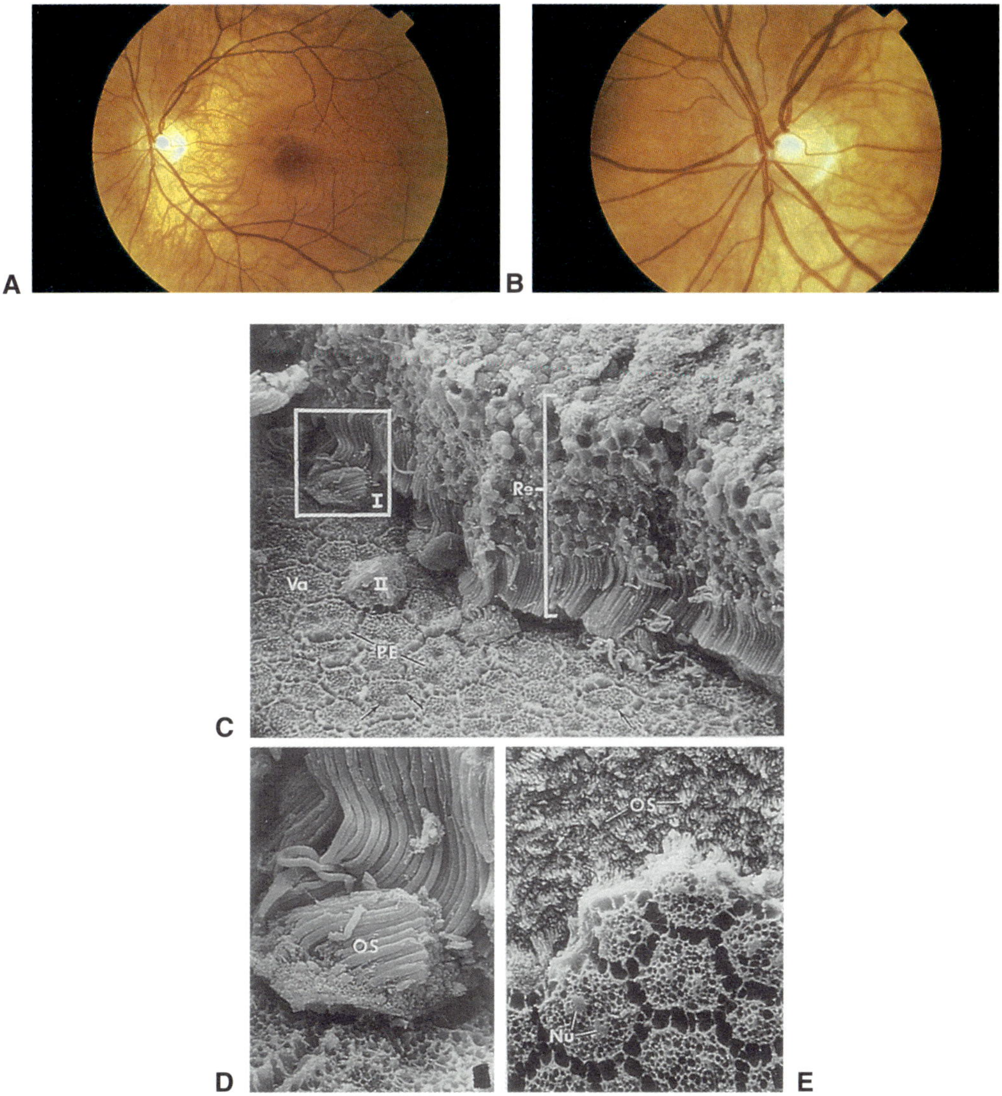

FIG III-29—**A**, Fundus photograph showing optic nerve head, fovea, and retinal vessels. **B**, Higher-power view of the optic nerve head, showing a normal optic cup and cilioretinal vessel. **C**, Scanning electron micrograph of the retina *(Re)* in relation to the pigmented epithelium *(PE)*. The outer segments *(OS)* of the photoreceptors can be seen attaching to individual pigmented cells at *I* and *II*. The vacuolated appearance *(Va)* of the epithelium is caused by the loss of pigment granules during tissue preparation. **D**, Enlargement of the area enclosed by rectangle in the top picture. **E**, View from the basal aspect of the pigmented epithelium with Bruch's membrane stripped away. Tissue shrinkage during preparation left a space separating the cell borders, and junctional complexes appear as threads bridging this space. Within the cells, nuclei *(Nu)* can be seen among the dissolved granules. (***C–E*** reproduced with permission from Kessel RG, Kardon RH. *Tissues and Organs: A Text-Atlas of Scanning Electron Microscopy*. New York: Freeman; 1979.)

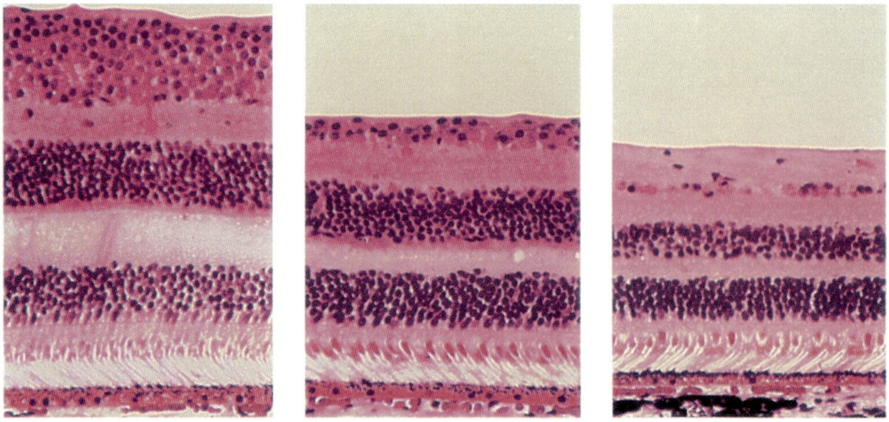

FIG III-30—Regional differences in the retina. From left to right: papillomacular bundle, macula, and peripheral retina (H&E, all same magnification). (Photographs courtesy of Thomas A. Weingeist, PhD, MD.)

Phagosomes are membrane-enclosed packets of disk outer segments that have been engulfed by the RPE. Several stages of disintegration are evident at any given time. In some species, shedding and degradation of the membranes of rod and cone outer segments follow a diurnal rhythm synchronized with daily fluctuations of environmental light.

Bok D. Retinal photoreceptor–pigment epithelium interactions. *Invest Ophthalmol Vis Sci.* 1985;26:1659–1694.

Young RW. The daily rhythm of shedding and degradation of rod and cone outer segment membranes in the chick retina. *Invest Ophthalmol Vis Sci.* 1978;17:105–116.

The cytoplasm of the RPE also contains numerous mitochondria (involved in aerobic metabolism), rough-surfaced endoplasmic reticulum, a Golgi apparatus, and a large round nucleus.

Throughout life, incompletely digested residual bodies, lipofuscin pigment, phagosomes, and other material are excreted beneath the basal lamina of the RPE. These contribute to the formation of *drusen.* Commonly classified by funduscopic appearance as either hard or soft, drusen are typically located between the basement membrane of the RPE cells and the inner collagenous zone of Bruch's membrane. They are thought to undergo a cycle of formation and regression, depending on the health of the overlying pigment epithelium. Hard drusen are thought to be a focal by-product of cell renewal and may thus be found even in young eyes. Softening, with the addition of membranous components, may indicate a general decline in cellular function. Drusen are presumed to be derived from the pigment epithelium, but other sources may exist.

Farkas TG. Drusen of the retinal pigment epithelium. *Surv Ophthalmol.* 1971;16:75–87.

Sarks SH. Aging and degeneration in the macular region: a clinico-pathologic study. *Br J Ophthalmol.* 1976;60:324–341.

Sarks SH, Sarks JP. Age-related macular degeneration. In: Ryan SJ, ed. *Retina.* 2nd ed. St Louis: Mosby; 1994;2:152–165.

Neurosensory Retina

The *neurosensory retina* is composed of neuronal, glial, and vascular elements (Figs III-31, III-32).

Neuronal elements The *photoreceptor layer* consists of highly specialized neuroepithelial cells called *rods* and *cones.* Each photoreceptor cell consists of an outer and an inner segment. The outer segments, surrounded by a mucopolysaccharide matrix, make contact with the apical processes of the RPE. Tight junctions or other intercellular connections do not exist between the photoreceptor cell outer segments and the RPE. The factors responsible for keeping these layers in apposition are poorly understood but probably involve active transport.

The *rod photoreceptor* consists of an *outer segment* containing multiple laminated disks resembling a stack of coins and a central connecting *cilium* that has a "9 plus 0" cross-sectional configuration rather than the "9 plus 2" configuration found in motile cilia. The rod *inner segment* is subdivided into two additional elements: an outer *ellipsoid* containing a large number of mitochondria and an inner *myoid* containing a large amount of glycogen; the myoid is continuous with the main cell body, where the nucleus is located (Fig III-33). The inner portion of the cell contains the synaptic body, or *spherule,* of the rod, which is formed by a single invagination that accommodates two horizontal cell processes and one or more central bipolar dendrites (Fig III-34).

The *extrafoveal cone photoreceptors* of the retina have conical ellipsoids and myoids, and their nuclei tend to be closer to the external limiting membrane than the nuclei of the rods. Although the structure of the outer segments of the rods and cones is similar, at least one important difference exists. Rod disks are not attached to the cell membrane; they are discrete structures. Cone disks are attached to the cell membrane and are thought to be renewed by membranous replacement. The cone synaptic body, or *pedicle,* is more complex than the rod spherule. Cone pedicles synapse with other rods and cones as well as with horizontal and bipolar cell processes. *Foveal cones* have cylindrical inner segments like rods, but otherwise are cytologically identical to extrafoveal cones. *Horizontal cells* make synaptic connections with many rod spherules and cone pedicles, and horizontal cells extend cell processes horizontally throughout the outer plexiform layer. *Bipolar cells* are oriented vertically. Their dendrites synapse with either rod or cone synaptic bodies, and their axons make synaptic contact with ganglion cells and amacrine cells in the inner plexiform layer.

The axons of the *ganglion cells* bend to become parallel to the inner surface of the retina, where they form the *nerve fiber layer* and later the axons of the optic nerve. Each optic nerve has more than 1 million optic nerve fibers. The nerve fibers from the temporal retina follow an arcuate course around the macula to enter the superior and inferior poles of the optic disc. The papillomacular fibers travel straight to the optic nerve from the fovea. The nasal axons also pursue a radial course. The visibility of the nerve fibers is enhanced when they are viewed ophthalmoscopically using green (red-free) illumination.

The neuronal elements and their connections in the retina are highly complex. Many types of bipolar, amacrine, and ganglion cells exist. The neuronal elements of more than 120 million rods and 6 million cones are interconnected, and signal processing within the neurosensory retina is significant.

Dowling JE. *The Retina: An Approachable Part of the Brain.* Cambridge, MA: Harvard University Press; 1987.

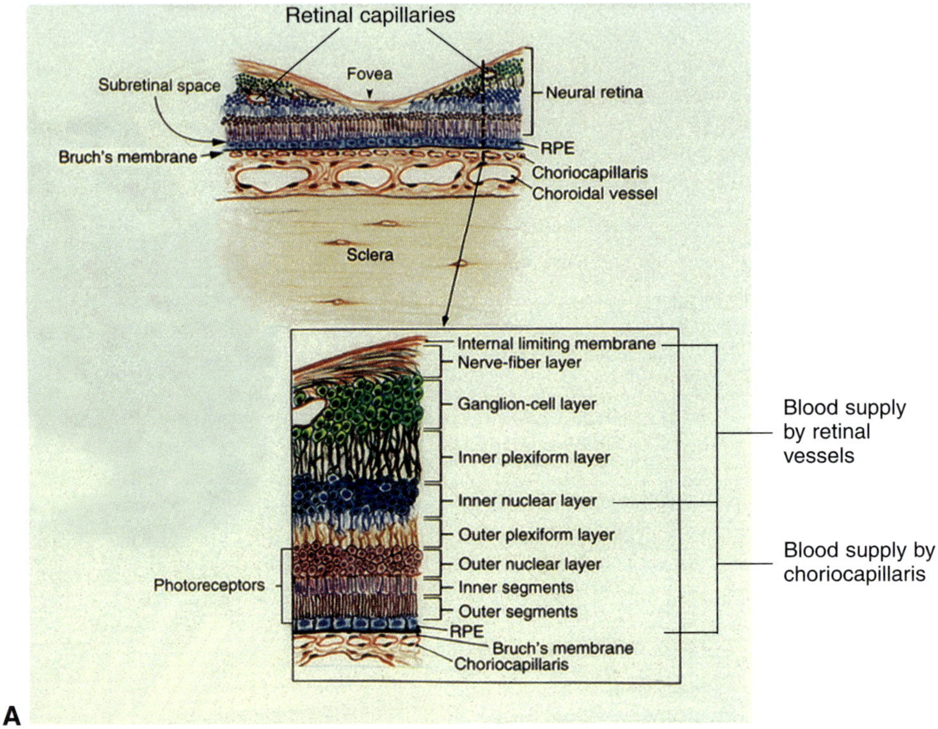

FIG III-31— **A**, Schematic cross section of retina demonstrating layers of retina and approximate location of blood supply to these layers. (Modified with permission from D'Amico DJ. Diseases of the retina. *N Engl J Med.* 1994;331:95–106.) **B**, Schematic diagram of cell types and histologic layers in the human retina. The basic relationship between rod *(R)* and cone *(C)* photoreceptors as well as bipolar *(B)*, horizontal *(H)*, amacrine *(A)*, inner plexiform cell *(I)*, and ganglion *(G)* neurons is depicted. Note that the Müller cell *(M)* almost extends across the whole thickness of the retina; the apical processes of Müller's cells form the external limiting membrane; the foot processes of Müller's cells partially form the internal limiting membrane. (Illustration by Christine Gralapp.) **C**, Optical coherence tomography (OCT) in vivo images of the normal human fovea at conventional *(top)* and ultrahigh *(bottom)* resolutions. Axial resolution is 10 mm *(top)* and 3 mm *(bottom)*. Several layers can be resolved in the ultrahigh-resolution image, including the nerve fiber layer, ganglion cell layer, inner plexiform layer, inner nuclear layer, outer nuclear layer, and retinal pigment epithelium. (Courtesy of James Fujimoto and Joel Schuman.)

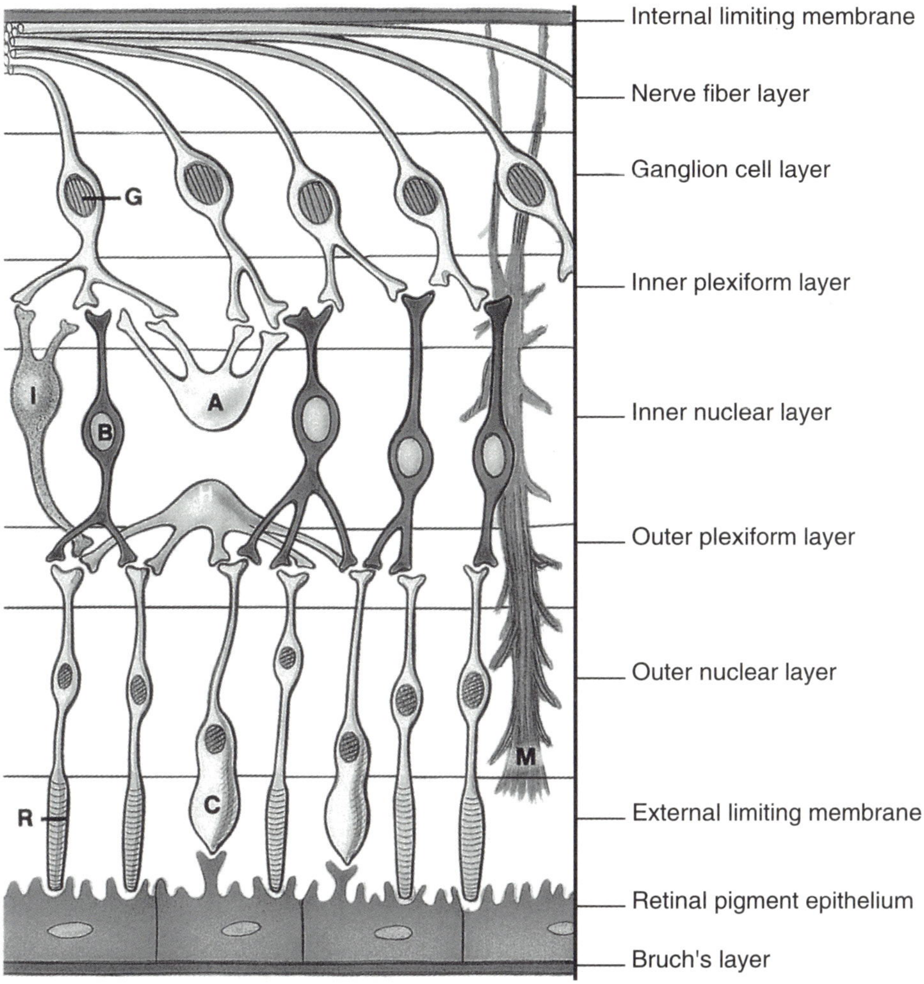
Internal limiting membrane
Nerve fiber layer
Ganglion cell layer
Inner plexiform layer
Inner nuclear layer
Outer plexiform layer
Outer nuclear layer
External limiting membrane
Retinal pigment epithelium
Bruch's layer
G
I
B
A
H
M
R
C

B

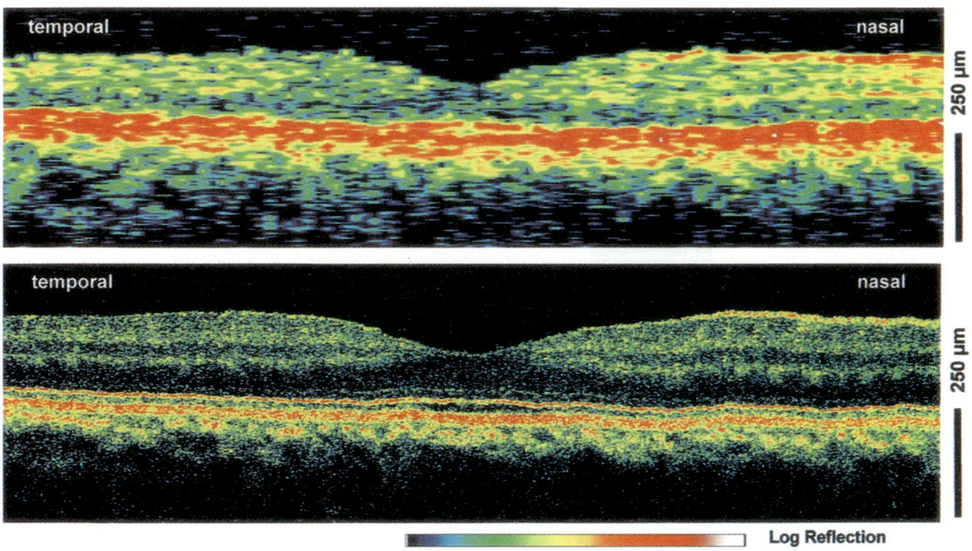
temporal
nasal
250 µm
temporal
nasal
250 µm
Log Reflection

C

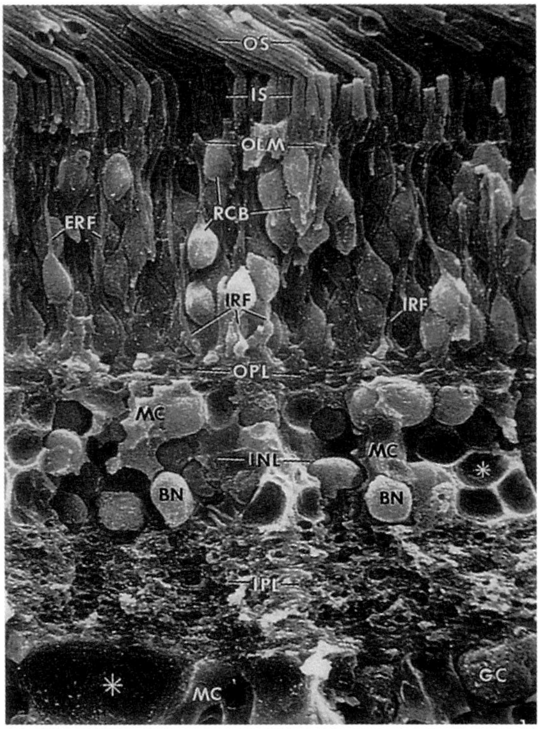

FIG III-32—Scanning electron micrograph of the retinal layers diagrammed in Fig III-31B, but oriented 180° in the anatomical position with the photoreceptors at the top. *OS,* outer segments of photoreceptors; *IS,* inner segments of photoreceptors; *OLM,* outer limiting membrane; *RCB,* rod cell bodies; *ERF,* external rod fibers dendrites; *IRF,* internal rod fibers axons; *OPL,* outer plexiform layer; *MC,* Müller's cell; *INL,* inner nuclear layer; *BN,* bipolar cell nuclei; *,* indentation in Müller's cell left by bipolar cells and ganglion cells after removal during tissue processing; *IPL,* inner plexiform layer; *GC,* ganglion cell. (Reproduced with permission from Kessel RG, Kardon RH. *Tissues and Organs: A Text-Atlas of Scanning Electron Microscopy.* New York: Freeman; 1979).

Glial elements *Müller cells* are glial cells that extend vertically from *the external limiting membrane* inward to the *internal limiting membrane*. Their nuclei are located in the inner nuclear layer. Müller cells provide structural support and nutrition to the retina along with the other glial elements, the *fibrous and protoplasmic astrocytes and microglia.*

Recent studies have provided evidence for the importance of Müller cells in retinal development and metabolism. Immunohistochemistry has shown that these cells contain cellular retinaldehyde–binding proteins, glutamine, taurine, and glutamine synthetase. Müller cells have also been shown to be involved in degradation of the neurotransmitters glutamate and gamma-aminobutyric acid. The presence of messenger RNA coding for carbonic anhydrase II implies that these cells also are important in buffering carbon dioxide liberated into the extracellular space by neurosensory elements of the retina. The production of insulin and growth factors by these cells may also be important in retinal metabolism.

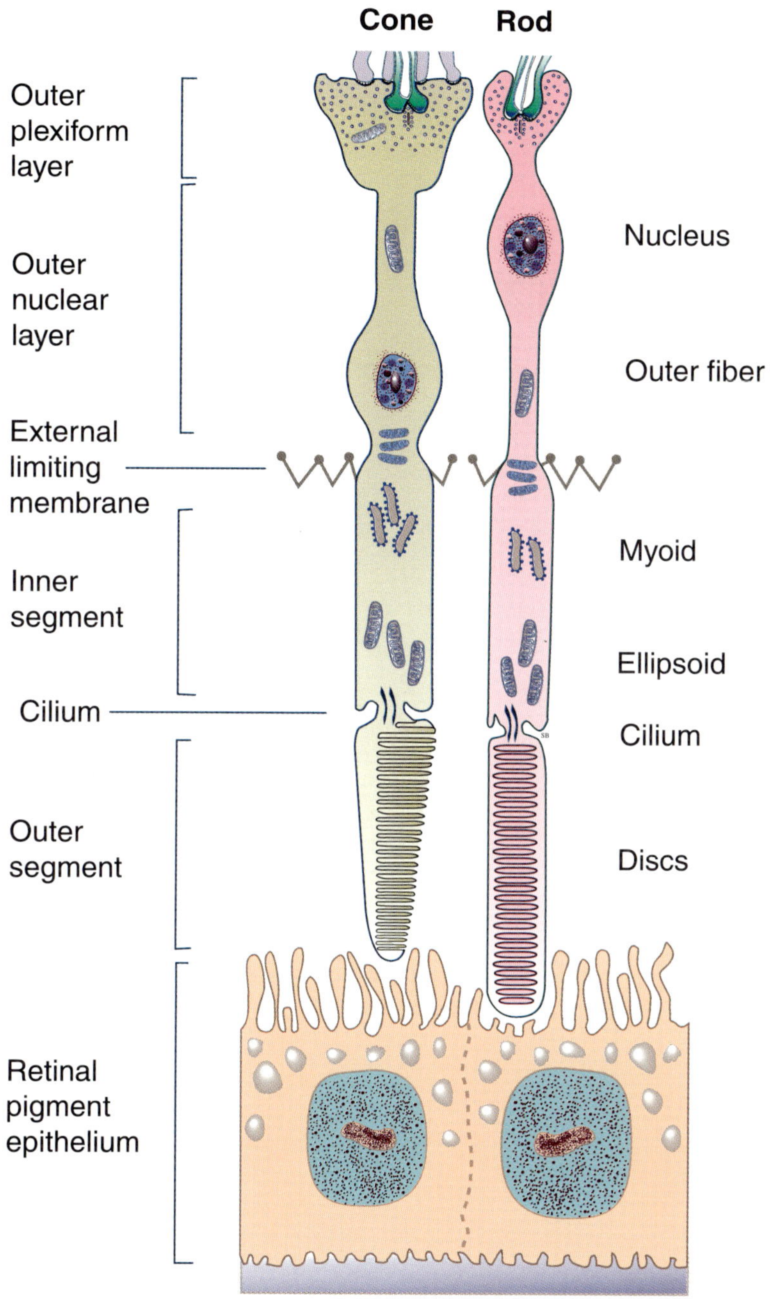

FIG III-33—Rod and cone photoreceptor cells. (Illustration by Sylvia Barker.)

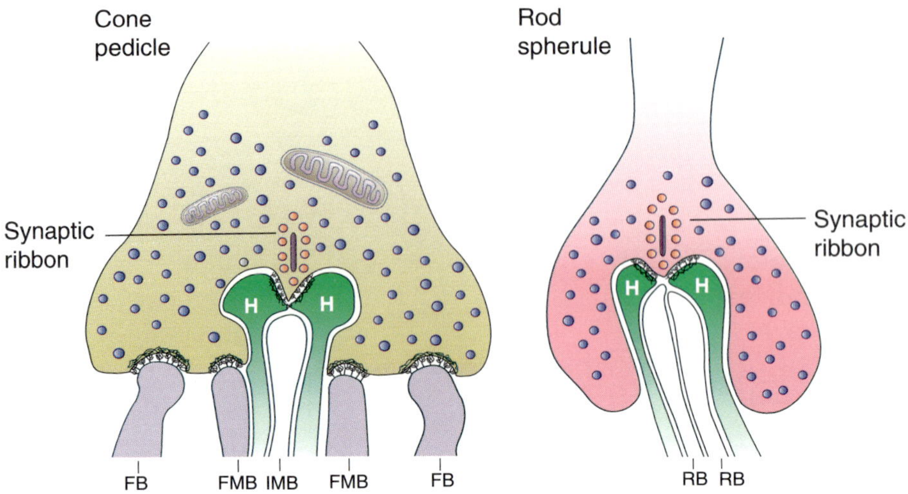

FIG III-34—Synaptic bodies cone pedicle, rod spherule; *H,* horizontal cell process; *B,* bipolar cell process. (Illustration by Sylvia Barker.)

The retina. In: Bron AJ, Tripathi RC, Tripathi BJ, eds. *Wolff's Anatomy of the Eye and Orbit.* 8th ed. London: Chapman & Hall; 1997.

Vascular elements The inner portion of the retina is perfused by branches of the central retinal artery. In 30% of eyes and 50% of persons, a cilioretinal artery also supplies part of the inner retina. A cilioretinal artery contributes to some portion of the macular circulation in approximately 15% of persons, but this artery may supply any portion of the retina.

Justice J Jr, Lehman RP. Cilioretinal arteries: a study based on review of stereo fundus photographs and fluorescein angiographic findings. *Arch Ophthalmol.* 1976;94: 1355–1358.

The retinal blood vessels are analogous to the cerebral blood vessels and maintain the inner blood–retina barrier. This physiologic barrier is due to the single layer of *nonfenestrated endothelial cells,* whose tight junctions are impervious to tracer substances such as fluorescein and horseradish peroxidase. A basal lamina covers the outer surface of the endothelium. The basement membrane contains an interrupted layer of *pericytes,* or *mural cells,* surrounded by their own basement membrane material.

Müller cells and other glial elements are generally attached to the basal lamina of retinal blood vessels. Retinal blood vessels lack an internal elastic lamina and the continuous layer of smooth muscle cells found in other vessels in the body. Smooth muscle cells are occasionally present in vessels near the optic nerve head. They become a more discontinuous layer as the retinal arterioles pass farther out to the peripheral retina. The retinal blood vessels do not ordinarily extend deeper than the middle limiting membrane. Where venules and arterioles cross, they share a common

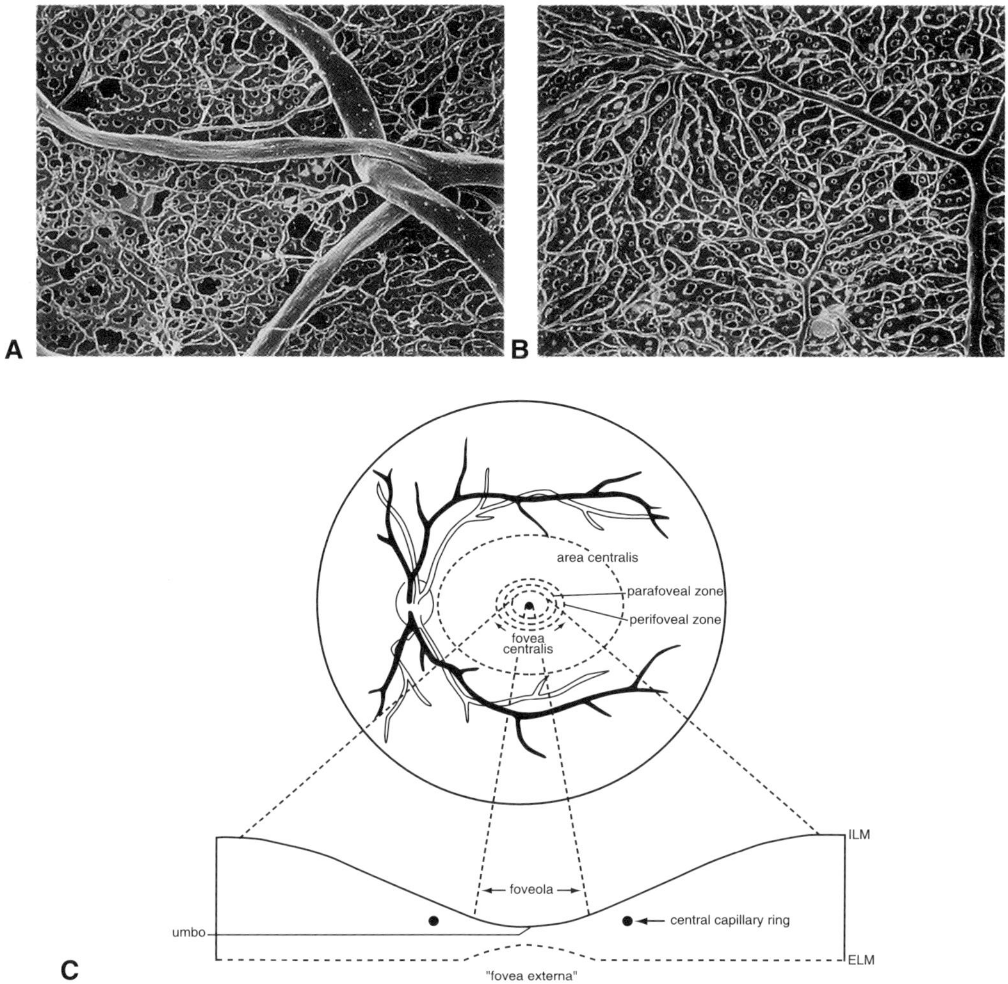

FIG III-35—**A** and **B**, Scanning electron micrographs of vascular cast of inner retinal circulation. **A**, An arteriovenous crossing with intervening capillary plexus and underlying choriocapillaris. **B**, End-arteriole supplying the inner retinal capillary plexus. **C**, Anatomic macula, also called *area centralis*. Anatomic fovea and foveola contained within center of macula. *ILM,* internal limiting membrane; *ELM,* external limiting membrane.

basement membrane. Venous occlusive disorders are common at an arteriovenous crossing (Fig III-35).

Hogan MJ, Feeney L. The ultrastructure of the retinal blood vessels. I, The large vessels. II, The small vessels. *J Ultrastruc Res.* 1963;9:10–46.

Stratification of the Neurosensory Retina

The neurosensory retina can be subdivided into several layers (see Figs III-31 and III-32).

The *external limiting membrane (ELM)* is formed by the attachment sites of adjacent photoreceptors and Müller cells. The ELM is therefore not a true membrane. In the peripheral retina, the ELM fuses with the pigment epithelium at the ora serrata. It is highly fenestrated.

The *outer plexiform layer (OPL)* is made up of the interconnections between the photoreceptor synaptic bodies and the horizontal and bipolar cells. The OPL is thicker and contains more fibers in the macular region, because the axons of the rods and cones become longer and more oblique as they deviate from the fovea. The OPL in this region is known as the *fiber layer of Henle.* At the edge of the foveola, it lies almost parallel with the internal limiting membrane. Accumulation of lipid and other blood products within the fiber layer of Henle accounts for the star pattern observed ophthalmoscopically at the macula in some cases of systemic hypertension.

The *inner nuclear layer (INL)* contains nuclei of bipolar, Müller, horizontal, and amacrine cells.

The next region is formed by a zone of desmosome-like attachments in the region of the synaptic bodies of the photoreceptor cells. The retinal blood vessels ordinarily do not extend beyond this point.

The *inner plexiform layer (IPL)* consists of axons of the bipolar and amacrine cells and dendrites of the ganglion cells and their synapses.

The *ganglion cell layer (GCL)* is made up of the cell bodies of the ganglion cells that lie near the inner surface of the retina.

The *nerve fiber layer (NFL)* is formed by axons of the ganglion cells. Normally, they do not become myelinated until after they pass through the lamina cribrosa of the optic nerve.

The *internal limiting membrane (ILM)* is also not a true membrane. It is formed by the footplates of the Müller cells and attachments to the basal lamina. The basal lamina of the retina is smooth on the vitreal side but appears undulating on the retinal side, where it follows the contour of the Müller cells. The thickness of the basal lamina varies.

Foos RY. Vitreoretinal juncture: topographical variations. *Invest Ophthalmol.* 1972; 11:801–808.

Huang D, Swanson EA, Lin CP, et al. Optical coherence tomography. *Science.* 1991; 254:1178–1181.

Drexler W, Morgner U, Ghanta RK, et al. Ultrahigh resolution ophthalmic optical coherence tomography. *Nature Med.* 2001;7:502–507.

Overall, cells and their processes in the retina are oriented perpendicular to the plane of the RPE in the middle and outer layers but parallel to the retinal surface in the inner layers. For this reason, deposits of blood or exudates tend to form round blots in the outer layers (where small capillaries are found) and linear or flame-shaped patterns in the nerve fiber layer. At the fovea, the outer layers also tend to be parallel to the surface (Henle's layer). As a result, radial or star-shaped patterns may arise when these extracellular spaces are filled with serum and exudate.

Macula

The terms *macula lutea, macula, posterior pole, area centralis, fovea,* and *foveola* have created confusion among both anatomists and clinicians. Clinical retina specialists tend to regard the *macula* as the area within the temporal vascular arcades. Histologically, it is the region with more than one layer of ganglion cell nuclei (Figs III-35 through III-37).

Orth DH, Fine BS, Fagman W, et al. Clarification of foveomacular nomenclature and grid for quantitation of macular disorders. *Trans Am Acad Ophthalmol Otolaryngol.* 1977;83:506–514.

The name *macula lutea* ("yellow spot") derives from the yellow color of the central retina in dissected cadaver eyes; this color is due to the presence of carotenoid pigments, chiefly located in Henle's layer.

Two major pigments have been identified—*zeaxanthin* and *lutein*—whose proportions vary with distance from the fovea: the lutein to zeaxanthin ratio is 1:2.4 in the central area (0.25 mm from the fovea) and greater than 2:1 in the periphery (2.2–8.7 mm from the fovea). This variation in pigment ratio corresponds to the rod to cone ratio. Lutein is more concentrated in rod-dense areas of retina, and zeaxanthin is more concentrated in cone-dense areas. Lipofuscin, the yellow age pigment, has been observed in the cytoplasm of the perifoveal ganglion cells by electron microscopy.

Bone RA, Landrum JT, Fernandez L, et al. Analysis of the macular pigment by HPLC: retinal distribution and age study. *Invest Ophthalmol Vis Sci.* 1988;29:843–849.

Nussbaum JJ, Pruett RC, Delori FC. Historic perspectives. Macular yellow pigment. The first 200 years. *Retina.* 1982;1:296–310.

The *fovea* is a concave central retinal depression approximately 1.5 mm in diameter; it is comparable in size to the optic nerve head (see Fig III-37). Its margins are clinically inexact, but in younger subjects the fovea is evident ophthalmoscopically as an elliptical light reflex that arises from the slope of the thickened internal limiting membrane of the retina. From this point inward, the basal lamina rapidly decreases in thickness as it dives down the slopes of the fovea toward the depths of the foveola, where it is barely visible even by electron microscopy.

Around the fovea is the *parafovea,* 0.5 mm wide, where the GCL, the INL, and the OPL are thickest; surrounding this zone is the most peripheral region of the macula, the *perifovea,* 1.5 mm wide.

The masking of choroidal fluorescence observed in the macula during fundus fluorescein angiography is caused partly by xanthophyll pigment and partly by the higher melanin pigment content of the foveal RPE.

The *foveola* is a central depression within the fovea, located approximately 4.0 mm temporal and 0.8 mm inferior to the center of the optic disc. It is approximately 0.35 mm across and 0.10 mm in thickness at its center. The borders of the foveola merge imperceptibly with the fovea. The nuclei of the photoreceptor cells in the region of the foveola bow forward toward the internal limiting membrane to form the *fovea externa.* Usually, only photoreceptors, Müller cells, and other glial cells are present in this area. Occasionally, light microscopy reveals ganglion cell nuclei just below the internal limiting membrane.

The photoreceptor layer of the foveola is made up entirely of cones, whose close packing accounts for the high visual acuity of this small area. The foveal cones

FIG III-36—Light micrograph of the macula. Compare with Figure III-37. (Photograph courtesy of Thomas A. Weingeist, PhD, MD.)

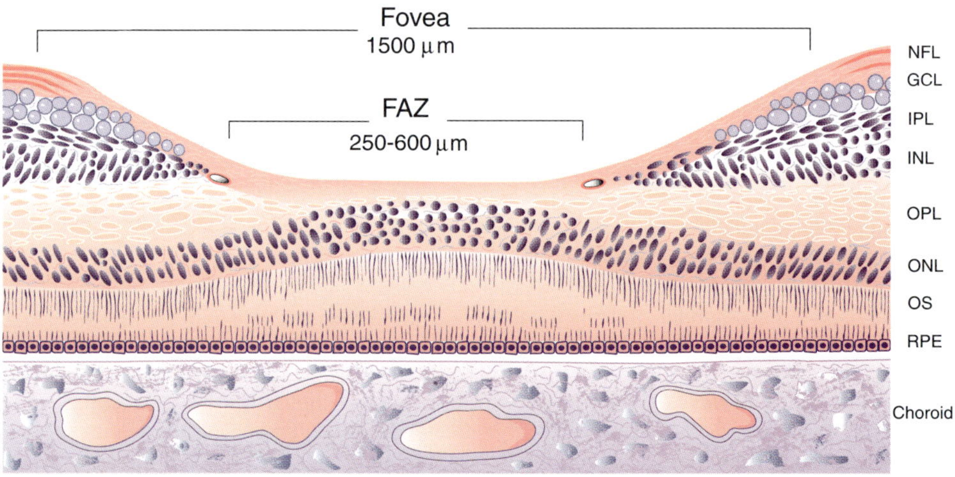

FIG III-37—Schematic section through the fovea. *FAZ,* foveal avascular zone; *GCL,* ganglion cell layer; *INL,* inner nuclear layer; *IPL,* inner plexiform layer/Henle's fiber layer; *ONL,* outer nuclear layer; *OPL,* outer plexiform layer; *OS,* outer segments of the photoreceptors; *RPE,* retinal pigment epithelium. (Illustration by Sylvia Barker.)

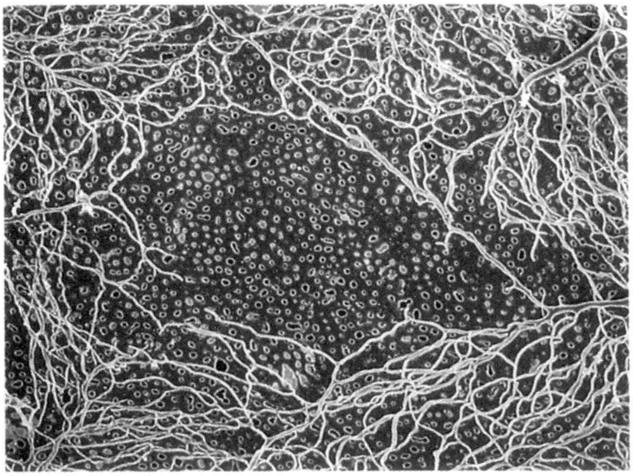

FIG III-38—Scanning electron micrograph of a retinal vascular cast at the fovea showing the foveal avascular zone and underlying choriocapillaris.

are shaped like rods but possess all the cytologic characteristics of extramacular cones. The outer segments are oriented parallel to the visual axis and perpendicular to the plane of the RPE. In contrast, the peripheral photoreceptor cell outer segments are tilted toward the entrance pupil.

The *foveal avascular zone (FAZ)*, or *capillary-free zone* (Fig III-38), is an important clinical landmark in the treatment of subretinal neovascular membranes by laser photocoagulation. The location of the FAZ is approximately that of the foveola, and its appearance in fundus fluorescein angiograms varies greatly. The diameter of the FAZ varies from 250 to 600 μm or more; often, a truly avascular or capillary-free zone cannot be identified.

Bird AC, Weale RA. On the retinal vasculature of the human fovea. *Exp Eye Res.* 1974;19:409–417.

Ora Serrata

The *ora serrata* is the boundary between the retina and the pars plana. Its distance from Schwalbe's line is between 5.75 mm nasally and 6.50 mm temporally. In myopia, this distance is greater; in hyperopia, it is smaller. Bruch's membrane extends anteriorly, beyond the ora serrata, but is modified since there is no choriocapillaris in the ciliary body.

At the ora serrata, the diameter of the eye is 20 mm and the circumference 63 mm; at the equator, the diameter is 24 mm and the circumference 75 mm. Topographically, the ora serrata is relatively smooth temporally and serrated nasally. Retinal blood vessels end in loops before reaching the ora serrata.

The ora serrata is in a watershed zone between the anterior and posterior vascular system, which may in part explain why peripheral retinal degeneration is relatively common. The peripheral retina in the region of the ora serrata is markedly

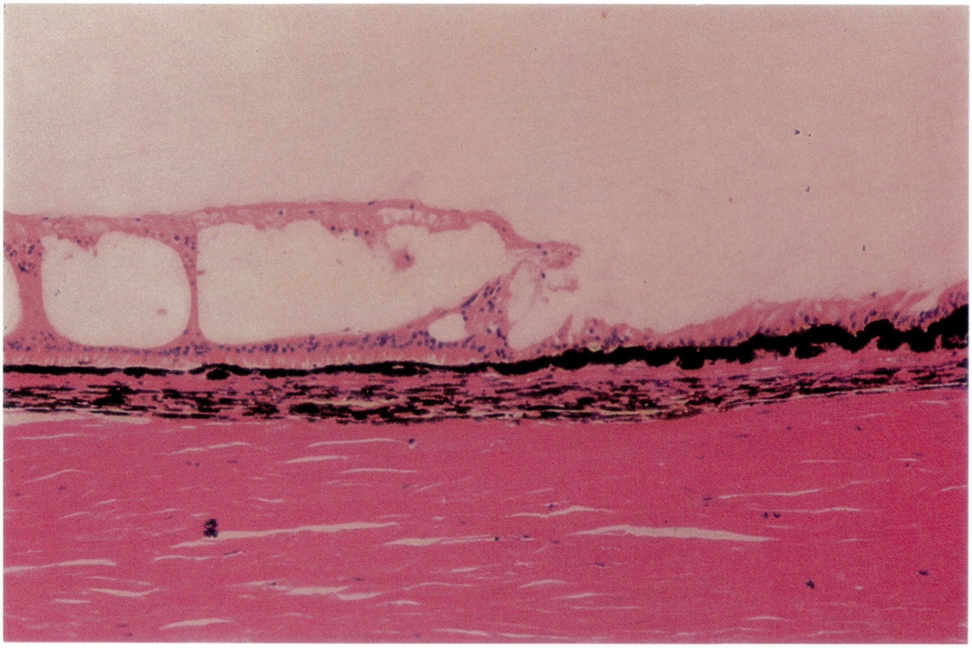

FIG III-39—Ora serrata. Note the malformed appearance of the peripheral retina and the cystic changes at the junction between the pars plana and the retina (H&E ×32). (Photograph courtesy of Thomas A. Weingeist, PhD, MD.)

attenuated. The photoreceptors are malformed, and the overlying retina frequently appears cystic in paraffin sections *(Blessig-Iwanoff cysts)* (Fig III-39).

Vitreous

The *vitreous* cavity occupies four fifths of the volume of the globe. The transparent vitreous humor is important to the metabolism of the intraocular tissues because it provides a route for metabolites used by the lens, ciliary body, and retina. Its volume is close to 4.0 ml. Although it has a gel-like structure, the vitreous is 99% water. Its viscosity is approximately twice that of water, mainly because of the presence of the mucopolysaccharide hyaluronic acid (Fig III-40).

At the ultrastructural level, fine collagen fibrils (chiefly type II) and cells have been identified in the vitreous. The origin and function of these cells is unknown. They have been termed *hyalocytes* and probably represent modified histiocytes, glial cells, or fibroblasts. The fibrils at the vitreous base merge with the basal lamina of the nonpigmented epithelium of the pars plana and the internal limiting membrane of the retina.

Sebag J, Balazs EA. Morphology and ultrastructure of human vitreous fibers. *Invest Ophthalmol Vis Sci.* 1989;30:1867–1871.

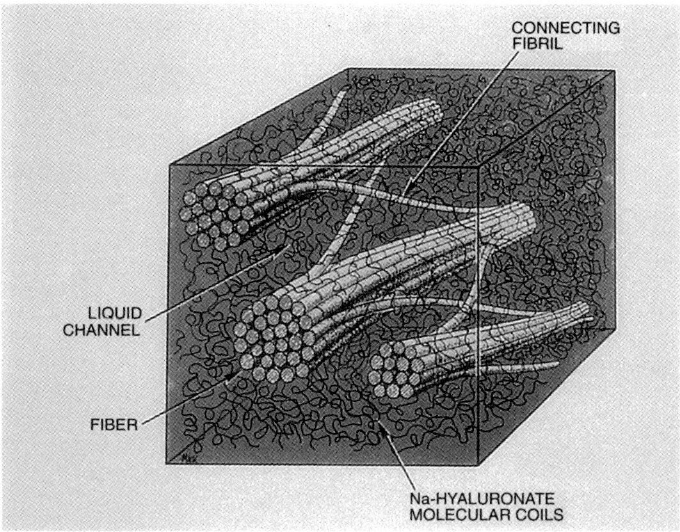

FIG III-40—Three-dimensional depiction of the molecular organization of the vitreous, showing the dissociation between hyaluronic acid molecules and collagen fibrils. The fibrils are packed into bundles, and the hyaluronic acid forms molecular "coils" that fill the intervening spaces to provide channels of liquid vitreous. (Reproduced with permission from Sebag J, Balazs EA. Morphology and ultrastructure of human vitreous fibers. *Invest Ophthalmol Vis Sci.* 1989;30:1867–1871.)

The vitreous adheres to the retina peripherally at the *vitreous base,* which extends from 2.0 mm anterior to the ora serrata to approximately 4.0 mm posterior to the ora serrata. Additional attachments exist at the disc margin, at the perimacular region, along the retinal vessels, and at the periphery of the posterior lens capsule. The vitreous becomes more fluid with age and frequently separates from the inner retina (*posterior vitreous detachment*—Fig III-41). The associated peripheral retinal traction is a potential cause of *rhegmatogenous retinal detachment* (Figs III-42 through III-44).

Foos RY. Posterior vitreous detachment. *Trans Am Acad Ophthalmol Otolaryngol.* 1972;76:480–497.

Michels RG, Wilkinson CP, Rice TA, eds. *Retinal Detachment.* St Louis: Mosby; 1990.

Sebag J. *The Vitreous: Structure, Function, and Pathobiology.* New York: Springer-Verlag; 1989.

Tolentino FI, Schepens CL, Freeman HM. *Vitreoretinal Disorders: Diagnosis and Management.* Philadelphia: Saunders; 1976:130–154.

During embryonic development, regression of the hyaloid vasculature results in the formation of an S-shaped channel *(Cloquet's canal),* which passes sinuously from a point slightly nasal to the posterior pole of the lens *(Mittendorf's dot)* to the margin of the optic nerve head. Remnants of this fetal vasculature may be observed clinically on the nerve head in the adult (*vascular loops* and *Bergmeister's papilla*).

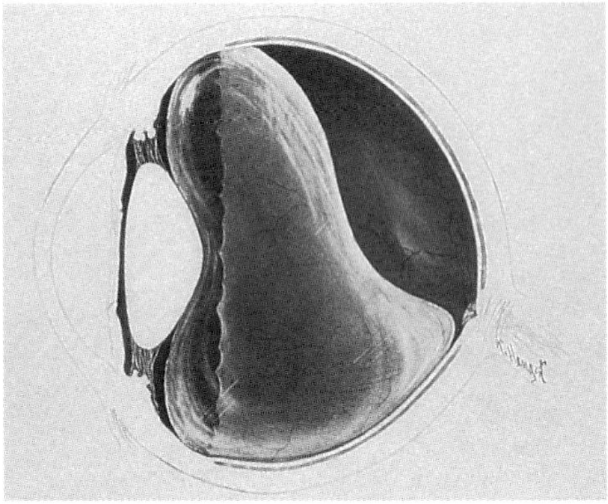

FIG III-41—Typical posterior vitreous detachment. The cortical vitreous initially separates from the retina in the posterior pole and the superior quadrants. The detachment may then progress farther anteriorly until reaching the posterior margin of the vitreous base in the inferior quadrants. (Reproduced with permission from Michels RG, Wilkinson CP, Rice TA, eds. *Retinal Detachment.* St Louis: Mosby; 1990.)

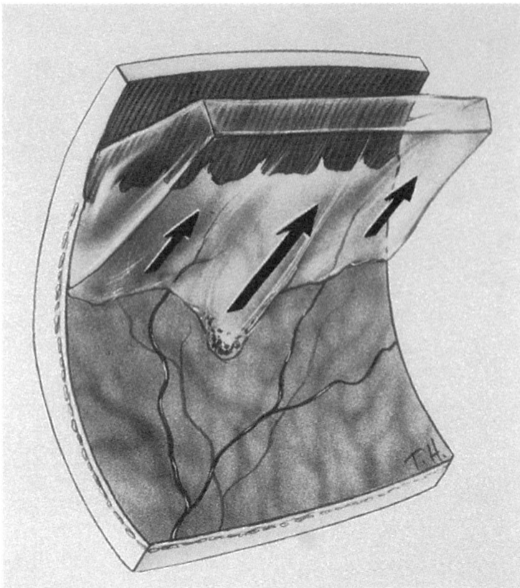

FIG III-42—Localized posterior extension of the vitreous base with firm underlying area of vitreoretinal attachment may result in greater traction in that area *(large arrow)* than along the adjacent vitreous base *(smaller arrows)*. (Reproduced with permission from Michels RG, Wilkinson CP, Rice TA, eds. *Retinal Detachment.* St Louis: Mosby; 1990.)

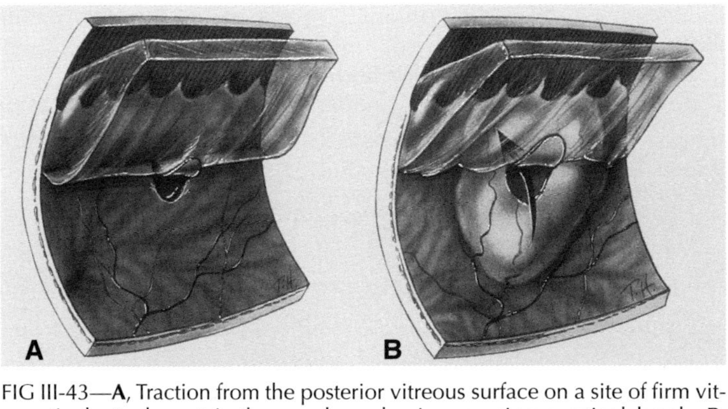

FIG III-43—**A**, Traction from the posterior vitreous surface on a site of firm vitreoretinal attachment is the usual mechanism causing a retinal break. **B**, Persistent traction on the flap of the retinal tear and fluid currents in the vitreous cavity contribute to retinal detachment. (Reproduced with permission from Michels RG, Wilkinson CP, Rice TA, eds. *Retinal Detachment.* St Louis: Mosby; 1990.)

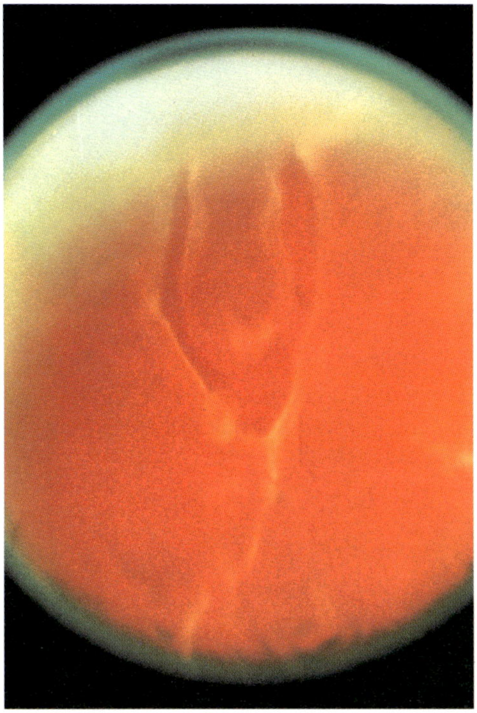

FIG III-44—Fundus photo of a flap retinal tear with associated retinal detachment. (Photograph courtesy of James Folk, MD.)

CHAPTER IV

Cranial Nerves: Central and Peripheral Connections

Cranial nerves (CN) I–VI are depicted in Figure IV-1 in relation to the bony canals and arteries at the base of the skull. The reader may find it useful to refer back to this figure as each of the cranial nerves is discussed. For further study, BCSC Section 5, *Neuro-Ophthalmology,* describes the cranial nerves and their function and dysfunction in detail.

Cranial Nerve I (Olfactory)

Cranial nerve I originates from small olfactory receptors in the mucous membrane of the nose. Unmyelinated CN I fibers pass from these receptors in the nasal cavity through the *cribriform plate* of the ethmoid bone and enter the ventral surface of the *olfactory bulb,* where they form the nerve.

The *olfactory tract* runs posteriorly from the bulb, beneath the frontal lobe of the brain in a groove (or *sulcus*) and lateral to the gyrus rectus (Fig IV-2). The *gyrus rectus* forms the anterolateral border of the suprasellar cistern. Meningiomas arising from the arachnoid cells in this area can produce important ophthalmic signs and symptoms associated with loss of olfaction.

Optic Nerve (Cranial Nerve II)

The *optic nerve* consists of more than 1 million axons that originate in the ganglion cell layer of the retina and extend toward the occipital cortex. The optic nerve may be divided into the following topographic areas:

- Intraocular portion of the optic nerve: optic disc, or nerve head; prelaminar; and laminar portions
- Intraorbital portion (located within the muscle cone)
- Intracanalicular portion (located within the optic canal)
- Intracranial portion (ending in the optic chiasm)

See Table IV-1 for a summary of regional differences; these are illustrated in the magnetic resonance scans shown in Figures IV-3 through IV-5.

The organization of the optic nerve is similar to that of the white matter of the brain. Developmentally, the optic nerve is part of the brain, and its fibers are surrounded by glial (and not Schwann cell) sheaths. The optic nerve varies in length from 35 to 55 mm and averages 40 mm. Part of the intraocular portion of the optic nerve is visible ophthalmoscopically as the *optic nerve head,* or *optic disc.* The optic nerve head is oval and measures approximately 1.5 mm horizontally and 1.75 mm vertically.

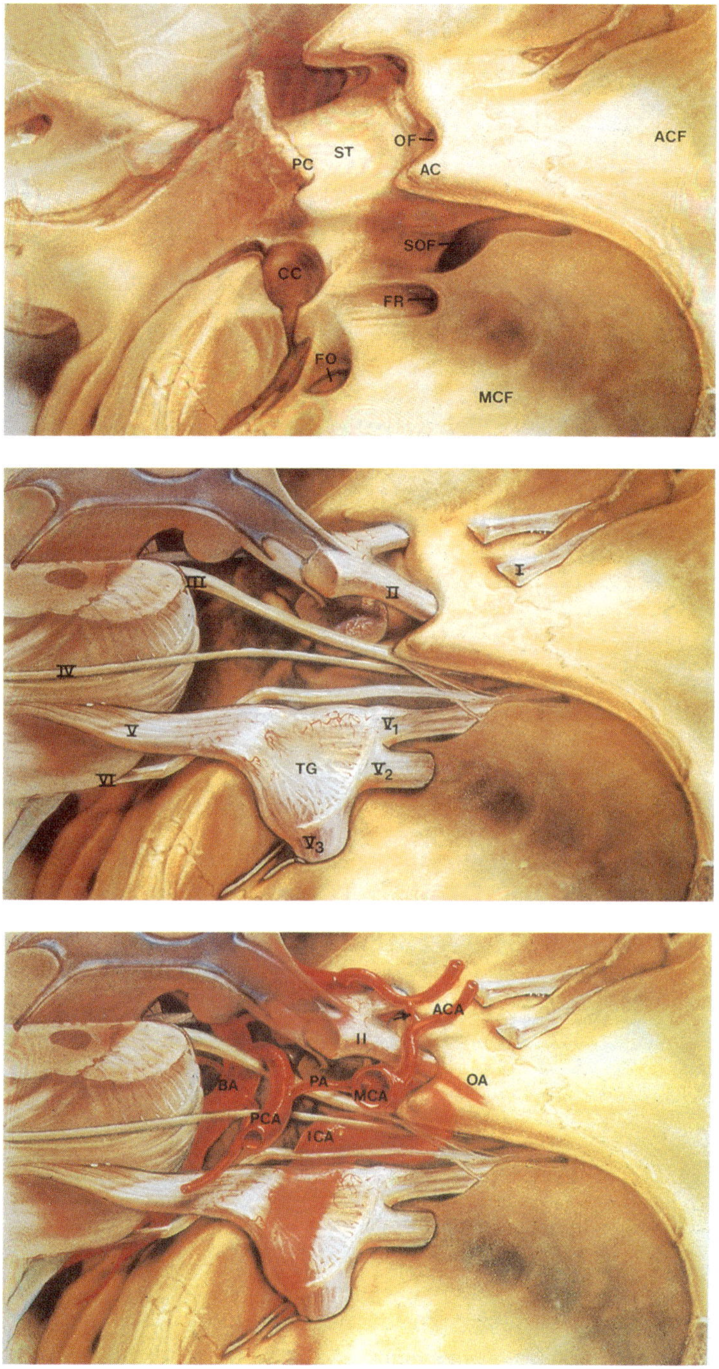

FIG IV-1—View from the right parietal bone looking downward into the skull base showing the relationship between the bony canals (**A**), nerves (**B**), and arteries (**C**) at the base of the skull. The orbits are located to the right, out of the picture (the roof of the orbits is just visible). The floor of the right middle cranial fossa is in the lower part. **A**, *PC*, posterior clinoid; *AC*, anterior clinoid; *ST*, sella turcica; *OF*, optic foramen; *CC*, carotid canal; *SOF*, superior orbital fissure; *FR*, foramen rotundum; *FO*, foramen ovale; *ACF*, anterior cranial fossa; *MCF*, middle cranial fossa. **B**, *I*, olfactory nerve; *II*, optic nerve; *III*, oculomotor nerve; *IV*, trochlear nerve; *V*, trigeminal nerve with ophthalmic V_1, maxillary V_2, and mandibular V_3 divisions; *VI*, abducens nerve; *TG*, trigeminal gasserian ganglion. **C**, *BA*, basilar artery; *PCA*, posterior cerebral artery; *PA*, posterior communicating artery; *MCA*, middle cerebral artery; *ACA*, anterior communicating artery; *OA*, ophthalmic artery; *arrow*, anterior communicating artery; *ICA*, internal carotid artery. (Reproduced with permission from Zide BM, Jelks GW, eds. *Surgical Anatomy of the Orbit.* New York: Raven; 1985.)

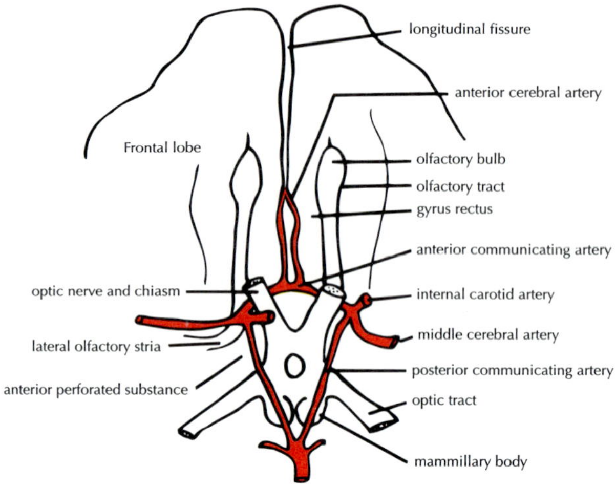

FIG IV-2—Inferior surface of the brain, depicting cranial nerves I and II and surrounding structures. (Illustration by Thomas A. Weingeist, PhD, MD.)

TABLE IV-1

REGIONAL DIFFERENCES IN THE OPTIC NERVE

SEGMENT	LENGTH mm	DIAMETER mm	BLOOD SUPPLY
Intraocular	1.0		Retinal arterioles
optic disc		1.5 × 1.75	Branches of posterior ciliary arteries
prelaminar			
laminar			
Intraorbital	25	3–4	Intraneural branches of central retinal artery; pial branches from CRA and choroid
Intracanalicular	4–10		Ophthalmic artery
Intracranial	10	4–7	Branches of internal carotid and ophthalmic artery

The intraorbital portion is 25–30 mm long, which is greater than the distance between the back of the globe and the optic canal (18 mm). For this reason, when the eye is in the primary position, the optic nerve runs a sinuous course.

Quigley HA, Addicks EM. Regional differences in the structure of the lamina cribrosa and their relation to glaucomatous optic nerve damage. *Arch Ophthalmol.* 1981; 99:137–143.

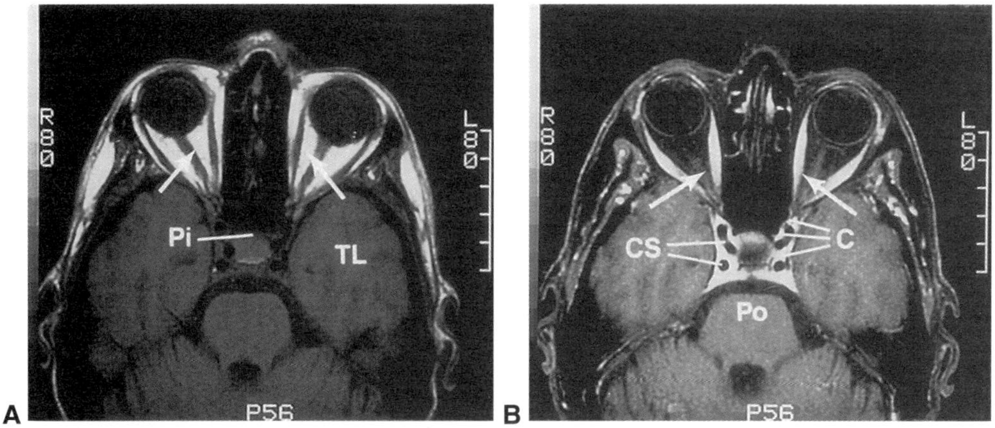

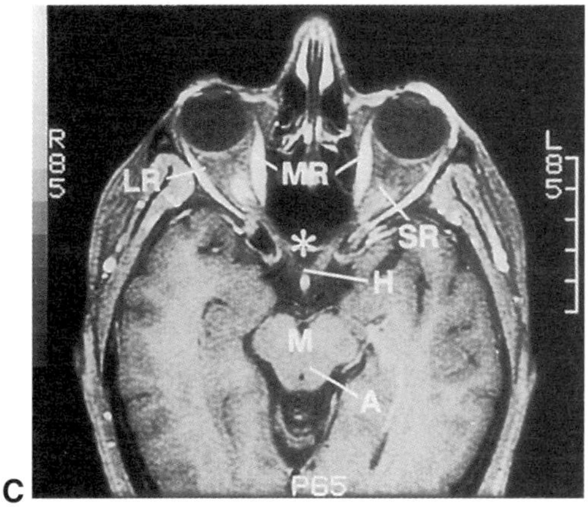

FIG IV-3—Axial MRIs showing course of the optic nerve *(arrows)* as it passes from the orbit to the chiasm *(*)* (C). **A**, Image made without fat suppression; ocular muscles and optic nerve appear dark against the bright fat signal in the orbit. **B**, Image made with fat suppression and gadolinium, causing the fat to appear dark with bright, enhancing ocular muscles. Normal optic nerves do not enhance. *SR,* superior rectus cut obliquely; *MR,* medial rectus; *LR,* lateral rectus; *C,* carotid arteries; *CS,* cavernous sinus; *Po,* pons; *M,* midbrain; *A,* aqueduct of Sylvius; *H,* hypothalamic stalk; *TL,* temporal lobe; *Pi,* pituitary.

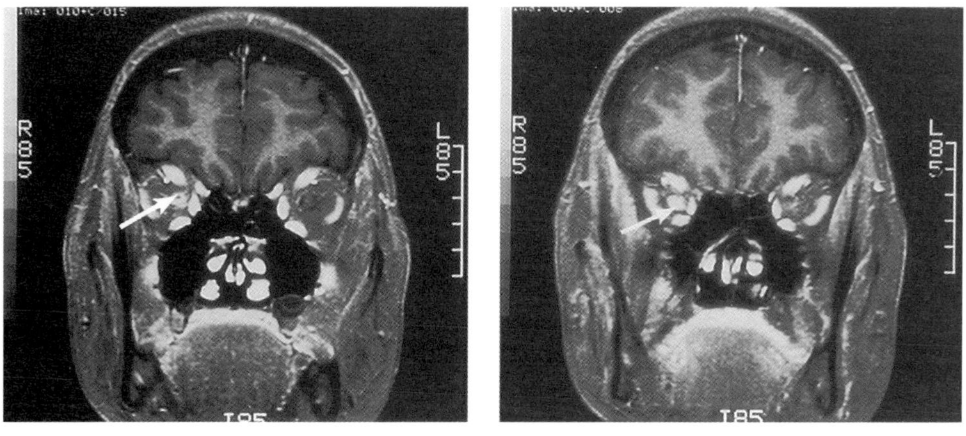

FIG IV-4—Coronal orbital MRI images with fat suppression and gadolinium enhancement in a patient with a meningioma of the right optic nerve. From left to right, images show successive cuts moving posteriorly. Note the abnormal enhancement of the pia of the right optic nerve *(arrow)* compared to the normal left nerve, and the relationship of the nerve to the extraocular muscles in which enhancement is normal.

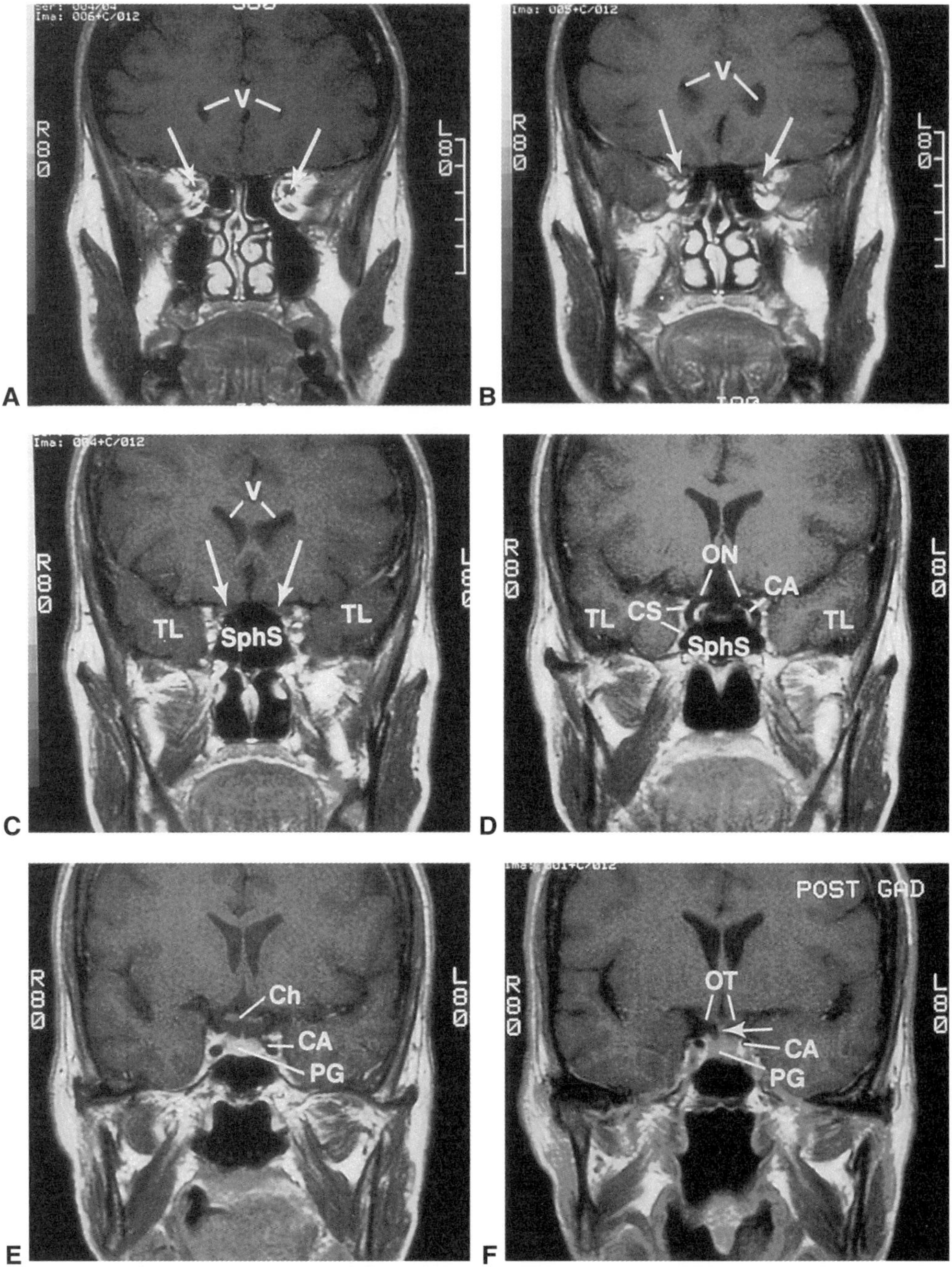

FIG IV-5—Series of coronal MRIs without fat suppression of a normal patient, starting at the posterior orbit and moving caudally toward the optic chiasm and optic tracts. The images move more posteriorly from **A** to **F**. The white surrounding the central dark optic nerve and peripheral dark muscles represents orbital fat at the orbital apex. **A**, Posterior orbit at muscle cone. *Arrows,* optic nerves; *V,* ventricles. **B**, Orbital apex, intracanalicular region. **C**, Intracranial region. *TL,* temporal lobe; *SphS,* sphenoid sinus. **D**, Intracranial portion of optic nerves. *ON,* optic nerve; *CS,* cavernous sinus; *CA,* carotid arteries. **E**, Plane of section is at chiasm. *Ch,* chiasm; *PG,* pituitary gland. **F**, Plane of section is at the optic tracts. *OT,* optic tracts; *arrow,* infundibulum.

Intraocular Portion

The optic nerve head is the principal site of many congenital and acquired ocular diseases; therefore, detailed knowledge of its anatomy is important for the practicing ophthalmologist. Its anterior surface is visible ophthalmoscopically as the *optic disc,* a 1.5 × 1.75 mm oval structure with a cup-shaped depression, the *physiologic cup,* located slightly temporal to its geometric center. The main branches of the central retinal artery and vein pass through the center of the cup. The optic nerve head can be described in four parts:

- Superficial nerve fiber layer
- Prelaminar region
- Laminar region
- Retrolaminar region

> Apple DJ, Rabb MF, Walsh PM. Congenital anomalies of the optic disc. *Surv Ophthalmol.* 1982;27:3–41.

Superficial nerve fiber layer As the nonmyelinated ganglion cell axons enter the nerve head, they retain their retinotopic organization, with fibers from the upper retina above and those from the lower retina below. Fibers from the temporal retina are lateral; those from the nasal side are medial. *Macular fibers,* constituting about one third of the nerve, are laterally placed. *Foveal fibers* are located peripherally, and *peripapillary fibers* centrally, in the nerve head. Patches of myelinated nerve fibers are occasionally seen in the retina. These patches result from the migration of oligodendrocytes beyond the lamina cribrosa along ganglion cell axons.

Prelaminar portion The ganglion cell axons that enter the nerve head are supported by a "wicker basket" of astrocytic glial cells and segregated into bundles, or *fascicles,* that pass through the lamina cribrosa. These *astrocytes* invest the optic nerve and form continuous circular tubes that enclose groups of nerve fibers throughout their intraocular and intraorbital course, separating them from connective tissue elements at all sites. There are no Müller's cells in the nerve head, but the astrocytes form an internal limiting membrane that covers the surface of the nerve head and is continuous with that of the retina. Astrocytes make up 10% of the nerve head volume.

When the optic nerve is damaged, axons and supporting glial elements can be lost, resulting in pathologic enlargement of the optic cup. This cupping may be the first objective sign of damage from glaucoma. Technologies such as confocal scanning laser ophthalmoscopy and scanning laser polarimetry can help quantify the loss of tissue at the optic nerve head. Such methods may be useful in diagnosing pathologic changes at this location and determining whether damage has progressed (Figs IV-6, IV-7).

The retinal layers terminate as they approach the edge of the optic disc. As already noted, the Müller cells that make up the internal limiting membrane are replaced by astrocytes. The pigment epithelium may be exposed at the temporal margin of the disc to form a narrow pigmented crescent. When the pigment epithelium and choroid fail to reach the temporal margin, crescents of partial or absent pigmentation can be seen. In the presence of chronic glaucoma, large crescents are predictive of greater visual field loss.

> Jonas JB, Naumann GOH. The anatomical structure of the normal and glaucomatous optic nerve. In: Kriegelstein GK, ed: *Glaucoma Update IV.* New York: Springer-Verlag; 1991:66–73.

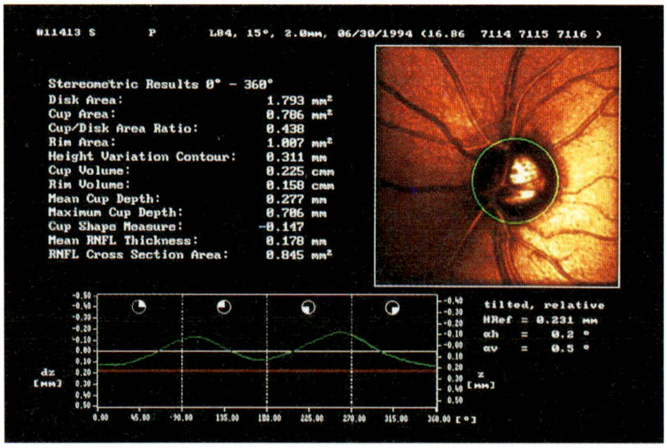

A

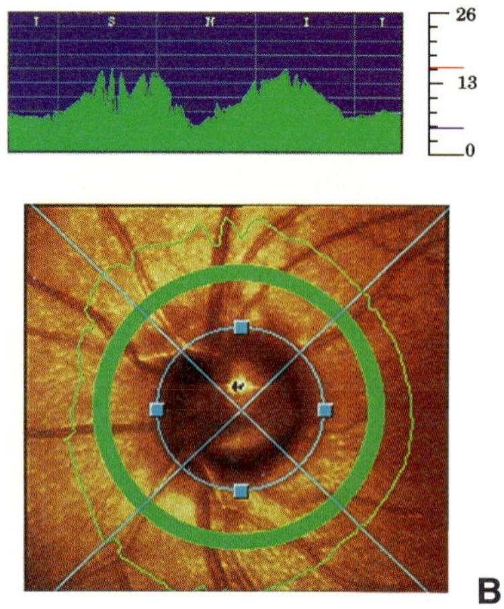

B

FIG IV-6—**A**, Confocal scanning laser ophthalmoscopy image (Heidelberg Retinal Tomograph) of the normal optic nerve head. Height-variation diagram *(bottom)* shows normal double-hump pattern corresponding to higher nerve fiber layer superiorly and inferiorly. **B**, Scanning laser polarimetry (Nerve Fiber Analyzer II) image and height-variation cross-sectional diagram *(top)* shows same double-hump pattern corresponding to thicker nerve fiber layer in these regions. (Photographs courtesy of Robert Weinreb, MD.)

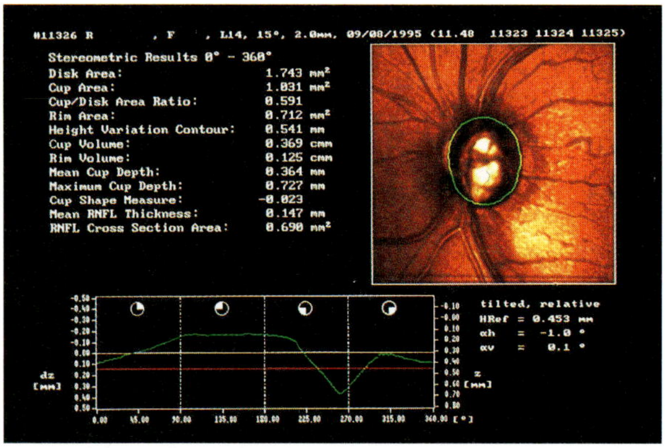

A

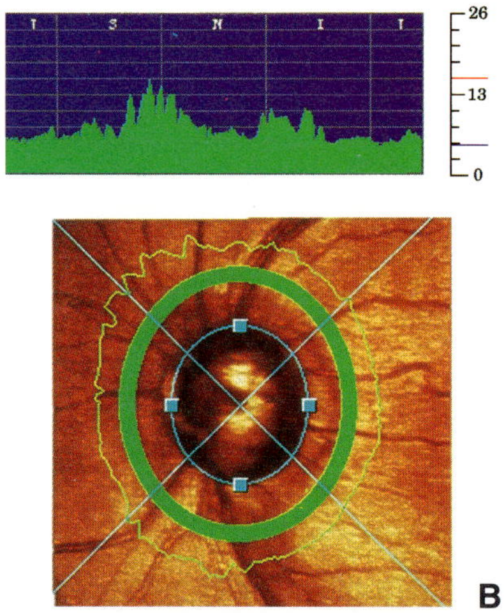

B

FIG IV-7— **A**, Confocal scanning laser ophthalmoscopy image (Heidelberg Retinal Tomograph) of the optic nerve from a 58-year-old female patient with glaucoma. Height-variation diagram shows depression in the inferotemporal region corresponding to notching of the neuroretinal rim and wedge-shaped defect of the nerve fiber layer. **B**, Scanning laser polarimetry (Nerve Fiber Analyzer II) image of the same eye shows similar depression in the inferotemporal region corresponding to thinning of the nerve fiber layer in this region. (Photographs courtesy of Robert Weinreb, MD.)

The relationship of the choroid to the prelaminar portion of the optic nerve partly accounts for the staining of the disc normally observed in late phases of fluorescein fundus angiography. The disc vessels do not leak, but the choroidal capillaries are freely permeable to fluorescein, which can therefore diffuse into the lamina.

Laminar portion The *lamina cribrosa* comprises about ten connective tissue plates, which are integrated with the sclera and whose pores transmit the axon bundles. The openings are wider above than below, which may imply lesser protection from the mechanical effects of pressure in glaucoma. The lamina contains type I and type III collagens, abundant elastin, and laminin and fibronectin. Astrocytes surround the axon bundles, and small blood vessels are present. The lamina cribrosa serves the following functions:

- Scaffold for the optic nerve axons
- Point of fixation for the central retinal artery and vein
- Reinforcement of the posterior segment of the globe

Retrolaminar portion Behind the lamina cribrosa, the optic nerve increases to 3 mm in diameter as a result of myelination of the nerve fibers and the presence of oligodendroglia and the surrounding meningeal sheaths (pia, arachnoid, and dura) (Fig IV-8). The *retrolaminar nerve* continues proximally (as the intraorbital part of the optic nerve) to the apex of the orbit. The axoplasm of the neurons contains neurofilaments, microtubules, mitochondria, and smooth endoplasmic reticulum.

Intraorbital Portion

Annulus of Zinn The intraorbital part of the optic nerve lies within the muscle cone. Before passing into the optic canal, the nerve is surrounded by the *annulus of Zinn,* formed by the origins of the rectus muscles. The *superior rectus* and the *medial rectus* partially originate from the sheath of the optic nerve. This connection may partly explain why patients with retrobulbar neuritis complain of pain on eye movement.

At the optic canal, the dural sheath of the nerve fuses to the periosteum, completely immobilizing the nerve. Blunt trauma, especially to the brow area, may transmit forces to this area; this can cause shearing between the dural sheath and its attachment to the periosteum, an interruption of the blood vessels traversing the periosteum and dural sheath, and severe nerve damage.

Meningeal sheaths: pia, arachnoid, dura mater The *pia mater* is the innermost layer of the optic nerve sheath. It is a vascular connective tissue coat, covered with meningothelial cells, which sends numerous septa into the optic nerve, dividing its axons into bundles. The septa continue throughout the intraorbital and intracanalicular portions of the nerve and end just before the chiasm. They contain collagen, elastic tissue, fibroblasts, nerves, and small arterioles and venules (Fig IV-9). They provide mechanical support for the nerve bundles and nutrition to the axons and glial cells. A mantle of astrocytic glial cells prevents the pia and septa from direct contact with nerve axons.

The *arachnoid mater* lines the dura mater and is connected to the pia across the subarachnoid space by vascular trabeculae. The *subarachnoid space* ends anteriorly at the level of the lamina cribrosa. Posteriorly, it is usually continuous with the subarachnoid space of the brain. Since the central retinal vessels cross this space, a rise

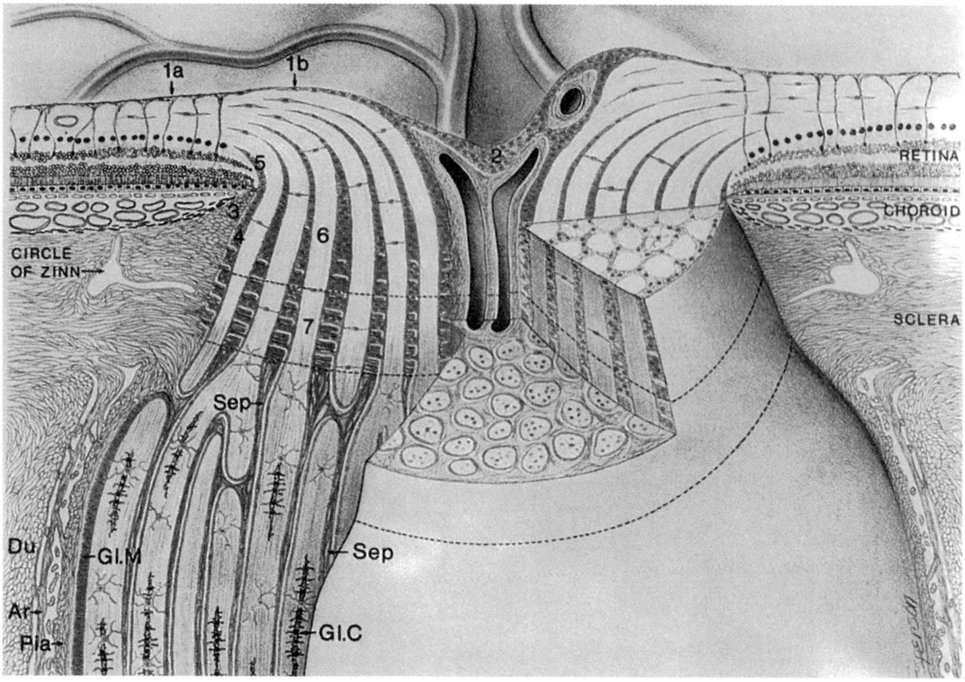

FIG IV-8—Three-dimensional drawing of the optic nerve head. Where the retina terminates at the optic disc edge, the Müller cells *(1a)* are continuous with the astrocytes, forming the internal limiting membrane *(1b)*. At the posterior termination of the choroid on the temporal side, the border tissue of Elschnig *(3)* lies between the astrocytes surrounding the optic nerve canal *(4)* and the stroma of the choroid. On the nasal side, the choroidal stroma is directly adjacent to the astrocytes surrounding the nerve. This collection of astrocytes surrounding the canal is known as the border tissue, which is continuous with a similar glial lining *(5)* at the termination of the retina. The nerve fibers of the retina are segregated into approximately 1000 fascicles by astrocytes *(6)*. Upon reaching the lamina cribrosa, or cribriform plate *(upper dotted line)*, the nerve fascicles *(7)* and their surrounding astrocytes are separated from each other by connective tissue. The cribriform plate is an extension of scleral collagen and elastic fibers through the nerve. The external choroid also sends some connective tissue to the anterior part of the lamina. At the external part of the lamina cribrosa *(lower dotted line)*, the nerve fibers become myelinated, and columns of oligodendrocytes and a few astrocytes are present within the nerve fascicles. The bundles continue to be separated by connective tissue septa all the way to the chiasm *(Sep)*. The septa are derived from the pia mater. This connective tissue is derived from the pia mater and is known as the septal tissue. A mantle of astrocytes *(Gl.M)*, continuous anteriorly with the border tissue, surrounds the nerve along its orbital course. The dura *(Du)*, arachnoid *(Ar)*, and pia mater *(Pia)* are shown. The nerve fibers are myelinated. Within the bundles, the cell bodies of astrocytes and oligodendrocytes form a column of nuclei *(Gl.C)*. The central retinal vessels are surrounded by a perivascular connective tissue throughout its course in the nerve. This connective tissue blends with the connective tissue of the lamina cribrosa and is called the central supporting connective tissue strand here. (Reproduced with permission from Anderson DR, Hoyt WF. Ultrastructure of intraorbital portion of human and monkey optic nerve. *Arch Ophthalmol.* 1969;82:507.)

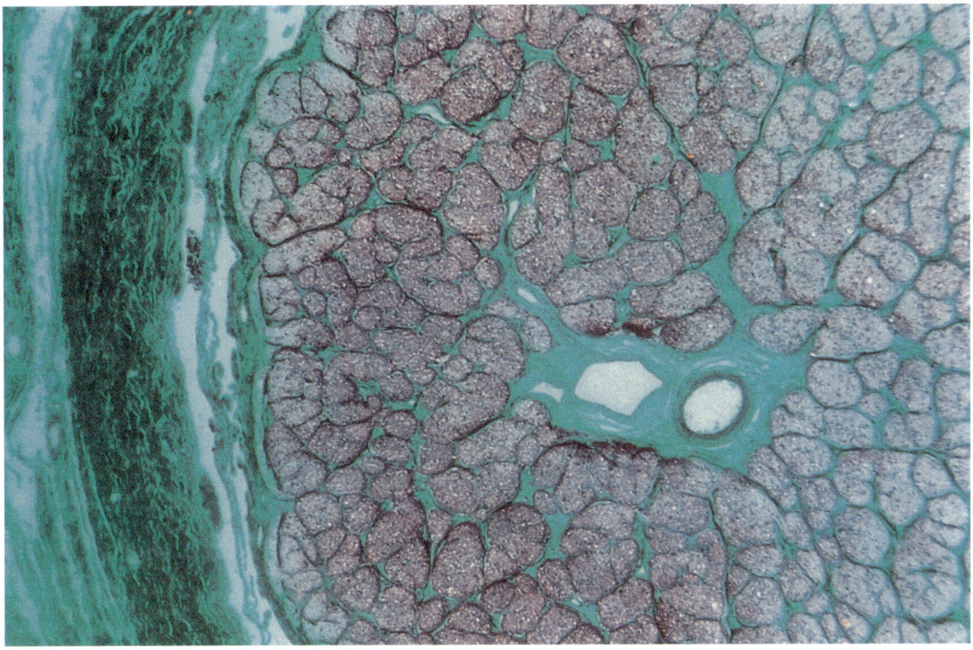

FIG IV-9—Meningeal sheaths. The dura mater, the outer layer, is composed of collagenous connective tissue. The arachnoid sheath is the middle layer made up of fine collagenous fibers arranged in a loose meshwork lined by endothelial cells. The innermost layer, the pia, is made up of fine collagenous and elastic fibers and is highly vascularized. Elements from both the arachnoid and the pia are continuous with the optic nerve septa (Masson trichrome ×64). (Photograph courtesy of Thomas A. Weingeist, PhD, MD.)

in intracranial pressure can compress the retinal vein and raise the venous pressure within the retina above the intraocular pressure. This situation causes the loss of spontaneous venous pulsation at the nerve head. Such an absence of pulsation may clinically indicate raised intracranial pressure. The arachnoid mater is composed of collagenous tissue, small amounts of elastic tissue, and meningothelial cells, which may give rise to *corpora arenacea.*

The thick *dura mater* encases the brain and makes up the outer layer of the meningeal sheath of the optic nerve. It is 0.3–0.5 mm thick and consists of dense bundles of collagen and elastic tissue that fuse anteriorly with the outer layers of the sclera.

The meninges of the optic nerve are supplied by sensory nerve fibers, which account in part for the pain experienced by patients with retrobulbar neuritis and other inflammatory optic nerve diseases.

Intracanalicular Portion

Within the *optic canal,* the blood supply of the optic nerve is derived from pial vessels originating from the ophthalmic artery. The optic nerve and surrounding arachnoid are tethered to the periosteum of the bony canal within the intracanalicular portion. Blunt trauma, particularly over the eyebrow, can transmit the force of injury to the intracanalicular portion, causing shearing and interruption of the blood supply to the nerve in this area. In addition, optic nerve edema in this area can produce a compartment syndrome, further compromising the function of the optic nerve within the confined space of the optic canal.

Intracranial Portion

After passing through the optic canals, the two optic nerves lie above the ophthalmic arteries, above and medial to the internal carotid arteries. The anterior cerebral arteries cross over the optic nerves and are connected by the anterior communicating artery, which completes the anterior portion of the *circle of Willis.* The optic nerves then pass posteriorly over the cavernous sinus to join in the optic chiasm. The chiasm then divides into right and left optic tracts, which end in their respective lateral geniculate bodies. From these bodies arise the *geniculocalcarine pathways* (or *visual radiations*), which pass to each primary visual cortex. Lesions at different locations along the visual pathway produce characteristic visual field defects that help to localize the site of damage (Fig IV-10).

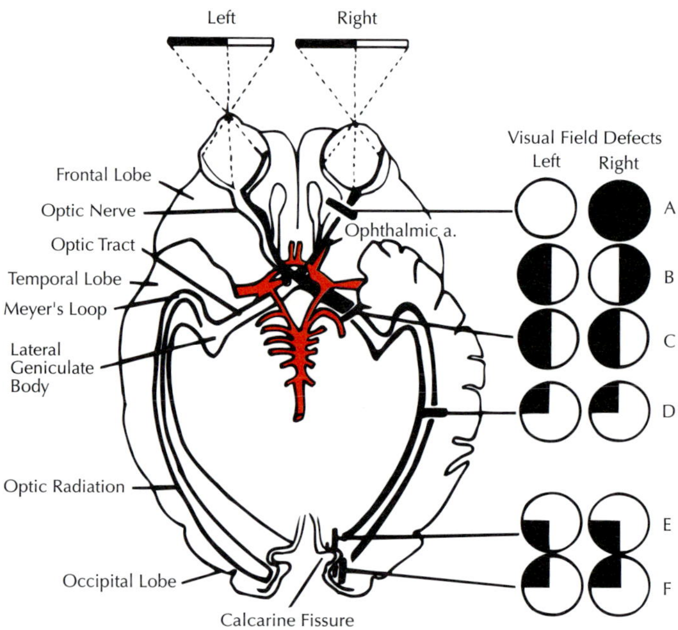

FIG IV-10—The visual pathway and the circle of Willis. (Illustration by Thomas A. Weingeist, PhD, MD.)

Chiasm The *optic chiasm* makes up part of the anterior inferior floor of the third ventricle. It is richly vascularized and, except in the intracranial portion, is surrounded by pia and arachnoid. The chiasm is approximately 12 mm wide, 8 mm long in the anteroposterior direction, and 4 mm thick. The extramacular fibers from the inferonasal retina cross anteriorly in the chiasm at *Wilbrand's knee* before passing into the optic tract. Extramacular superonasal fibers cross directly to the opposite tract. Extramacular temporal fibers remain uncrossed in the chiasm and optic tract. The *macular projections* are located centrally in the optic nerve and constitute 80%–90% of the total volume of the optic nerve and the chiasmal fibers. The temporal macular fibers pursue a direct course through the chiasm as a bundle of uncrossed fibers. Nasal macular fibers cross in the posterior part of the chiasm. Approximately 53% of the optic nerve fibers are crossed, and 47% are uncrossed.

Optic tract Each *optic tract* contains ipsilateral temporal and contralateral nasal fibers from the optic nerves (Fig IV-11). Fibers (both crossed and uncrossed) from the upper retinal projections travel medially in the optic tract; lower projections move laterally. The macular fibers adopt a dorsolateral orientation as they course toward the lateral geniculate body.

Lateral geniculate body The *lateral geniculate body,* or *nucleus,* is the synaptic zone for the higher visual projections. It is an oval, caplike structure that receives approximately 70% of the optic tract fibers within its six alternating layers of gray and white matter. Layers 1, 4, and 6 of the lateral geniculate body contain axons from the contralateral optic nerve. Layers 2, 3, and 5 arise from the ipsilateral optic nerve. The six layers, numbered consecutively from below upward, give rise to the optic radiations.

Optic radiations The *optic radiations* connect the lateral geniculate body with the cortex of the occipital lobe. The fibers of the optic radiations leave the lateral geniculate body and run around the temporal horn of the lateral ventricle, approaching the anterior tip of the temporal lobe (the so-called *loop of Meyer*). They then sweep backward toward the visual area of the occipital lobe. Damage to the visual radiation in the anterior temporal lobe gives rise to a wedge-shaped, upper homonymous "pie in the sky" visual field defect.

Visual cortex The *visual cortex,* the thinnest area of the human cerebral cortex, has six cellular layers and occupies the superior and inferior lips of the calcarine fissure on the posterior and medial surfaces of the occipital lobes. Macular function is extremely well represented in the visual cortex and occupies the most posterior position at the tip of the occipital lobe. The most anterior portion of the calcarine fissure is occupied by contralateral nasal retinal fibers only. The posterior cerebral artery, a branch of the basilar artery, supplies the visual cortex almost exclusively. The blood supply to the occipital lobe does show anatomical variation, however, with the middle cerebral artery making a contribution in some persons.

Blood Supply of the Optic Nerve

The *ophthalmic artery* lies below the optic nerve. The central retinal artery and, usually, two long posterior ciliary arteries branch off from the ophthalmic artery once it has entered the muscle cone at the annulus of Zinn.

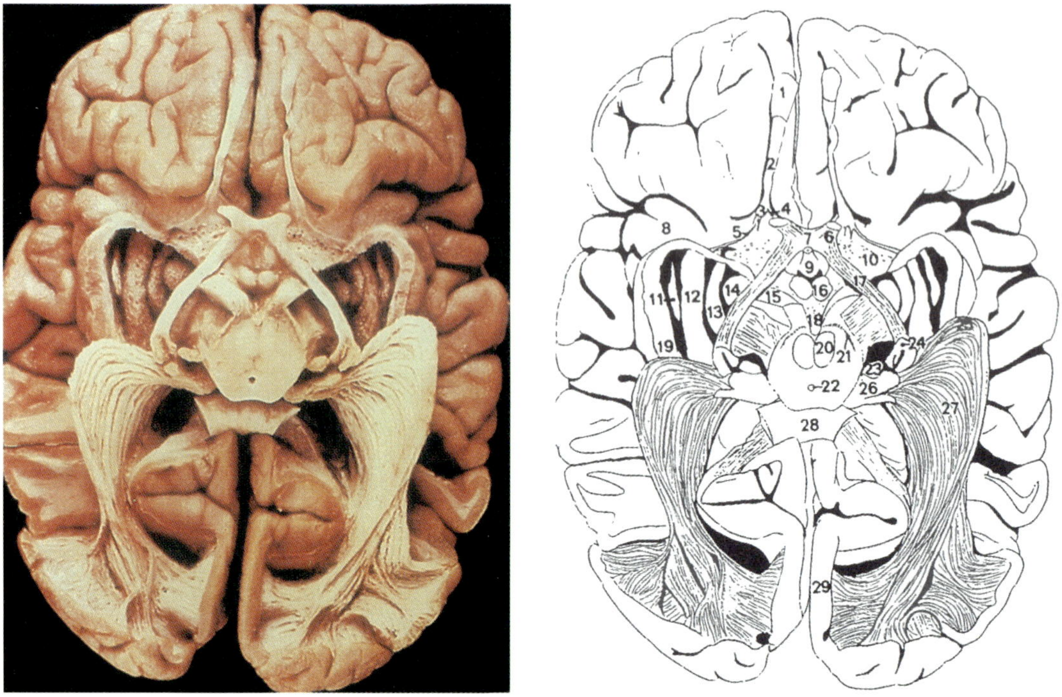

FIG IV-11—Anatomic dissection of the visual radiations. (Reproduced with permission from Gluhbegovic N, Williams TH. *The Human Brain: A Photographic Guide.* Hagerstown, MD: Harper & Row; 1980.)

1. Olfactory bulb
2. Olfactory tract
3. Olfactory trigone
4. Medial olfactory stria
5. Lateral olfactory stria
6. Optic nerve
7. Optic chiasma
8. Limen insulae
9. Tuber cinereum with infundibulum
10. Anterior (rostral) perforated substance
11. Claustrum
12. Putamen
13. Lateral part of globus pallidus
14. Medial part of globus pallidus
15. Basis pedunculi
16. Mamillary body
17. Optic tract
18. Posterior (interpeduncular) perforated substance
19. Cortex of insula
20. Superior cerebellar peduncle
21. Substantia nigra
22. Mesencephalic (cerebral) aqueduct
23. Medial geniculate nucleus
24. Lateral geniculate body
25. Temporal genu of optic radiation
26. Pulvinar of thalamus
27. Sagittal stratum
28. Splenium of corpus callosum
29. Upper lip of calcarine sulcus

The blood supply of the optic nerve varies from one segment of the nerve to another. Although the blood supply can vary widely, a basic pattern has emerged from a multitude of studies.

Hayreh SS. Blood supply and vascular disorders of the optic nerve. In: Cant JS, ed. *The Optic Nerve: Proceedings of the Second William Mackenzie Memorial Symposium Held in Glasgow, 7th–10th September 1971.* London: Kimpton; 1972:59–67.

Hayreh SS. Structure and blood supply to the optic nerve. In: Heilmann K, Richardson KT, Aulhorn E, eds. *Glaucoma: Conceptions of a Disease.* Stuttgart: Thieme; 1978: 78–96.

The arterial supply of the optic nerve head is as follows: The *retrolaminar nerve* is supplied chiefly by pial and short posterior ciliary vessels, with some help from the central retinal and septal arteries and sometimes recurrent choroidal arteries. The *lamina* is supplied by short posterior ciliary arteries or by branches of the arterial *circle of Haller and Zinn.* This circle arises from the paraoptic branches of the short posterior ciliary arteries and is usually embedded in the sclera around the nerve head. It is often incomplete and may be divided into a superior and an inferior half. There is no supply from the central retinal artery in this region.

The *prelaminar nerve* is supplied by the short posterior ciliary arteries (cilioretinal arteries, if present) and recurrent choroidal arteries, although their relative contribution is debated. The nerve fiber layer is supplied by the central retinal artery (Figs IV-12 through IV-14). The posterior ciliary arteries are *terminal arteries,* and the area where the respective capillary beds from each artery meet has been termed the *watershed zone.* When perfusion pressure drops, the tissue lying within this area is the most vulnerable to ischemia. Consequences can be significant when the entire optic nerve head or a portion of it lies within the watershed zone, as shown by fluorescein angiograms taken during acute anterior ischemic optic neuropathy (Fig IV-15).

The *intraorbital portion* of the optic nerve is supplied proximally by the pial vascular network and by neighboring branches of the ophthalmic artery. Distally, it is also supplied by intraneural branches of the central retinal artery. Most anteriorly, short posterior ciliary arteries and occasional peripapillary choroidal arteries contribute.

The *intracanalicular portion* of the optic nerve is supplied almost exclusively by the ophthalmic artery.

The *intracranial portion* of the optic nerve is supplied primarily by branches of both the internal carotid artery and the ophthalmic artery.

Central retinal artery and vein The lumen of the *central retinal artery* is surrounded by nonfenestrated endothelial cells with typical zonulae occludentes like those in retinal vessels. The central retinal artery, however, differs from retinal arterioles in that it contains a fenestrated internal elastic lamina and an outer layer of smooth muscle cells surrounded by a thin basement membrane. The retinal arterioles have no internal elastic lamina and lose their smooth muscle cells shortly after entering the retina. The central retinal vein consists of endothelial cells, a thin basal lamina, and a thick collagenous adventitia.

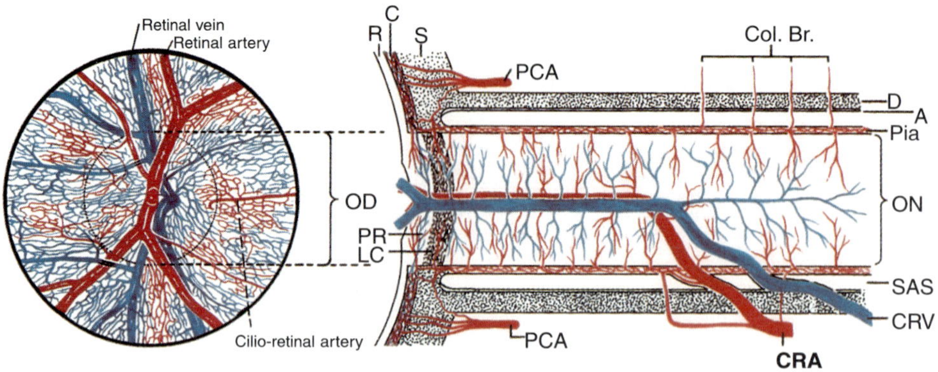

FIG IV-12—Diagram of blood supply of the optic nerve head and intraorbital optic nerve. *A,* arachnoid; *C,* choroid; *CRA,* central retinal artery; *Col Br,* collateral branches; *CRV,* central retinal vein; *D,* dura; *LC,* lamina cribrosa; *OD,* optic disc; *ON,* optic nerve; *PCA,* posterior ciliary arteries; *PR,* prelaminar region; *R,* retina; *S,* sclera; *SAS,* subarachnoid space. (Reproduced with permission from Hayreh SS. Anatomy and physiology of the optic nerve head. *Trans Am Acad Ophthalmol Otolaryngol.* 1974;78:240–254.)

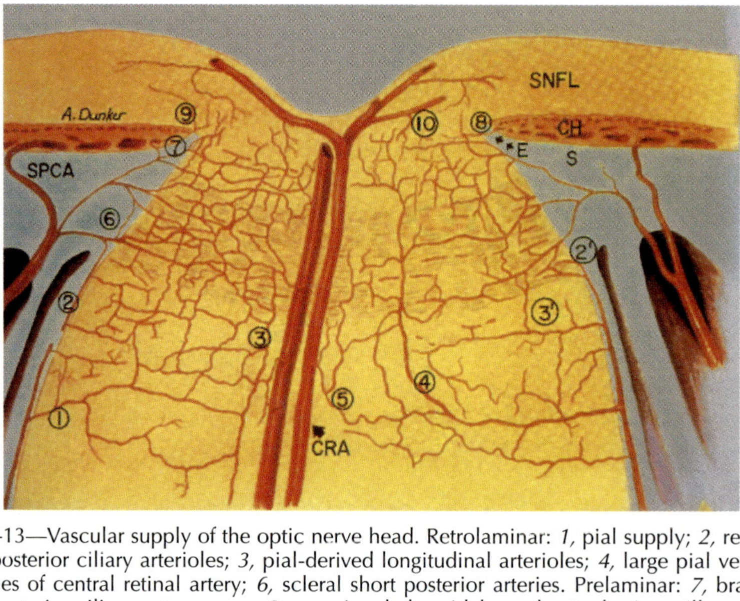

FIG IV-13—Vascular supply of the optic nerve head. Retrolaminar: *1,* pial supply; *2,* recurrent short posterior ciliary arterioles; *3,* pial-derived longitudinal arterioles; *4,* large pial vessels; *r,* branches of central retinal artery; *6,* scleral short posterior arteries. Prelaminar: *7,* branch of short posterior ciliary enters nerve; *8,* occasional choroidal supply; *9,* choriocapillary anastomosis, epi- and peripapillary branches of the central retinal artery. (Reprinted from Lieberman MF, Maumenee AE, Green WR. Histologic studies of the vasculature of the anterior optic nerve. *Am J Ophthalmol.* 1976;82:405–423, © 1976, with permission from Elsevier Science.)

FIG IV-14—Vascular cast of optic nerve surface capillaries that are supplied by branches of the central retinal artery. The branches of the central retinal artery and vein are shown originating in the center. The retinal surface capillary plexus appears continuous with the capillaries of the optic nerve surface. In this specimen, the vasculature was filled with a polymer and allowed to harden; the tissue was then digested with alkali; and the cast was viewed by scanning electron microscopy (magnification ×50).

Cranial Nerve III (Oculomotor)

Although *cranial nerve III* contains only 24,000 fibers, it supplies all the extraocular muscles except the superior oblique and the lateral rectus. It also carries cholinergic innervation to the pupillary sphincter and the ciliary muscle.

CN III arises from a complex group of cells in the rostral midbrain, or *mesencephalon,* at the level of the superior colliculus. This nuclear complex lies ventral to the periaqueductal gray matter, is immediately rostral to the CN IV nuclear complex, and is bounded inferolaterally by the medial *longitudinal fasciculus.*

The CN III nucleus consists of several distinct, large motor cell subnuclei, each of which subserves the extraocular muscle it innervates (Fig IV-16). Except for a single central caudal nucleus that serves both levators, the cell groups are paired. Fibers from the superior rectus cross in the caudal aspect of the nucleus and therefore supply the contralateral superior rectus muscles.

The *Edinger-Westphal nucleus* is cephalad and dorsomedial in location. It provides the parasympathetic preganglionic efferent innervation to the ciliary muscle and pupillary sphincter. The most ventral subnuclei supply the medial rectus muscles. A subnucleus for ocular convergence has been described but is not found consistently in primates.

Carpenter MB. *Core Text of Neuroanatomy.* 4th ed. Baltimore: Williams & Wilkins; 1991.

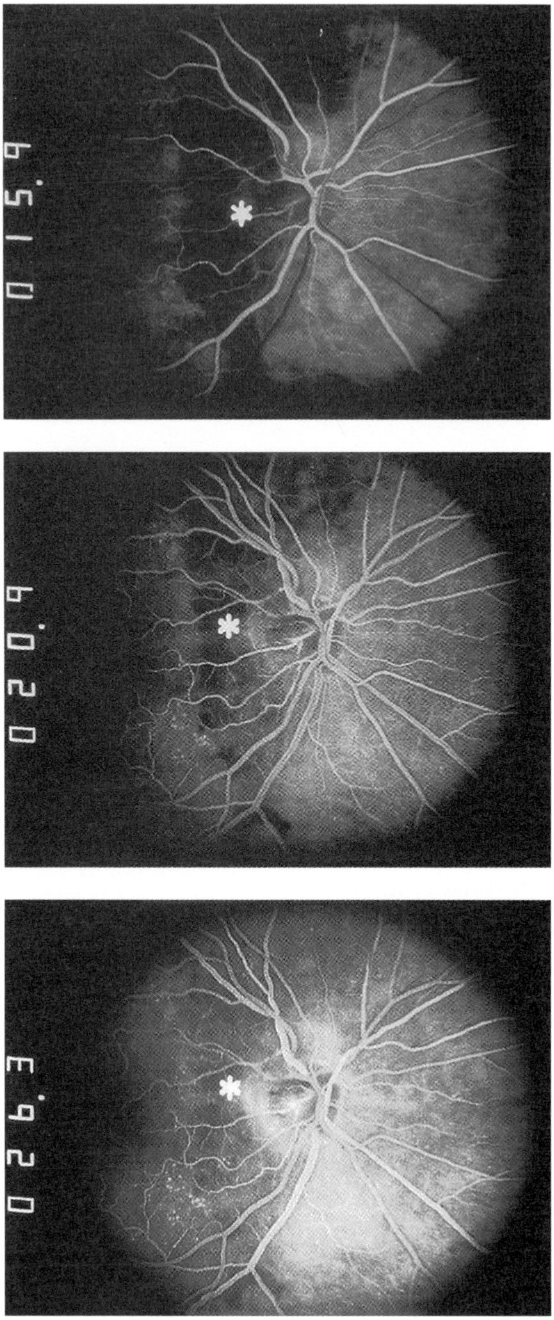

FIG IV-15—Fluorescein angiogram shows sequence of early phase of filling in watershed zone of choroidal filling *(*)* between the areas of choroid supplied by two posterior ciliary arteries during acute anterior ischemic optic neuropathy. Note that the optic nerve lies within the watershed zone and thus may be prone to ischemic damage from drops in perfusion pressure. The lobular filling pattern of the choroid is also shown.

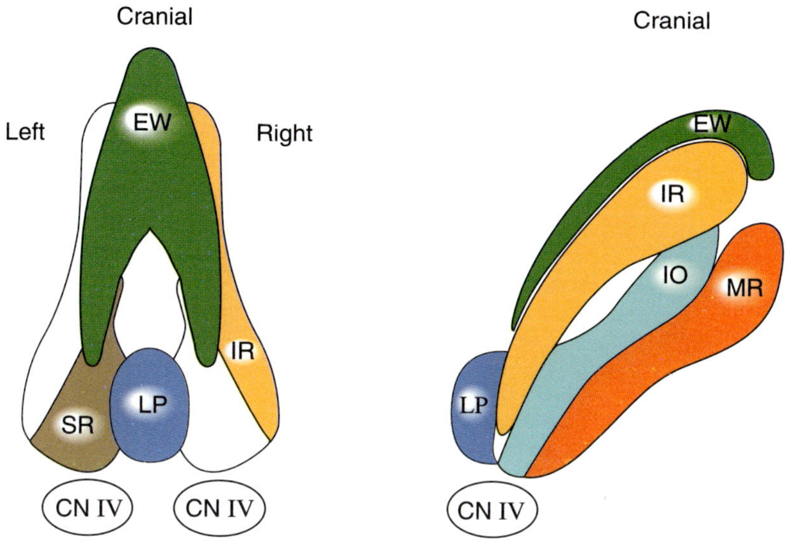

FIG IV-16—Diagram of the oculomotor nuclear complex (cranial nerve III), supplied by different subnuclei. Note that a central caudal nucleus supplies *both* levator muscles and that the nucleus for each superior rectus supplies the *contralateral* muscle. *EW,* Edinger-Westphal nucleus; *IO,* nucleus to the inferior oblique muscle, *IR;* nucleus to the inferior rectus muscle; *LP,* nucleus to the levator palpebrae muscle; *MR,* nucleus to the medial rectus muscle; *SR,* nucleus to the superior rectus muscle. (Illustration by Sylvia Barker.)

The fascicular portion of CN III travels ventrally from the nuclear complex, through the red nucleus, between the medial aspects of the cerebral peduncles, and through the corticospinal fibers. It exits in the interpeduncular space. In the subarachnoid space, CN III passes below the posterior cerebral artery and above the superior cerebellar artery, the two major branches of the basilar artery (Fig IV-17). The nerve travels forward in the interpeduncular cistern lateral to the posterior communicating artery and penetrates the arachnoid between the free and attached borders of the tentorium cerebelli. Tumors of the oculomotor nerve, such as gliomas, cause an enlargement and thickening that can be easily identified in magnetic resonance images (Fig IV-18).

Aneurysms that affect CN III commonly occur at the junction of the posterior communicating and internal carotid arteries. The nerve pierces the dura on the lateral side of the posterior clinoid process, initially traversing the roof of the cavernous sinus. It runs along the lateral wall of the cavernous sinus and above CN IV and enters the orbit through the superior orbital fissure.

CN III usually divides into superior and inferior divisions after passing through the annulus of Zinn in the orbit. Alternatively, it may divide within the anterior cavernous sinus. CN III maintains a topographic organization even in the midbrain, so that lesions almost anywhere along its course may cause a divisional nerve palsy.

Kardon RH, Traynelis VC, Biller J. Inferior division paresis of the oculomotor nerve caused by basilar artery aneurysm. *Cerebrovasc Dis.* 1991;1:171–176.

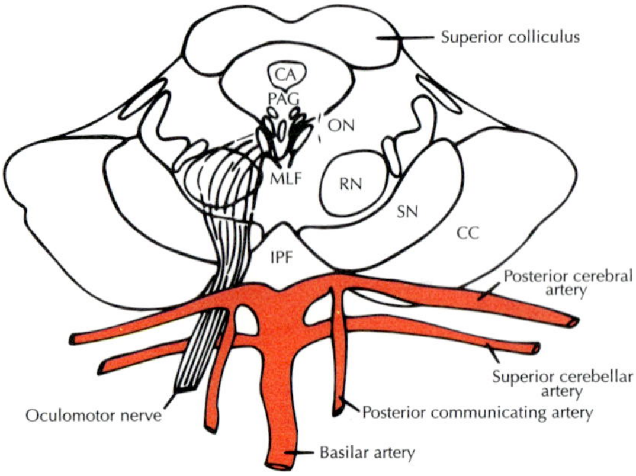

FIG IV-17—Cross section through the midbrain at the level of the third-nerve nucleus. Note relationship between cranial nerve III (oculomotor) and posterior cerebral, superior cerebellar, and posterior communicating arteries. *CA,* cerebral aqueduct; *PAG,* periaqueductal gray; *ON,* oculomotor nucleus; *MLF,* medial longitudinal fasciculus; *RN,* red nucleus; *SN,* substantia nigra; *CC,* crus cerebri (includes corticospinal tract); *IPF,* interpeduncular fossa.

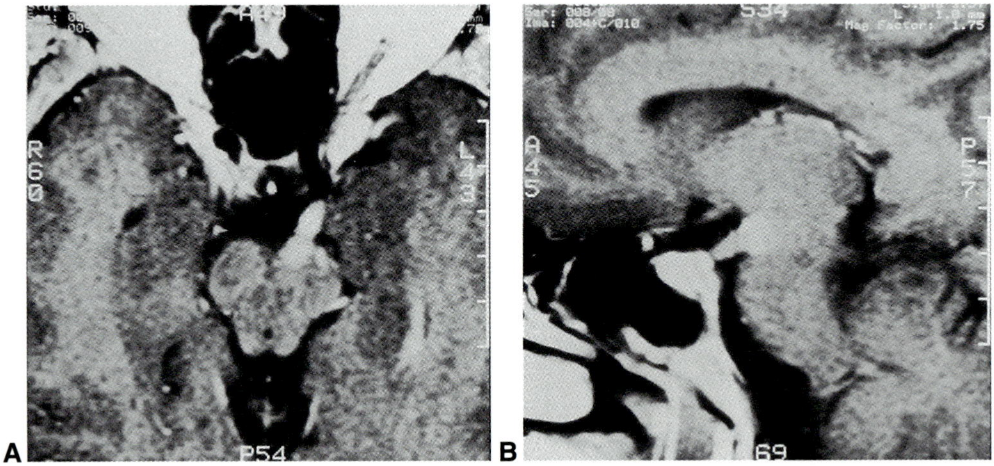

FIG IV-18—MRI of an (**A**) axial and (**B**) sagittal section of a patient with a glioma of the left oculomotor nerve, causing a thickening of the nerve as it exits the midbrain. The course of the enlarged nerve can thus be easily discerned. (Courtesy of Randy H. Kardon, MD, PhD.)

The *superior division* of CN III innervates the superior rectus and levator palpebrae muscles. The larger *inferior division* splits into three branches to supply the medial and inferior rectus muscles and the inferior oblique.

The *parasympathetic fibers* wind around the periphery of the nerve, enter the inferior division, and course through the branch that supplies the inferior oblique muscle. They join the *ciliary ganglion,* where they synapse with the *postganglionic fibers,* which emerge as many *short ciliary nerves.* These pierce the sclera and travel through the choroid to innervate the pupillary sphincter and the ciliary muscle. The superficial location of these fibers makes them more vulnerable to compression than to ischemia. Pupillary dilation is a sensitive (and commonly early) sign of compression.

Pathways for the Pupil Reflexes

Light Reflex

The *light reflex* consists of a simultaneous and equal constriction of the pupils in response to illumination of one or the other eye. The *afferent* pupillary pathway coincides with that of the visual pathway and includes a decussation of nasal fibers in the chiasm. At the posterior part of the optic tract, the pupillary fibers leave the visual fibers and pass to the lateral side of the midbrain to reach the *pretectal nuclei* at the level of the superior colliculus. Here, *efferent* fibers arise and pass to the Edinger-Westphal nuclei, decussating partially (both ventral to the aqueduct and dorsally, in the *posterior commissure*). Preganglionic parasympathetic fibers leave each Edinger-Westphal nucleus and run in the oculomotor nerve as it leaves the brain stem. The fibers spiral downward to lie medially in the nerve at the level of the petroclinoid ligament and inferiorly in the inferior division of the third nerve as it enters the orbit. These fibers synapse in the ciliary ganglion and give rise to postganglionic myelinated short ciliary nerves, about 3%–5% of which are pupillomotor. The rest are designated for the ciliary muscle and are concerned with the near reflex.

Near Reflex

The *near reflex* is a *synkinesis* that occurs when attention is changed from distance to near. This reflex includes *accommodation, pupil constriction,* and *convergence.* The reflex is initiated in the *occipital association cortex,* from which impulses descend along corticofugal pathways to relay in pretectal and possibly tegmental areas. From these relays, fibers pass to the Edinger-Westphal nuclei, the motor nuclei of the medial rectus muscles, and the nuclei of the sixth cranial nerve. Fibers for the near reflex approach the pretectal nucleus from the ventral aspect, so that with compressive dorsal lesions of the optic tectum there is sparing of the near pupil reflex relative to the light reflex *(light-near dissociation).* Efferent fibers for accommodation follow the same general pathway as those for the light reflex, but their final distribution (via the short ciliary nerves) is to the ciliary muscle.

Cranial Nerve IV (Trochlear)

Cranial nerve IV contains the fewest nerve fibers of any cranial nerve (approximately 3400), but it has the longest intracranial course (75 mm). The nerve nucleus is located in the caudal mesencephalon at the level of the inferior colliculus near the periaqueductal gray matter, ventral to the aqueduct of Sylvius. It is continuous with

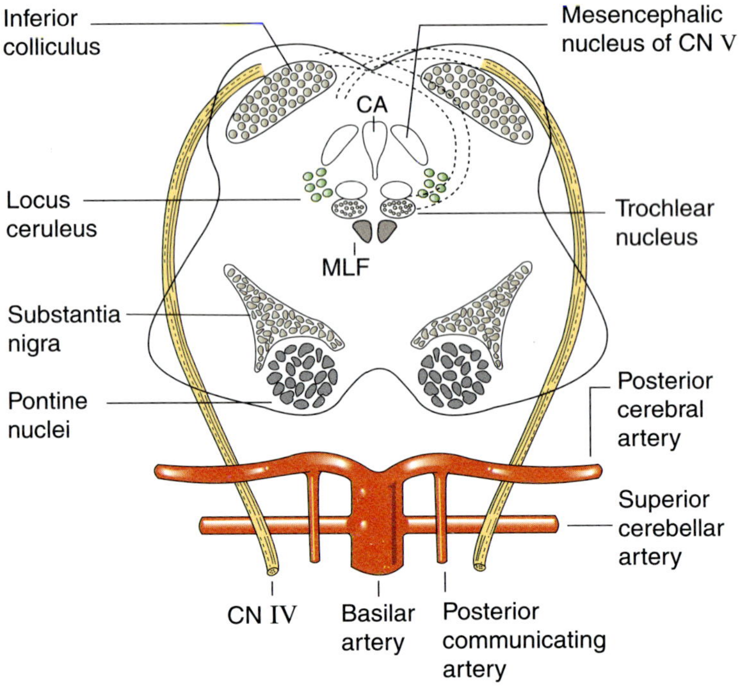

FIG IV-19—Cross section of the midbrain at the level of the inferior colliculi and cranial nerve IV. The dorsal exit of the trochlear nerves is *dotted* to signify that they decussate caudal to this level in the superior medullary velum. *CA*, cerebral aqueduct; *MLF*, medial longitudinal fasciculus. (Illustration by Sylvia Barker.)

the caudal end of the CN III nucleus and differs histologically only in the smaller size of its cells. Like the CN III nucleus, it is bounded ventrolaterally by the medial longitudinal fasciculus (Fig IV-19).

The fascicles of CN IV curve dorsocaudally around the periaqueductal gray matter and decussate completely in the *superior medullary velum*. The nerves exit the brain stem just beneath the inferior colliculus. Thus, CN IV is the only cranial nerve that is completely decussated and the only motor nerve to exit dorsally from the nervous system. As it curves around the brain stem in the ambient cistern, CN IV runs from beneath the free edge of the tentorium, passes between the posterior cerebral and superior cerebellar arteries, and then pierces the dura mater to enter the cavernous sinus.

Brodal A. *Neurological Anatomy in Relation to Clinical Medicine*. 3rd ed. New York: Oxford University Press; 1981.

CN IV travels beneath CN III and above the ophthalmic division of CN V in the lateral wall of the cavernous sinus. On occasion, communications occur between CN IV and the ophthalmic division of CN V, the sympathetics of the paracarotid plexus, or the lacrimal branch of CN V. These communications are of no known clinical significance. CN IV enters the orbit through the superior orbital fissure outside the annulus of Zinn and runs superiorly to innervate the superior oblique muscle. Because of its location outside the muscle cone, CN IV is usually not affected by injection of retrobulbar anesthetics.

Brodal A. *Neurological Anatomy in Relation to Clinical Medicine.* 3rd ed. New York: Oxford University Press; 1981: 533.

Cranial Nerve V (Trigeminal)

Cranial nerve V, the largest cranial nerve, possesses both sensory and motor divisions. The *sensory portion* subserves the greater part of the scalp, forehead, face, eyelids, eye, lacrimal gland, extraocular muscles, ear, dura mater, and tongue. The *motor portion* innervates the muscles of mastication through branches of the mandibular division.

The CN V nuclear complex extends from the midbrain to the upper cervical segments, often as caudal as C4. It consists of the following four nuclei, from above downward:

- Mesencephalic nucleus
- Main sensory nucleus
- Spinal nucleus and tract
- Motor nucleus located in the pons

Important interconnections exist between the different subdivisions of the fifth-nerve sensory nuclei and the reticular formation (Fig IV-20).

Mesencephalic nucleus The *mesencephalic nucleus* mediates proprioception and deep sensation from the masticatory, facial, and extraocular muscles. The nucleus extends inferiorly into the posterior pons as far as the main sensory nucleus.

Main sensory nucleus The *main sensory nucleus* lies in the pons, lateral to the motor nucleus. It is continuous with the mesencephalic nucleus (above) and with the spinal nucleus (below). It receives its input from ascending branches of the sensory root, and it serves light touch from the skin and mucous membranes. The *sensory root* of cranial nerve V, upon entering the pons, divides into an ascending tract and a descending tract. The former terminates in the main sensory nucleus, and the latter ends in the spinal nucleus.

Spinal nucleus and tract The *spinal nucleus and tract* extend through the medulla to C4. The nucleus receives pain and temperature afferents from the descending spinal tract, which also carries cutaneous components of cranial nerves VII, IX, and X that serve sensations from the ear and external auditory meatus. The sensory fibers from the *ophthalmic division of CN V* (V_1) terminate in the most ventral portion of the spinal nucleus and tract. Fibers from the *maxillary division* (V_2) end in the midportion of the spinal nucleus (in a ventral-dorsal plane). The fibers from the *mandibular division* (V_3) end in the dorsal parts of the nucleus.

The cutaneous territory of each of the CN V divisions is represented in the spinal nucleus and tract in a rostral-caudal direction. Fibers from the midfacial region (perioral and perinasal) are thought to terminate most rostrally in the nucleus; fibers from the peripheral face and scalp end in the caudal portion. The zone between them, the midface, is projected onto the central portion of the nucleus. This "onionskin" pattern of cutaneous sensation has been derived from clinical studies in patients with

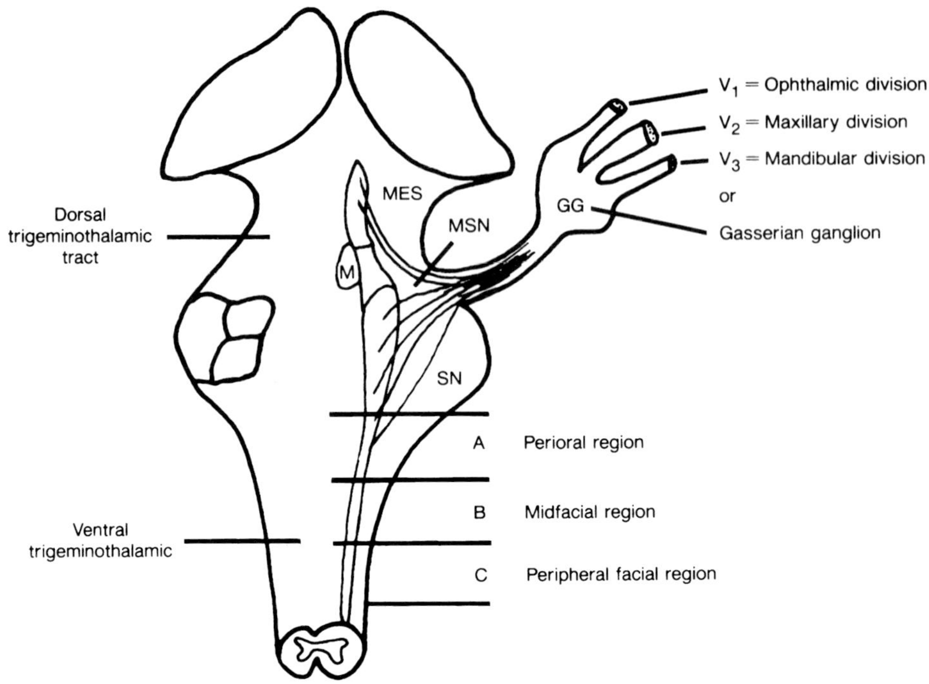

FIG IV-20—Cranial nerve V complex (dorsal view of brain stem). *GG,* gasserian ganglion; *MES,* mesencephalic nucleus; *MSN,* main sensory nucleus; *M,* motor nucleus; *SN,* spinal nucleus and tract. *A, B,* and *C* are portions of the caudal spinal nucleus that correspond to concentric areas of the face: *A,* perioral; *B,* midface, including eyes; *C,* peripheral face and scalp.

damage to the spinal nucleus and tract (Fig IV-21). In other words, damage to the trigeminal sensory nucleus at the level of the brain stem causes bilateral sensory loss in concentric areas of the face, with the sensory area surrounding the mouth in the center. When a patient verifies this distribution of sensory loss, the lesion is in the brain stem. Conversely, when the sensory loss follows the peripheral distribution of the trigeminal sensory divisions (ophthalmic, maxillary, and mandibular), the lesion lies in the fifth nerve after it exits the brain stem.

Axons from the main sensory, spinal, and portions of the mesencephalic nuclei relay sensory information to higher sensory areas of the brain. The axons cross the midline in the pons and ascend to the thalamus along the ventral and dorsal trigeminothalamic tracts. They terminate in the nerve cells of the ventral posteromedial nucleus of the thalamus. These cells in turn send axons through the internal capsule to the postcentral gyrus of the cerebral cortex.

Motor nucleus The *motor nucleus* is located medial to the main sensory nucleus in the pons. It receives fibers from both cerebral hemispheres, the reticular formation, the red nucleus, the tectum, the medial longitudinal fasciculus, and the mesencephalic nucleus. A monosynaptic reflex arc is formed by cells from the mesencephalic nucleus and the motor nucleus. The motor nucleus sends off axons that

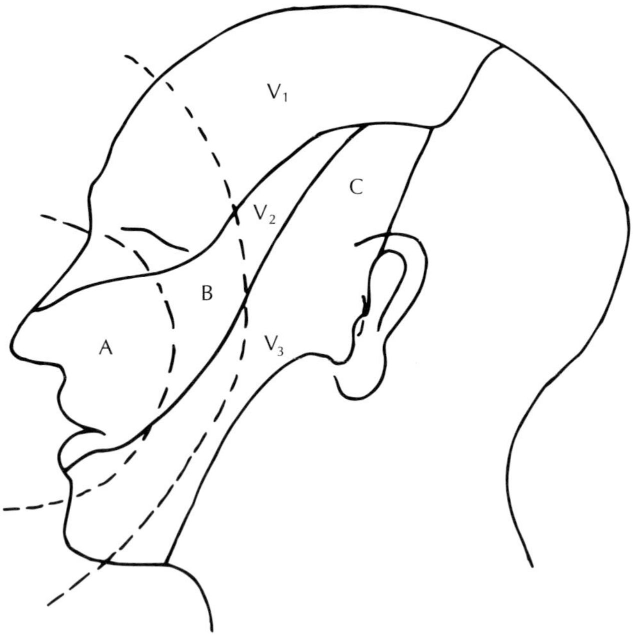

FIG IV-21—Cranial nerve V (trigeminal): pattern of facial sensation. "Onionskin" distribution from lesions of the trigeminal sensory nucleus in the brain stem *(A, B,* and *C)* is delineated by *dotted lines.* Lesions of the trigeminal sensory nucleus in the brain stem result in an onionskin distribution of altered sensation (dotted lines *A, B,* and *C*); lesions of the ophthalmic (V_1), maxillary (V_2), and mandibular (V_3) nerves result in the pattern of sensory loss delineated by the solid lines. For in-depth discussion, see Brodal A. *Neurological Anatomy in Relation to Clinical Medicine.* 3rd ed. New York: Oxford; 1981:524–529. (Illustration by Thomas A. Weingeist, PhD, MD.)

form the motor root, which eventually supplies the muscles of mastication (pterygoid, masseter, temporalis), the tensor tympani, the tensor veli palatini, the mylohyoid, and the anterior belly of the digastric.

The intracranial fifth nerve emerges from the upper lateral portion of the ventral pons, passes over the petrous apex, forms the gasserian ganglion, and then divides into three branches. The *gasserian ganglion,* also called the *semilunar* (or *trigeminal) ganglion,* contains the cells of origin of all the fifth-nerve sensory axons. The crescent-shaped ganglion occupies a recess in the dura mater posterolateral to the cavernous sinus. This recess, called *Meckel's cave,* is near the apex of the petrous part of the temporal bone in the middle cranial fossa. Medially, the gasserian ganglion is close to the internal carotid artery and the posterior cavernous sinus.

Divisions of Cranial Nerve V

The three divisions of cranial nerve V are the ophthalmic (V_1), the maxillary (V_2), and the mandibular (V_3).

CN V_1 The *ophthalmic division* enters the cavernous sinus lateral to the internal carotid artery and courses beneath cranial nerves III and IV. It gives off a tentorial-dural branch within the sinus, which supplies sensation to the cerebral vessels, the dura mater of the anterior fossa, the cavernous sinus, the sphenoid wing, the petrous apex, Meckel's cave, the tentorium cerebelli, the falx cerebri, and the dural venous sinuses. CN V_1 passes into the orbit through the superior orbital fissure and divides into three branches: frontal, lacrimal, and nasociliary.

The *frontal nerve* divides into the supraorbital and the supratrochlear nerves, which provide sensation for the medial portion of the upper eyelid and the conjunctiva, forehead, scalp, frontal sinuses, and side of the nose.

The *lacrimal nerve* innervates the lacrimal gland and the neighboring conjunctiva and skin. It was once suggested that postganglionic parasympathetic lacrimal secretory fibers, arising in the pterygopalatine ganglion, were carried to the lacrimal gland via a zygomaticotemporal connection with the lacrimal nerve. However, it is now thought more likely that the gland receives its parasympathetic supply directly, from the retro-orbital plexus (see below).

The *nasociliary nerve* supplies sensation through nasal branches to the middle and inferior turbinates, septum, lateral nasal wall, and tip of the nose. The infratrochlear branch serves the lacrimal drainage system, the conjunctiva, and the skin of the medial canthal region. Long ciliary nerves carry sensory fibers from the ciliary body, the iris, and the cornea and provide the sympathetic innervation to the dilator muscle of the iris. Sensation from the globe is carried by short ciliary nerves. The fifth-nerve fibers pass through the ciliary ganglion to join the nasociliary nerve. The ciliary nerves also contain postganglionic parasympathetic fibers from the ganglion to the pupillary sphincter and the ciliary muscle.

CN V_2 The *maxillary division* leaves the gasserian ganglion to exit the skull through the foramen rotundum, which lies below the superior orbital fissure. CN V_2 courses through the pterygopalatine fossa into the inferior orbital fissure, then runs through the infraorbital canal as the *infraorbital nerve.* After exiting the infraorbital foramen, CN V_2 divides into an inferior palpebral branch supplying the lower eyelid, a nasal branch for the side of the nose, and a superior labial branch for the upper lip. The teeth, maxillary sinus, roof of the mouth, and soft palate are also innervated by branches of the maxillary division.

CN V_3 The *mandibular division* contains both sensory and motor fibers. It exits the skull through the foramen ovale and provides motor input for the masticatory muscles. Sensation is supplied to the mucosa and skin of the mandible, lower lip, tongue, external ear, and tympanum.

Williams PL, Warwick R. *Gray's Anatomy.* 38th ed. Edinburgh: Churchill Livingstone; 1995.

Cranial Nerve VI (Abducens)

The nucleus of *cranial nerve VI* is situated in the floor of the fourth ventricle, beneath the *facial colliculus,* in the *caudal pons.* Fibers of CN VII pass over or loop around

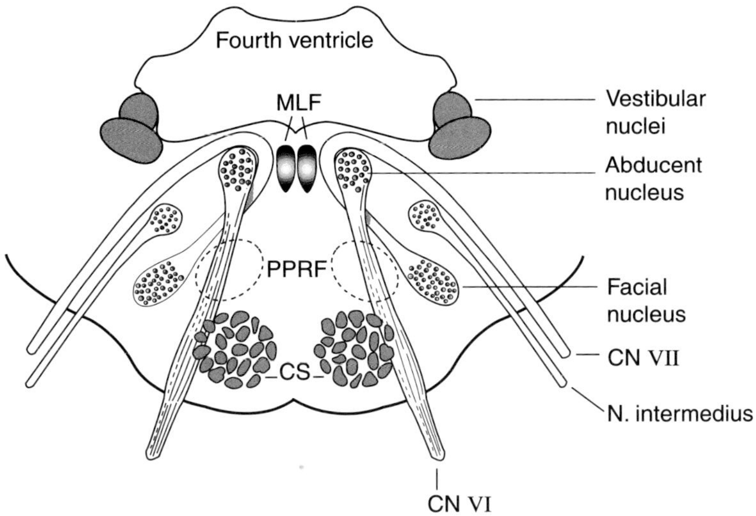

FIG IV-22—Cross section of the pons at the level of the cranial nerve VI (abducens) nucleus. *MLF,* medial longitudinal fasciculus; *PPRF,* pontine paramedian reticular formation; *CS,* corticospinal tract. (Illustration by Sylvia Barker.)

the sixth-nerve nucleus and exit in the *cerebellopontine angle.* The medial longitudinal fasciculus lies medial to the sixth-nerve nucleus. The fascicular portion of the nerve runs ventrally through the *paramedian pontine reticular formation* and the *pyramidal tract* and leaves the brain stem in the *pontomedullary junction* (Fig IV-22).

CN VI takes a vertical course along the ventral face of the pons and is crossed by the *anterior inferior cerebellar artery.* It ascends farther through the subarachnoid space along the surface of the *clivus,* surrounded by *Batson's venous plexus,* to perforate the dura mater below the crest of the petrous portion of the temporal bone about 2 cm below the *posterior clinoid process.* It then passes intradurally through or around the inferior petrosal sinus and beneath the *petroclinoid (Gruber's) ligament* through *Dorello's canal,* where it enters the cavernous sinus. In the cavernous sinus, CN VI runs below and lateral to the carotid artery and may transiently carry sympathetic fibers from the *carotid plexus.* It passes through the superior orbital fissure within the annulus of Zinn and innervates the lateral rectus muscle on its ocular surface.

Williams PL, Warwick R. *Gray's Anatomy.* 38th ed. Edinburgh: Churchill Livingstone; 1995.

Cranial Nerve VII (Facial)

Cranial nerve VII is a complex mixed sensory and motor nerve. The motor root contains special visceral efferent fibers that innervate the muscles of facial expression. The so-called *sensory root* of cranial nerve VII is the *nervus intermedius,* which contains special visceral afferent, general somatic afferent, and general visceral efferent fibers. The special *visceral afferent fibers,* which convey the sense of taste from the anterior two thirds of the tongue, terminate centrally in the nucleus of the tractus solitarius. The general *somatic afferent fibers* convey sensation from the external auditory meatus and the retroauricular skin; centrally, they enter the spinal nucleus of CN V. The general *visceral efferent fibers* provide preganglionic parasympathetic innervation by way of the sphenopalatine and submandibular ganglia to the lacrimal, submaxillary, and sublingual glands.

The *motor nucleus of CN VII* is a cigar-shaped column, 4 mm long, located in the caudal third of the pons. It is ventrolateral to the sixth-nerve nucleus, ventromedial to the spinal nucleus of CN V, and dorsal to the superior olive. Four distinct subgroups within the nucleus innervate specific facial muscles; the ventral portion of the intermediate group probably supplies axons to the orbicularis oculi. The part of the nucleus supplying the upper half of the face receives corticobulbar input from both cerebral hemispheres. The lower half of the face is influenced by corticobulbar fibers from the opposite cerebral hemisphere.

Fibers from the motor nucleus course dorsomedially to approach the floor of the fourth ventricle and then ascend immediately dorsal to the sixth-nerve nucleus. At the rostral end of the sixth-nerve nucleus, the main facial motor fibers arch over its dorsal surface (forming the internal genu of CN VII) and then pass ventrolaterally between the spinal nucleus of the fifth nerve and the seventh-nerve nucleus to exit the brain stem at the pontomedullary junction. The bulge formed by the seventh-nerve genu in the floor of the fourth ventricle is the facial colliculus (Fig IV-23).

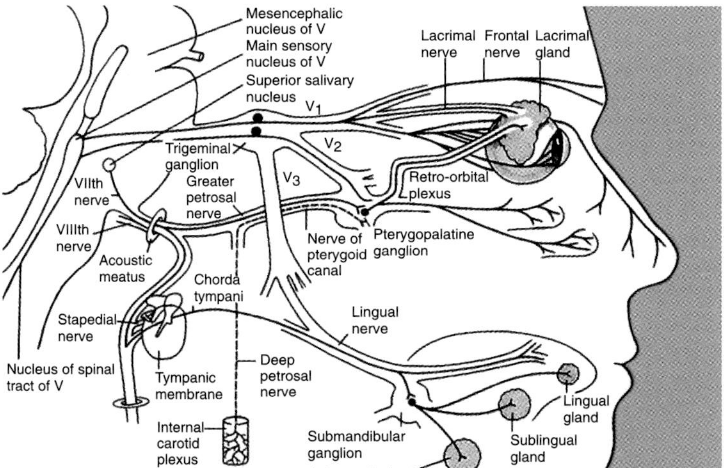

FIG IV-23—Lacrimal reflex arc (after Kurihashi). The afferent pathway is provided by the first and second divisions of the trigeminal nerve. The efferent path proceeds from the lacrimal nucleus (close to the superior salivary nucleus) via the facial nerve (nervus intermedius), through the geniculate ganglion, the greater superficial petrosal nerve, and the nerve of the pterygoid canal (where it is joined by sympathetic fibers from the deep petrosal nerve). The nerve passes to the pterygopalatine ganglion, where it synapses with postganglionic fibers. These reach the lacrimal gland directly and via the retro-orbital plexus of nerves. They carry cholinergic and vasoactive intestinal polypeptide (VIP)-ergic fibers to the gland. (From Bron AJ, Tripathi RC, Tripathi BJ. *Wolff's Anatomy of the Eye and Orbit.* 8th ed. London, New York: Chapman & Hall; 1997.)

Carpenter MB. *Core Text of Neuroanatomy.* 4th ed. Baltimore: Williams & Wilkins; 1991.

The *sensory nucleus* of CN VII is the rostral portion of the tractus solitarius, sometimes known as the *gustatory nucleus.* It lies lateral to the motor and parasympathetic nuclei in the caudal pons. Sensations of taste from the anterior two thirds of the tongue are carried by special visceral afferent fibers to this nucleus. The impulses travel along the lingual nerve and chorda tympani; the cell bodies for these impulses are located in the geniculate ganglion. They eventually reach the brain through the nervus intermedius.

Cranial nerve VII, the nervus intermedius, and cranial nerve VIII (acoustic) pass together through the *lateral pontine cistern* in the cerebellopontine angle and enter the *internal auditory meatus* in a common meningeal sheath. The seventh nerve and the intermedius nerve then enter the fallopian canal, the longest bony canal traversed by any cranial nerve (30 mm).

CN VII can be divided into three segments in its course through this canal. After passing anterolaterally for a short distance known as the *labyrinthine segment,* the nerves bend sharply at the *geniculate ganglion* and are then directed dorsolaterally past the *tympanic cavity.* This 90° bend, known as the *tympanic segment,* is the external genu of CN VII. Two parasympathetic branches from the superior salivatory and lacrimal nuclei leave the nerve at the tympanic segment: the greater superficial petrosal nerve and a small filament that joins the inferior petrosal nerve.

The *third (mastoid) segment* of the nerve is directed straight down toward the base of the skull. The stapedius nerve leaves and the chorda tympani joins CN VII in the mastoid segment. The seventh-nerve trunk then exits the skull at the stylomastoid foramen and separates into a large temporofacial and a small cervicofacial division between the superficial and deep lobes of the parotid gland. This area of branching is known as the *pes anserinus.*

The *temporofacial division* gives rise to the temporal, zygomatic, and buccal branches. The *cervicofacial division* is the origin of the marginal mandibular and colli branches. However, anastomoses and branching patterns are numerous. Commonly, the temporal branch supplies the upper half of the orbicularis oculi and the zygomatic branch supplies the lower half. The frontalis, corrugator supercilii, and pyramidalis muscles are usually innervated by the temporal branch.

The *parasympathetic outflow* originates in the superior salivatory nucleus and the lacrimal nucleus, which lie posterolateral to the motor nucleus. Both probably receive afferent fibers from the hypothalamus. The superior salivatory nucleus also receives input from the olfactory system. The hypothalamic fibers reaching the lacrimal nucleus may mediate emotional tearing, and there is supranuclear input from the cortex and the limbic system. Reflex lacrimation is controlled by afferents from the sensory nuclei of CN V. These preganglionic parasympathetic fibers pass peripherally as part of the nervus intermedius and divide into two groups near the external genu of CN VII. The lacrimal group of fibers pass to the pterygopalatine ganglion in the greater superficial petrosal nerve. The salivatory group of fibers projects through the *chorda tympani* nerve to the *submandibular ganglion* to innervate the submandibular and sublingual salivary glands.

The *greater superficial petrosal nerve* passes forward on the anterior surface of the petrous temporal bone to join the deep petrosal nerve (sympathetic) and form the nerve of the pterygoid canal. This nerve enters the pterygopalatine fossa; joins the *pterygopalatine ganglion*; and gives rise to unmyelinated postganglionic fibers that

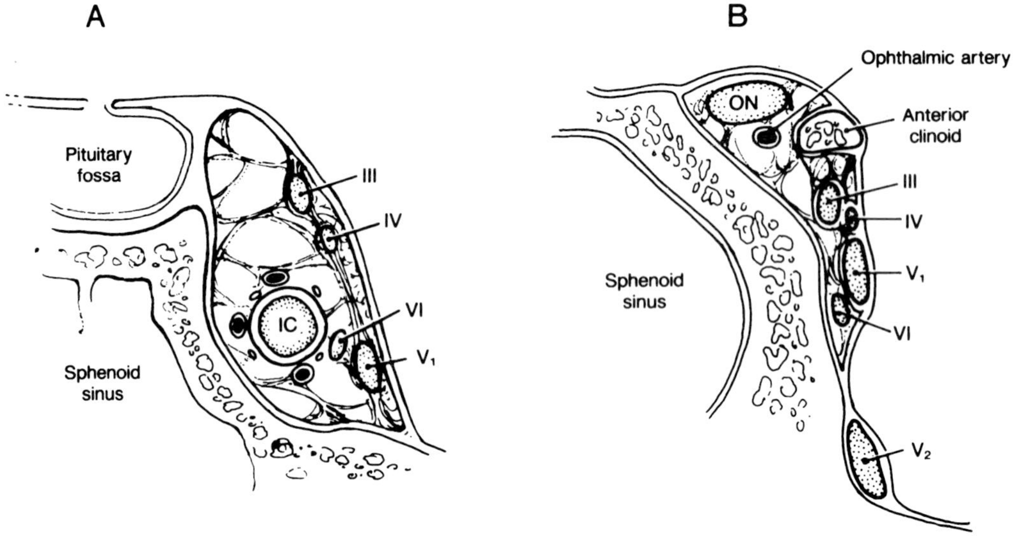

FIG IV-24—Cavernous sinus, coronal sections. **A**, At the level of the pituitary fossa. **B**, At the level of the anterior clinoid process. *ON*, optic nerve; *IC*, internal carotid artery. (Reproduced by permission from Doxanas MT, Anderson RL. *Clinical Orbital Anatomy.* Baltimore: Williams & Wilkins; 1984.)

innervate the globe, lacrimal gland, glands of the palate, and nose. Those parasympathetic fibers destined for the orbit enter it via the inferior orbital fissure. Here, they are joined by sympathetic fibers from the carotid plexus and form a *retro-orbital plexus* of nerves, whose *rami oculares* supply orbital vessels or enter the globe to supply the choroid and anterior segment structures. Some of these fibers enter the globe directly; others enter via connections with the short ciliary nerves. The rami oculares supply the lacrimal gland directly and not via zygomaticotemporal connections with the lacrimal nerve, as is usually stated. These parasympathetic fibers, unlike those of CN III, carry nitrergic and VIP-ergic axons in addition to cholinergic fibers. They are vasodilatory to the choroid and supply structures in the drainage angle. Whether they are all postganglionic is unclear, since scattered accessory ganglia are found in the choroid associated with nitrergic and VIP-ergic fibers.

Cavernous Sinus

The *cavernous sinus* is an interconnected series of venous channels located just posterior to the orbital apex and lateral to the sphenoidal air sinus and pituitary fossa (Fig IV-24). The following structures are located within the venous cavity:

- The internal carotid artery, surrounded by the sympathetic carotid plexus
- Cranial nerves III, IV, and VI
- The ophthalmic and maxillary divisions of cranial nerve V

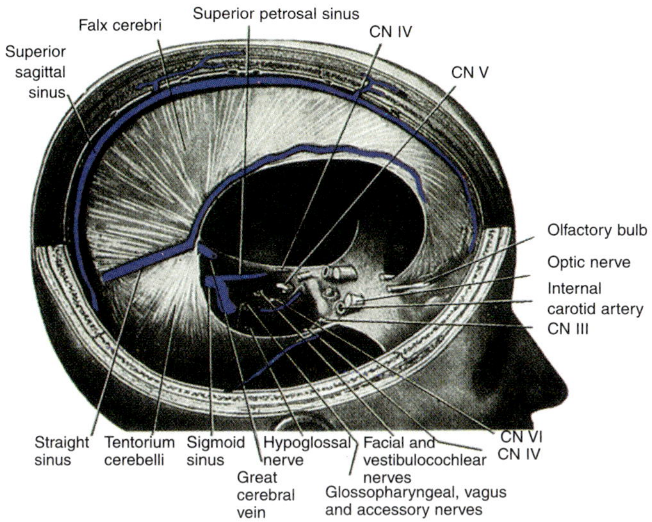

FIG IV-25—Three-dimensional drawings of the venous sinuses of the brain, their interconnections, and the relationship to the dura. (Reproduced with permission from Williams PL, Warwick R. *Gray's Anatomy.* 38th ed. Edinburgh: Churchill Livingstone; 1995.)

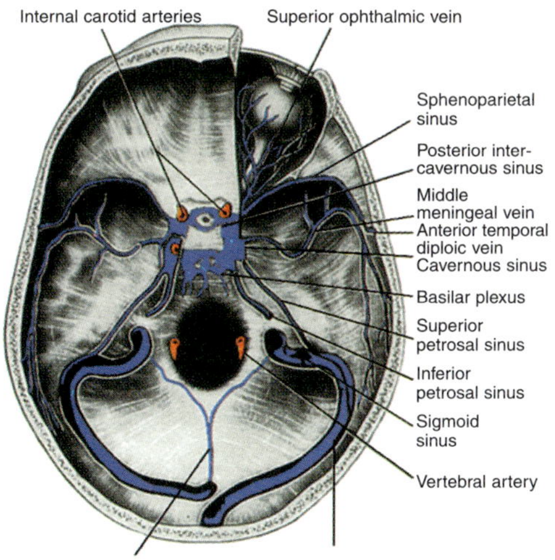

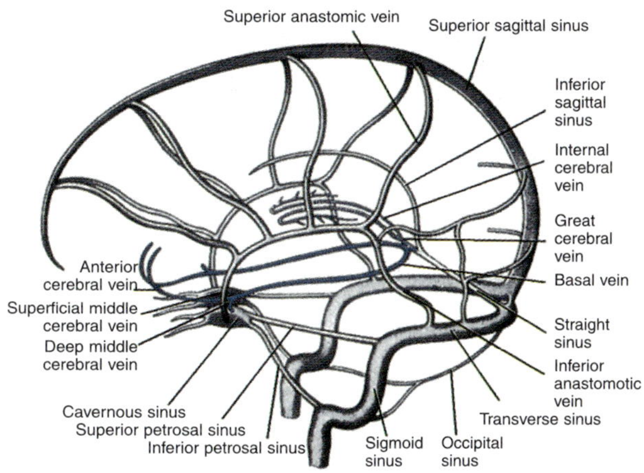

Other Venous Sinuses

The cavernous sinus is only one part of an interconnecting series of venous channels that carry blood away from the brain and drain into the *internal jugular veins.* Other venous sinuses include the superior sagittal, transverse, straight, sigmoid, and petrosal. The various components of the venous system are depicted in Figure IV-25. Thrombosis in any portion of the venous sinuses can lead to increased venous pressure and may cause intracranial hypertension and papilledema.

Circle of Willis

The major arteries supplying the brain are the right and left internal carotid arteries (which distribute blood primarily to the rostral portion of the brain) and the right and left vertebral arteries (which join to form the basilar artery). The basilar artery primarily distributes blood to the brain stem and posterior portion of the brain. These arteries interconnect at the base of the brain at the circle of Willis (Fig IV-26). These interconnections help to distribute blood to all regions of the brain even when a portion of the system becomes occluded.

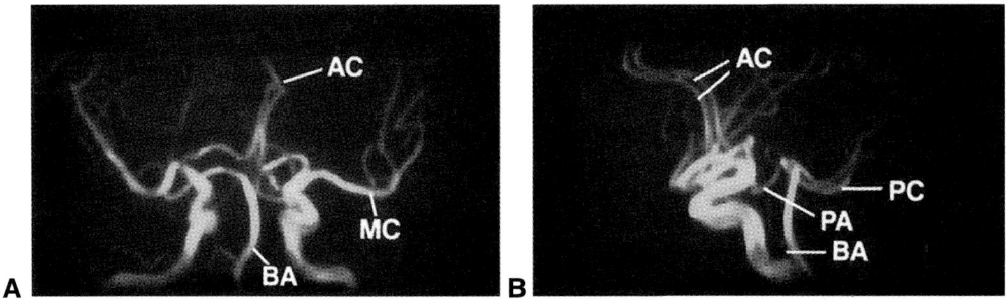

FIG IV-26—**A**, Magnetic resonance angiogram showing circle of Willis in an anteroposterior view. **B**, Same patient shown with an oblique view. *BA*, basilar artery; *PA*, posterior communicating artery; *MC*, middle cerebral artery; *AC*, anterior cerebral artery; *PC*, posterior cerebral artery. (Photographs courtesy of T. Talli, MD, and W. Yuh, MD.)

PART 2

EMBRYOLOGY

Glossary

Agenesis Absence of an organ resulting from failure of its primordium to appear in embryonic development. Aplasia.

Animal cap Cells of the blastula that originated as a result of regional differences in the fertilized ovum and are destined normally to form epidermis.

Aniridia Absence of the iris.

Anisocoria Inequality in diameter of the pupils.

Anlage Primordium. Primitive tissue from which an organ or part develops (plural, *anlagen*).

Anophthalmos A developmental defect characterized by absence of the eye.

Aplasia Lack of development of an organ; frequently used to describe complete suppression or the failure of development of a structure from the embryonic primordium.

Apoptosis Intrinsically programmed cell death that is characterized by distinctive morphologic changes, especially in the nucleus, and is responsible for physiologic deletion of cells.

Choristoma A mass of tissue histologically normal for an organ or part of an organ other than the site where it appears.

Coloboma A fissure of a part or parts of the eye; failure of fusion and subsequent development. May be congenital or acquired.

Cryptophthalmos A developmental anomaly in which the skin is continuous over the eyeballs without any indication of eyelid formation.

Cyclopia A developmental anomaly characterized by a single orbital fossa with the globe absent, rudimentary, or single and the nose absent or present as a tubular appendage located above the orbit.

Dermoid Congenital inclusion of epidermal and associated connective tissue at the line of closure of the fetal cleft.

Dysgenesis Defective development.

Dysplasia Abnormal development of tissues or cells with retention of some features resembling normal structures.

Ectoderm The outermost of the three primary germ layers of the embryo. The epidermis and the epidermal tissues develop from surface ectoderm as the nasal, hair, and skin glands; the nervous system and the external sense organs develop from neuroectoderm. Neural crest cells arise from primitive ectoderm.

Embryogenesis The earliest stages of development of a new individual from a fertilized ovum.

Evagination An outpouching of a layer or part. Evagination of the optic anlage forms the optic vesicle.

Fissure Cleft or groove formed as the optic cup invaginates.

Growth factor A protein that binds to its receptor and induces up-regulation or down-regulation of specific cell activities.

Holoprosencephaly Lack of (or defect in) development of the ventral brain.

Homeobox A sequence of base pairs in the DNA that is very similar in genes of many different species.

Hypertelorism An abnormal increased distance between two organs.

Hypoplasia Defective, limited, or incomplete development.

Induction The stimulatory process whereby one tissue directs a second tissue to develop in a specific direction.

Invagination The infolding of one part within another; the optic vesicle infolds to form the double-layered optic cup.

Leukoma A dense white opacity of the cornea.

Mesenchyme A dispersed population of undifferentiated embryonic cells, stellate-shaped and arranged loosely, that have their embryonic origin from either mesoderm or neural crest cells.

Mesoderm The middle layer of the three primary germ layers of the embryo, lying between the ectoderm and the endoderm. In the orbit, it gives rise to the vascular endothelium and the myoblasts of the extraocular muscles.

Messenger RNA (mRNA) Ribonucleic acid that provides the information as triplicate sequences of nucleotides (codons) for the synthesis of protein.

Microphthalmos A small, disorganized eye.

Nanophthalmos A rare developmental anomaly in which the eyeballs are abnormally small but are without other deformities.

Neural crest cells A cellular mass of primitive ectoderm that arises at the peak of the neural folds as they close.

Neural folds The paired folds, one lying on either side of the neural plate, that form the neural tube.

Neural tube The epithelial tube that develops from the neural plate and forms the central nervous system of the embryo.

Neurocristopathy Clinical disorder that results from abnormal induction, proliferation, migration, differentiation, or regression of neural crest cells.

Neuroectoderm The region of the ectoderm destined to become the neural tube.

Optic cup The cup-shaped structure of the primitive eye formed by invagination of the optic vesicle.

Optic pits Indentation of optic primordia at the stage of neural folds.

Optic sulcus A groove or furrow in the neural plate and folds that is the first identifiable structure destined to become the eye.

Optic vesicle Saccular structure of the embryonic eye formed by evagination of the optic pit.

Organogenesis The development or growth of organs; period of development after embryogenesis; begins about the fourth week in ocular development.

Placode A platelike structure, especially a thickened plate of ectoderm in the early embryo from which an organ, lens, or glands develop.

Pluripotent The capacity to differentiate into any one of several cell types.

Posterior embryotoxon A congenital opacity of the margin of the cornea; also called *arcus juvenilis*.

Primordium The earliest discernible indication during embryonic development of an organ or part. Also called *anlage* or *rudiment*.

Progenitor cell An ancestral or parent cell.

Somite One of a pair of mesodermal cell masses situated on either side of the embryonic neural tube.

Synophthalmia Form of cyclopia in which parts of the two eyes are joined at the midline.

Teratogen An agent or factor that produces physical defects in the developing embryo.

Teratoma A true neoplasm made up of various types of tissue, none of which is native to the area in which it occurs.

Tunica vasculosa lentis The vascular envelope that encloses and nourishes the developing lens of the fetus. Normally, this structure regresses and disappears shortly before birth.

CHAPTER V

Ocular Development

Introduction

Experimental studies conducted in recent decades have revolutionized our understanding of ocular development. Consequently, the original treatises on the growth and differentiation of the eye have been modified. The classic *germ layer theory* depicted the epithelium of the cornea, the retina, and the neural components of the uveal tract as derived from ectoderm, and the remainder of the ocular structures as derived from mesoderm.

Although this general schema is still used, it is currently recognized that the embryonic and fetal development of the human eye involves a series of sequential events, including inductive interactions and morphogenetic movement of cells from distant regions of the embryo. The primary tissues involved in these processes are the *head epidermis, neuroectoderm,* and *mesenchyme.* Three elements have been identified as making important (if not pivotal) contributions to the genesis of the eye:

- Growth factors
- Homeobox genes
- Neural crest cells

Each is described separately below, but interaction among these elements is also crucial. Part 3 of this section, Genetics, also discusses some of the concepts reviewed in Chapters V and VI.

Growth Factors

The process of *induction* is mediated by tissue communication by way of macromolecules that act as chemical signals. *Growth factors* are now known to be active in the earliest stages of embryonic development. They are a class of trophic substances that participate in the control of normal development by modulating the migration, proliferation, and differentiation of cells. These molecules act at nanomolar concentrations by binding with high affinity to specific receptor sites localized in the plasma membrane of the target cell.

The embryonic genome is not transcribed until the stage of midblastula transition, which takes place several hours after fertilization. The messenger RNAs for the growth factors involved in the earliest aspects of the growth and differentiation of the fertilized egg are endogenous and are supplied from maternal sources until the embryonic tissues become able to synthesize them de novo. These growth factors include:

- Fibroblast growth factor (FGF)
- Transforming growth factor β (TGF-β)
- Insulin-like growth factor I (IGF-I)

Experimental studies have revealed that when cells of the animal cap are exposed to FGF, they are induced to differentiate into posterior mesoderm that is destined to form tissues of the caudal region. However, TGF-β induces animal cap cells to differentiate into mesoderm that forms structures in the head region, including the eye.

Growth factors also regulate the levels of expression of *homeobox genes*, which function as a mechanism for controlling the establishment of the overall arrangement of the eye as an organ. Visual acuity requires a precise spatial arrangement of the tissues of the eye; thus, it is critical that the homeobox genes be expressed at the appropriate level and time.

Some growth factors, especially TGF-β, are crucial in directing the migration and developmental patterns of cranial neural crest cells by influencing the synthesis and degradation of the extracellular matrix. Various components of the extracellular matrix act as morphogenetic factors that facilitate a complex series of integrated tissue interactions, movements, and shape changes, especially during the earliest stages of morphogenesis of the optic vesicle and lens.

Differentiation of the various ocular tissues appears to be at least partly controlled by a variety of growth factors. For example, the FGFs induce the neuroectodermal cells that line the inner wall of the optic cup to develop as neural retina; FGF is also responsible for certain aspects of differentiation of the lens epithelial cells into lens fibers. However, the differentiation of lens epithelial cells immediately anterior to the equator—as well as their mitotic activity—is promoted by the IGFs. The synergistic action of multiple trophic factors appears to be a significant regulatory tool for initiating cellular activities and for limiting abnormal development.

Tripathi BJ, Tripathi RC, Livingston AM, et al. The role of growth factors in the embryogenesis and differentiation of the eye. *Am J Anat.* 1991;192:442–471.

*Homeobox Genes**

Homeobox genes contain a distinctive segment of DNA, 180 base pairs in length, that shows similarity in the sequence of the nucleotides. This region is termed the *homeobox* (from the Greek *homoios* ["like, resembling"] and *box* [the extent of the conserved sequence]). The homeobox encodes an almost identical sequence of approximately 60 amino acids, the *homeodomain*, in the protein products of these genes. Because they control the activity of many subordinate genes, homeobox genes are considered "master" genes. Conserved evolutionarily, these genes are present throughout the plant and animal kingdoms. Identification of individual homeobox genes is based on their original characterization in the fruit fly—for example, the paired box, or *PAX*, genes. Homologues of the fruit fly genes are also recognized; for example, the antennapedia genes are designated *HOX* genes in mammals.

The function of homeobox genes is mediated by the homeodomain, which recognizes and binds to specific DNA sequences in the subordinate genes, thereby activating or repressing their expression. Thus, these genes act as transcription factors. On the basis of the pattern of homeobox gene expression, which is restricted both spatially and temporally during the earliest stages of development, the vertebrate

* Because many genes contain a conserved sequence of nucleotides, or *homeobox region*, those responsible for the segmental development of the embryo are more correctly called *homeotic genes*. However, the terms *homeobox* and *homeotic* are used interchangeably in the literature.

embryo can be subdivided anteriorly to posteriorly into fields of cells that have different developmental capacities. This organizational plan precedes the formation of any specific organ or structures. The fact that homeobox genes are located on the chromosomes in the same order as they are expressed along the anteroposterior axis of the embryo indicates that they are activated sequentially.

The same homeobox genes are expressed again later in embryogenesis, apparently to specify the identity of a particular cell. Experimental evidence suggests that homeobox genes are activated not only by growth factors—especially FGFs and TGF-βs—but also by retinoic acid.

Investigations in several vertebrate species have revealed the involvement of specific homeobox genes in the development of the eye. For example, expression of the *PAX6* gene marks the location of the lens-competent region in the head ectoderm before the optic vesicle can be recognized. During the early stages of eye development, two *HOX* genes are expressed with a distinct spatial and temporal relationship: The *HOX8.1* gene is expressed in the surface ectoderm in a region destined to form the corneal epithelium and in the optic vesicle, where the retina will differentiate before invagination occurs. *HOX7.1*, which is expressed after the formation of the optic cup, marks the region of the future ciliary body. Subsequently, the *PAX6* gene has a role in the expression of tissue-specific genes in the eye; for example, it induces both differentiation of progenitor cells into neurons in the retina and the expression of zeta crystallins in lens epithelial cells.

Grindley JC, Davidson DR, Hill RE. The role of Pax-6 in eye and nasal development. *Development.* 1995;121:1433–1442.

Li HS, Yang JM, Jacobson RD, et al. Pax-6 is first expressed in a region of ectoderm anterior to the early neural plate: implications for stepwise determination of the lens. *Dev Biol.* 1994;162:181–194.

Matsuo T. The genes involved in the morphogenesis of the eye. *Jpn J Ophthalmol.* 1993;37:215–251.

Monaghan AP, Davidson DR, Sime C, et al. The Msh-like homeobox genes define domains in the developing vertebrate eye. *Development.* 1991;112:1053–1061.

Neural Crest Cells

Neural crest cells arise from neuroectoderm located at the crest of the neural folds at approximately the same time as the folds fuse to form the *neural tube.* They are a transient population of cells: After they migrate to different regions of the embryo, differentiation occurs. The contribution made by crest cells to the tissues of the developing embryo was recognized in experimental studies. Grafts of neural crest primordia or paraxial mesoderm from Japanese quail were transplanted into host chick embryos, or vice versa. The presence of a heterochromatin condensation in the cell nucleus of the quail, which replicates with each mitotic division, provided a marker not possessed by the chick cells and allowed the origins of specific tissues to be identified throughout development. Studies in mice using tritiated thymidine as a nuclear marker have confirmed the observations made in chick/quail chimeras. Although exact labeling is not possible in human embryos, the fact that basic developmental patterns are similar among vertebrates supports the conclusion that comparable events occur.

Most mesenchymal cells of the facial primordia are derived from the neural crest. Crest cells do not arise from the region of the forebrain. However, neural crest cells from the diencephalic, mesencephalic, and rhombencephalic regions migrate anteriorly along the dorsum of the embryo. Crest cells that originate from the posterior midbrain form the *maxillary primordia,* and those from the hindbrain form the *mandibular primordia.* The crest cells from the diencephalon contribute to the tissue of the frontonasal mass; later, they are joined by cells from the anterior midbrain that migrate to and settle around the optic vesicles. The anterior flexure of the embryo aids the migration of neural crest cells ventrad and cephalad.

The extracellular matrix has a significant role in directing the migration of neural crest cells. Molecules such as fibronectin promote migration, whereas others such as proteoglycans are inhibitory. The positive and negative cues that the crest cells encounter during migration appear to guide the cells along the correct pathways to the appropriate destination. Because the synthesis and secretion of extracellular matrix molecules such as collagen, fibronectin, and proteoglycans can be influenced by growth factors, especially TGF-β, cytokines also have a role in regulating the migration of crest cells.

Early in development, neural crest cells are pluripotent, and their final differentiation is considerably influenced by local factors. Crest cells from the hindbrain normally form the connective tissue of the visceral arch and contribute to the formation of the cranial sensory ganglia. However, if these hindbrain neural crest cells are grafted in place of the posterior diencephalic and mesencephalic crest population, they differentiate appropriately into ocular, orbital, and facial tissues.

Neural crest cells make a major contribution to the connective tissue components of the eye and orbit. Notable exceptions include the striated fibers of the extraocular muscles and the endothelial cells that line all blood vessels of the eye and orbit. Both of these exceptions arise from mesoderm (Table V-1).

Neurocristopathy Congenital and developmental anomalies that involve cells derived from the neural crest have been grouped together under the term *neurocristopathies.* Most of these abnormalities result from defects in either the migration of neural crest cells or their terminal differentiation. Several conditions are common in combination with ocular defects, especially those of the anterior segment:

- Craniofacial and dental malformations
- Middle-ear deafness
- Malformations of the skull, shoulder girdle, and upper spine

Primary cleft palate can be produced by extirpation of the neural folds prior to crest cell migration. The median face malformation (severe orbital hypertelorism) is thought to result from an impaired midline coalescence of the frontonasal process.

Johnston MC, Noden DM, Hazelton RD, et al. Origins of avian ocular and periocular tissues. *Exp Eye Res.* 1979;29:27–43.

Lallier T, Leblanc G, Artinger KB, et al. Cranial and trunk neural crest cells use different mechanisms for attachment to extracellular matrices. *Development.* 1992;16: 531–541.

Noden DM. Periocular mesenchyme: neural crest and mesodermal interactions. In: Jakobiec FA, ed. *Ocular Anatomy, Embryology, and Teratology.* Philadelphia: Harper & Row; 1982.

Tripathi BJ, Tripathi RC, Wisdom JE. Embryology of the anterior segment of the human eye. In: Ritch R, Shields MB, Krupin T, eds. *The Glaucomas.* 2nd ed. St Louis: Mosby; 1996:1.

Table V-1

Derivatives of Embryonic Tissues

ECTODERM
Neuroectoderm
Neurosensory retina
Retinal pigment epithelium
Pigmented ciliary epithelium
Nonpigmented ciliary epithelium
Pigmented iris epithelium
Sphincter and dilator muscles of iris
Optic nerve, axons, and glia
Vitreous
Cranial Neural Crest Cells
Corneal stroma and endothelium
Sclera (see also mesoderm)
Trabecular meshwork
Sheaths and tendons of extraocular muscles
Connective tissues of iris
Ciliary muscles
Choroidal stroma
Melanocytes (uveal and epithelial)
Meningeal sheaths of the optic nerve
Schwann cells of ciliary nerves
Ciliary ganglion
All midline and inferior orbital bones, as well as parts of orbital roof and lateral rim
Cartilage
Connective tissue of orbit
Muscular layer and connective tissue sheaths of all ocular and orbital vessels
Surface Ectoderm
Epithelium, glands, cilia of skin of eyelids and caruncle
Conjunctival epithelium
Lens
Lacrimal gland
Lacrimal drainage system
Vitreous
MESODERM
Fibers of extraocular muscles
Endothelial lining of all orbital and ocular blood vessels
Temporal portion of sclera
Vitreous

Embryogenesis

In the 2 weeks after fertilization, the impregnated ovum undergoes a series of repeated cell divisions and, through repositioning and reorientation of the cells, becomes sequentially *morula, blastula,* and *gastrula* (Fig V-1). Only the inner cell mass, a small number of cells derived from the fertilized ovum, differentiates subsequently into the embryo. The outer cell mass, or *trophoblast,* forms the placenta and support tissues. The formation of the *epiblast* and *hypoblast* from the inner cell mass pre-

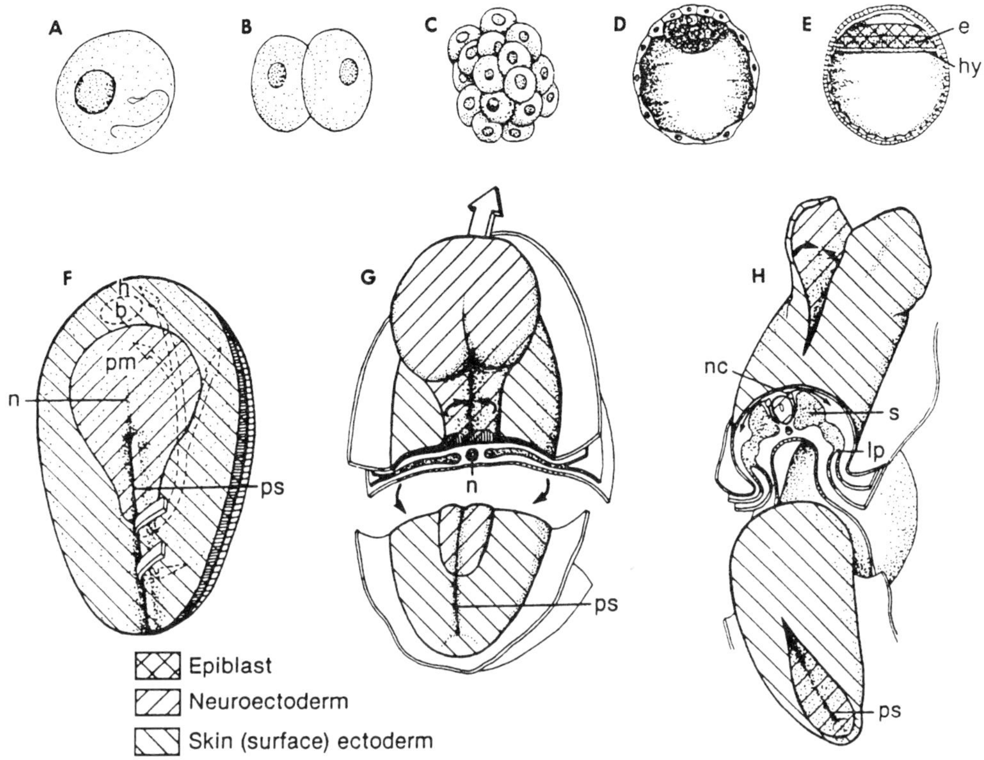

FIG V-1—Early stages of human embryonic development. **A–C**, Fertilization and earliest cell divisions to morula stage. **D**, Sectioned blastocyst. A fluid-filled cavity has formed, and cells that will form the embryo (darker area indicates inner cell mass) are distinct from other cells that will develop into support tissues (eg, the placenta). **E**, Embryo-forming cells have now separated into two layers: the epiblast (*e*) and hypoblast (*hy*). **F**, Dorsal view of an embryo slightly more advanced than the sectioned embryo illustrated in **E**. Gastrulation movements (*arrows*) bring cells from the upper layer through the primitive streak (*ps*) into the potential space between the two layers to form the middle germ layer (*mesoderm*). Mesodermal cells fail to penetrate between the ectoderm and endoderm at the oral plate (*b,* buccopharyngeal membrane), which later forms the embryonic partition between the oral and pharyngeal cavities. At this stage, the heart primordium (*h*) lies anterior to the oral plate. The notochord (*n*) is formed from the anterior (cephalic) end of the primitive streak. The prochordal mesoderm (*pm*) is subjacent to the neural plate on the region between *n* and *b*. **G**, Early stages of neural tube folding and closure and folding of the lateral body walls (*solid arrows*). The anterior neural plate has begun to "overgrow" (*open arrow*) the heart primordium and future oral region, including the buccopharyngeal membrane. **H**, Embryo folding is nearing completion. Migration of cranial neural crest cells (*nc*) in the hindbrain region has been initiated. In contrast to the trunk crest cells, most of those forming in the head region migrate laterally—under the surface ectoderm but superficial to the somites (*s*) and the lateral plate (*lp*) of the mesoderm. (Reproduced with permission from Serafin D, Georgiade NG. *Pediatric Plastic Surgery.* St Louis: Mosby; 1984.)

cedes gastrulation, a process that results in the establishment of the three primary germ layers: *ectoderm, mesoderm,* and *endoderm* (Fig V-2).

Cells of the epiblast in the medial region of the embryonic disk begin to proliferate at the caudal end, which causes the development of a thickening known as the *primitive streak.* The cells of the primitive streak migrate both laterally and cephalad

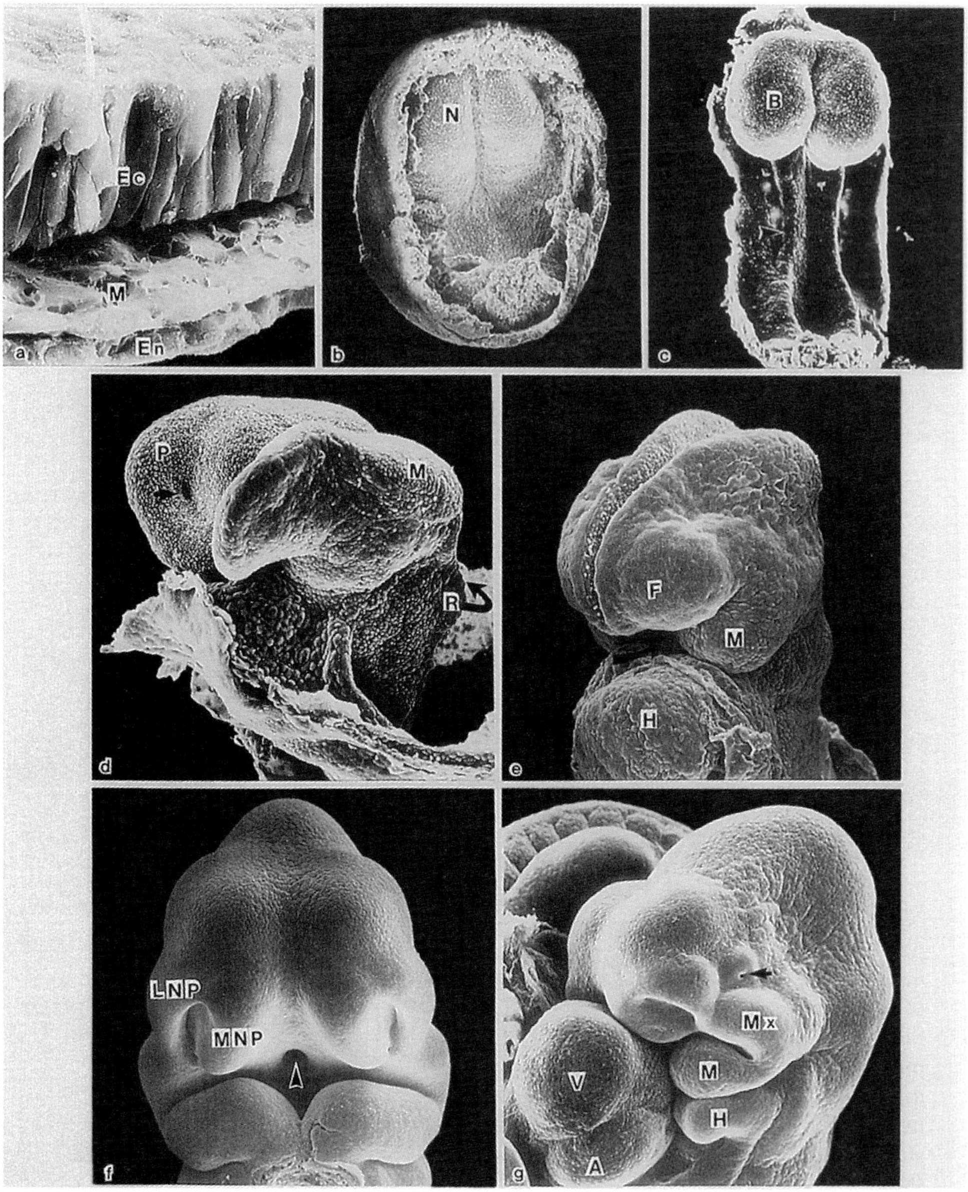

FIG V-2—Normal craniofacial development. **a**, A parasagittal section through the cranial aspect of a gastrulation-stage mouse embryo. The cells of the three germ layers, ectoderm (*Ec*), mesoderm (*M*), and endoderm (*En*), have distinct morphologies. **b**, The developing neural plate (*N*) is apparent in a dorsal view of this presomite mouse embryo. **c**, Neural folds (*arrow*) can be seen in the developing spinal cord region. The lateral aspects of the brain (*B*) region have not yet begun to elevate in this mouse embryo in the head-fold stage. **d**, Three regions of the brain can be distinguished at this six-somite stage: prosencephalon (*P*), mesencephalon (*M*), and rhombencephalon (*R, curved arrow*). Optic sulci (*small arrow*) are seen as evaginations from the prosencephalon. **e**, The neural tube has not yet fused in this 12-somite embryo. The stomodeum (*arrow*), or primitive oral cavity, is bordered by the frontonasal prominence (*F*), the first visceral arch (mandibular arch, *M*), and the developing heart (*H*). **f**, Medial and lateral nasal prominences (*MNP, LNP*) surround olfactory pits in 36-somite mouse embryo. Rathke's pouch (*arrow*) can be distinguished in the roof of the stomodeum. **g**, In this lateral view of a 36-somite mouse embryo, the first and second (hyoid, *H*) visceral arches are apparent. The region of the first arch consists of maxillary (*Mx*) and mandibular (*M*) components. Note the presence of the eye with its invaginating lens (*arrow*). Atrial (*A*) and ventricular (*V*) heart chambers can be distinguished. (Reproduced with permission from Sulik KK. Embryonic origin of holoprosencephaly: interrelationship of the developing brain and face. *Scan Electron Microsc.* 1982;1:311.)

beneath the epiblast, where they give rise to the mesenchymal cells of the intraembryonic mesoderm. The cells that remain in the epiblast are now recognized as the embryonic ectoderm. Some cells of the primitive streak invade the hypoblast and laterally displace most of these cells to give rise to the embryonic endoderm. The primitive streak elongates by the addition of cells at the caudal end and thus establishes the axial orientation of the embryo. The cranial end of the primitive streak enlarges as the primitive node. Mesenchymal cells that migrate cranially from this site form the medial notochordal process, which develops into the primitive mesenchymal axial skeleton (or *notochord*) of the embryo. The development of the notochord induces the overlying ectoderm to differentiate into neuroectoderm that becomes identified as the *neural plate*. The brain and the eye develop from the most anterior region of the neural plate.

Growth of the lateral part of the neural plate results in folds that develop upward and outward, parallel to the neural groove from the head to the caudal region. At this stage, neuroectoderm lines the inner folds, and surface ectoderm covers the outer surface of the folds. The neuroectodermal cells at the apex of the folds proliferate and produce a population of neural crest cells, which contribute extensively to the tissues of the eye (Fig V-3). The cephalic neural folds grow and expand markedly. At the end of the third week after conception, the neural folds begin to close to form the neural tube. This process starts in the midregion of the embryo and proceeds anteriorly and posteriorly at the same time. As the neural tube closes, three events important to the development of the eye and orbit occur simultaneously (Fig V-4):

- Optic pits develop from the optic sulci, small depressions present in the cephalic neuroectoderm.
- Neural crest cells begin migration.
- As the anterior neural tube closes, it flexes ventrally.

Organogenesis of the Eye

The chronology of ocular development is provided in Table V-2. The *optic sulci* are first recognizable as slight, curved indentations in the widest part of each neural fold just internal to the peak of the ridge (Fig V-5). The long axis of the depression is roughly parallel to that of the neural groove. The *optic pits*, formed of a single layer of neuroectoderm, develop from the continued evagination of the sulci. As the neural tube closes, the pits deepen and become optic vesicles, which appear as symmetrical, hollow, hemispherical outgrowths on the lateral sides of what is now the forebrain vesicle. The *optic vesicles* remain attached to, and continuous with, the neural tube by *optic stalks* composed of neuroectodermal cells (Fig V-6). The expansion and ballooning that take place in the hollow optic vesicle do not occur in the stalk, which remains as a tubular link from the cavity of the vesicle to that of the diencephalon.

As the optic vesicle approaches the outer wall of the embryo, a focal thickening of the cells, the *lens placode*, develops in the surface ectoderm, which has been primed by lens-bias signals during earlier embryogenesis. In the fourth week, invagination of the lens placode leads to formation of the *lens vesicle*, which initially remains attached to the surface ectoderm by the lens stalk. Simultaneously, differential growth and movement of the cells of the optic vesicle result in the invagination of its temporal and lower walls and the formation of the *optic cup*.

The outer layer of the optic cup will evolve as a monolayer of cells, the retinal pigment epithelium. The inner, invaginated layer will differentiate into the neuro-

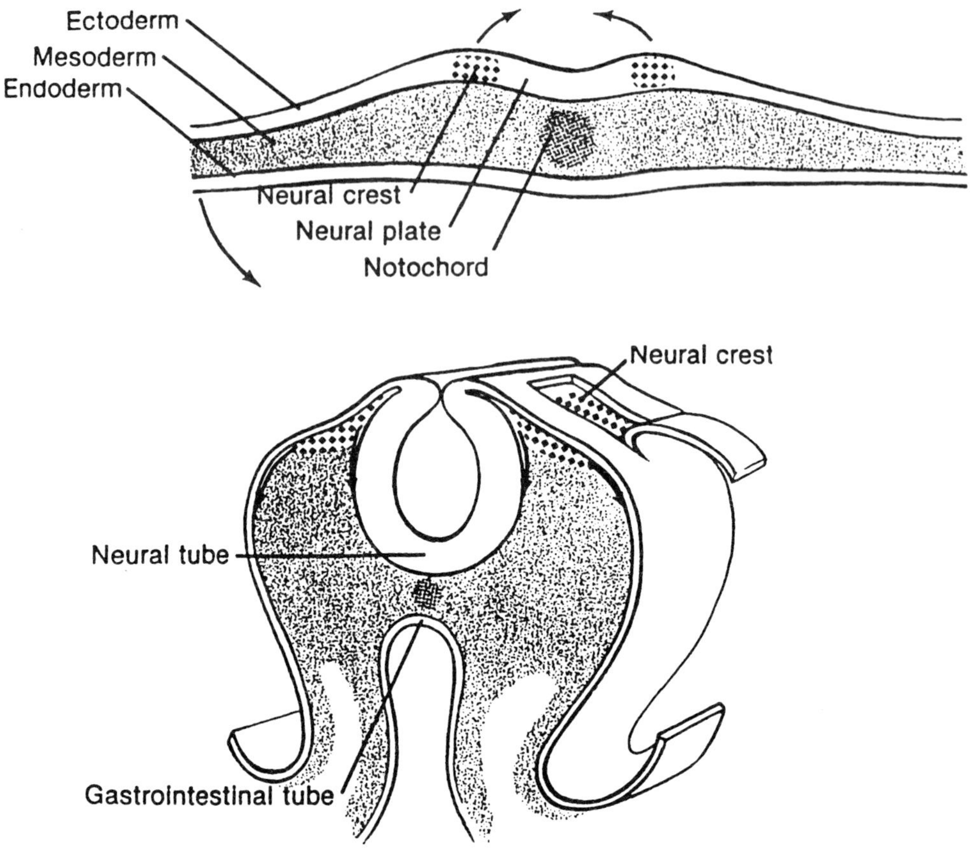

FIG V-3—Cross sections through embryos before (**top**) and after (**bottom**) the onset of migration of crest cells (*diamond pattern*). The ectoderm has been peeled back in the lower figure to show the underlying neural crest cells. (Reproduced with permission from Johnston MC, Sulik KK. Development of face and oral cavity. In: Bhaskar SN. *Orban's Oral Histology and Embryology.* 9th ed. St Louis: Mosby; 1980.)

sensory retina. Initially, the cup is incomplete in its inferior portion (Fig V-7). The indentation, or fissure, between the folds or margins of the cup is called the *embryonic fissure* (previously the *choroidal,* or *fetal, fissure*). Invagination pushes the neuroectodermal cells originally near the surface deep into the cup near the outer layer. Invagination produces a fold in the neuroectoderm anteriorly, adjacent to the lens, called the *rim* of the optic cup. At first, the two layers of the developing cup have a small space between them, the *optic ventricle*; however, as invagination proceeds and the inner and outer layers become juxtaposed, the cavity of the optic ventricle progressively narrows. Basement membrane lines both the outer and the inner layers of the cup. The apices of the cells in both layers meet end to end as the ventricle narrows. The ventricle cavity remains throughout life as a potential space, the *subretinal space.*

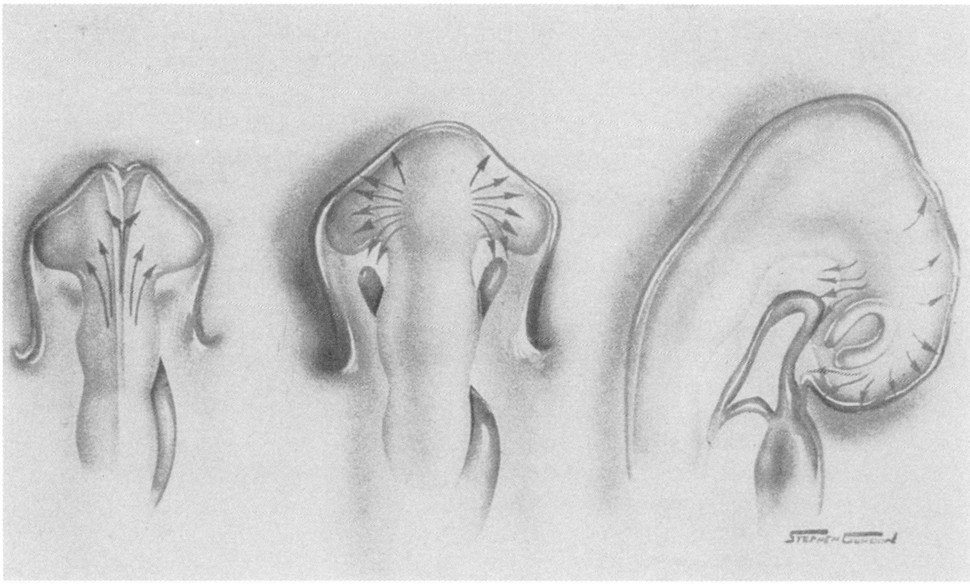

FIG V-4—Migration of cranial neural crest cells from dorsal diencephalic and mesencephalic regions. *Left,* Cells begin migration anteriorly as tube closes. *Center,* Crest cells move in waves around optic vesicle and lose continuity with the surface cells. *Right,* Neural tube flexes ventrally, carrying optic cup and crest cells ventrally. (Redrawn from M. Johnston, 1966.)

The embryonic fissure extends from the rim of the cup near the lens to the distal optic stalk. This fissure allows vessels of the hyaloid system to be incorporated within the eye (Fig V-8). To complete the entire wall of the globe, the two lips of the embryonic fissure meet and fuse. Closure begins in the midregion of the cup near the equator of the globe and proceeds anteriorly to the rim and posteriorly down the stalk, enclosing the hyaloid artery. This process occasionally gives rise to a bridge coloboma with a posterior and peripheral component separated by a band of normal tissue (see Chapter VI, Fig VI-18D). The inner and outer layers of the cup meet end to end. Because the primitive cells are still labile, they seal the fissure without evidence of a seam or scar. Incomplete or inadequate closure produces a coloboma of the iris, ciliary body, choroid, or optic disc, depending on the extent of the failed closure and secondary attempts to close the defect (Fig V-9).

Neurosensory Retina

The *neurosensory,* or *neural, retina* arises from the inner layer of neuroectodermal cells of the optic cup (Fig V-10). Differentiation of this cell layer commences early, and within 1 month of fertilization mitotic activity has produced three to four compact rows of cells that rapidly increase in numbers. The nuclei segregate at the outer two thirds of the primordial retina toward the outer layer of the optic cup. This region is recognized as the *primitive zone*. The ciliated apices of the cells are directed outward into the rapidly shrinking cavity of the optic ventricle. The inner third of the

TABLE V-2

CHRONOLOGY OF EMBRYONIC AND FETAL DEVELOPMENT OF THE EYE

22 days	Optic primordium appears in neural folds (1.5–3.0 mm).
25 days	Optic vesicle evaginates. Neural crest cells migrate to surround vesicle.
28 days	Vesicle induces lens placode.
Second month	Invagination of optic and lens vesicles. Hyaloid artery fills embryonic fissure. Closure of embryonic fissure begins. Pigment granules appear in retinal pigment epithelium. Primordia of lateral rectus and superior oblique muscles grow anteriorly. Eyelid folds appear. Retinal differentiation begins with nuclear and marginal zones. Migration of retinal cells begins. Neural crest cells of corneal endothelium migrate centrally. Corneal stroma follows. Cavity of lens vesicle is obliterated. Secondary vitreous surrounds hyaloid system. Choroidal vasculature develops. Axons from ganglion cells migrate to optic nerve. Glial lamina cribrosa forms. Bruch's membrane appears.
Third month	Precursors of rods and cones differentiate. Anterior rim of optic vesicle grows forward and ciliary body starts to develop. Sclera condenses. Vortex veins pierce sclera. Eyelid folds meet and fuse.
Fourth month	Retinal vessels grow into nerve fiber layer near optic disc. Folds of ciliary processes appear. Iris sphincter develops. Descemet's membrane forms. Schlemm's canal appears. Hyaloid system starts to regress. Glands and cilia develop.
Fifth month	Photoreceptors develop inner segments. Choroidal vessels form layers. Iris stroma is vascularized. Eyelids begin to separate.
Sixth month	Ganglion cells thicken in macula. Recurrent arterial branches join the choroidal vessels. Dilator muscle of iris forms.
Seventh month	Outer segments of photoreceptors differentiate. Central fovea starts to thin. Fibrous lamina cribrosa forms. Choroidal melanocytes produce pigment. Circular muscle forms in ciliary body.

TABLE V-2 (Continued)

CHRONOLOGY OF EMBRYONIC AND FETAL DEVELOPMENT OF THE EYE

Eighth month	Chamber angle completes formation. Hyaloid system disappears.
Ninth month	Retinal vessels reach the periphery. Myelination of fibers of optic nerve is complete to lamina cribrosa. Pupillary membrane disappears.

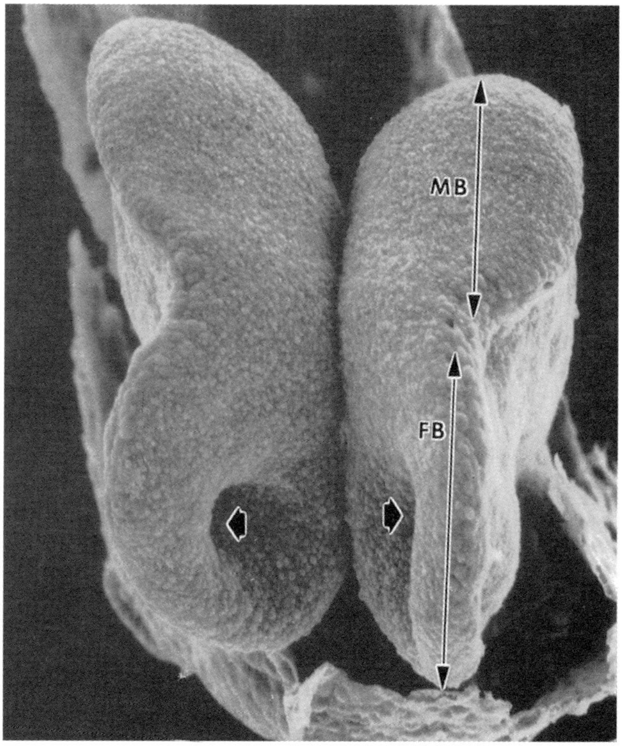

FIG V-5—Scanning electron microscopy of normal mouse embryo on day 8 of gestation; five somite pairs. The bilateral optic sulci (*short arrows*) form as evaginations of the forebrain (*FB*). *MB,* midbrain (×250). (Reproduced with permission from Cook CS, Sulik KK. Keratolenticular dysgenesis [Peters anomaly] as a result of acute embryonic insult during gastrulation. *J Pediatr Ophthalmol Strabismus.* 1988;25:60–66.)

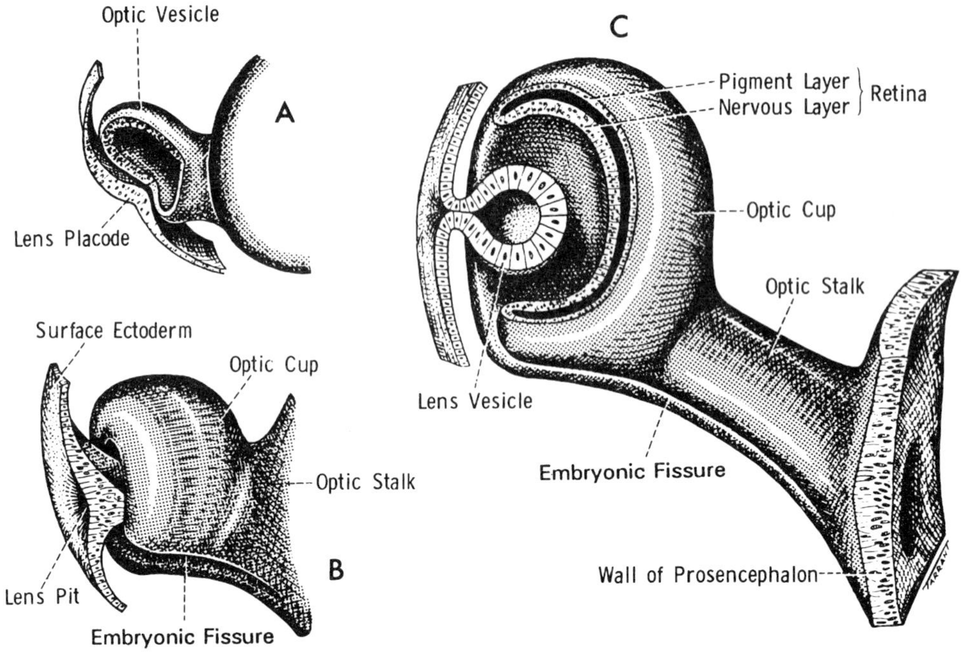

FIG V-6—Diagram of the development of the human optic cup. The optic vesicle and cup have been partly cut away in **A** and **C**, and the lens vesicle is sectioned for clarity. **A**, 4.5 mm embryo (27 days). **B**, 5.5 mm embryo. **C**, 7.5 mm embryo (28 days). (Reproduced from Tripathi RC, Tripathi BJ. Comparative physiology and anatomy of the aqueous outflow pathway. In: Davson H, ed. *The Eye.* 3rd ed. Orlando: Academic Press; 1984.)

developing retina is initially devoid of nuclei and is termed the *inner marginal zone;* it eventually differentiates into the nerve fiber layer. The primitive and marginal zones are recognizable only until the seventh week of gestation. Little is known about the stimuli that initiate and direct the complex migration and subsequent differentiation of the primitive neuroepithelial cells into the retina.

Differentiation of the retina begins in the center of the optic cup and gradually extends peripherally toward its rim. Neural and glial cells develop simultaneously. By 5 weeks of gestation, the putative ganglion and Müller cells have migrated from the outer neuroepithelial layers toward the vitreous cavity. As a result, the nuclei of the neuroblastic cells become segregated as two distinct layers, the *inner and outer neuroblastic layers.* These two layers are separated by a region of tangled cell processes known as the *transient nerve fiber layer of Chievitz,* which becomes the definitive inner plexiform layer between weeks 9 and 12 of gestation (except in the macula, where it persists until birth). At 9–12 weeks, the four major horizontal layers of the retina become distinguishable.

The *ganglion cells* are the first cells of the retina to become clearly differentiated. Their axonal processes and dendritic trees begin to develop at about the sixth week of gestation. Axons from ganglion cells nearest the posterior pole are the first

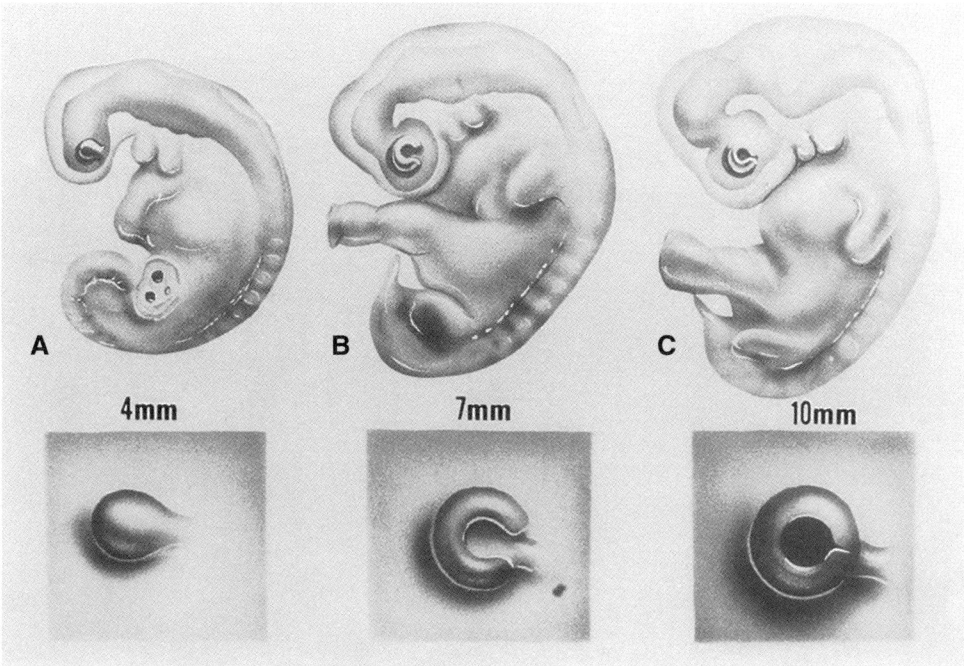

FIG V-7—Ocular and somatic development. *A,* Flexion of neural tube and ballooning of optic vesicle. *B,* Upper limb buds appear as optic cup and embryonic fissure emerge. *C,* Completion of optic cup with closure of fissure. Convolutions appear in the brain and leg buds appear. The size of the fetus is noted. *Lower sequence:* optic vesicle; optic cup with open embryonic fissure; cup with fissure closing.

to enter the optic stalk and induce formation of the optic nerve. The number of ganglion cells increases rapidly between weeks 15 and 17 of gestation, then decreases between weeks 18 and 30 because of apoptosis. The ganglion-cell somas grow larger with advancing gestational age.

The processes of the Müller cells extend from the inner basal lamina of the optic vesicle toward the optic ventricle. As soon as the photoreceptors enlarge and become morphologically distinct as cones, the development of junctional complexes on adjacent lateral surfaces of these cells and of the Müller cell processes gives rise to the *external limiting membrane.*

Photoreceptors arise from the outermost layer of neuroblastic cells. Mitotic activity, abundant in the outer neuroblastic layers in weeks 4–12, ceases in the central retina by week 15 of gestation, and differentiation of the cones begins in the region of the putative fovea. The cilia on the apices of the cells that had invaginated the adjacent retinal pigment epithelium disappear, and precursors of outer segments gradually develop. Initially, cylindrical cytoplasmic processes extend toward the apical region of the retinal pigment epithelial cells.

Differentiation of cone outer segments begins at 5 months, when multiple infoldings develop in the plasma membrane of the processes. The folds separate from the plasma membrane, and their orientation as flattened, lamellar disks parallels the

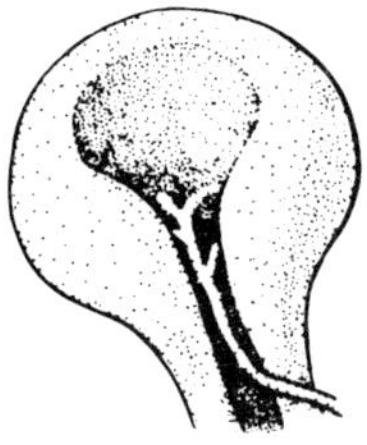

FIG V-8—Optic cup and stalk with open embryonic fissure below. Hyaloid artery from dorsal ophthalmic artery enters cavity through posterior aspect of the embryonic fissure. Rim of optic cup is above. Lens is not pictured.

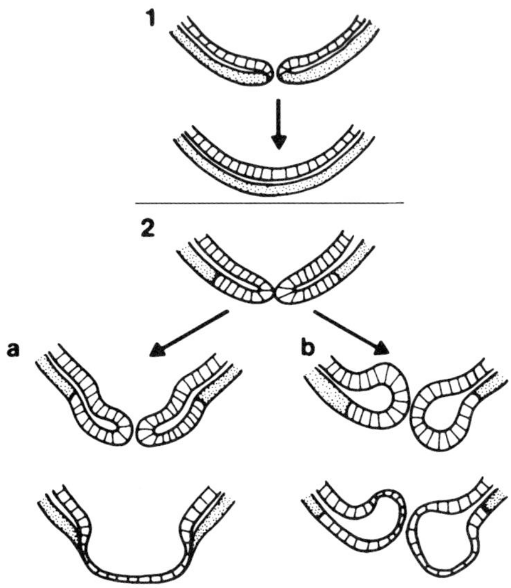

FIG V-9—Closure of lips of embryonic fissure. **1**, Normal closure: Inner layers (neurosensory retina) and outer layers (*dotted area*, retinal pigment epithelium) meet and merge. Basement membrane forms on both surfaces. **2**, Coloboma formation: Ectropion of the inner retina at the lips of the fissure results in imperfect fusion; pigment epithelium is displaced laterally by cells of neurosensory retina. **a**, A simple coloboma results in defective retina and retinal pigment epithelium. Uvea and sclera (not shown) are thin and dysgenic. **b**, In a cystic coloboma, the primary vesicular cavity enlarges adjacent to the point of defective closure.

development of the horizontal cells. The cell bodies of the rods are dispersed among the cones and are first recognizable by their dark nuclei with condensed peripheral chromatin. The rod outer segments develop during the seventh month of gestation.

Amacrine cells are identified by their large, round, pale-staining nuclei. They are first seen scattered at the inner border of the outer neuroblastic layer by week 14 of gestation. Bipolar cells do not differentiate until week 23. The bipolar dendrites extend to the outer plexiform layer by week 25, at which time the horizontal cells probably differentiate.

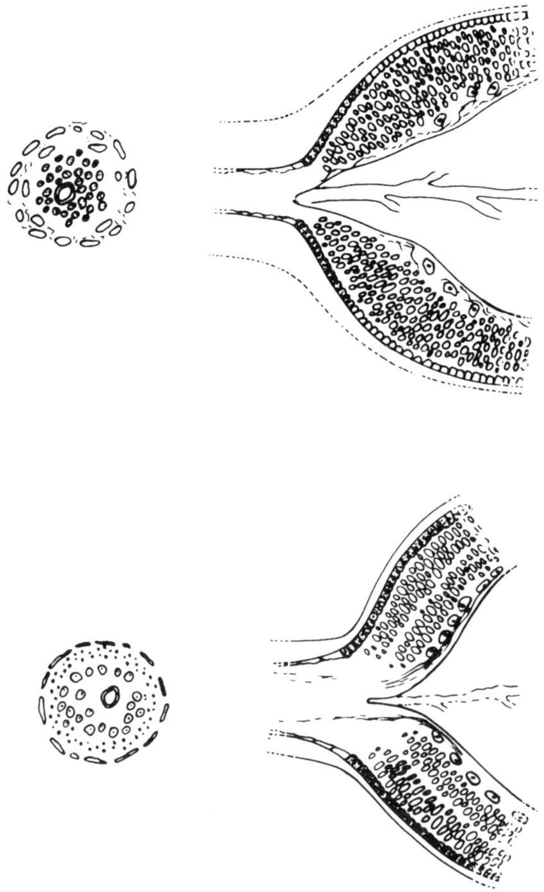

FIG V-10—Development of retina and optic nerve. **Top**, At right, fetal neurosensory retina develops from neuroectoderm as ganglion cells migrate from outer primitive zone of closely packed nuclei to inner marginal zone of fibrils. A few axons from ganglion cells grow toward the optic nerve. Retinal pigment epithelium begins melanization in posterior pole. At left, cross section shows fetal optic nerve with center of vacuolating primitive cells through which axons from ganglion cells will grow toward the brain. Neural crest cells as mesenchyme loosely ring the nerve. Hyaloid artery enters vitreous (fifth week). **Bottom**, At right, migration of nuclei results in three nuclear layers and plexiform layers. Cross section of optic nerve on left shows axons of ganglion cells (*black dots*) migrating through vacuolating cells, first in periphery of the nerve. Neural crest cells condense to meningeal sheaths of optic nerve (seventh week).

Fovea Differentiation of the neurons, photoreceptors, and glial cells in the fovea occurs early because this region is the focal point for the centraperipheral development of the retina. The different cell types, as well as many synapses and intercellular junctions, are already established by 15 weeks of gestation. Thinning of the ganglion cell and inner nuclear layers begins at 24–26 weeks of gestation and gives rise to the earliest recognizable depression in the area of the macula.

The foveal pit becomes more prominent by the seventh month as a result of the marked thinning of the inner nuclear layer. An acellular fibrous zone is now present on both the nasal and temporal sides of the fovea. By this time, major changes have occurred in the cones: the width of the inner segments is decreased, whereas their length is increased along with the length of the fibers of the fiber layer of Henle. Only two layers of ganglion cells remain at 8 months, and the inner nuclear layer at the foveola is reduced to three rows or fewer because of lateral displacement. At birth, axons of bipolar cells that pass to the inner plexiform layers constitute the prominent transient layer of Chievitz. Relocation of all layers to the periphery of the foveal slope, which leaves the nuclei of the cones uncovered in the foveola, occurs by 4 months after birth. However, remodeling of the elements of the fovea continues until nearly 4 years of age, at which time the transient layer of Chievitz is lost completely.

Hendrickson AE, Yuodelis C. The morphological development of the human fovea. *Ophthalmology*. 1984;91:603–612.

Rhodes RH. A light microscopic study of the developing human neural retina. *Am J Anat.* 1979;154:195–209.

Tripathi BJ, Tripathi RC. Development of the human eye. In: Bron AJ, Tripathi RC, Tripathi BJ, eds. *Wolff's Anatomy of the Eye and Orbit.* 8th ed. London: Chapman and Hall; 1997.

Yuodelis C, Hendrickson AE. A qualitative and quantitative analysis of the human fovea during development. *Vision Res.* 1986;26:847–855.

Retinal Pigment Epithelium (RPE)

Mitotic activity continues in the pseudostratified, columnar epithelial cells that constitute the outer wall of the optic cup up to the sixth week of gestation. The apical borders of adjacent cells are already joined by zonulae occludens and zonulae adherens junctional complexes. At week 6, melanogenesis begins; concurrently, the cilia that had been present on the inner cell surface (ie, adjacent to the developing neurosensory retina) disappear. The RPE cells are the first in the body to produce melanin. Whether in the retina or in the choroid, the stages of melanin production are the same: Premelanosomes gradually become melanosomes.

Differentiation of the RPE begins at the posterior pole and proceeds anteriorly, so that by 8 weeks of gestation the RPE is organized as a single layer of hexagonal columnar cells located posteriorly. The cells become tall and cuboidal during the third and fourth months, and the terminal web becomes well established at the lateral apical borders. The RPE is thought to be fully functional at this stage. The increase in surface area of the RPE (which takes place after birth to accommodate the subsequent growth of the globe) is achieved by enlargement and expansion of individual cells.

The basement membrane of the RPE becomes the inner portion of *Bruch's membrane*; the outer layer of Bruch's membrane, also basement membrane, is laid down by the choriocapillaris layer. The embryonic pigment epithelial cells have a profound inductive influence on the development of the choroid, sclera, and neurosensory retina. In areas where pigment epithelium does not form, as sometimes happens along the line of closure of the embryonic fissure, the underlying choroid, sclera, and retina are hypoplastic (see Fig V-9). The nature of the inductive stimulus is not known.

Oguni M, Tanaka O, Shinohara H, et al. Ultrastructural study on the retinal pigment epithelium of human embryos, with special reference to quantitative study on the development of melanin granules. *Acta Anat (Basel).* 1991;140:335–342.

Optic Nerve

The optic nerve develops from the optic stalk, the original connection between the optic vesicle and the forebrain. Initially, the stalk is composed of an inner zone of closely packed neuroectodermal cells surrounded by a less compact layer of undifferentiated neural crest cells. Late in the sixth week of gestation, some cells of the inner region vacuolate and degenerate, and nerve fibers from the ganglion cells migrate through the spaces thus created. Other cells of the inner zone differentiate as glial cells. By the seventh week, the optic disc contains the hyaloid artery, which is surrounded by axons and covered by a mantle of glial cells, many of which disappear by the seventh month. The glial cells also give rise to the glial elements of the lamina cribrosa during the eighth week of gestation. Differentiation of the neural crest cells into the pia, arachnoid, and dura mater of the optic nerve begins in the seventh week, but the sheaths become well defined only after the fourth month.

The number of axons increases rapidly: By 10–12 weeks of gestation, some 1.9 million axons are present in the optic nerve, rising to 3.7 million by 16 weeks. Later, attrition of axons causes the number of fibers to drop to approximately 1.1 million, which establishes the adult condition by 33 weeks. The loss of axons parallels the degeneration of ganglion cells in the fetal retina and may be related to the segregation of terminals as discrete laminae in the dorsal lateral geniculate body.

As the axons grow toward the lateral geniculate body, partial crossover occurs at the optic chiasm. Cells located at the chiasm midline (probably radial glial cells) express certain repulsive or inhibitory molecules that provide a guidance cue by acting specifically on ipsilateral projecting axons.

Myelination starts in the chiasm at the seventh month of gestation, proceeds toward the eye, and ceases at the lamina cribrosa by about 1 month after birth. Occasionally, medullated fibers develop in the retina. They appear on ophthalmoscopic examination as a flat, serrated white patch on the inner surface of the retina. The medullation is usually interrupted at the lamina, but occasionally it is continuous across the lamina from nerve to retina.

Some fetuses demonstrate a response to light as early as the eighth week of gestation, which indicates that at least some central nervous system pathways are established. By 5 months, 50% of the growth of the optic nerve and disc has occurred; by birth, 75%; and before 1 year of age, 95%.

Provis JM, van Driel D, Billson FA, et al. Human fetal optic nerve: overproduction and elimination of retinal axons during development. *J Comp Neurol.* 1985;238:92–100.

Rimmer S, Keating C, Chou T, et al. Growth of the human optic disk and nerve during gestation, childhood, and early adulthood. *Am J Ophthalmol.* 1993;116:748–753.

Wizenmann A, Thanos S, Boxberg Y, et al. Differential reaction of crossing and noncrossing rat retinal axons on cell membrane preparations from the chiasm midline: an in vitro study. *Development.* 1993;117:725–735.

Lens

One of the earliest events in embryogenesis is determination of lens development. The interaction that takes place between the surface ectoderm and the underlying chordamesoderm during midgastrulation imparts a lens-forming bias on an extensive region of head ectoderm. Next, the anlage of the eye conveys an inductive signal to the ectoderm, which determines the region of the presumptive lens. The mesoderm beneath the putative lens ectoderm transmits another signal late in gastrulation (but when the neural plate is still open) that potentiates the fate of the tissue that will become the lens. Finally, by invoking the final phase of determination and enhancing differentiation during neurulation, the optic vesicle designates the specific region of the head ectoderm that will become the lens. The surface ectoderm can respond to the influence of the optic vesicle only during a precise period of development.

The lens is first apparent at about 27 days' gestation as a disk-shaped thickening of surface epithelial cells over the optic vesicle. This lens placode and its thin basal lamina are separated from the basal lamina of the optic vesicle by a narrow space containing fine filaments that have a role in the gradual invagination of the lens placode to form the lens vesicle. Initially, the vesicle, consisting of a single layer of cells with apices directed inward, is covered by a basal lamina that seals anteriorly to complete the formation of the lens capsule (Fig V-11). Ultimately, the lens vesicle separates from the surface epithelium at about 33 days' gestation. The area of lens separation from the surface ectoderm heals without residuum. The epithelial cells deposit additional basal lamina material, which forms the lens capsule. Initially, the posterior capsule is more prominent than the anterior capsule. The lens capsule isolates the lens constituents immunologically within the globe.

During closure of the lens vesicle, DNA synthesis decreases in the cells that form the posterior half of the lens; simultaneously, specific lens proteins *(crystallins)* are synthesized. By day 45 of gestation, the posterior cells, or *primary lens fibers,* have lengthened to fill the cavity of the vesicle from posterior to anterior. The posterior cells of the vesicle account for most of the growth of the lens during the first 2 months of embryogenesis. The primary fibers form the compact core of the lens, known as the *embryonic nucleus.*

The pre-equatorial epithelial cells retain their mitotic activity throughout life, producing secondary lens fibers. These fibers are displaced inward between the capsule and the embryonic nucleus and meet on the vertical planes, the *lens sutures.* The first suture marking the fetal nucleus is shaped like a Y anteriorly and an inverted Y posteriorly. The basic anatomy of the lens is established after the first layer of secondary fibers has been laid down at the seventh week of gestation.

At first, the lens is spherical, but it becomes ellipsoid with the addition of secondary fibers. As secondary fibers are added, the sutures become more complex and dendriform. In the third month, the innermost fibers mature; cytoplasmic fibrillar material increases and cellular organelles decrease. The nuclei of the deeper cells, at first homogeneous and dense, are lost; the chromatin and ribosomes disintegrate. The equatorial diameter of the unfixed human lens measures 2 mm at 12 weeks and 6 mm at 35 weeks. Both the growth and maturation of lenticular fibers continue throughout life. BCSC Section 11, *Lens and Cataract,* discusses the development of the lens in detail in its Chapter IV, Embryology.

The zonular apparatus begins to develop after the tertiary vitreous has formed. The ciliary epithelial cells then synthesize collagen fibrils of the zonular fibers. As they increase in number, strength, and coarseness, the fibers reach the lens and merge with the anterior and posterior capsule by the fifth month of gestation.

Grainger RM, Henry JJ, Saha MS, et al. Recent progress on the mechanisms of embryonic lens formation. *Eye.* 1992;6:117–122.

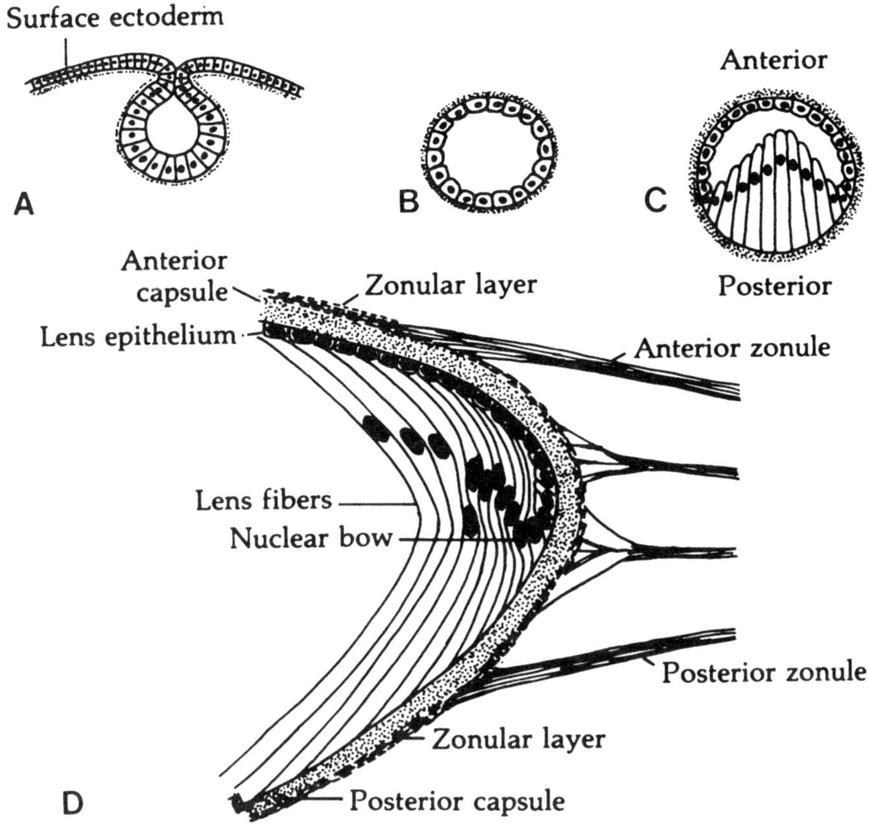

FIG V-11—Diagram of stages in the development of the lens and its capsule. **A**, Formation of lens vesicle from invagination of surface ectoderm together with its basal lamina in an embryo corresponding to 32 days of gestation. **B**, Separation of the vesicle from the surface ectoderm and its surrounding basal lamina. **C**, Obliteration of lens vesicle cavity by elongation of posterior cells at about 35 days' gestation. **D**, Equatorial region of the fully formed lens. Attachment of zonular fibers to the anterior, posterior, and equatorial regions of the lens periphery becomes apparent at approximately 5½ weeks' gestation. Note the change in polarity of cells from anterior to posterior regions of the lens. (Reproduced from Tripathi RC, Tripathi BJ. Anatomy of the human eye. In: Davson H, ed. *The Eye.* 3rd ed. Orlando: Academic Press; 1984.)

Saha MS, Spann CL, Grainger RM. Embryonic lens induction: more than meets the optic vesicle. *Cell Differ Dev.* 1989;28:153–171.

Vitreous (Fig V-12)

Between the fourth and fifth weeks of gestation, the space between the lens vesicle and the inner layer of the optic cup becomes filled with fibrils, mesenchymal cells, and vascular channels of the hyaloid system. Together, these elements constitute the *primary vitreous.* Initially, the fibrillar content is of ectodermal origin, being derived

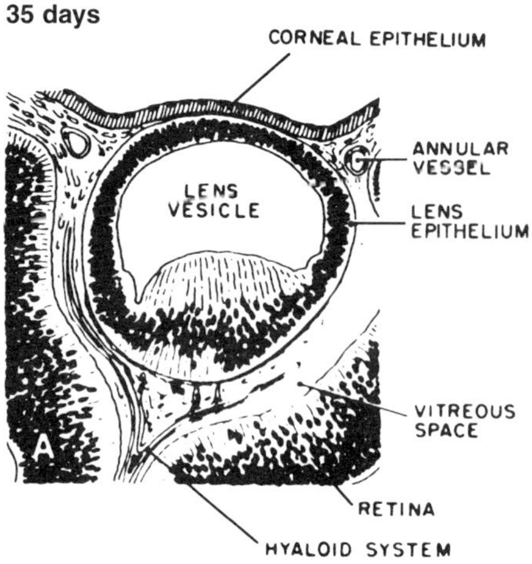

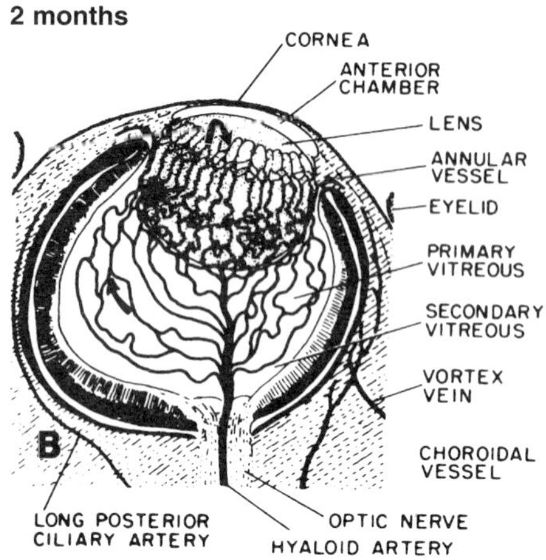

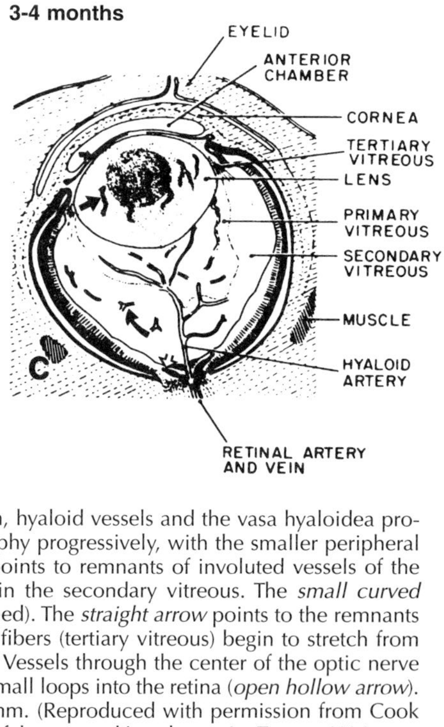

FIG V-12—Main features in vitreous development and the regression of the hyaloid system shown in drawings of sagittal sections. **A**, At 35 days, hyaloid vessels and their branches, the vasa hyaloidea propria, occupy the space between the lens and the neural ectoderm. A capillary net joins the capsula perilenticularis fibrosa, which is composed of ectodermal fibrils associated with vasoformative mesenchyme from the periphery. The ground substance of the primary vitreous is finely fibrillar. **B**, By the second month, the vascular primary vitreous reaches its greatest extent. Arborization of the vasa hyaloidea propria (*curved arrow*) fills the retrolental area and is embedded in collagen fibrils. An avascular secondary vitreous of more finely fibrillar composition forms a narrow zone between the peripheral (outer) branches of the vasa hyaloidea propria and the retina. The *hooked arrow* points to the vessel of the pupillary membrane. The drawing is a composite of embryos at 15–30 mm. **C**, During the fourth month, hyaloid vessels and the vasa hyaloidea propria, together with the tunica vasculosa lentis, atrophy progressively, with the smaller peripheral channels regressing first. The *large curved arrow* points to remnants of involuted vessels of the superficial portion of the vasa hyaloidea propria in the secondary vitreous. The *small curved arrow* indicates the pupillary membrane (not sketched). The *straight arrow* points to the remnants of the atrophied capsulopupillary vessels. Zonular fibers (tertiary vitreous) begin to stretch from the growing ciliary region toward the lens capsule. Vessels through the center of the optic nerve connect with the hyaloid artery and vein and send small loops into the retina (*open hollow arrow*). The drawing is a composite of fetuses at 75–110 mm. (Reproduced with permission from Cook CS, Ozanics V, Jakobiec FA. Prenatal development of the eye and its adnexa. In: Tasman W, Jaeger EA, eds. *Duane's Foundations of Clinical Ophthalmology.* Philadelphia: Lippincott; 1991.)

from the fibrils already in place between the invaginating lens placode and the inner layer of the optic cup. The mesenchymal cells are mostly mesodermal in origin, having invaded the cavity of the optic cup with the hyaloid vessel through the patent optic fissure. However, some mesenchymal cells are derived from neural crest cells that migrated over the rim of the cup. The vascular primary vitreous attains its maximum development by 2 months of gestation.

The development of the *secondary vitreous* begins soon after the primary vitreous is established. The secondary vitreous is avascular and consists of type II collagen fibrils and hyalocytes, which are presumed to be derived from mesenchymal cells of the primary vitreous that differentiated into monocytes. The content of hyaluronic acid in the vitreous is very low during the prenatal period but increases after birth. Initially, the secondary vitreous occupies only a narrow space between the retina and the posterior limit of the primary vitreous. The continued development of the secondary vitreous, until the end of the third month, is related to the regression of the hyaloid system and the simultaneous retraction of the primary vitreous. Remnants of the atrophied hyaloid system and primary vitreous remain throughout life as *Cloquet's canal.*

Between the third and fourth months of gestation, collagen fibrils of the secondary vitreous condense and become attached to the internal limiting membrane at the rim of the optic cup. The condensation of fibrils extends to the lens equator and constitutes the *tertiary vitreous.* The zonular apparatus of the lens ultimately develops anterior to these collagen fibrils.

Choroid

The development of the choroid begins at the anterior region of the optic cup and proceeds posteriorly toward the optic stalk. Choroidal development is associated with the condensation of neural crest cells around the cup that differentiate into cells of the choroidal stroma. Endothelium-lined blood spaces appear in this mesenchymal tissue and first coalesce as the embryonic annular vessel at the rim of the optic cup. During the fourth and fifth weeks of gestation, the *choriocapillaris* begins to differentiate. The choriocapillary network is formed by mesodermal cells that come in contact with the RPE, which is differentiating simultaneously.

The embryonic eye is completely invested with a primitive layer of capillaries at the beginning of the sixth week of gestation. Adjacent endothelial cells are joined by punctate junctional complexes and zonulae occludens. Characteristic diaphragmed fenestrations develop in the endothelium between the seventh and ninth weeks. At the same time, the basal lamina becomes defined as a continuous layer of extracellular material surrounding the capillaries. Toward the RPE, this basal lamina constitutes the fifth layer of Bruch's membrane.

The network of vascular channels is supplied by vessels from the internal carotid artery and, later, by the primitive ophthalmic arteries. The channels drain into two main blood spaces, the superior orbital and inferior orbital venous plexuses, and from there into what will become the cavernous sinuses. By the end of the second month of gestation, short ciliary arteries enter the capillary coat. Arteries can be distinguished by narrow lumina and walls two or more cells thick; veins are larger and lined only by endothelium.

Definite layering of the choroidal vasculature begins in the third month, when the outer layer of large vessels develops. Mainly venous, this layer receives small efferent branches of the choriocapillaris and connects with the vortex veins, which eventually perforate the neighboring sclera. During the fourth month of gestation, the anterior ciliary and long posterior ciliary arteries form the major arterial circle of the iris. Recurrent branches extend from this vessel into the ciliary body by the end of the fifth month. (However, the final anastomosis with the arterial circulation of the choroid is not established until the eighth month.) During the fifth month of gestation, the third layer of medium-sized arterioles develops between the choriocapillaris and the outer layer of large vessels. This layer is initially confined to the level of the equator and reaches the developing ciliary body only at the sixth month.

The choroidal stroma is demarcated by the sclera at the end of the third month of gestation. Initially, the stroma consists of a loosely organized framework of collagen fibrils and abundant fibroblasts. Elastic tissue is laid down during the fourth month. Melanosomes appear between weeks 24 and 27 of gestation, most notably in the melanocytes of the outer choroid and suprachoroid. Melanocytes differentiate from neural crest cells. Melanogenesis proceeds anteriorly from the optic disc to the ora serrata. A few immature melanosomes can be found in the choroidal melanocytes at birth.

Heimann K. The development of the choroid in man. *Ophthalmic Res.* 1972;3:257–273.

Sellheyer K. Development of the choroid and related structures. *Eye.* 1990;4:255–261.

Cornea and Sclera

The separation of the lens vesicle from the surface ectoderm initiates the development of the cornea (Fig V-13). By the end of the fifth week of gestation, the ectoderm consists of two layers of epithelial cells that rest on a thin basal lamina (Fig V-14). Detachment of the lens vesicle induces the basal layer of epithelial cells to secrete collagen fibrils and glycosaminoglycans, which occupy the space between the lens and the corneal epithelium and constitute the primary stroma. Mesenchymal cells migrate from the margins of the rim of the optic cup along the posterior surface of the primary stroma. The first of three successive waves of ingrowth, these neural crest–derived cells form the corneal endothelium.

At 5–6 weeks of gestation, the cornea consists of the following:

- A superficial squamous and a basal cuboidal layer of epithelial cells
- A primary stroma
- A double layer of endothelial cells posteriorly

Further development of the stroma is preceded by the ingrowth of another wave of mesenchymal cells from the rim of the optic cup, which proceeds in two directions. The cells of the posterior extension grow between the lens epithelium and the corneal endothelium and are destined to form the primary pupillary membrane. Concurrently, hydration of the hyaluronic acid component of the primary stroma causes swelling that seems to make space available for the next migratory wave of cells. At approximately 7 weeks' gestation, the anterior extension of mesenchymal cells migrates into the corneal stroma. These cells differentiate into *keratocytes* that secrete type I collagen fibrils and form the matrix of the mature (or secondary) corneal stroma.

The stroma attains its maximum width, which is approximately double the normal postembryonic width. Dehydration (especially of hyaluronic acid) and compression of the connective tissue cause the later reduction in thickness. Morphogenesis of keratocytes begins in the posterior stroma and proceeds anteriorly. The cells synthesize proteoglycans and collagen fibrils, which are organized as lamellae. Each lamella continues to grow by the formation of additional fibrils *(interstitial growth)*;

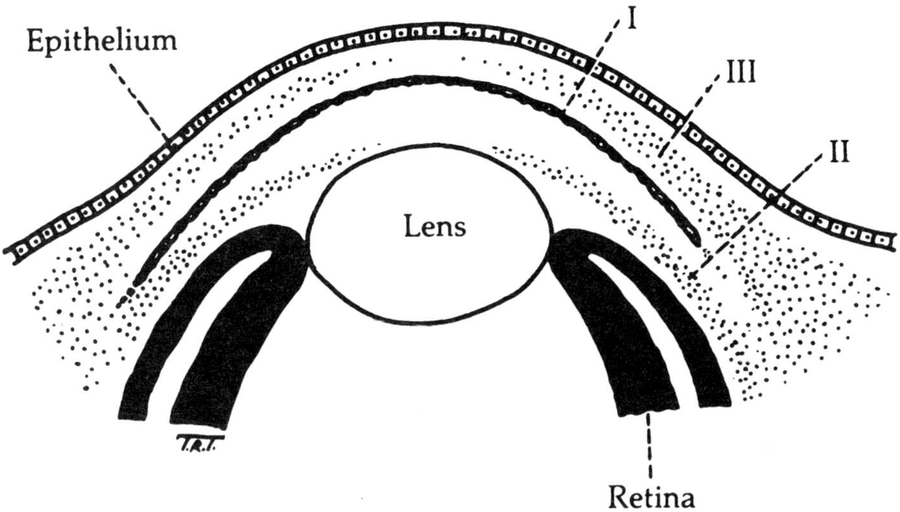

FIG V-13—Three successive waves of ingrowth of neural crest cells associated with differentiation of anterior chambers. *I,* First wave forms corneal endothelium. *II,* Second wave forms iris and part of the pupillary membrane. *III,* Third wave forms keratocytes.

simultaneously, successive layers of lamellae are added *(appositional growth).* As the lamellae increase in length and width, the diameter and thickness of the cornea enlarge.

The endothelium in the central region of the cornea becomes a single layer of flattened cells by the third month of gestation. The cells rest on an interrupted basal lamina, which is the future *Descemet's membrane.* At this stage in development, Descemet's membrane consists of two zones: the *lamina densa* toward the stroma and the *lamina lucida* adjacent to the endothelium. Subsequent growth of Descemet's membrane forms a unique organization that is recognized as the fetal banded zone, which attains a maximum thickness of about 3 μm at birth. In postnatal life, the posterior nonbanded zone of Descemet's membrane is composed of a homogeneous, fibrillogranular material. This region continues to thicken with age.

By the middle of the fourth month of gestation, the apices of adjacent endothelial cells are joined by zonulae occludens. This development corresponds with the onset of aqueous humor production by the ciliary processes. Late in the fourth month, the acellular *Bowman's zone* of the anterior stroma is formed (Fig V-15). It is thought that the most superficial keratocytes synthesize and lay down the collagen fibrils and ground substance as they migrate somewhat posteriorly in the stroma.

The diameter of the unfixed cornea measures 2 mm at 12 weeks' gestation, 4.5 mm at 17 weeks, and 9.3 mm at 35 weeks.

The *sclera* is formed by mesenchymal cells that condense around the optic cup. Most of these cells are derived from the neural crest. However, those in the caudal region of the sclera are probably derived from paraxial mesoderm that lies juxtaposed to the caudomedial surface of the optic cup throughout the period of crest cell migration. The sclera develops anteriorly before the seventh week of gestation

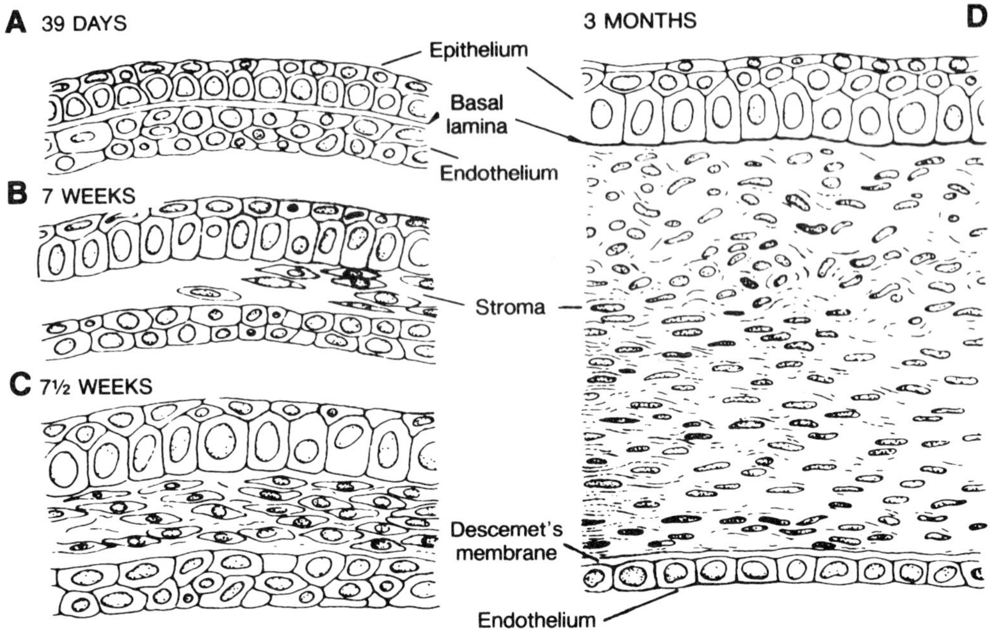

FIG V-14—Development of cornea in central region. **A**, At day 39, two-layered epithelium rests on basal lamina and is separated from endothelium (two to three layers) by narrow acellular space. **B**, At week 7, mesenchymal cells from periphery migrate into space between epithelium and endothelium. **C**, Mesenchymal cells (future keratocytes) are arranged in four to five incomplete layers by 7½ weeks, and a few collagen fibrils are present among cells. **D**, By 3 months, the epithelium has two to three layers of cells, and the stroma has about 25–30 layers of keratocytes that are arranged more regularly in the posterior half. Thin, uneven Descemet's membrane lies between the most posterior keratocytes and the now single layer of endothelium. (Reproduced with permission from Cook CS, Ozanics V, Jakobiec FA. Prenatal development of the eye and its adnexa. In: Tasman W, Jaeger EA, ed. *Duane's Foundations of Clinical Ophthalmology.* Philadelphia: Lippincott; 1991.)

and gradually extends posteriorly. The alignment of cells into parallel layers and the deposition of collagen fibrils are evidence of differentiation. Deposits of elastin and glycosaminoglycans are added to the extracellular matrix at a later stage. By the third month of gestation, some undifferentiated mesenchymal cells have migrated between the nerve fibers in the optic nerve. These cells become oriented transversely and synthesize extracellular matrix materials to form the lamina cribrosa.

Sellheyer K, Spitznas M. Development of the human sclera. A morphological study. *Graefes Arch Clin Exp Ophthalmol.* 1988;226:89–100.

Tripathi BJ, Tripathi RC. Development of the human eye. In: Bron AJ, Tripathi RC, Tripathi BJ, eds. *Wolff's Anatomy of the Eye and Orbit.* 8th ed. London: Chapman and Hall; 1997.

Tripathi BJ, Tripathi RC, Wisdom JE. Embryology of the anterior segment of the human eye. In: Ritch R, Shields MB, Krupin T, eds. *The Glaucomas.* 2nd ed. St Louis: Mosby; 1996:1.

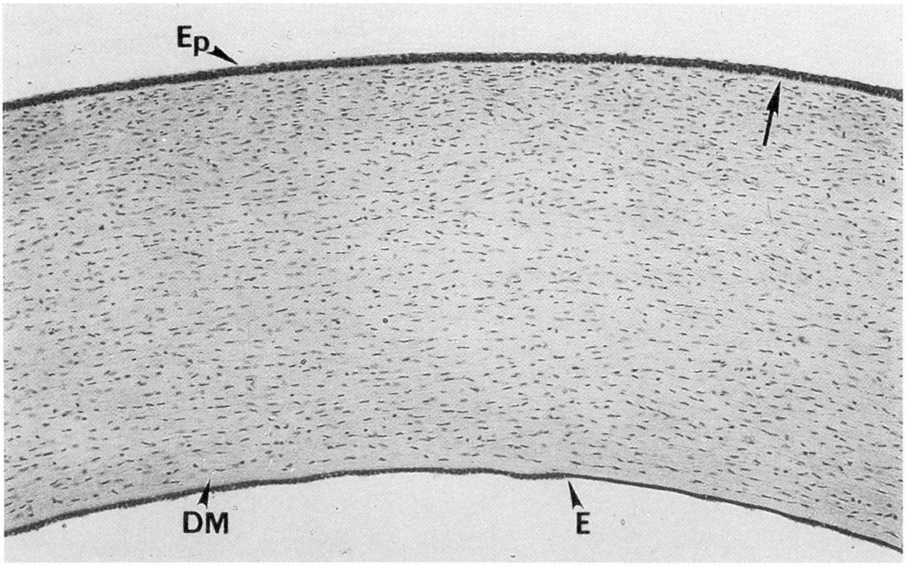

FIG V-15—Light micrograph of central cornea in a 6-month fetus. Epithelium (*Ep*) consists of two to three cell layers. Stroma contains numerous keratocytes, and there is an indistinct Bowman's zone (*arrow*). Descemet's membrane (*DM*) is clearly demarcated. *E,* endothelium. (Original magnification ×120.) (Reproduced with permission from Tripathi BJ, Tripathi RC, Wisdom J. Embryology of the anterior segment. In: Ritch R, Shields MB, Krupin T, eds. *The Glaucomas.* 2nd ed. St Louis: Mosby; 1996.)

Anterior Chamber, Angle, Iris, and Ciliary Body

The anterior chamber is first recognizable as the slitlike space that results after the ingrowth of the first wave of mesenchymal cells and the posterior extension of the second wave. By approximately 7 weeks' gestation, the angle of the anterior chamber is occupied by a nest of loosely organized mesenchymal cells of neural crest origin. These cells will develop into the *trabecular meshwork.* At the posterior aspect of the angle, mesodermal cells are developing into the vascular channels of the pupillary membrane. Loosely organized mesenchymal cells and the pigment epithelium of the forward-growing optic cup are also present in this region.

Anteriorly, cells that resemble the corneal endothelium form a layer that extends to the angle recess; these cells meet the anterior surface of the developing iris, thus demarcating the angle of the anterior chamber by week 15 of gestation (Fig V-16). Beginning at the third month of gestation and continuing for a considerable time after birth (up to the age of 4 years), the angle recess progressively deepens. It also appears to be repositioned posteriorly because of the differential growth rate of adjacent tissues.

Initially, no demarcation exists between the mesenchymal cells that will form the trabecular meshwork and those that will differentiate into the ciliary muscle. The extracellular matrix of the trabecular beams is synthesized and deposited by the differentiating trabecular cells beginning at week 15 and continuing up to the eighth

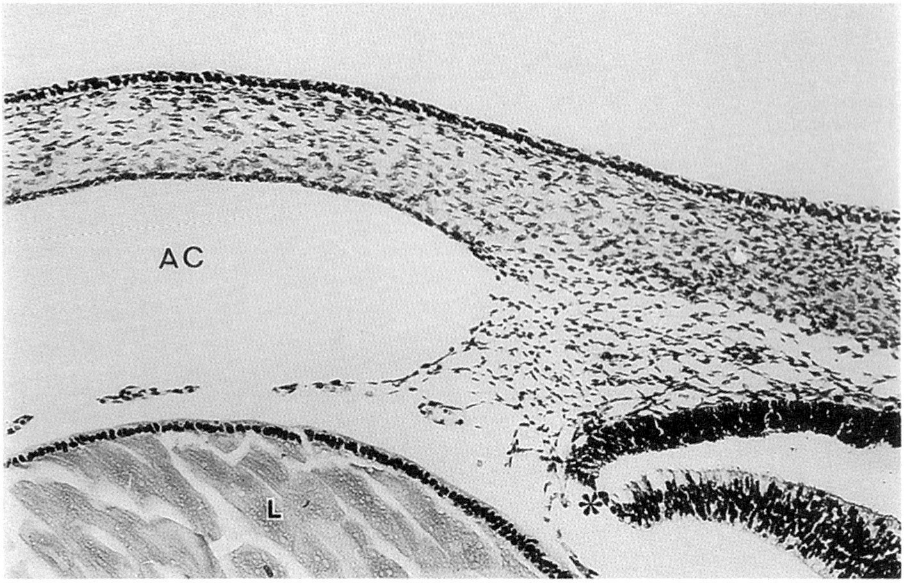

FIG V-16—Light micrograph of the eye of an 11-week fetus in meridional section. The angular region is poorly defined at this stage and is occupied by loosely arranged, spindle-shaped cells. Schlemm's canal is unrecognizable, and ciliary muscles and ciliary processes are not yet formed; the latter are derived from neural ectodermal fold (*asterisk*). Corneal endothelium appears continuous with cellular covering of primitive iris. *AC,* anterior chamber. *L,* lens. (Original magnification ×230.) (Reproduced with permission from Tripathi RC, Tripathi BC. Functional anatomy of the anterior chamber angle. In: Tasman W, Jaeger EA, eds. *Duane's Foundations of Clinical Ophthalmology.* Philadelphia: Lippincott; 1991.)

month of gestation. Even as early as 12–14 weeks' gestation, the cellular layer that lines the trabecular meshwork on its anterior chamber aspect is perforated by gaps of 2–8 μm in diameter. As development proceeds, these gaps become larger, and eventually the open spaces of the meshwork directly communicate with the anterior chamber.

Schlemm's canal develops from a small plexus of venous canaliculi by the end of the third month of gestation. Derived from mesodermal mesenchyme, these channels function initially as blood vessels. Other mesenchymal cells surround the canal during the fourth month of gestation. These cells and their secreted extracellular matrix materials will form the juxtacanalicular tissue. Characteristic vacuolar configurations begin to appear in the endothelial cells that line Schlemm's canal at about the beginning of the fifth month. Their development corresponds to the onset of aqueous humor circulation. The canal begins to function as an aqueous sinus rather than as a blood vessel.

Differentiation of the ciliary epithelium occurs in the two layers of neuroectoderm just behind the advancing optic cup. Late in the third month, longitudinal indentations appear in the outer pigmented layer. Between the third and fourth months, the inner nonpigmented layer starts to follow the contour and adhere to the pigmented layer. These radial folds, approximately 75 in number, are the beginning of the *ciliary processes.*

At week 10 of gestation, precursor *ciliary muscle* cells are identified as an accumulation of mesenchymal cells between the primitive ciliary epithelium and the anterior sclera condensation at the margin of the optic cup. Differentiation, which begins in the outermost (sclerad) cells during week 12, is evident from the myofilaments that surround plaques of dense bodies along the plasmalemma. The meridional portion of the muscle becomes organized during the fifth month, followed by the circular and radial components. The circular muscle continues to develop for at least 1 year after birth.

The development of the *iris* is associated with the formation of the anterior portion of the tunica vasculosa lentis. At about the sixth week of gestation, vascular channels of this embryonic structure are present as blind outgrowths from the annular vessel that encircles the rim of the optic cup. The developing vessels extend into the mesenchymal cells that cover the anterior lens surface and will ultimately be incorporated into the iris stroma. The most anterior region of the tunica vasculosa lentis is replaced subsequently by the pupillary membrane. At the end of the third month of gestation, after the future ciliary processes have formed, both walls of the optic cup (at its margin) grow forward beneath the pupillary membrane and mesenchymal cells. The mesenchymal tissue of the iris differentiates earlier than does the neuroectoderm. Cells in the developing stroma become fibroblast-like and secrete collagen fibrils and other components of the extracellular matrix.

The earliest differentiation of the *sphincter muscle* from the anterior layer of epithelium (the forward extension of the RPE) occurs at 3 months of gestation. However, myofibrils are not synthesized until the fifth month, and the muscle does not come to lie free in the stroma until the eighth month of gestation. The dilator muscle is not apparent until the sixth month, and differentiation of the myoepithelial cells continues after birth.

Pigmentation of the posterior epithelial layer of the iris, which is a continuation of the nonpigmented layer of the ciliary body and hence of the neurosensory retina, commences at the pupillary margin at midterm and proceeds toward the periphery. It ceases at the iris root by the end of the seventh month.

The pupillary portion of the tunica vasculosa lentis is resorbed during the sixth month of gestation. The remains of an incomplete arteriovenous anastomosis at the ciliary end of the sphincter muscle demarcates the collarette. The pupillary membrane atrophies near term.

The iris is still immature at birth. Much of the extracellular matrix is yet to be laid down in the stroma. The collarette is closer to the pupil in the newborn than it is in the adult eye.

McMenamin PG. A quantitative study of the prenatal development of the aqueous outflow system in the human eye. *Exp Eye Res.* 1991;53:507–517.

Reme C, d'Epinay SL. Periods of development of the normal human chamber angle. *Doc Ophthalmol.* 1981;51:241–268.

Sellheyer K, Spitznas M. Differentiation of the ciliary muscle in the human embryo and fetus. *Graefes Arch Clin Exp Ophthalmol.* 1988;226:281–287.

Strek W, Strek P, Nowogrodzka-Zagorska M, et al. Hyaloid vessels of the human fetal eye. A scanning electron microscopic study of corrosion casts. *Arch Ophthalmol.* 1993;111:1573–1577.

Tripathi BJ, Tripathi RC, Wisdom JE. Embryology of the anterior segment of the human eye. In: Ritch R, Shields MB, Krupin T, eds. *The Glaucomas.* 2nd ed. St Louis: Mosby; 1996:1.

Vascular System

The development of the vascular system of the eye and orbit is complex. Many vessels are transitory, arising and regressing in response to the changing needs of the embryonic eye. Vascular channels from the internal carotid artery develop in the mesenchyme around the optic vesicle late in the fourth week. Primitive dorsal and ventral ophthalmic arteries bud inward from the carotid and join a loose reticulum of capillaries around the optic vesicle. The system is drained into the future cavernous sinuses by way of plexuses. The early vessels are primarily ocular. A transient vessel, the stapedial artery, arises from the carotid to supply the expanding orbit. Later, the distal part of the stapedial artery is annexed to the ophthalmic artery.

The hyaloid artery is a branch of the primitive dorsal ophthalmic artery that arises at the juncture of the optic stalk and the optic cup at the time of closure of the embryonic fissure. The annular vessel that develops at the rim of the optic cup is supplied by the dorsal and ventral arteries. When incorporated in the optic cup, the hyaloid system extends toward and around the lens to join the annular vessel. Together with the tunica vasculosa lentis, the hyaloid system nourishes the interior of the developing eye.

The primitive dorsal ophthalmic artery becomes the definitive ophthalmic artery of the orbit at the sixth week of gestation. It supplies the temporal long posterior ciliary artery, the short posterior ciliary arteries, and the central retinal artery. The primitive ventral ophthalmic artery almost disappears; only a portion remains as the long posterior nasal ciliary artery.

The major arterial circle of the iris develops in the mesenchyme that surrounds the optic cup. It is located slightly lateral and anterior to the annular vessel and is formed by a coalescence of branches from the long ciliary arteries. Vascular twigs with little connective tissue grow from both the annular vessel and the major arterial circle to form the pupillary membrane, a system of radial vascular loops over the surface of the iris and lens. The pupillary arcades disappear centrally but remain peripherally as the minor circle of the iris. They provide vessels of the mature iris. Because the tissues that demarcate the anterior chamber angle are repositioned during development, the major arterial circle of the iris is ultimately located in the ciliary body.

At the fourth month of gestation, spindle-shaped mesenchymal cells arise from the hyaloid artery at the optic disc. These cells infiltrate the inner layers of the retina as solid cords of undifferentiated cells. Lumina develop, initially as slitlike openings behind the advancing edge of the invading mesenchymal cells. Retinal vascularization proceeds centripetally, and a boundary zone consisting of undifferentiated cells distinguishes the avascular and vascular retina. Endothelial cells differentiate first; adjacent cells are joined by zonulae occludens and gap junctions.

Vascularization of the nasal retina is complete before that of the temporal retina because of the shorter distance from the optic disc to the nasal ora serrata. By the fifth month, patent vessels have extended superiorly and inferiorly on the temporal aspect of the retina, sparing the region of the putative macula. Small blood vessels begin to develop in the ganglion cell layer of the foveal slope at the sixth month. The adult pattern of arterioles, veins, and capillaries is established through a process of remodeling and retraction of the primitive capillary network. Although capillaries

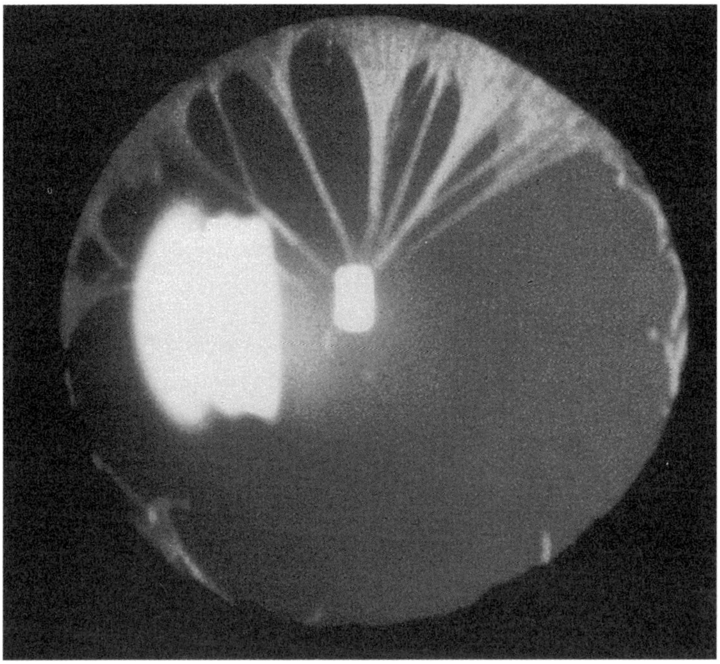

FIG V-17—Persistent pupillary membrane.

reach the ora serrata by the eighth month, the mature pattern of vascularization is not achieved until 3 months after birth.

The hyaloid system and the tunica vasculosa lentis atrophy in the third trimester. Occasionally, either system may persist after birth (Fig V-17).

Penfold PL, Provis JM, Madigan MC, et al. Angiogenesis in normal human retinal development: the involvement of astrocytes and macrophages. *Graefes Arch Clin Exp Ophthalmol.* 1990;228:255–263.

Periocular Tissues and Eyelids

The frontonasal and maxillary processes of neural crest cells occupy the space that surrounds the optic cups by the fourth week of gestation. The bones, cartilage, fat, and connective tissues of the orbit develop from these cells. All bones of the orbit are membranous except the sphenoid, which is initially cartilaginous. Ossification begins during the third month of gestation, and fusion occurs between the sixth and seventh months.

The extraocular muscles arise from myotomic cells of the pre-otic mesodermal somites that have shifted cranially. These cells become located within the neural crest mesenchyme that is situated on the dorsal and caudal aspects of the developing eye. Although the extraocular muscles were once thought to begin developing at

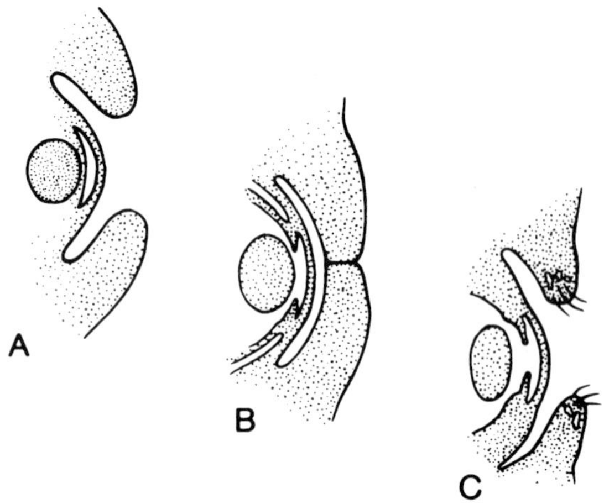

FIG V-18—Development of the eyelids. **A**, Seventh week: upper and lower eyelid folds grow over the eye. **B**, Eyelids fuse during the eighth week. Fusion starts along the nasal margin. **C**, As cilia and glandular structures develop, eyelids gradually open from the fifth to the seventh month.

the primitive muscle cone that surrounds the optic nerve in the fifth week of gestation, recent evidence suggests that the muscles arise in situ. Myoblasts with myofibrils and immature Z bands are distinguishable by the fifth week of gestation. At about 7 weeks, the dorsomedial aspect of the superior rectus muscle gives rise to the levator muscle, which grows laterally and over the superior rectus toward the eyelid. The tendons of the extraocular muscles fuse with the sclera in the vicinity of the equator late in the third month.

The upper eyelid first develops as a proliferation of surface ectoderm in the region of the future outer canthus at 4–5 weeks of gestation (Fig V-18). During the second month, both the upper and the lower eyelids are discernible as undifferentiated skin folds that surround mesenchyme of neural crest origin. Later, mesodermal mesenchyme infiltrates the eyelids and differentiates into the palpebral musculature. The eyelid folds grow toward each other as well as laterally. Starting near the inner canthus, the margins of the folds fuse at approximately 10 weeks' gestation. As the folds adhere to each other, evolution of cilia and glands continues. The orbicularis muscle condenses in the fold in week 12. The eyelid adhesions gradually break down late in the fifth month, coincident with the secretion of sebum from the sebaceous glands and cornification of the surface epithelium.

The lacrimal gland begins to develop between the sixth and seventh weeks of gestation. Solid cords of epithelial cells proliferate from the basal cell layer of the conjunctiva in the temporal region of the fornix. Neural crest–derived mesenchymal cells aggregate at the tips of the cords and differentiate into acini. Ducts of the gland are formed at approximately 3 months by vacuolation of the cord cells and the development of lumina. Lacrimal gland (reflex) tear production does not begin until 20 or more days after birth. Hence, newborn infants cry without tears.

Between the third and sixth months of gestation, eyelid appendages and pilosebaceous units develop from invaginations of epithelial cells into the underlying mesenchyme.

Oguni M, Setogawa T, Matsui H, et al. Timing and sequence of the events in the development of extraocular muscles in staged human embryos: ultrastructural and histochemical study. *Acta Anat (Basel)*. 1992;143:195–198.

Sevel D. The origins and insertions of the extraocular muscles: development, histologic features, and clinical significance. *Trans Am Ophthalmol* Soc. 1986;84:488–526.

Tripathi BJ, Tripathi RC. Development of the human eye. In: Bron AJ, Tripathi RC, Tripathi BJ, eds. *Wolff's Anatomy of the Eye and Orbit.* 8th ed. London: Chapman and Hall; 1997.

Realignment of the Globe

Initially, the axes of the two optic cups and the optic stalks form an angle of 180°. At 3 months' gestation, this angle has decreased to 105°. With continued enlargement; remodeling; and repositioning of the head, face, and brain throughout gestation, the eyes become oriented in their anterior position. At birth, the axes form an angle of 71°. However, the adult orientation of 68° is not achieved until the age of 3 years.

CHAPTER VI

Congenital Anomalies

Congenital anomalies are defects present at birth. They result both from genetic influences and from a variety of local and systemic environmental effects. A *teratogen* is an agent that produces or increases the incidence of congenital malformation. Because exposure to a teratogen can occur at any stage of embryonic development, its effects differ according to the time, duration, and intensity of exposure.

A teratogen acting in the first trimester on primordial cells produces severe damage to the ocular primordium and its derivative tissues. Major ocular developmental disasters involve the entire globe and are often associated with defects in the orbital, cerebral, and facial tissues, although defects may be limited to one or more tissues in the eye (Fig VI-1). Because cellular development continues after organogenesis, genetic influences or environmental effects at a later time may lead to dysfunction without gross structural abnormalities.

Genetic Influences

Microphthalmos and coloboma formation are associated with many different chromosomal anomalies, which suggests that the same morphologic defect can be caused by different genetic influences occurring at different stages in the evolution of neural crest cells.

Warburg M, Friedrich U. Coloboma and microphthalmos in chromosomal aberrations. Chromosomal aberrations and neural crest cell developmental field. *Ophthalmic Pediatr Genet.* 1987;8:105–118.

Mutations in homeobox genes are known to produce congenital ocular abnormalities. Often, the mutation results from the deletion or insertion of a single nucleotide that causes a frameshift in the coding region. Several congenital ocular anomalies have been matched to specific mutations—for example, aniridia, posterior embryotoxon, Peters anomaly, Axenfeld anomaly, and congenital cataract are produced by mutations in *PAX6*; oculorenal syndrome and coloboma of the optic nerve by *PAX2* mutations; and cyclopia by Sonic hedgehog mutations. A mutation in different homeobox genes may produce the same clinical manifestations (eg, *RIEG1* mutation also results in Peters anomaly), or a different mutation in the same gene may produce a similar but distinct phenotype (eg, *RIEG1* mutation causes Rieger anomaly). Future investigations will identify additional roles for homeobox genes in normal development and their aberrant expression in abnormal ocular development.

Sanyanusin F, Schimmenti LA, McNoe LA, et al. Mutation of the PAX2 gene in a family with optic nerve colobomas, renal anomalies and vesicoureteral reflux. *Nat Genet.* 1995;9:358–364.

Traboulsi EL. Developmental genes and ocular malformation syndromes. *Am J Ophthalmol.* 1993;115:105–107.

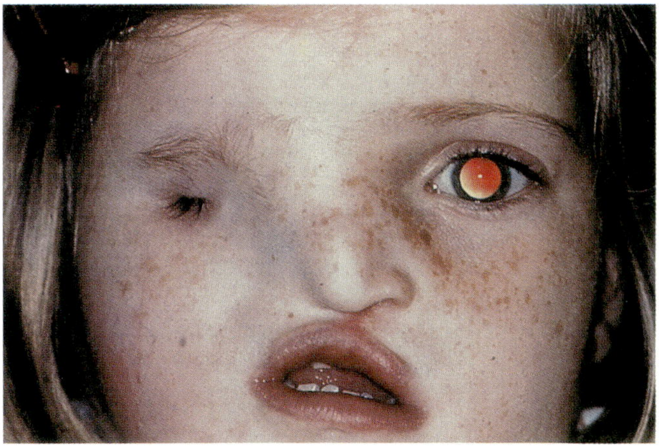

FIG VI-1—Microphthalmic eye with defects in eyelids, brow, nose, and mouth on same side as deformed eye.

Traboulsi EL. Ocular malformations and developmental genes. *J AAPOS.* 1998;2: 317–323.

Nongenetic Teratogens

Nongenetic teratogens include:

- Toxins
- Maternal infections
- Nutritional deficiencies
- Radiation
- Drugs
- Developmental failures
- Traumatic insult

Early studies showed that an excess or deficiency of vitamin A during development causes eye abnormalities such as coloboma and lens defects. However, it is now apparent that the acid form of this vitamin—retinoic acid—is critical not only to early ocular development but also to the induction of congenital anomalies. Exposure of the developing human embryo to excess amounts of retinoic acid causes malformation of the retina and optic nerve by affecting the expression of homeobox genes such as *PAX2, MSH-C,* and Sonic hedgehog.

Hyatt GA, Dowling JE. Retinoic acid. A key molecule for eye and photoreceptor development. *Invest Ophthalmol Vis Sci.* 1997;38:1471–1475.

Sulik KK, Dehart DB, Rogers JM, et al. Teratogenicity of low doses of all-trans retinoic acid in presomite mouse embryos. *Teratology.* 1995;51:398–403.

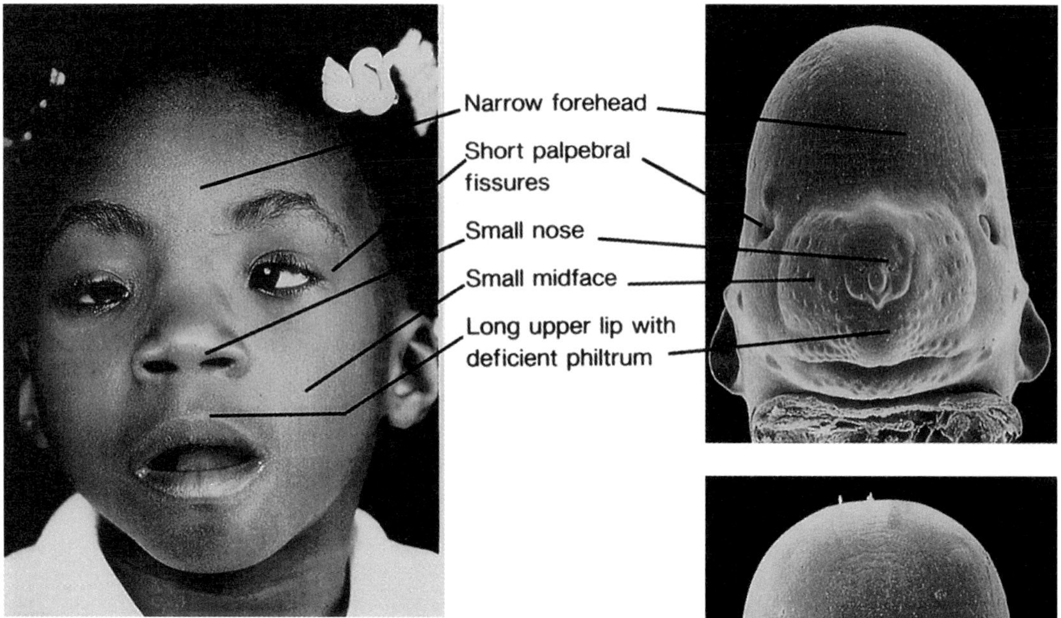

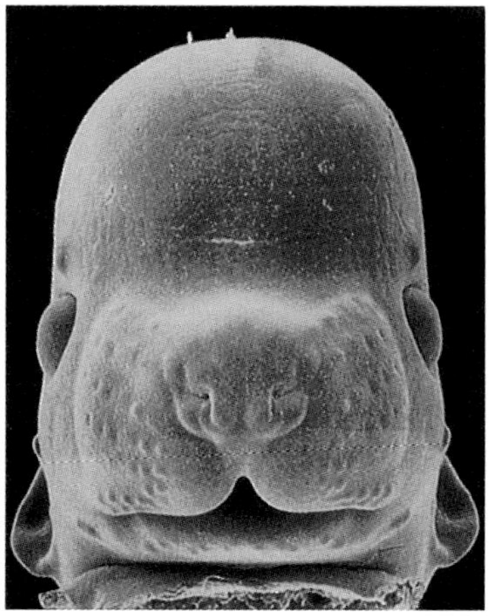

FIG VI-2—**Upper right**, A fetus removed from a mouse to which alcohol (ethanol) had been administered early in pregnancy (gastrulation stage) shows numerous facial characteristics similar to those of a child with fetal alcohol syndrome **(left)**. **Below**, A control mouse fetus. (Left photograph courtesy of Marilyn Miller, MD. Right photographs reproduced with permission from Kathleen Sulik, PhD. In: Serafin D, Georgiade NG. *Pediatric Plastic Surgery.* St Louis: Mosby; 1984.)

Prenatal exposure to alcohol results in a distinct pattern of delayed growth, mental retardation, and abnormal behavior patterns, as well as multiple congenital malformations that is recognized as fetal alcohol syndrome (FAS). Various ocular abnormalities include (Fig VI-2):

- Anomalies of the adnexa (strabismus, blepharoptosis, epicanthus)
- Intraocular defects (cataract, glaucoma, coloboma of uvea, persistent hyperplastic primary vitreous, dysmorphogenesis of the retina, and optic nerve hypoplasia)

The majority of children with FAS have optic nerve hypoplasia which, in experimental models, results from reduced densities of ganglion cells and their axons as well as damage to glial cells and myelin sheaths in the optic nerve.

The mouse model of FAS shows that the teratogenic effect of alcohol starts with the insult to the optic primordia during the gastrula stage. A small optic vesicle results in a deficient lens vesicle, which manifests as microphakia. Delay in lens

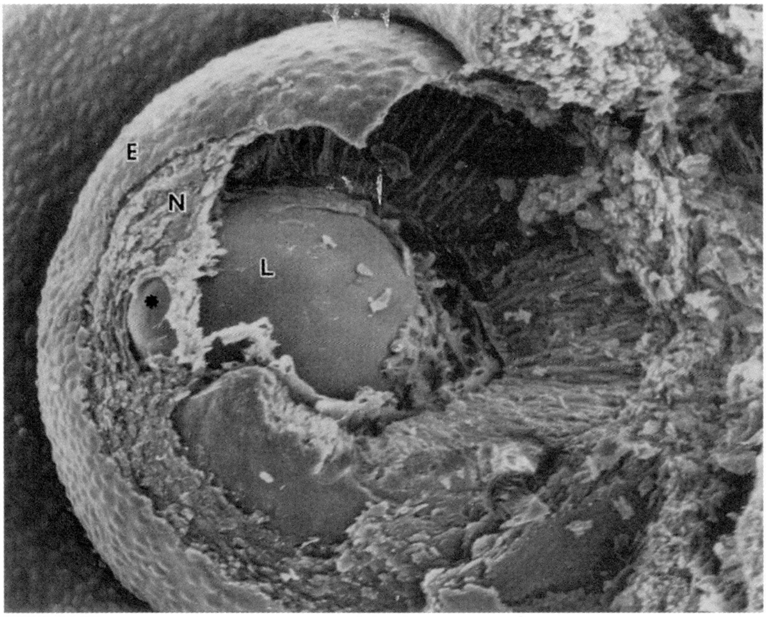

FIG VI-3—Eye of day-14 fetus exposed to ethanol on day 7 of gestation. Anterior lenticonus *(*)* as a result of delayed lens detachment prevents migration of neural crest (*N*) to form the axial corneal stroma and endothelium. *E,* surface ectoderm; *L,* lens. (Freeze fracture, scanning electron micrograph ×195.) (Reproduced with permission from Cook CS, Sulik KK. Keratolenticular dysgenesis [Peters anomaly] as a result of acute embryonic insult during gastrulation. *J Pediatr Ophthalmol Strabismus.* 1988;25:60–66.)

detachment from surface ectoderm leads to myriad anterior segment anomalies that resemble Peters and Axenfeld anomalies; such conditions are explained by impairment of the migration of the neural crest cells that should have formed the corneal stroma, endothelium, and iris (Figs VI-3, VI-4). Together, they cause microphthalmos with a secondary persistence of primary vitreous.

Persistence of the embryonic fissure, which leads to a failure in maintaining intraocular pressure, is one explanation for microphthalmos and coloboma formation. In the mouse model, these malformations appear to be primary events caused by faulty induction produced by the early insult of alcohol on the gastrula forebrain, which gives rise to the evaginating optic primordia.

Cook CS, Nowotny AZ, Sulik KK. Fetal alcohol syndrome. Eye malformations in a mouse model. *Arch Ophthalmol.* 1987;105:1576–1581.

Pinazo-Duran MD, Renau-Piqueras J, Guerri C, et al. Optic nerve hypoplasia in fetal alcohol syndrome: an update. *Eur J Ophthalmol.* 1997;7:262–270.

Stromland K, Hellstrom A. Fetal alcohol syndrome—an ophthalmological and socioeducational prospective study. *Pediatrics.* 1996;97(6 Part 1):845–850.

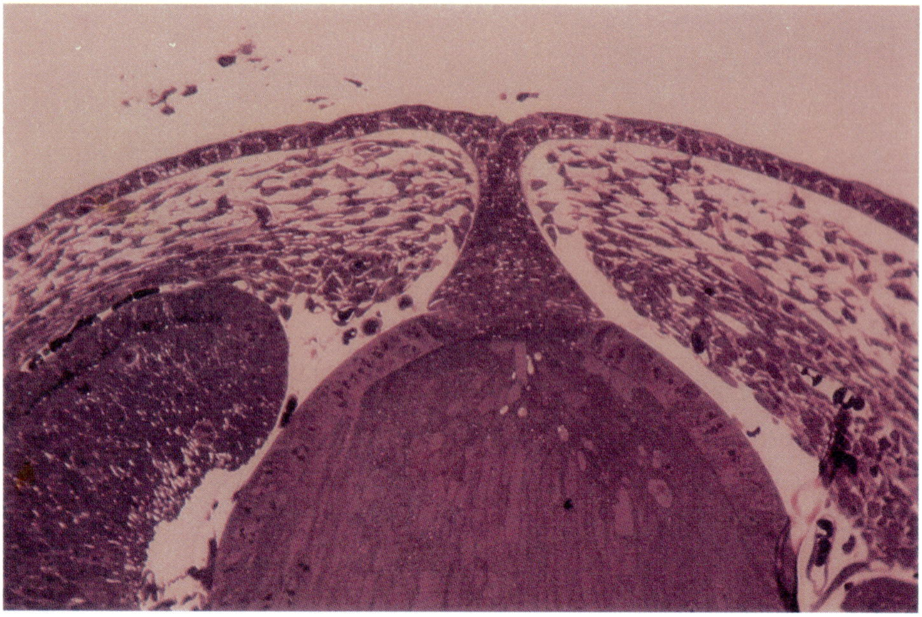

FIG VI-4—Histologic section of isotretinoin-exposed fetus. The lens stalk forms a barrier to neural crest migration (toluidine blue, ×100). (Reproduced with permission from Cook CS, Sulik KK. Keratolenticular dysgenesis [Peters anomaly] as a result of acute embryonic insult during gastrulation. *J Pediatr Ophthalmol Strabismus.* 1988;25:60–66.)

Anophthalmos, Microphthalmos, and Nanophthalmos

Anophthalmos is the total absence of ocular tissues (Figs VI-5 through VI-7). *Microphthalmos* is the presence of a small, often disorganized globe. True anophthalmos is extremely rare, and the diagnosis can be confirmed only when the orbital contents are examined histologically and no evidence of ocular tissues can be identified. Most instances of clinical anophthalmos are actually severe microphthalmos (Fig VI-8).

Primary anophthalmos can occur in an otherwise normal child. It is usually bilateral and isolated, implying complete agenesis of the primitive ocular anlage. *Secondary anophthalmos,* a result of complete suppression of the development of the forebrain, is lethal. *Consecutive anophthalmos* implies the initial development of an ocular structure that undergoes secondary degeneration or destruction.

Microphthalmos results from a variety of conditions, notably trisomy 13. Defects in the microphthalmic eye include the following:

- Corneal leukomas
- Immature chamber angles
- Colobomas with or without cysts
- Cataract
- Keratolenticular adhesions

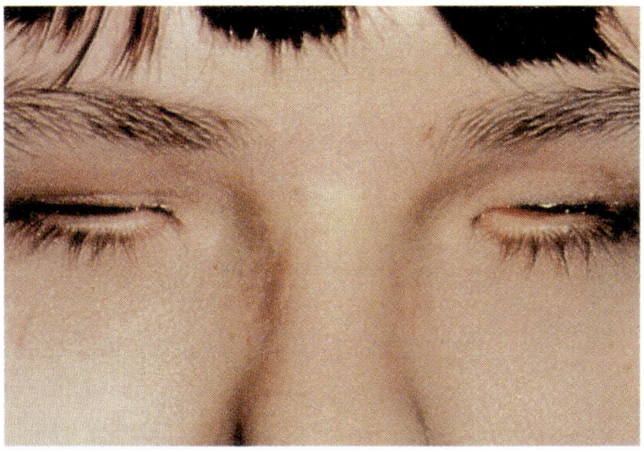

FIG VI-5—Bilateral clinical anophthalmos. Eyelids, brows, and orbits are well formed but small.

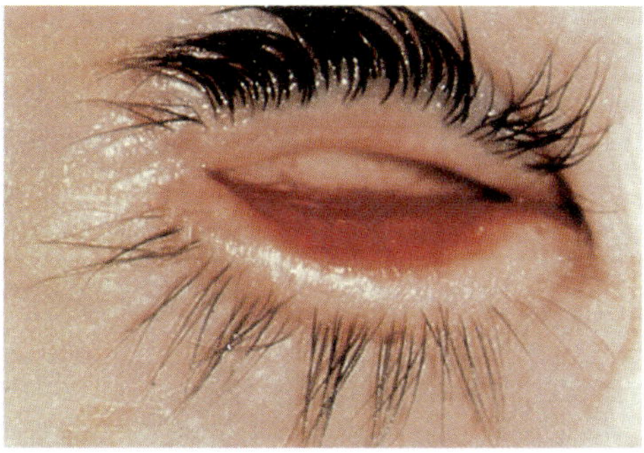

FIG VI-6—Partially opened small eyelids of anophthalmic orbit.

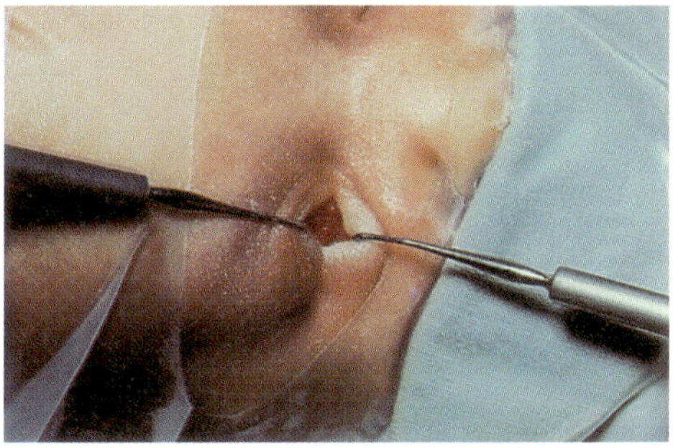

FIG VI-7—Anophthalmos. Palpebral fissure is fully opened, and no eye tissue is evident.

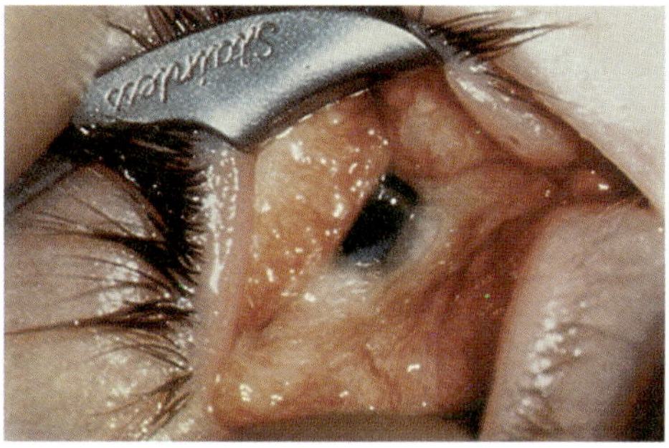

FIG VI-8—Severely microphthalmic eye is almost buried by redundant eyelid tissue.

- Persistent hyperplastic primary vitreous (persistent fetal vasculature)
- Retinal dysplasia
- Bizarre migrations of the pigment epithelium
- Hypoplasia of the optic nerve

Other defects may also be associated with microphthalmos.

The *nanophthalmic* globe is a small eye without major internal disorganization. Associated anomalies may include narrow palpebral fissures, deep-set eyes, high hyperopia, thickened sclera, and occasionally glaucoma. Nanophthalmos can be inherited as either an autosomal dominant or an autosomal recessive trait.

Cyclopia and Synophthalmia

Cyclopia is a lethal condition in which a single ocular structure is present in the upper, median portion of the face. It is associated with dramatic, symmetric deformities

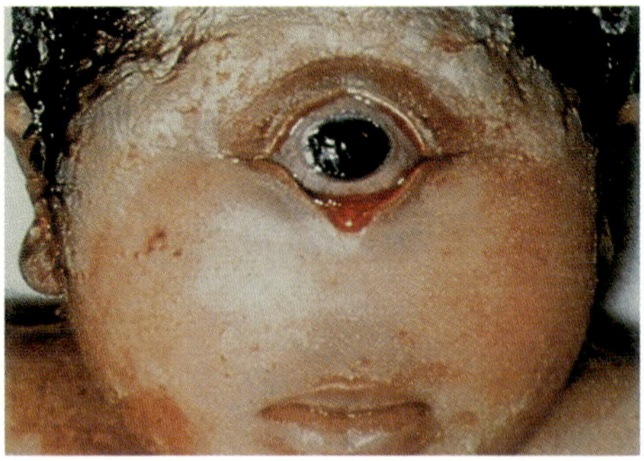

FIG VI-9—True cyclopia with large midline eye, hypoplastic philtrum. No proboscis. (Photograph courtesy of Ruiz and Biomedical Foundation.)

of the nose, skull, orbits, and brain (Fig VI-9). These include:

- Holoprosencephalic (nonhemispheric) brain
- Single median orbit
- Proboscis (primitive dysplastic nose) above the eye
- Absence of nasal bones and upper nasal passages
- Hypoplastic oral cavity
- General midline hypoplasia

The cyclopean eye consists of a single eyelid opening, beneath which there may be two small eyeballs that show partial fusion at the posterior portion within a single orbit. The eyes may be reasonably well formed, or they may consist of various rudimentary dysplastic ocular tissues. The single optic nerve, composed predominantly of glial cells, emerges from the midline of the posterior region of the fused globes.

A defect in normal midline patterning of tissues is responsible for the fusion of the two eyes in a single orbit. Recent evidence indicates that failure to develop midline structures of the neural plate results from either genetic or environmental mutations in the homeobox gene Sonic hedgehog. The transient loss in signaling of this gene probably involves, at least in part, members of the TGF-β family of growth factors, and it inhibits development of the frontonasal bud primordia from local paraxial mesoderm (which lies juxtaposed to the caudomedial surface of the optic cup) and the neural crest cells.

Hu D, Helms JA. The role of sonic hedgehog in normal and abnormal craniofacial morphogenesis. *Development.* 1999;126:4873–4884.

Kakita A, Wakabayashi K, Sekizuka N, et al. Cyclopia: histogenesis of the single optic nerve. *Acta Neuropathol (Berl).* 1997;94:509–513.

Synophthalmia is a less severe condition, but it is also lethal. In this condition, portions of two incomplete, usually symmetrical globes are joined in the midline as a single median ocular structure. Most infants born with cyclopia or synophthalmia do not show other somatic defects or chromosomal abnormalities.

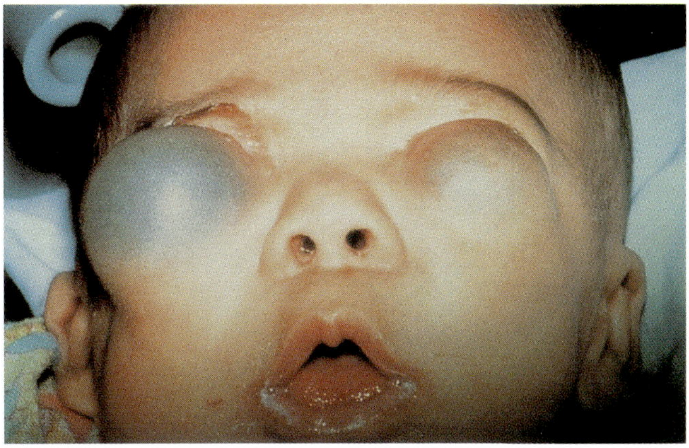

FIG VI-10—Orbital cysts replacing both globes.

Orbital Cyst and Cryptophthalmos

Several conditions may appear as a cystic bulge within the palpebral fissure when the eyelids are separated (Fig VI-10). They include congenital cystic eye, an eye with a cystic coloboma, a teratoma with cystic changes, orbital encephalocele, and congenital rhabdomyosarcoma. Rhabdomyosarcoma is discussed in BCSC Section 4, *Ophthalmic Pathology and Intraocular Tumors*; Section 6, *Pediatric Ophthalmology and Strabismus*; and Section 7, *Orbit, Eyelids, and Lacrimal System*.

In the true cystic eye, the optic vesicles remain in an embryonic state. This extremely rare condition results from failure of invagination of the lens placode and optic vesicle. Differentiation of all ocular tissues is limited. Clinically smooth, undivided skin without eyelid differentiation may cover the bulging structure and appear as cryptophthalmos.

A *cystic coloboma* results from faulty closure of the fetal fissure. Ectropion of the inner layer of the optic cup occurs as the fissure closes. The two layers of the cup are not perfectly apposed to each other, leaving a remnant of the primary optic vesicle. With closure of the fissure, the remaining vesicular space expands, developing a cystlike structure external to the globe. The eye may be smaller than the cystic coloboma.

Orbital teratomas are rare and contain tissues representative of all primitive germ layers. The glandular tissues produce secretions in closed sacs, leading to the formation of either a single cyst or multiple cystic spaces.

Encephaloceles are soft and pulsatile. Anterior orbital encephaloceles occur at the inner angle of the orbit, with cerebral tissues bulging through any of the sutures involving the frontal, ethmoidal, lacrimal, or maxillary bones. The eye may be displaced laterally or downward. Anterior encephaloceles are usually evident at birth. Posterior encephaloceles are less obvious. Although the bony defect in the posterior orbit is present at birth, proptosis of the globe may evolve slowly over months or years, and this congenital defect may not be diagnosed until clinical symptoms present. Radiographs, CT, and MRI may be diagnostic. A defect in the sphenoidal bone may result in proptosis from a posterior encephalocele in neurofibromatosis.

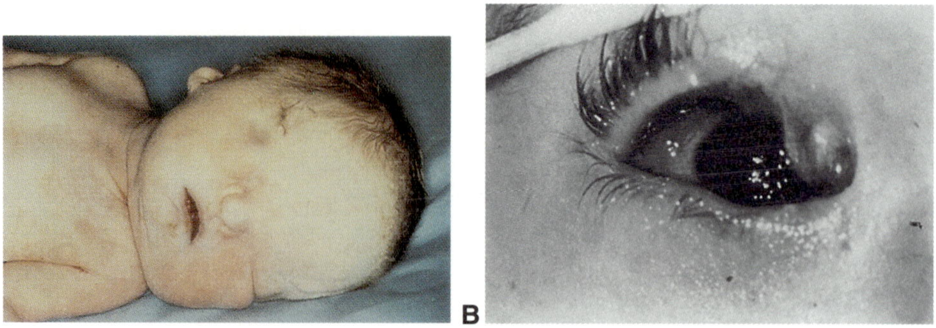

A B

FIG VI-11—**A**, Complete cryptophthalmos, both eyes. **B**, Incomplete cryptophthalmos of right eye with eyelid fused to cornea superonasally.

Cryptophthalmos, the hidden or buried eye, is a failure of normal eyelid formation in which folds do not form and the skin forms a single continuous layer over the anterior surface of the eye, extending from brow to cheek (Fig VI-11). A small linear dimpling may represent an aborted eyelid fold. Unlike cystic eyes, the lens in cryptophthalmos has usually formed and sits in a highly anomalous anterior chamber without iris or ciliary body. There is no cornea. The inward migration of the first, second, and third wave of neural crest cells did not take place. The retina and optic nerve are usually normal.

In *limited cryptophthalmos,* bands of skin remain adherent to the eye, usually superiorly, and eyelid formation is incomplete. Other defects associated with cryptophthalmos include absence of the eyebrows, microphthalmos, cleft palate, cleft lip, facial fissure, malformed ears, malformations of the genitourinary tract, and syndactyly of the fingers and toes.

The cosmetic appearance in each of these conditions is usually unsatisfactory. Echographic and CT scanning provide useful information, not only about the orbit but also about the intracranial variations. True diagnosis may be possible only with histologic examination of the excised structure. In some cases, a multidisciplinary surgical team, including a neurosurgeon, is needed to extirpate or repair these defects, especially in the case of encephaloceles.

Developmental Anomalies of the Cornea

The corneas of infants appear large relative to the rest of the face. At age 1 year, a cornea measuring less than 9 mm horizontally is too small *(microcornea)* and one more than 12 mm is too large *(megalocornea).* Microcornea (Fig VI-12) is common with microphthalmos. In *sclerocornea,* the scleral rim extends anteriorly to replace the peripheral cornea, leaving 3–8 mm of clear cornea centrally. Because of the opacity, the cornea appears small but may be of normal dimensions.

In megalocornea (Fig VI-13), the diameter of the cornea may be up to 18 mm. It must be differentiated from the enlarged cornea of congenital glaucoma.

Leukomas are discrete, white, opacified areas in the central or peripheral cornea. Opacification varies and sometimes occurs only at the edge of a defect. The

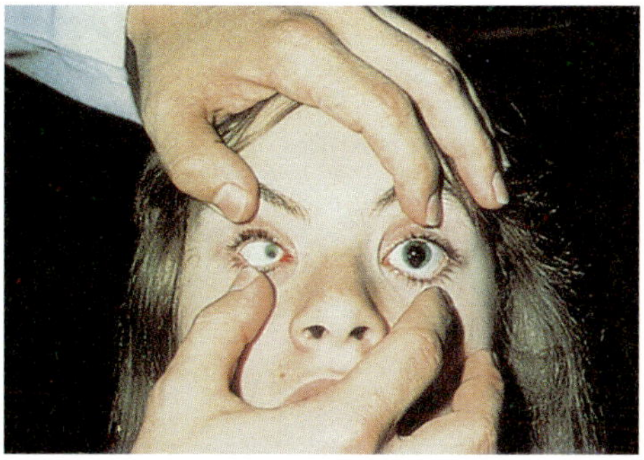

FIG VI-12—Severe microcornea and microphthalmos OD. Both irides are colobomatous.

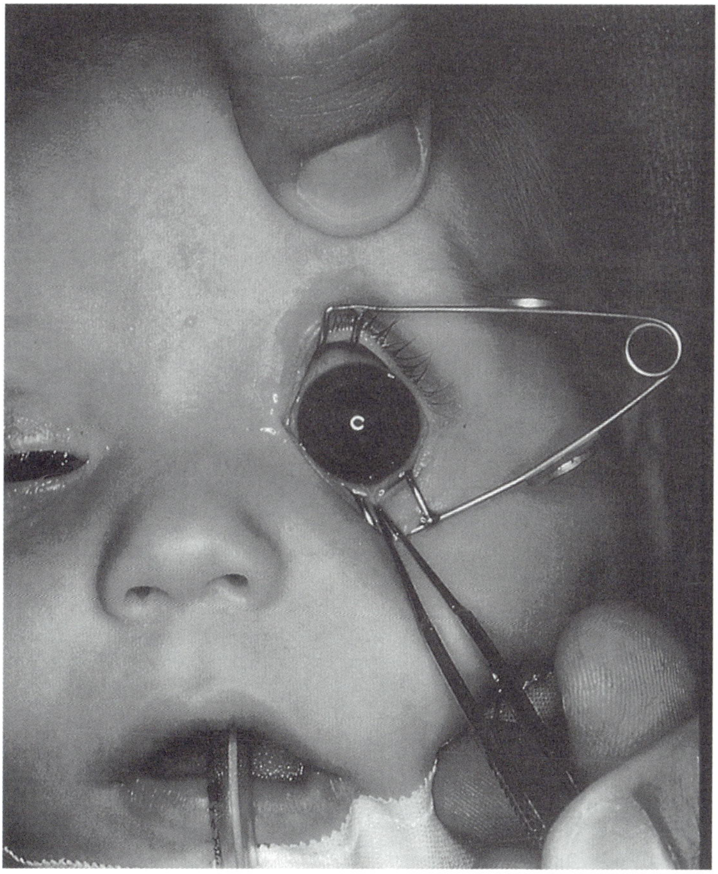

FIG VI-13—Megalocornea.

opacities may be round, oval, arcuate, or irregular. Congenital leukomas may be associated with other anomalies of anterior segment dysgenesis that once were (mistakenly) called *mesodermal dysgenesis.* These anomalies, including congenital glaucoma, are now known as *neurocristopathies.* They are thought to result from defective terminal induction or migration of tissues derived from neural crest cells that form the cornea and chamber angle. Synonyms used clinically can be helpful, however, as they describe specific clinical entities. These other terms include posterior embryotoxon, posterior keratoconus, and the anomalies of Rieger, Axenfeld, and Peters.

Bahn CF, Falls HF, Varley GA, et al. Classification of corneal endothelial disorders based on neural crest origin. *Ophthalmology.* 1984;91:558–563.

The usual dermoid or dermolipoma is found at the limbus and only rarely is entirely corneal. Dermoids may also be found in the conjunctiva and orbit. *Dermoids* are choristomas that are thought to represent arrests or inclusions of epidermal and connective tissues during closure of fetal clefts. The dermoid is covered by squamous epithelium and may contain hair follicles, sebaceous glands, nerve bundles, fat, or glandular tissue. *Dermolipomas* are usually solid, consisting entirely of fatty and fibrous tissue, and are generally located between the lateral and superior rectus muscles. Dermoids that cover the entire eye usually herald a poorly formed globe. Epibulbar dermoids are found in *Goldenhar syndrome (oculoauriculovertebral dysplasia),* one of the neural crest cell–derived, first brachial arch, facial midline–clefting anomalies.

Congenital Anterior Chamber Anomalies

Anomalies of the normal anatomy of the anterior chamber result from inadequate regression of tissues or failure of cell differentiation. *Posterior embryotoxon,* the most common and benign of these anomalies, is an exaggerated, prominent, and centrally displaced Schwalbe's line (Fig VI-14), present in 3%–15% of normal eyes. Commonly inherited as an autosomal dominant trait, it may be noticed on biomicroscopic examination but is best seen gonioscopically.

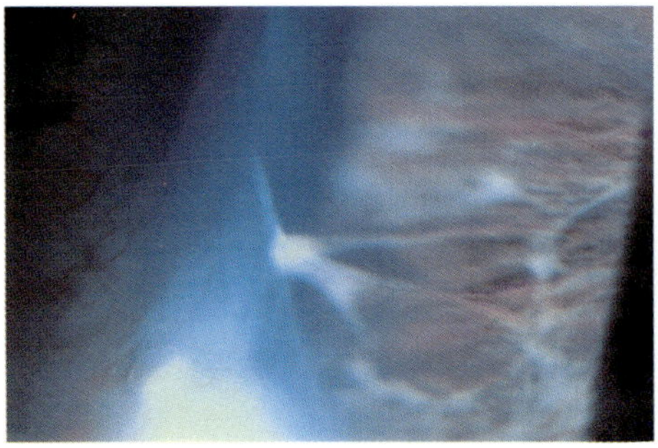

FIG VI-14—Posterior embryotoxon on biomicroscopic examination.

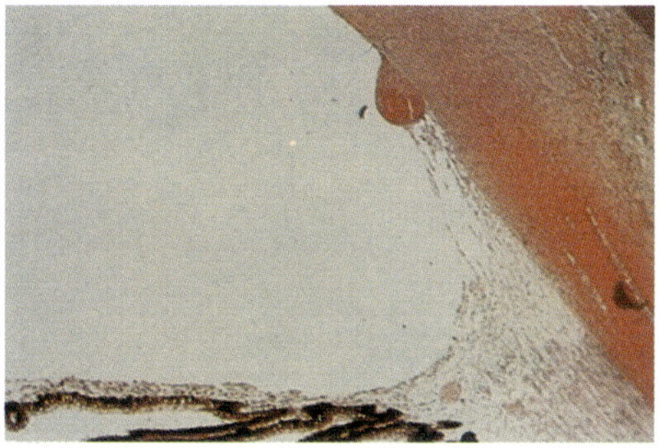

FIG VI-15—Axenfeld anomaly with prominent Schwalbe's line and iris strands.

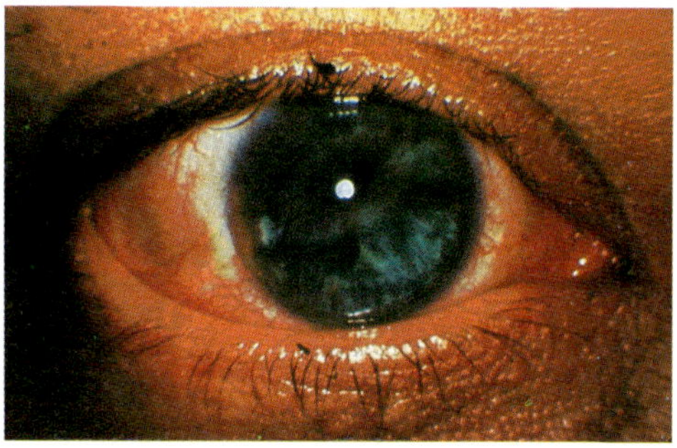

FIG VI-16—Rieger anomaly with stromal hypoplasia, visible sphincter muscle, oval pupil, and prominent Schwalbe's line.

Axenfeld-Rieger syndrome has been introduced recently as the term for a spectrum of congenital anterior segment defects previously called *mesodermal dysgenesis* and a variety of eponyms. All the conditions listed in the following two paragraphs are now subsumed under the single name Axenfeld-Rieger syndrome.

Axenfeld anomaly includes both posterior embryotoxon and iris processes that cross the angle and insert into Schwalbe's line (Fig VI-15). *Rieger anomaly* includes posterior embryotoxon, iris strands, and hypoplasia of the iris stroma (Fig VI-16).

In *Rieger syndrome*, anomalies of the face (maxillary hypoplasia, hypertelorism, hypodontia, oligodontia, or anodontia) and umbilicus are added to the anomalies of the anterior ocular segment. Strabismus, macrocornea, microcornea, cataracts, corectopia, and pseudopolycoria have been recorded. Rieger syndrome is inherited as an autosomal dominant trait with a high degree of penetrance.

Peters anomaly (Fig VI-17) includes a central defect in Descemet's membrane and absence of endothelium. Often, iris strands are attached to the edges of the defect. Clinically, the defective central cornea is opacified. The lens may or may not be cataractous, and it may adhere to the defect in Descemet's membrane by a keratolenticular stalk. The condition may be bilateral or unilateral and can be dominantly inherited. An isolated presentation, often accompanied by chromosomal and other systemic anomalies, is common. Experimental studies have revealed that exposure of the embryo to ethanol or isotretinoin can cause structural abnormalities suggestive of Peters anomaly.

Cook CS, Sulik KK. Laminin and fibronectin in retinoid-induced keratolenticular dysgenesis. *Invest Ophthalmol Vis Sci.* 1990;31:751–757.

Posterior keratoconus is a localized or diffuse, central or paracentral, crater-like depression in the posterior cornea associated with a deep stromal opacity. Unlike the case with Peters anomaly, both the endothelium and Descemet's membrane are present, and no other ocular abnormality is apparent. Because the anterior corneal curvature is unaffected, good vision is common. The term *internal ulcer of von Hippel* signifies a secondary infectious or inflammatory condition, but some cases may actually fall within the spectrum of Peters anomaly and posterior keratoconus.

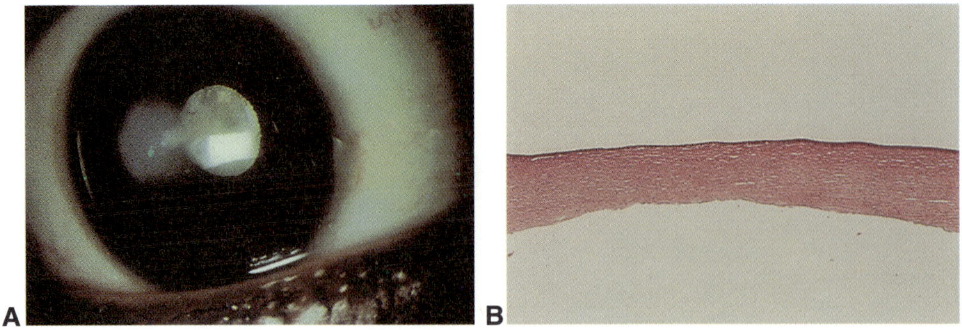

FIG VI-17—**A**, Clinical appearance of anterior segment in Peters anomaly. The lens is connected to the central posterior region of the cornea by a keratolenticular stalk. **B**, Histologic section of cornea in Peters anomaly. The central cornea is thinned and lacks Descemet's membrane and endothelium. Bowman's zone of the anterior stroma in this region is degenerate. (Original magnification ×10.) (Reproduced from Tripathi RC, Tripathi BJ, Gaster RN. Clinicopathologic study of Peters' anomaly. In: Trevor-Roper P, ed. *The Cornea in Health and Disease.* The Royal Society of Medicine International Congress Symposium Series no. 40. London: Academic Press; 1981.)

Abnormalities of the Iris

Anisocoria, a difference in the size of the two pupils of 0.5–2.0 mm, occurs in 2% of the normal population. It may be transmitted as an autosomal dominant trait with variable expressivity. *Polycoria,* the presence of many openings in the iris, may result from local hypoplasia or hyperplasia of the stroma and pigment epithelium. *Ectopia* (or *displaced pupil*) and *corectopia* (or *deformed pupil*) may be caused by unequal growth in parts of the iris or by defective contributions of neural crest cells to the stroma. Both pigmented and nonpigmented congenital cysts of the iris may be present, sometimes in the pupillary area and sometimes buried beneath the peripheral iris.

Aniridia implies total absence of the iris, but some iris tissue usually remains. Rudimentary iris stumps occur around the entire circumference of the base of the iris. Dilator and sphincter muscles are usually absent. Retroillumination shows the pupil extending almost to the periphery of the cornea. Aniridic infants are highly photophobic. Aniridia can be inherited as an autosomal dominant or recessive trait. Nonfamilial aniridia is associated with Wilms tumor, ambiguous genitalia and other genitourinary anomalies, and mental retardation. A small but variable interstitial deletion of the short arm of chromosome 11 (11p–) is usually present. Aniridia has never been reported in families with autosomal dominant Wilms tumor.

Colobomas

The term *coloboma* implies the absence of part or all of a tissue. It may result from developmental aberrations, surgery, or injury. Congenital colobomas of the eye are classified as typical or atypical varieties (Fig VI-18). Typical colobomas are caused by the failure of the embryonic fissure to close in the fifth week. Atypical colobomas appear in locations outside the region of the embryonic fissure. Colobomas may be partial or complete. When they occur in the iris, they have the appearance of an obliquely placed, inverted teardrop; the ballooned area is uppermost, and the

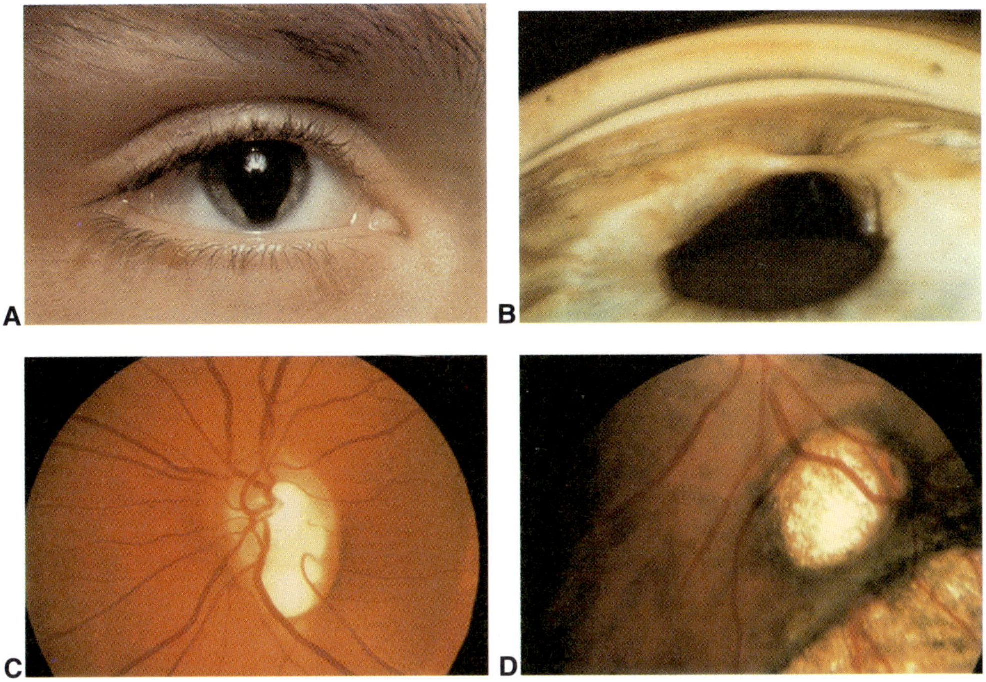

FIG VI-18—**A**, Iridic coloboma showing typical location with teardrop directed inferonasally. **B**, Gonioscopic view showing colobomatous dimple near iris root separated from main coloboma of iris by bridge of tissue. (Photograph courtesy of Alan R. Liss, Inc.) **C**, Small coloboma of optic disc inferonasally in left eye. **D**, Coloboma of fundus with skip areas. Note healthy tissue between optic disc and small colobomas and bridge of tissue above large inferior coloboma.

tapered end of the drop is directed inferonasally. *Keyhole pupil* is a common lay term.

Closure of the embryonic fissure starts at the equator of the globe and proceeds anteriorly and posteriorly. Areas of normal fusion may alternate with areas of defective closure, resulting in partial or incomplete colobomas. The coloboma may extend completely from the optic disc through the pupillary margin, or it may be represented by one or more defects along the normal fusion line. Occasionally, only one tissue is involved. Thus, colobomas of the optic disc may or may not be associated with iridic colobomas, and vice versa.

More than half of typical colobomas are bilateral (although often asymmetrical). Typical colobomas may be inherited as an autosomal dominant trait with incomplete penetrance and expressivity. Although colobomas are caused by defective fusion of the lips of the embryonic fissure, the failure of retinal pigment epithelium to develop results in hypoplasia of the underlying choroid and sclera (and overlying retina) and in ectasia or staphyloma of the globe. A coloboma of the retina results in an absolute scotoma corresponding to the defect. In cases with a large posterior pole coloboma, visual acuity may be subnormal, with secondary strabismus and nystagmus. Lens colobomas usually result from lack of zonular pull in that region of the lens because of hypoplasia of the corresponding ciliary body and zonular fibers.

Abnormalities of the Lens

Spherophakia is a spherical lens that is occasionally associated with other defects, as in Marfan syndrome. In later life, the lens often dislocates superiorly. In *microspherophakia,* which is associated with Weill-Marchesani syndrome, the lens is both small and spherical, and it may dislocate into the anterior chamber. The habitus of persons with these two syndromes is profoundly different: the patient with Marfan syndrome is tall and unusually lean with arachnodactyly; the patient with Weill-Marchesani syndrome is short (usually less than 5 feet tall), muscular, and stocky.

In Lowe syndrome, developmental delay, neuromuscular hypotonia, and aberrations of amino acid metabolism are associated with congenital cataracts and, frequently, glaucoma. The lenses are small plaques reduced in both the anteroposterior and the equatorial diameters. Lowe syndrome is found almost exclusively in males; the condition has a recessive X-linked pattern of inheritance and a characteristic lens opacity in carrier females.

A *cataract* is an opacity in the lens, whether or not visual impairment results. Congenital cataracts that impair vision are relatively uncommon, but many types of opacities in the newborn lens—including all manner of dots, clumps, lines, clefts, deposits, colors, membranes, and disks—have been described. Lenticonus is a central outpouching of the anterior or posterior surface of the lens. High myopia is characteristic.

Scattered opacities and various embryonic sutural cataracts are compatible with good visual acuity because only part of the lens is involved and may not require surgery. The potential for increased vision in an infant with congenital cataracts must be carefully assessed and weighed against the hazards of surgery and the problems of aphakia, including appropriate optical correction and orthoptic rehabilitation. With total bilateral congenital cataracts, the prognosis is worse because of the high incidence of other congenital anomalies, especially microcornea, poorly dilating pupils, and development of glaucoma. The management of cataract and aphakia in infants and children is discussed in BCSC Section 11, *Lens and Cataract*; BCSC Section 6, *Pediatric Ophthalmology and Strabismus*; and in Focal Points, *Childhood Cataracts*, vol. XIV, no. 1, 1996.

Cataracts that result from congenital rubella are dense, white, pearly opacities in the nucleus. The virus invades the developing lens and damages it. The virus becomes sequestered in the lens and may remain there for months after birth. If maternal infection occurred in the first trimester, the infant is more likely to be severely damaged than if the infection occurred later. Many organs in the body can be damaged by rubella. The classic triad for an infant infected with rubella in the first trimester is ocular (cataracts), auditory (deafness), and systemic (cardiac). Glaucoma occasionally occurs in a child with rubella but is usually not concomitant with rubella cataracts. A diffuse, generalized, irregular pigmentation called *salt-and-pepper fundus* may be the only residuum of a maternal rubella infection during the second trimester.

Abnormalities of the Retina

Folds, cysts, septa, and congenital nonattachment may be found in the retina. Dysplasia of the retina implies aberrant differentiation in any of the cellular layers and is often associated with the formation of *rosettes*—circular or oval groupings of dysplastic retinal cells that may be found in any retinal layer. In retinoblastoma, the rosettes are uniform in size in a given specimen; in dysplasia, the rosettes are highly variable in size and shape. Rosettes may be an incidental finding in eyes enucleated for unrelated conditions later in life, or they may replace the entire retina. In cases of severe dysplasia, total retinal detachment is often an associated finding. Profound retinal dysplasia, as in trisomy 13 and trisomy 18, may not be evident clinically, because the cornea and other ocular media may be too cloudy to allow ophthalmoscopic examination of the fundus. Abnormal electroretinogram results can reveal retinal dysplasia not detectable by clinical inspection.

Retinal Pigment Epithelium

Variations in the RPE may be evident at birth. Localized hypertrophy of the RPE is the most common. These flat, heavily pigmented spots are sharply delineated from the surrounding normal pigment epithelium. They occur as an isolated lesion but may be clustered throughout the entire posterior pole and are uncommonly bilateral. When they are bilateral, Gardner syndrome (a familial cancer syndrome) must be considered. Patches are smaller proximal to the disc and generally increase in size toward the periphery. Several patches grouped together are called *congenital grouped pigmentations,* or *bear tracks.*

Lewis RA, Crowder WE, Eierman LA, et al. The Gardner syndrome. Significance of ocular features. *Ophthalmology.* 1984;91:916–925.

The *macular coloboma* is so called because it appears as a punched-out macula lacking retina and choroid. Generally ascribed to congenital toxoplasmosis, it is bounded by dense irregular clumps of pigment, measuring 2–4 disc diameters and associated with temporal disc pallor. Such lesions are occasionally reported in siblings.

Abnormalities of the Optic Disc and Nerve

Aplasia of the optic nerve is extremely rare. Hypoplasia of the optic disc is more common (Figs VI-19, VI-20). The condition may be associated with microphthalmos and malformations of the forebrain. Bilateral optic nerve hypoplasia and absent septum pellucidum indicate de Morsier syndrome. Unilateral hypoplasia is common in normal persons.

Colobomas and optic pits are found in the optic nerve. The coloboma may be limited to the nerve or may be continuous with a total uveal coloboma. A pit is a deep, circular depression in the nerve head, usually located temporally, inferotemporally, or inferiorly. During organogenesis, a small pouch of retinal tissue, including glia, nerve fibers, and RPE, pouches downward between the sclera and the nerve under the dural sheath. The lamina cribrosa is defective, but the adjacent choroid and sclera are normal. The abnormal saccular protrusion is thought to allow direct continuity of the subdural space with the subretinal space. In later life, cerebrospinal fluid percolates through into the subretinal space as a serous detachment of the retina. Arcuate scotomata may be associated with optic pits.

Remnants of the hyaloid system may persist as white, fibrous, or glial strands, veils, or collarettes on the disc *(Bergmeister's papilla).* Remnants of hyaloid vessels may extend into the vitreous and are occasionally patent *(persistent posterior hyaloid artery)* (Figs VI-21, VI-22). *Mittendorf's dot,* a small white dot of tissue posterior to the lens on the nasal side, is a minute remnant of the anterior hyaloid artery (Fig VI-23).

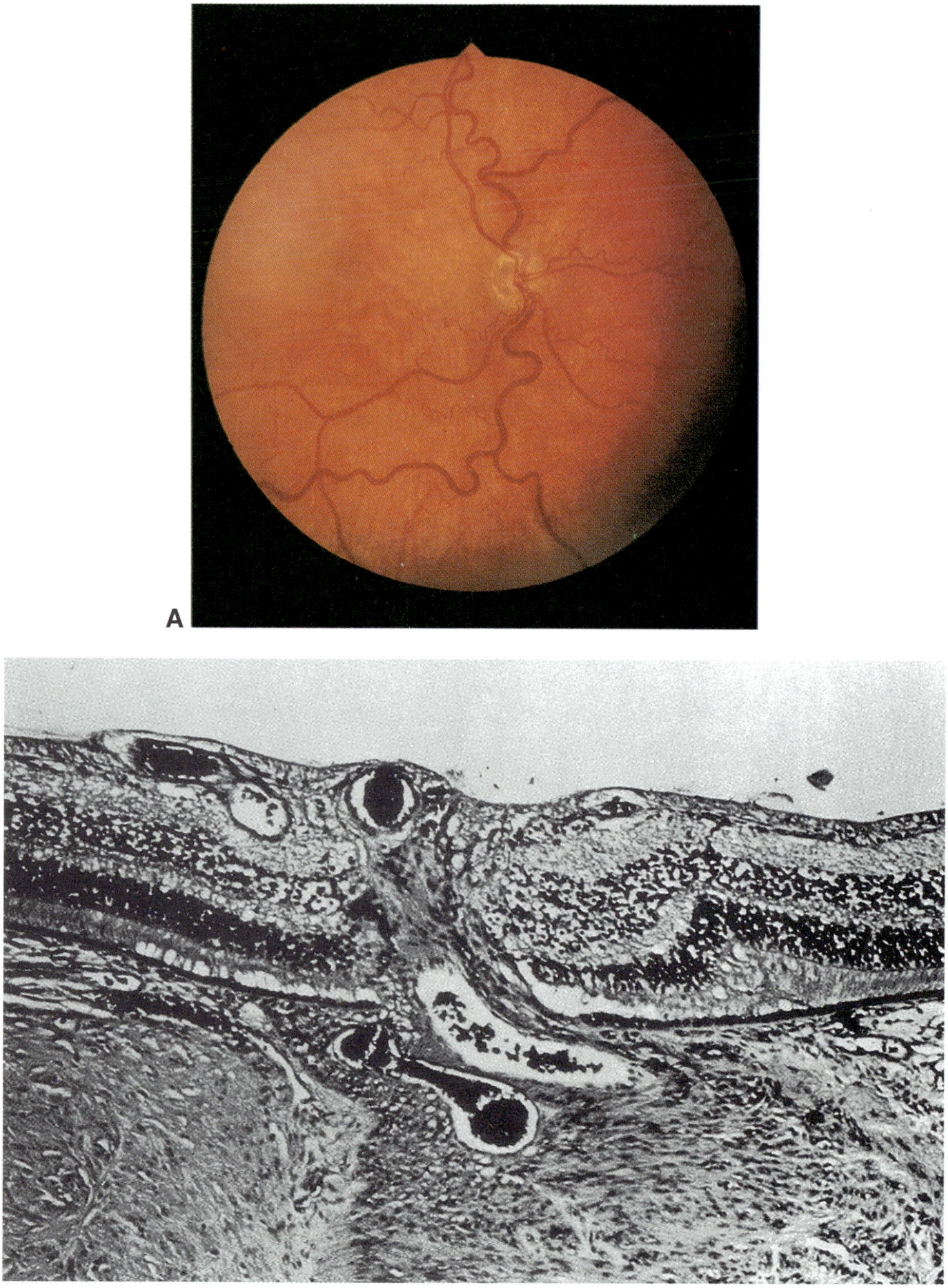

FIG VI-19—**A**, Optic nerve hypoplasia (ONH). A normal nerve of 1.5 mm diameter is 10–15 central retinal vessel widths across. This disc is far smaller. Part of what seems to be disc is a scleral rim (halo sign). **B**, Histologic section through a hypoplastic optic disc that shows a defect in retinal pigment epithelium with small central retinal artery (H&E ×125).

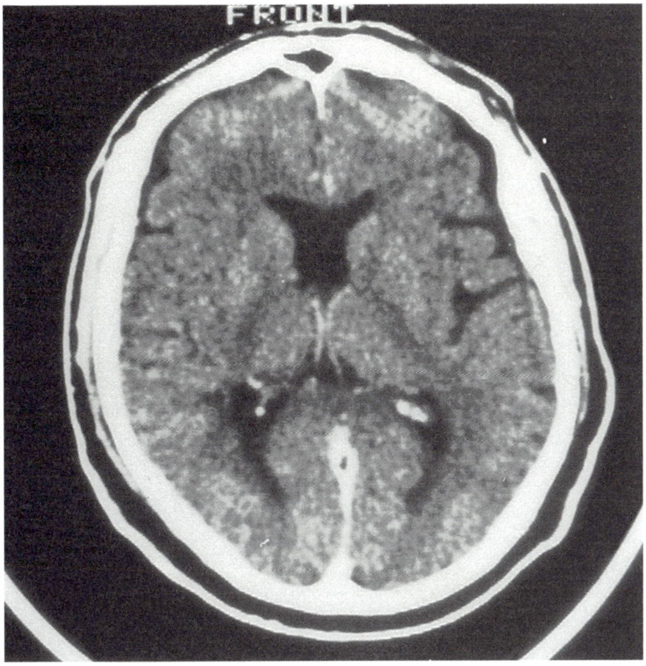

FIG VI-20—CT scan of absent septum pellucidum with bilateral ONH (de Morsier syndrome).

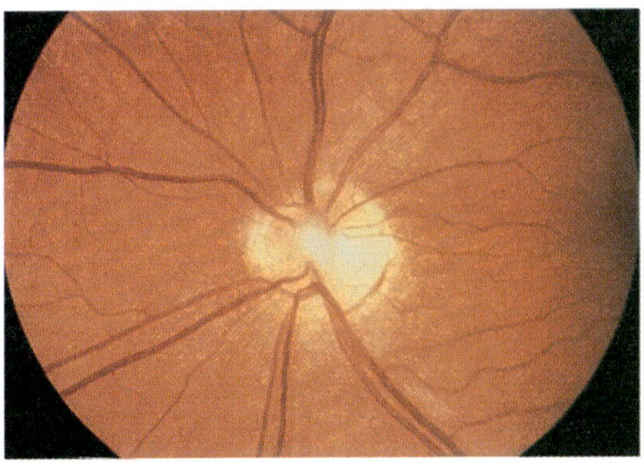

FIG VI-21—Remnant of glial tissue of hyaloid artery on optic disc.

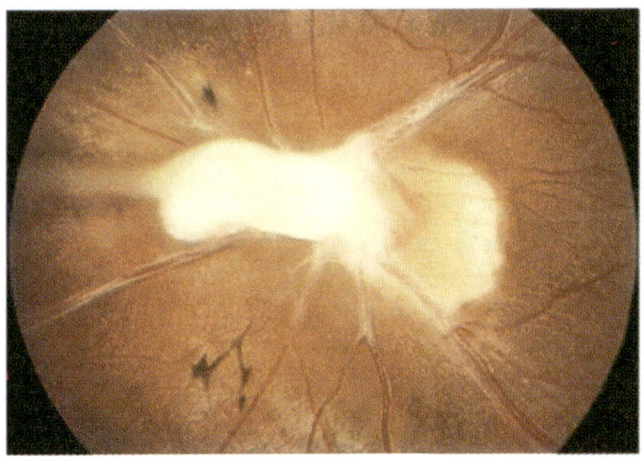

FIG VI-22—Atrophic stalk of hyaloid artery from optic disc into vitreous cavity.

Persistent Hyperplastic Primary Vitreous

Persistent hyperplastic primary vitreous (persistent fetal vasculature) is included in the differential diagnosis of leukocoria. Visual potential is limited. The condition is almost invariably unilateral and associated with microphthalmos. With growth of the child, the microphthalmos becomes accentuated because the involved globe grows more slowly, if at all. The anterior chamber is shallow; the lens may be either clear

FIG VI-23—Clinical appearance of Mittendorf's dot *(arrow)*, a retrolenticular remnant of anterior hyaloid.

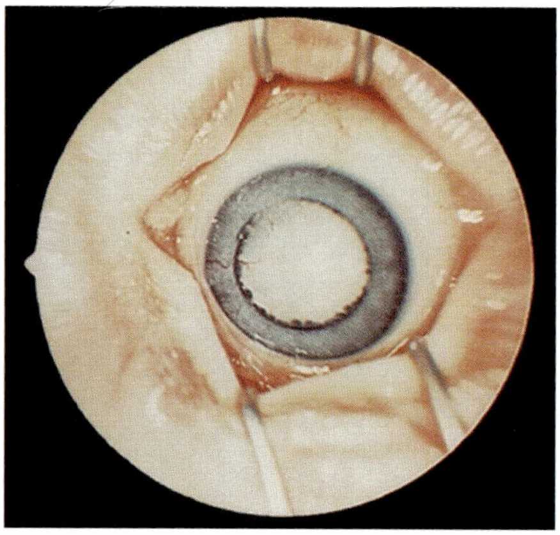

FIG VI-24—Persistent hyperplastic primary vitreous. Note ciliary processes visible in pupil.

or cataractous. A fibrovascular sheath forms behind the lens, contracts, and elongates the ciliary processes, which appear as dark reflections when the pupil is dilated (Fig VI-24). Vessels may occasionally grow through a tear in the posterior capsule. As the retrolenticular membrane contracts, the retina may detach. Removal of the lens and vitreous through the pars plana often prevents secondary glaucoma, but it may not improve the visual outcome because of retinal dysplasia and amblyopia.

Abnormalities of Pigmentation

The skin and the eye may be hyperpigmented or hypopigmented. *Oculodermal melanocytosis (nevus of Ota)* is a unilateral congenital condition with hyperpigmentation of the melanocytes in the periocular skin. The hyperpigmentation may also involve the uveal tract and the sclera. Open-angle glaucoma is occasionally associated with oculodermal melanocytosis.

Nevi that are pigmented at birth are sometimes found on the skin of the eyelids, conjunctiva, caruncle, iris, and choroid. Pigmentation developing at the time of puberty may cause concern. Malignant melanomas may arise from nevi, but usually only later in life. Malignant melanomas of ocular or periocular tissues are extremely rare in childhood.

Albinism is the result of a series of defects in the synthesis of melanin pigment. In *oculocutaneous albinism,* all tissues of the body lack normal melanin (Fig VI-25). In so-called *ocular albinism,* the globe is the most obvious clinical site of involvement; however, recent histologic evaluations show that the skin is also a target of morphologic changes in the X-linked form, in both affected males and carrier females. Visual acuity in tyrosinase-negative persons is poor but stable throughout life. Nystagmus is almost invariably present. With the several tyrosinase-positive variants, visual acuity may be poor in the first decade of life, but it improves as some uveal pigmentation develops with age. Pigment deficiency in the pigment epithelium will be apparent as variable degrees of transillumination of the irides, but stromal iris pigmentation may occur in all forms of albinism. Foveal hypoplasia is characteristic.

Well-demarcated, highly pigmented spots may be found in one or more quadrants of the fundus. These areas, formed by hypertrophic RPE cells, are congenital and do not enlarge. The spots may be single (Fig VI-26) or multiple.

Orbital Dermoid Cyst

A dermoid cyst is a choristoma that may be found in several locations in the orbit, most frequently in the superior and temporal aspects. An enlarging tumefaction displaces the globe. The tumors are congenital but may not be noted at birth because of their small size. Many become evident only in the second and third decades of life. Dermoid cysts produce a characteristic radiographic appearance of a well-circumscribed defect in the orbital bone with an increased bone density at the tumor margin. Disfiguring orbital dermoids may be removed.

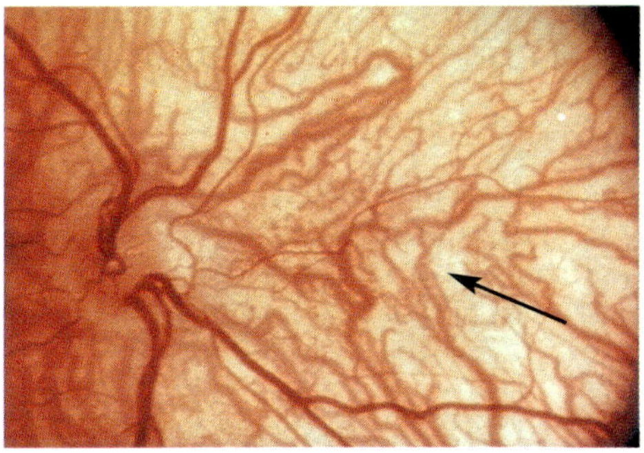

FIG VI-25—Fundus of albino. Optic disc is at left. Note also the associated hypoplasia of the fovea *(arrow).*

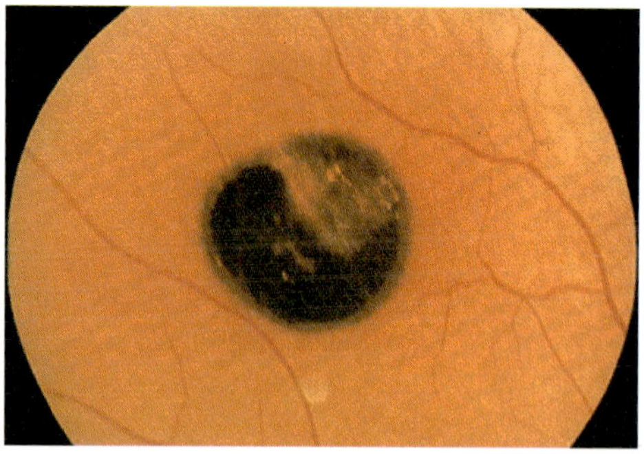

FIG VI-26—Congenital hypertrophy of retinal pigment epithelium.

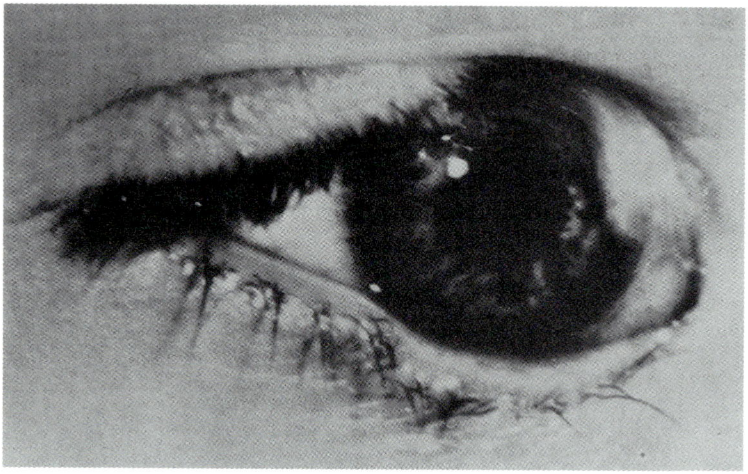

FIG VI-27—Moderately large coloboma of upper eyelid.

Eyelid Colobomas

Colobomas of the eyelids occur infrequently. The upper eyelid is more commonly affected (Fig VI-27). Colobomas range in size from a small notch to the absence of most of the eyelid. Faulty fusion of the facial processes, which are the embryonic progenitors of much of the eyelid tissue, results in a coloboma.

PART 3

GENETICS

CHAPTER VII

Introduction

Genetics is the study of human variability. Although genetics is a relatively new science compared to such disciplines as anatomy and physiology, its significance in the overall understanding of human life cannot be overstated. Genetic knowledge can enhance our understanding of the processes of cellular function, embryology, and development and our concepts of what is and is not a genetic disease. Many researchers think that as much as 90% of medical disease either has a major genetic component or involves genetic factors that may significantly influence the disease.

The discovery of unknown genes has opened new areas of understanding of physiology at the cellular or tissue level. One important example is the discovery of *homeotic selector genes* (eg, the *HOX* and *PAX* gene families) that regulate, guide, and coordinate early embryologic development and differentiation. (These genes, also called *homeobox genes,* are also discussed in Part 2, Embryology.) Another example is the identification of the genes that appear to be transcribed as initiating events in the process of *apoptosis,* or programmed cell death, which itself appears critical for normal embryogenesis.

Genetic disorders affect about 5% of the liveborn infants in the United States. Approximately 50% of childhood blindness has a genetic cause. At the beginning of this millennium, more than 10,000 human gene loci were known by mendelian phenotypes and/or cellular and molecular genetic methods. In about 10%–15% of known genetic diseases, clinical findings are limited to the eye; a similar percentage includes systemic disorders with ocular manifestations.

Terminology

Vocabulary is one of the greatest impediments to the understanding of genetics and molecular biology. The reader is urged to refer to the definitions of unfamiliar terms in the glossary before proceeding through the text. See also the glossary for Part 2, Embryology.

Glossary

Acceptor splice site The junction between the 3′ or downstream end of an intron and the 5′ or upstream end of the next exon. The consensus sequence is $\frac{T}{G}N\frac{C}{T}AG/G$ for the intron–exon boundary, where *N* is a purine (G or A). See *donor splice site* and *splice junction site.*

Acrocentric Type of chromosome in which the centromere is located near one end—for example, chromosomes 13, 14, 15, 21, and 22.

Allele Alternative form of a gene or DNA sequence that may occupy a given locus on a pair of chromosomes. Clinical traits, gene products, or disorders are said to be *allelic* if they are determined to be at the same locus and *nonallelic* if they are determined to reside at different loci.

Allele-specific oligonucleotide (ASO) A synthetic segment of DNA about 20 nucleotides in length. When hybridized to an unknown DNA sample, the ASO will bind to and thus identify the complementary sequence or specific string of base pairs. Used in detecting disease mutation.

Allelic association See *linkage disequilibrium.*

Allelic heterogeneity When different alleles at the same locus are capable of producing an abnormal phenotype.

Alu repeat sequence A common short interspersed element (SINE), 300 base pairs long, that occurs 500,000 times scattered throughout the genome. Often involved in errors of duplication or in mutational events. Unique to primates.

Amber codon The primitive stop codon TAG, which is thought to become the consensus sequence for the exon–intron and intron–exon boundaries (or *splice junction sites*) with loss of the thymine. See *stop codon.*

Aneuploidy An abnormal number of chromosomes.

Anticipation The occurrence of a dominantly inherited disease at an earlier age (often with greater severity) in subsequent generations. Now known to occur with expansion of a trinucleotide repeat sequence. Seen, for example, in fragile X syndrome, myotonic dystrophy, and Huntington disease.

Anti-oncogene See *tumor-suppressor genes.*

Antisense strand of DNA That strand of double-stranded DNA that serves as template for RNA transcription. Also called the *noncoding,* or *transcribed,* strand. See *sense strand of DNA.*

Apoptosis The process by which internal or external messages trigger expression of specific genes and their products, resulting in the initiation of a series of cellular events that involve fragmentation of the cell nucleus, dissolution of cellular structure, and orderly cell death. Unlike traumatic cell death, apoptosis results in the death of individual cells rather than clusters of cells and does not lead to the release of inflammatory intracellular products. Also called *programmed cell death (PCD).*

Ascertainment The method of selecting families for inclusion in a genetic study.

Assortative mating Mating between persons with preference for or against a specific genotype; that is, nonrandom mating.

Autosome Any chromosome other than the sex (X or Y) chromosomes. The normal human has 22 pairs of autosomes.

BAC A *b*acterial *a*rtificial *c*hromosome used as a cloning vector. BACs can be used to clone up to 150 kb of DNA.

Bacteriophage A virus containing DNA or RNA whose host is bacteria. Bacteriophage can be used for the transduction or insertion of fragments of DNA into bacteria for cloning purposes.

Barr body Inactive X chromosome seen in the nucleus of some female somatic cells.

Base pair (bp) Two complementary nitrogen bases that are paired in double-stranded DNA. Used as a unit of physical distance or length of a sequence of nucleotides.

Carrier An individual who has a pair of genes consisting of one normal and one abnormal, or *mutant,* gene. Usually, such persons are by definition phenotypically "normal," although in certain disorders biochemical evidence of a deficient or defective gene product may be present. Not infrequently, carriers of an X-linked disorder may show partial expression of a genetic trait.

CCAAT box A sequence of nucleotides about 75–80 bp upstream from the transcription initiation site of many genes that is thought to play a role in promoter function.

cDNA clone A host cell that contains a vector containing a fragment of complementary DNA (cDNA) from another organism.

Centimorgan A measure of the crossover frequency between linked genes. One centimorgan equals 1% recombination and represents a physical distance of about 1 million bp. Abbreviated cM.

Centromere The constricted region of the chromosome that is associated with spindle fibers during mitosis and meiosis. It is important in the movement of chromosomes to the poles of the dividing cell.

Chorionic villus sampling (CVS) Transcervical procedure in which chorionic villi are retrieved with a flexible suction catheter for use in studies to establish a prenatal diagnosis.

Chromatid One of the duplicate arms (also called *sister chromatids*) of chromosomes that are created after DNA replication during mitosis or the first division of meiosis.

Chromatin The complex of DNA and proteins that is present in chromosomes.

Clinical heterogeneity Different mutations at the same locus producing different phenotypes. Examples include macular dystrophy and retinitis pigmentosa from differing mutations of peripherin/*RDS* and Crouzon, Pfeiffer, and Apert syndromes from mutations of *FGFR2*.

Cloning vector Any DNA molecule capable of autonomous replication within a host cell into which DNA can be inserted for amplification. Cloning vectors can be derived from plasmids, bacteriophages, viruses, and yeast. Examples of cloning vectors include YAC (yeast artificial chromosome), BAC (bacterial artificial chromosome), and PAC (P1 artificial chromosome).

Codominance Simultaneous expression of both alleles of a heterozygous locus (eg, ABO blood groups).

Codon The basic unit of the genetic code. The DNA molecule is a chain of nucleotide bases that is "read" in units of three bases *(triplets),* which will translate (through messenger RNA) to an amino acid. Thus, each triplet codon specifies a single amino acid.

Complementary DNA (cDNA) DNA created by the action of reverse transcriptase from messenger RNA. Such DNA does not have introns, as does genomic DNA.

Compound heterozygote Gene locus having two different, abnormal alleles.

Congenital Present at birth. The term has no implications about the origin of the feature.

Consanguinity Mating between blood relatives, or a genetic relationship by descent from a common ancestor.

Consensus sequence The most common or idealized sequence of base pairs (or encoded amino acids) for a given region of a gene. See *acceptor splice site* for an example.

Conservation A genetic sequence or nucleotide position is said to be conserved or show conservation if a similar sequence is present among different species at one gene or related genes of similar sequence.

Contig A set of overlapping clones, each containing a fragment of a specific region or DNA sequence, that collectively covers the region without an interruption.

Cosmid A self-replicating vector (hybrid bacteriophage) used for cloning of DNA fragments into bacteria. Cosmids accommodate a DNA sequence of about 40 kilobases (kb) and are useful for creation of gene libraries.

Crossing over A process in which homologous chromosomes (chromatids) exchange segments by breakage and the physical exchange of segments, followed by repair of the breaks. Crossing over is a regular event in meiosis but occurs only rarely in mitosis. Also termed *recombination.*

Degeneracy of the code The genetic code is termed *degenerate* because most of the 20 amino acids are encoded by more than one of the 64 possible triplet codons.

Digenic inheritance Simultaneous inheritance of two nonallelic mutant genes, giving rise to a genetic disorder wherein inheritance of only one of the two is insufficient to cause disease. An example is retinitis pigmentosa caused by simultaneous inheritance in the heterozygous state of otherwise tolerable mutations of both the *ROM1* and peripherin/*RDS* genes. The simplest form of polygenic inheritance.

Diploid The number of chromosomes in most somatic cells, which in humans is 46. The diploid number is twice the haploid number, which is the number of chromosomes in gametes.

Dispersed repetitive DNA Repeated sequences interspersed throughout the genome. These are either short interspersed elements (SINEs) such as the Alu sequence family (about 300 bp in length), or long interspersed elements (LINEs) up to 6000 bp in length such as the L1 family. These repetitive sequences can lead to mutational events through unequal recombination, causing disease or tumors.

DNA Deoxyribonucleic acid, the nucleic acid of chromosomes.

Dominant An allele that is expressed in the phenotype when inherited along with a normal allele. See *recessive.*

Dominant medical disorder A distinctive disease state that occurs in a (dominant) heterozygous genotype. Classically, normal dominant traits give the same phenotype in both the heterozygous and homozygous states. Homozygotes for dominant disease-producing alleles are rare and usually more severely affected than heterozygotes.

Dominant negative An autosomal dominant mutation that disrupts the function of the normal or wild-type allele in the heterozygous state, giving a phenotype approaching that of the homozygous mutant.

Donor splice site The junction between the 3′ end of an exon and the 5′ end of the next intron. The consensus sequence is $\frac{C}{A}\mathrm{AG/GT}\frac{A}{G}\mathrm{AGT}$ for the exon–intron boundary. See *acceptor splice site* and *splice junction site.*

Endonuclease A phosphodiester-cleaving enzyme, usually derived from bacteria, that cuts nucleic acids at internal positions. Restriction endonucleases cut at specific recognition sites determined by the occurrence of a specific sequence of four, five, or six base pairs. Endonuclease specificity may also be confined to substrate conformation, nucleic acid species (DNA, RNA), and the presence of modified nucleotides.

Enhancer Any sequence of DNA upstream or downstream of the coding region that acts in *cis* (ie, on the same chromosome) to increase (or, as a negative enhancer, decrease) the rate of transcription of a nearby gene. Enhancers may display tissue specificity and act over considerable distances.

Eukaryote Organisms with their DNA located within a nucleus (includes all multicellular and higher unicellular organisms). See *prokaryote.*

Exon Any segment of a gene that is represented in the mature mRNA product. See *intron.*

Expressed-sequence tag (EST) A partial sequence of a gene that uniquely identifies its message. These tags are useful, through reverse-transcriptase polymerase chain reaction (RT-PCR), for determining the expression of genes.

Expressivity The variation in clinical manifestation among individuals with a particular genotype, usually a dominant medical disorder. The variability may be a difference in either age of onset (manifestation) or severity. See *penetrance.*

Fragile site Reproducible sites of secondary constrictions, gaps, or breaks in chromatids. Fragile sites are transmitted as mendelian codominant traits and are usually not associated with abnormal phenotype. The most notable exceptions are the association of fragile X chromosomes and X-linked mental retardation and postpubertal macro-orchidism (fragile X syndrome). See *trinucleotide repeat expansion.*

Frame shift mutation Any mutation, usually a deletion or insertion of a nucleotide or a number of nucleotides not divisible by three, that results in a loss of the normal sequences of triplets, causing the new sequence to code for entirely different amino acids from the original. It usually leads to the eventual chance formation of a *stop codon.*

Galtonian inheritance The form of inheritance whereby traits appear to be passed from one generation to the next in a particulate fashion rather than as segregating traits, as is the case with mendelian traits. Disorders of mitochondrial DNA (mtDNA) exhibit galtonian rather than mendelian inheritance characteristics.

Gene The segment of DNA and its associated regulatory elements coding for a single trait, usually a single polypeptide or mRNA. Recently expanded to include any expressed sequence of nucleotides that has functional significance, including DNA sequences that govern the punctuation (promoter) or regulation (enhancer) of transcription.

Genetic Related to or produced by a gene.

Genocopy Different nonallelic genotypes that result in a similar phenotype (often a medical disorder).

Genome The sum total of the genetic material of a cell or of an organism.

Genomic clone A host cell that contains a vector containing a fragment of genomic DNA from another organism.

Genotype The genetic constitution of an organism. Also used to denote the specific set of two alleles inherited at a locus.

Germinal mosaicism The occurrence in an individual of two populations of gametes, one with a normal allele and the other with a disease-producing mutant gene. Of "new" cases of some autosomal dominant diseases (eg, osteogenesis imperfecta), 5%–10% are thought to result from germinal mosaicism; offspring of the affected parent are at significant risk for the same disease.

Haploid Half the number of chromosomes in most somatic cells, equal to the number of chromosomes in gametes. In humans, the haploid number is 23. Also used to denote the state in which only one of a pair or set of chromosomes is present. See *diploid.*

Haploid insufficiency (haploinsufficiency) The condition of dominant genetic disease caused by reduction in gene product to levels that are insufficient to produce the desired function of the protein. For example, aniridia and Waardenburg syndrome result from insufficiency of the single functional copy of the *PAX6* and *PAX3* genes, respectively, to activate transcription of the genes that they normally control.

Haplotype The combination of linked polymorphisms or marker alleles for a given region of DNA on a single chromosome.

Hemizygous (hemizygote) Having only one allele at a locus; usually refers to X-linked loci in males, who normally have only one set of X-linked genes. An individual who is missing an entire chromosome or a segment of one chromosome is considered hemizygous for the genes on the homologous chromosome.

Hereditary Genetically transmitted or capable of being genetically transmitted from parent to offspring. Not quite synonymous with *heritable,* which implies the ability to be transmitted to the next generation but does not intrinsically connote inheritance from the last generation. See *genetic.*

Heterogeneity (genetic heterogeneity) The production of a phenotype (or apparently similar phenotypes) by different genetic entities. Refers to genetic disorders that are found to be two or more fundamentally distinct entities. See *genocopy.*

Heteronuclear RNA (hnRNA) The messenger RNA from the initiator codon to the stop codon. About 25% of these represent immature RNAs prior to splicing out of the introns. The function of the other 75% is unknown. Also called *heterogeneous nuclear RNA.*

Heteroplasmy The presence of two or more different populations of mitochondria within a cell, each carrying a different allele (or the presence or absence of a mutation) at a given locus.

Heterozygous (heterozygote) Having two unlike alleles at a particular locus. See *hemizygous, homozygous.*

Holandric Inheritance through genes on the Y chromosome.

Homeobox A conserved 180 bp sequence of DNA, first detected within homeotic selector genes, that helps determine the cell's fate.

Homeotic selector genes Genes that appear to regulate the activity or expression of other genes, eventually guiding the embryonic development of cells into body segments, body parts, and specialized organ systems. Examples are the *HOX* and *PAX* families of developmental genes. The *HOX* family represents 38 homeobox genes that are linearly arranged in four independent complexes termed *HOX1, HOX2, HOX3,* and *HOX4.* These gene clusters reside on chromosomes 7, 17, 12, and 2, respectively. Whereas *HOX* genes are involved in early body plan organization, *PAX* genes are involved in somewhat later organ and body part development. See the discussion of homeobox genes in Chapter V, Ocular Development.

Homologous chromosomes The two members of a matched pair of (sister) chromosomes, one derived from each parent, that have the same gene loci, but not necessarily the same alleles, in the same order.

Homoplasmy The presence of a single population of mitochondria within a cell, each carrying the same allele (or the same presence or absence of a mutation) at a given locus.

Homozygous (homozygote) Having two like or identical alleles at a particular locus in diploid genome. Sometimes misused to refer to *compound heterozygote* (see above).

Host cell In the context of recombinant genetics, the organism (usually a bacterium such as *Escherichia coli*) into which is inserted the vector (usually a plasmid or bacteriophage) containing the foreign DNA. Hosts are used to propagate the vector and, hence, the cloned DNA segment.

Hybridization The bonding (by Watson-Crick base-pairing) of single-stranded DNA or RNA into double-stranded DNA or RNA. The ability of stretches of DNA or RNA to hybridize with each other is highly dependent on the similarity or identity of the base-pair sequence.

Illegitimate transcripts Rare transcripts isolated from white blood cells for genes that are not generally expressed in these cells. Reverse-transcriptase polymerase chain reaction (RT-PCR) can be used to generate cDNA from these transcripts for diagnostic purposes.

Imprinting The reversible marking or inactivation of an allele by inheritance (through either the maternal or the paternal lineage), which may significantly alter gene expression. The imprinting is reversed if the gene is passed through subsequent generations through the opposite parental line. This phenomenon occurs in Prader-Willi and Angelman syndromes; it may also occur with mutations of the Wilms tumor gene. One mechanism of imprinting is thought to involve methylation of 5′ elements of the gene.

Initiator codon The triplet code that, when coded into messenger RNA (mRNA), initiates translation of the mRNA by causing binding of a special type of transfer RNA called *initiator tRNA*. In prokaryotes (bacteria), either AUG or GUG can act as an initiator codon. In eukaryotes, AUG is the only initiator codon and codes for methionine.

Intervening sequence Intron.

Intron A segment of DNA that is transcribed into RNA but is ultimately removed from the transcript by splicing together the sequences on either side of it (exons).

Isochromosome An abnormal chromosome created by deletion of one arm and duplication of the other arm, such that the chromosome has two equal-length arms of the same loci sequence extending in opposite directions from the centromere.

Karyotype A photographic record or a computer printout of an individual's chromosome set arranged in a standard pattern in pairs by size, shape, band pattern, and other identifiable physical features.

Kilobase (kb) 1000 bp of DNA or 1000 bases of single-stranded RNA.

L1 repeat element A repetitive long interspersed DNA element (LINE)—about 10,000 copies up to 6000 bp in length—dispersed throughout the genome. Thought to be involved in mutational events.

Liability With reference to polygenic or multifactorial inheritance, the graded continuum of increasing susceptibility to a disease or trait.

Library A complete set of clones presumably including all genetic material of interest from an organism, tissue, or specific cell type at a specified stage of development. A *genomic library* contains cloned DNA fragments from the entire genome; a cDNA library contains fragments of cloned DNAs generated by reverse transcription from mRNA. Genomic libraries are useful sources to search for genes, whereas cDNA libraries give information about expression within the source cell or tissue.

Linkage A concept that refers to loci rather than to the alleles that reside on those loci. Exists when the loci of two genes or DNA sequences are physically close enough to each other on the same chromosome that alleles at the two loci do not assort independently at meiosis but tend to be inherited together.

Linkage disequilibrium The state in which alleles that reside at loci close together in the genome remain inherited together through many generations because the close physical distance makes crossover between the loci extremely unlikely. Thus, alleles that are in linkage disequilibrium are present in subpopulations of individuals (eg, those with a given disease) in greater-than-expected frequencies. Also called *allelic association.*

Locus The physical site on a chromosome occupied by a particular gene; often colloquially used interchangeably with *gene.*

Locus heterogeneity The situation when mutations at different loci produce a similar phenotype. An example would be X-linked retinitis pigmentosa from *RP2* at Xp11 and *RP3* at Xp21.

LOD score (*lo*garithm of *od*ds, or **log of the likelihood ratio)** A statistical method that tests whether a set of linkage data indicates that two loci are linked or unlinked. The LOD score is the logarithm to the base 10 of the odds favoring linkage. By convention, an LOD score of 3 (1000:1 odds in favor of linkage) is generally accepted as proof of linkage.

Lyonization Inactivation of genes on either the maternally or the paternally derived X chromosome in somatic cells, occurring at about the time of implantation. First proposed by Mary Lyon.

Meiosis The special form of cell division that occurs in germ cells by which gametes of haploid chromosomal number are created. Each of the chromatids, which are clearly visible by prophase, contains a long double helix of DNA associated with histones and other chromosomal proteins. At anaphase, the chromatids separate at the centromere and migrate to each half of the dividing cell; thus, each daughter cell receives an identical set of chromatids (which become the chromosomes for that cell). During the first, or *reduction,* division of meiosis, the chromatids of homologous chromosomes undergo crossover (during the diplotene phase), and the number of chromosomes is reduced to the haploid number by the separation of homologous chromosomes (with duplicate chromatids) to each daughter cell. During the second division of meiosis, the sister chromatids separate to form the haploid set of chromosomes of each gamete.

Mendelian disorder (single-gene disorder) A trait or medical disorder that follows patterns of inheritance that suggest the state is determined by a gene at a single locus.

Microsatellite (eg, **dinucleotide** or **trinucleotide repeats)** Tandemly repeated segments scattered throughout the genome of varying numbers of two to four nucleotides in a row. For example, a stretch of consecutive CA combinations of bases (NNNCACACACACACACACACACANNN or $[CA]_{10}$, where N is any base) in a DNA strand. The highly variable nature of the number of repeats provides information useful as markers for establishing linkage to disease loci. See *satellite DNA* and *short tandem repeats.*

Minisatellite Array of repeated, nested segments of the same sequence of multiple triplet codons, each segment (consensus repeat unit) varying between 14 and 100 bp. Minisatellites are extraordinarily polymorphic and extremely useful as markers for establishing linkage, because they are often situated upstream or downstream from genes. The repeats are inherently unstable and can undergo mutation at a rate of up to 10%. Defects of some minisatellites are associated with cancer and insulin-dependent diabetes mellitus. Other terms used are *variable number of tandem repeats (VNTRs)* or *variable tandem repeats (VTRs).* See *satellite DNA.*

Missense mutation A mutation, often the change of a single nucleotide, that results in the substitution of one amino acid for another in the final gene product.

Mitosis The ordinary form of cell division that results in daughter cells identical in chromosomal number to the parent cell.

Mosaic An individual or tissue with at least two cell lines of different genotype or distinctive chromosomal constitution that develop after the formation of the zygote.

Multifactorial inheritance The combined operation of several unspecified genetic and environmental factors in the inheritance of a particular trait or disease. See *polygenic inheritance.*

Mutation Any alteration of a gene or genetic material from its "natural" state, regardless of whether the change has a positive, neutral, or negative effect.

Nitrogen bases Nitrogen-containing compounds, either the *purines* guanine and adenine or the *pyrimidines* cytosine, thymine, and uracil. These bases are abbreviated as G, A, C, T, and U, respectively.

Nondisjunction Failure of two chromosomes to separate during meiosis or mitosis.

Nonsense mutation Any mutation that either results directly in formation of a stop codon or creates a stop codon in the downstream sequence after a frame shift mutation through creation of a frame shift.

Northern blot Imprint of an electrophoretic gel that separates fragments of mRNA according to their size and mobility. The fragments are identified by hybridization to cDNA probes.

Nucleoside The combination of a nitrogen-containing base and a five-carbon sugar. The five nucleosides are adenosine (A), guanosine (G), cytidine (C), uridine (U), and thymidine (T). Note that the abbreviations are the same as those for the nitrogen bases that characterize the nucleoside.

Nucleosome The primary unit of chromatin, consisting of a 146 bp sequence of DNA wrapped twice around a core composed of eight histone molecules.

Nucleotide The combination of a nitrogen-containing base, a five-carbon sugar, and one or more phosphate groups. The nucleotides are designated by three capital letters as follows: adenosine monophosphate (AMP), deoxyadenosine monophosphate (dAMP), uridine diphosphate (UDP), adenosine triphosphate (ATP), etc. Although nucleotides are linked together by phosphodiester linkage into long sequences known as *nucleic acids,* they also perform other important functions, such as carrying chemical energy (ATP), combining with other groups to form coenzymes (coenzyme A, or CoA), and acting as intracellular signaling molecules (cyclic AMP, or cAMP).

Oncogene A defective gene that is capable of transforming cells to a neoplastic phenotype characterized by loss of growth control and/or tumorigenesis in a suitable host or site. In many cases, cancer is caused by the growth-stimulating effects of increased expression, protein activation, or aberrant regulation of transcription factors required for normal growth. Certain oncogenes are produced by chromosomal translocations of normal transcription factor genes to other regions adjacent to more abundantly expressed genes, causing inappropriate excessive expression. See *tumor-suppressor genes.*

Open reading frame (ORF) Any part of the genome that could be translated into a protein sequence because of the absence of stop codons. An exon is an example of an ORF. See *exon.*

Origin of replication The site(s) on a chromosome where replication is initiated and proceeds bidirectionally. The site of binding of the origin replication complex (ORC). Also called the *replication origin.*

Origin replication complex (ORC) A series of proteins involved in DNA synthesis and replication that bind to the origin of replication as one of the initiating events of DNA replication.

p arm The short arm of a chromosome in relation to the centromere. From *petit.*

Penetrance The proportion of individuals of a given genotype who show any evidence of an associated phenotype. Usually refers to the proportion of individuals heterozygous for a dominant disease who show any evidence of the disease. Non-penetrance is the lack of phenotypic evidence of the genotype. See *expressivity.*

Pharmacogenetics The area of biochemical genetics concerned with genetically controlled variations in drug responses.

Phenocopy The occurrence of a particular clinical phenotype (often a medical disorder) as a result of nonmutagenic environmental factors (eg, exposure to a drug or virus), when the more usual basis for the phenotype is an altered genotype.

Phenotype The total observable nature of an individual, resulting from interaction of the genotype with the environment (in medicine, often a disease phenotype).

Plasmid Circular extrachromosomal DNA molecules in bacteria that can independently reproduce in a host. Plasmids were originally detected because of their ability to transfer antibiotic resistance genes to bacteria. They can be used as vectors in recombinant DNA research.

Pleiotropism Multiple end effects (in different organ systems) arising from a single (mutant) gene or gene pair.

Polygenic inheritance Determined by the operation of an unspecified number of genes with additive effects. See *multifactorial inheritance.*

Polymerase chain reaction (PCR) A procedure whereby segments of DNA or RNA can be amplified without resorting to the conventional techniques of molecular cloning by use of flanking oligonucleotides called *primers* and repeated cycles of amplification with DNA polymerase. The steps involve:

- Heating to separate the molecules into single-stranded DNA
- Repeated annealing to the complementary target DNA sequences or primers specifically designed to delimit the beginning and ending of the target segment
- Extension of the primer sequences with the enzyme DNA polymerase, creating double-stranded DNA
- Separation of the products into single-stranded DNA

In effect, the amount of DNA is doubled with each cycle. Often, 30 or more cycles are used to obtain sufficient amplification for further testing.

Polymorphism Two or more alleles with a frequency greater than 1% in a given population.

Post-translational modification Changes or modifications of gene products after translation, including removal of amino acids from the end of the peptide, addition or removal of sugars, and addition of lipid side chains or phosphate groups to specific sites in the protein. Often, such changes are essential for proper protein localization or function.

Proband The affected person whose disorder, or concern about a disorder, brings a family or pedigree to be genetically evaluated. Also called the *propositus* (male), *proposita* (female), or *index case.*

Prokaryote Single-cellular organisms, such as bacteria, with their DNA located within the cytoplasm, with no nucleus. See *eukaryote.*

Promoter That sequence of nucleotides upstream (5′) to the coding sequence of a gene that determines the site of binding of RNA polymerase and, hence, initiation of transcription. Different promoters for the same gene may exist and can result in alternately spliced gene products and tissue-specific expression. The promoter may contain the consensus DNA sequence TATA$\frac{A}{T}$A$\frac{A}{T}$ (the so-called TATA box) about 25–30 bp (5′) upstream from the transcription start site.

Proposita, propositus Same as *proband,* above.

Proto-oncogene A normal gene that is involved in cell division or proliferation. Abnormalities in expression or regulation can cause the gene to become activated to an oncogene, which can lead to cancer. Several proto-oncogenes are involved in intracellular signal transduction—the process by which external messages influence the machinery that governs growth and differentiation.

Pseudodominance The appearance of vertical transmission of a recessive genetic disorder from one generation to the next, usually through the mating of an affected homozygote with a heterozygote, which produces affected offspring.

Pseudogene A defective copy of a gene. It often lacks introns and is rarely, if ever, expressed. Some pseudogenes are thought to have arisen by reverse transcription of mRNA that has had the introns spliced out. Others, such as globin pseudogenes, have arisen from silencing of a tandem duplicate. Since they are released from conservation (the maintenance of essential DNA sequences necessary for function) through selection, pseudogenes (compared to the original functional gene) often contain numerous base-pair changes and other mutational events.

Purine Nitrogen-containing base: adenine (A) and guanine (G) in DNA or RNA.

Pyrimidine Nitrogen-containing base: thymine (T) and cytosine (C) in DNA or uracil (U) in RNA.

q arm The long arm of a chromosome. See *p arm.*

Recessive Classically, a gene that results in a phenotype only in the homozygous state. See *dominant.*

Recessive medical disorder A disease state whose occurrence requires a homozygous (or compound heterozygous) genotype—that is, a double dose of the mutant allele. Heterozygotes are essentially normal.

Recombinant An individual who has a combination of genes on a single chromosome unlike that in either parent. Usually applied to linkage analysis, wherein *recombinant* refers to a haplotype (a set of alleles on a specific chromosome) that is not present in either parent because of a recombination crossover.

Recombinant DNA DNA that has been cut out of one organism, reinserted into the DNA of a vector (plasmid or phage), and then reimplanted into a host cell. Also, any act of altering DNA for further use.

Recombination The formation of a new set of alleles on a single chromosome unlike that in either parent; due to crossover during meiosis.

Relatives, first-degree Individuals who share on average half of their genetic material with the proband: parents, siblings, offspring.

Relatives, second-degree Individuals who share on average one fourth of their genetic material with the proband: grandparents, aunts and uncles, nieces and nephews, grandchildren.

Replication Creation of a new linear DNA copy by the enzyme DNA polymerase, proceeding from the 5′ side of bound primer to the 3′ end of the DNA sequence. Replication of DNA occurs during chromosomal duplication.

Replication slippage An error of DNA replication or copying. Because of the similarity of repeated base-pair sequences, one or more repeats are skipped over and not represented in the copied DNA sequence.

Replicative segregation The process by which, through partitioning of copies of mtDNA to each daughter cell during division, some cells receive a preponderance of normal or mutant copies. Replicative segregation tends to result in conversion of heteroplasmy to homoplasmy with associated development of disease within the affected tissue, if the tissue becomes homoplasmic for the mutant mtDNA. This phenomenon explains the development of new organ system involvement in multisystem mitochondrial diseases.

Restriction fragment length polymorphisms (RFLPs) RFLPs represent the variation in the length of genomic DNA fragments created by the loss or gain of an endonuclease restriction site. They can be used to map genes or link specific physical or genetic traits.

Retrotransposition The insertion of a *retroposon* (a segment of DNA created by reverse transcription from an RNA template) into the genome. Because of the staggered cut made in the target DNA by the endonuclease involved in the recombination event, a short duplication (3–12 bp) of the target site sequence is created at the ends of the transposed element. Because the transposable element may contain transcriptional initiation and/or termination signals, this process is one mechanism by which fusion genes, such as those causing certain forms of leukemia or cancer formation, can arise.

Reverse transcription The process, performed by the enzyme reverse transcriptase, whereby messenger RNA is converted back to DNA. If the introns have already been spliced out of the precursor mRNA, the product of this process is complementary DNA (cDNA).

Satellite DNA Nuclear DNA that migrates at separate positions or bands from the bulk of DNA during CsCl gradient centrifugation. Satellite DNAs are long segments of DNA that consist of short DNA sequences repeated hundreds or thousands of times at a stretch in the genome. Satellite DNAs form the ends and centers of chromosomes. Telomeric DNA is a form of satellite DNA.

Segregation The separation of pairs of alleles at meiosis.

Sense strand of DNA The strand of double-stranded DNA that corresponds in its 5′ to 3′ sequence to the expressed mRNA. Also called the *coding,* or *nontranslated, strand.*

Sequence-tagged sites (STSs) Short unique sequences of DNA, usually 200–500 bp, scattered throughout the genome that serve as landmarks for the physical mapping of genes. The presence of a specific STS in any sample can be determined by the polymerase chain reaction. If the STS is detected, the sample has genomic material from the known region of that STS. Currently, the average distance between STSs is 100 kb.

Sex linked Genes on the X or Y (sex) chromosomes. Often used improperly to mean X linked.

Short tandem repeats (STRs) Sequences of repeated copies of 2–5 bp that occur every 10 kb in the human genome. The variation in number of copies within a given STR is highly polymorphic and thus useful for gene mapping. See *microsatellite* and *minisatellite.*

Simplex A term used to denote that only one individual is affected within a given family. For example, a single male or female with a genetic disease would be called a *simplex case.* This term implies no inheritance type.

Smallest region of overlap (SRO) The minimum chromosomal or nucleotide sequence that is deleted among all individuals who have a phenotype thought to be the result of a particular chromosomal deletion. This deleted region is presumed to contain the gene or genes that cause the phenotype.

Southern blot Imprint of an electrophoretic gel that separates fragments of DNA according to their size.

Splice junction site The DNA region that demarcates the boundaries between exons and introns. The specific sequence determines whether the site acts as a 5′ donor or a 3′ acceptor site during splicing. Single base-pair changes or mutations that involve splice junction sites may result in skipping of the following exon or incorporation of part of the adjacent intron into the mature mRNA. See *acceptor splice site* and *donor splice site* for the consensus sequences.

Spliceosome Multicomponent ribosomal ribonuclear protein complex (40S to 60S) that is involved in the removal of introns from heteronuclear, or *precursor messenger,* RNAs.

Splicing That process by which the introns are removed from the precursor messenger RNA and the exons are joined together as mature mRNA prior to translation. Takes place within spliceosomes.

Sporadic A trait that occurs in a single member of a kindred with no other affected individuals in the family. The term has been used by some geneticists to imply that the trait is nongenetic.

Stop codon (termination codon) The DNA triplet that causes translation to end when it is coded into messenger RNA. The DNA stop codons are TAG, TAA, and TGA. Expressed as mRNA, these are UAG, UAA, and UGA.

Synteny The presence of genes on the same chromosome, even if linkage cannot be demonstrated. Also used to denote homologous chromosomal locations between species.

TATA box A promoter element about 25–30 bp (5′) upstream from the transcription start site that contains the consensus sequence TATA$\frac{A}{T}$A$\frac{A}{T}$. The TATA box is recognized by transcription factors that bind to the region, and it is critical in the initiation of transcription.

Telomeric DNA A type of highly repetitive satellite DNA that forms the tips of chromosomes and prevents them from fraying or joining. It decreases in size as a concomitant of aging. Defects in the maintenance of telomeres may play a role in cancer formation.

Threshold In polygenic or multifactorial inheritance, a relatively sharp qualitative difference beyond which individuals are considered to be affected. The threshold is presumed to have been reached by the cumulative effects of the polygenic and multifactorial influences.

Transcription The synthesis as catalyzed by a DNA-dependent RNA polymerase of a single-stranded RNA molecule from the antisense strand of a double-stranded DNA template in the cell nucleus.

Translation The process by which a polypeptide is synthesized from a sequence of specific messenger RNA.

Translocation The transfer of a part of one chromosome to a nonhomologous chromosome.

Trinucleotide repeat expansion (contraction) The process by which long sequences of multiple triplet codons (see *minisatellite*) are lengthened or shortened in the process of gene replication. The process of expansion of trinucleotide repeats over consecutive generations results in the genetic phenomenon of *anticipation.* The underlying mechanisms for expansion (or contraction) appear to be replication slippage and unequal crossing over in the region of the repeats. Most disorders involving trinucleotide repeats are dominant in inheritance (eg, fragile X syndrome, myotonic dystrophy, Huntington disease, Kennedy disease), but one is autosomal recessive (Friedreich ataxia).

Tumor-suppressor genes Genes that must be present in one fully functional copy in order to keep cells from uncontrolled proliferation. Two "hits" (inactivations) of the gene, one for each allele, must occur in a given cell for tumor formation to occur. Examples include the genes for retinoblastoma, Wilms tumor, tuberous sclerosis, p53, ataxia-telangiectasia, and von Hippel–Lindau disease. Also called *anti-oncogenes.* See *oncogene.*

Unequal crossing over An error in the events of chromosomal duplication and cell division occurring during meiosis and, rarely, during mitosis. Probably because of similar sequences or repeated segments, chromosomal exchange occurs between nonhomologous regions of the chromosome, resulting in duplication and deletion of genetic material in the daughter cells.

Uniparental disomy The conveyance to a child of two copies of an abnormal gene or chromosome by only one parent (the other parent makes no contribution). The child can be affected with autosomal recessive disease even if only one of the parents is a carrier for the abnormal gene. This occurrence has been reported in cystic fibrosis and Prader-Willi and Angelman syndromes.

Untranslated region (UTR) The regions upstream (5′ UTR) and downstream (3′ UTR) of the open reading frame of a gene. The 5′ UTR contains the promoter and part or all of the regulatory regions of the gene. The 3′ UTR presumably also serves important functions in regulation and mRNA stability.

Vector A viral, bacteriophage, or plasmid DNA molecule into which a stretch of either genomic DNA or cDNA or a specific gene can be inserted. The λ-bacteriophage can accept segments of DNA up to 25 kb long. Cosmid vectors can accommodate a segment 40 kb long. BAC (bacterial artificial chromosomes) and YAC (yeast artificial chromosome) vectors can accept much larger fragments of DNA.

Western blot Imprint of an electrophoretic gel that separates proteins according to their size and mobility. The proteins are usually identified by immunologic methods.

Wild type A normal phenotype of an organism. Also, a normal allele as compared to a mutant allele.

X linked Term that refers to genes on the X chromosome.

Y linked Term that refers to genes on the Y chromosome.

Yeast artificial chromosome (YAC) A yeast artificial chromosome is used as a cloning vector. YACs can be used to clone very large segments of DNA (up to 1000 kb).

CHAPTER VIII

Molecular Genetics

This chapter provides a review of molecular genetics (with emphasis on basic concepts), an overview of the techniques for manipulating deoxyribonucleic acid (DNA) in the laboratory, and an appreciation of the power and implications of molecular investigations for the study of inherited diseases. Ophthalmic examples and applications are used whenever possible to illustrate concepts and techniques.

DNA, Genes, and Chromosomes

DNA

The genetic information encoded as DNA in each cell directs the development and function of complex organisms. All the information required to reproduce an organism—from conception through the stages of embryonic development to adulthood—is organized, stored in, and retrievable from either nuclear or, to a limited but very important degree, mitochondrial DNA. Also contained in DNA is all the information necessary to establish normal physiologic cellular mechanisms to sustain life and function.

Structure of DNA *Chromosomes* are composed primarily of proteins and DNA. The helical structure of DNA, first defined in 1953 by James Watson and Francis Crick, consists of a chain of deoxyribose (pentose, or five-carbon, sugar) molecules, linked by phosphate, with each carrying a single nitrogenous base. Covalent phosphodiester bonds join the 5′ carbon of one deoxyribose sugar to the 3′ carbon of the next sugar in the sequence. The nitrogenous base is either a *purine* (adenine or guanine) or one of the *pyrimidines* (thymine or cytosine) (Fig VIII-1). A sugar-phosphate unit together with a nitrogen base is called a *nucleotide*. Each nitrogen base can form a hydrogen bond with only one of the other bases to form a *base pair*. Thus, within double-stranded DNA, two chains entwine into counter-rotating helices. Adenine always pairs with thymine, and guanine always pairs with cytosine. Approximately 3.2 billion base pairs constitute human DNA.

RNA Ribonucleic acid (RNA) differs from DNA in the following ways:

- It is usually a single, rather than a double, strand
- It contains the sugar *ribose* instead of the sugar *deoxyribose*
- It incorporates uracil instead of thymine as one of the four nitrogenous bases

Complementary nature of DNA The two strands of DNA are complementary. Each strand contains the information in the reverse sequence of the other strand. If the sequence is known for one strand, the sequence for the other strand can always be inferred.

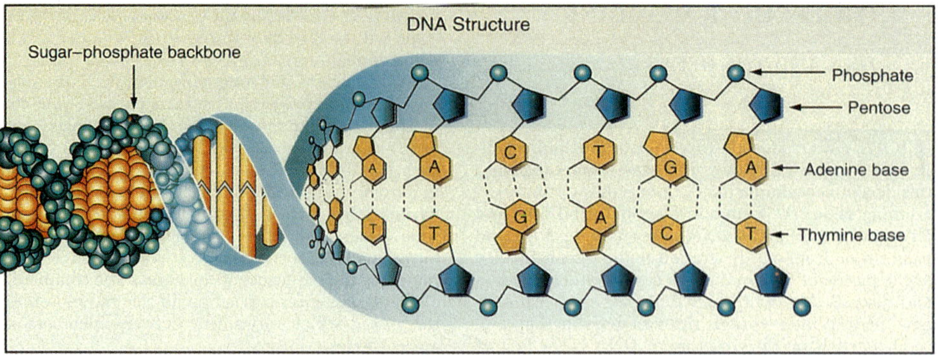

FIG VIII-1—The sequence of four bases (guanine, adenine, thymine, and cytosine, in *orange*) that determines the specificity of genetic information. The bases face inward from the sugar-phosphate backbone *(green)* and form pairs *(dashed lines)* with complementary bases on the opposing strand. (Reproduced with permission from Rosenthal N. Molecular medicine: DNA and the genetic code. *N Engl J Med.* 1994;331:40.)

Double-stranded DNA dissociates with high pH or temperature and anneals with cooling or other appropriate conditions to re-form. The property of single-stranded DNA to reassociate only with sequences of complementary DNA is the basis for much of recombinant genetics and many of the modern techniques of DNA manipulation.

Genetic code Each base pair codes a "letter," and each triplet *(codon)* represents a word in a sort of biological Morse code that passes on genetic information (Table VIII-1; Fig VIII-2).

The linear sequence of the four bases in RNA is a key factor in determining which of the 20 amino acids will make up a particular polypeptide chain. The four nucleotides in RNA—adenine (A), cytosine (C), guanine (G), and uracil (U)—combine linearly in groups of three, or *codons.* Each codon represents a start, an amino acid, or a stop. Of the 64 possible codons, 61 code for amino acids and 3 code for cessation of translation. Thus, the genetic code is redundant, with 18 amino acids coded by more than one triplet.

Four codons signal a start or stop of the construction of an amino acid chain. The AUG codon (which codes for methionine) initiates translation, whereas a UGA, UAA, or UAG codon stops translation. Translation is discussed later (see under Gene Translation).

Sense and antisense DNA Nuclear DNA in each chromosome exists in double-stranded helix conformation. Because of the complementary nature of DNA, all of the information is present on each strand. However, the information is read out as RNA by the transcription of only one of the single strands of DNA. The strand of DNA called the *sense,* or *coding,* strand is *not* translated by RNA polymerase, but it has the same 5′ to 3′ sequence as the messenger RNA (mRNA). The strand that actually is read is called the *antisense,* or *translated,* strand. This is the strand that is transcribed to mRNA, which in turn is processed and translated into the polypeptide.

TABLE VIII-1

SYMBOLS FOR AMINO ACIDS

AMINO ACID	THREE-LETTER SYMBOL	ONE-LETTER SYMBOL	AMINO ACID	THREE-LETTER SYMBOL	ONE-LETTER SYMBOL
Alanine	Ala	A	Isoleucine	Ile	I
Arginine	Arg	R	Leucine	Leu	L
Asparagine	Asn	N	Lysine	Lys	K
Aspartic acid	Asp	D	Methionine	Met	M
Asn and/or Asp	Asx	B	Phenylalanine	Phe	F
Cysteine	Cys	C	Proline	Pro	P
Glutamine	Gln	Q	Serine	Ser	S
Glutamic acid	Glu	E	Threonine	Thr	T
Gln and/or Glu	Glx	Z	Tryptophan	Trp	W
Glycine	Gly	G	Tyrosine	Tyr	Y
Histidine	His	H	Valine	Val	V

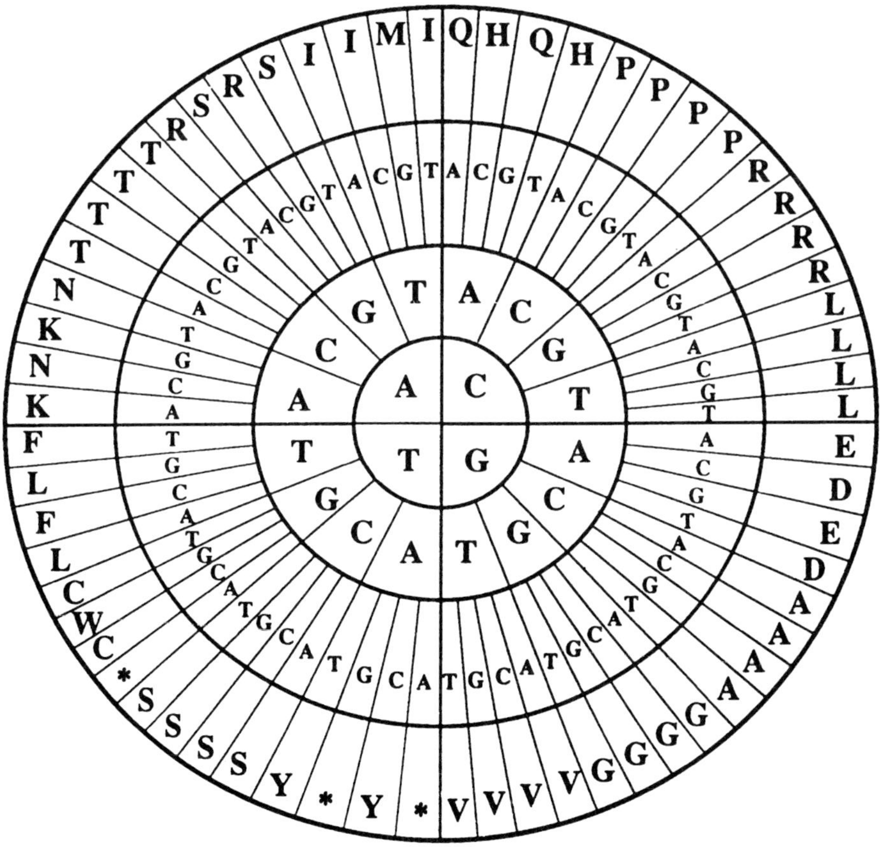

FIG VIII-2—The outermost circle represents the amino acid or termination code *(*)*; the trinucleotide encoding the amino acid or "stop" signal is given on the radial, starting with the first base of the codon in the center. *A,* adenine; *C,* cytosine; *G,* guanine; *T,* thymine. See Table VIII-1 for key to symbols. (Reproduced with permission from McKusick VA. *Mendelian Inheritance in Man: A Catalog of Human Genes and Genetic Disorders.* 11th ed. Baltimore: The Johns Hopkins University Press; 1994:xii.)

Therefore, the genetic code for the strand of DNA that is untranslated must be considered in understanding the sequence of amino acids in a given peptide, because the coding sequence of this strand is identical to the sequence of amino acids in the polypeptide finally released by the ribosomes.

One approach to treatment of autosomal dominant disease is to target the translated strand of the mutant allele by antisense DNA, a sequence of DNA designed to anneal to and block the processing or translation of the abnormal mRNA.

Gene structure A *gene* is the coding sequence for a protein, ribosomal RNA (rRNA), or other gene product and its associated regulatory sequences. Genes contain several regulatory (enhancer, repressor) and structural (coding introns and exons) components to ensure accurate transcription and translation at the appropriate levels and times.

Genes are arranged linearly on chromosomes. The genomic structure can be divided into the so-called *5′ untranslated region* (which contains the *promoter*) and *regulatory regions* (*enhancer, inducer,* and *inhibitor regions*) that serve as targets for *transcription factors,* which activate genes and regulate and help determine whether a gene is transcribed (Fig VIII-3). The promoter, which is under developmental and tissue-specific control, turns the gene on (ie, allows transcription and translation).

Following the initiation codon is the *open reading frame (ORF)* composed of *exons* (sequences that code for amino acids that will be present in the final protein) and *introns* (sequences that are spliced out during the processing of mRNA). Following the last exon is the *3′ untranslated region (3′ UTR).* The function of this region is known to be partly regulatory. Indeed, a mutation within the 3′ UTR region of the gene for the enzyme myotonin kinase is thought to cause myotonic dystrophy.

The development of introns in higher organisms may have had evolutionary benefits. Introns have allowed eukaryotes to evolve beyond the limits of genes seen in single-celled organisms, and they may have other roles as well. The compartmentalization of coding segments into exons may have allowed for more rapid evolution of proteins by allowing for alternative processing of precursor RNA (alternative splicing) and rearrangements of exons during gene duplication (exon shuffling). Encoded within some introns are so-called *small nucleolar RNAs (snoRNAs),* which are thought to play a role in ribosome assembly. Certain genes may be regulated by intron-encoded RNAs that bind either DNA or RNA. Introns may thus provide a previously unsuspected system for regulating gene expression and may also be involved in maintenance of genomic structure.

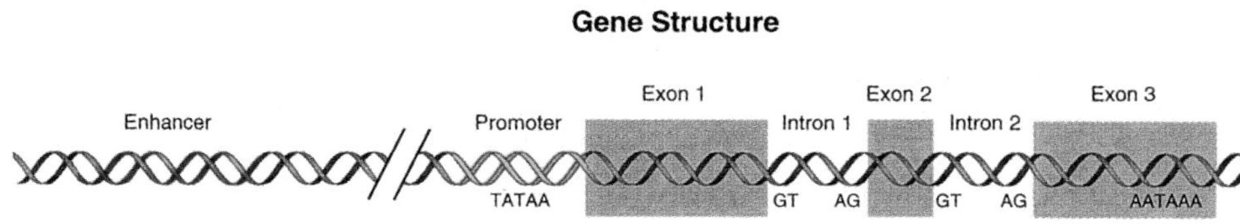

FIG VIII-3—The DNA sequences that are transcribed as RNA are collectively called the gene and include exons (expressed sequences) and introns (intervening sequences). Introns invariably begin with the nucleotide sequence GT and end with AG. An AT-rich sequence in the last exon forms a signal for processing the end of the RNA transcript. Regulatory sequences that make up the promoter and include the TATA box occur close to the site where transcription starts. Enhancer sequences are located at variable distances from the gene. (Reproduced with permission from Rosenthal N. Molecular medicine: regulation of gene expression. *N Engl J Med.* 1994;331:932.)

Some introns contain complete separate genes, and some of these may cause disease or influence expression of other genes. Expansion of unstable repeats within introns can cause abnormal splicing and result in genetic disease.

Untranslated and other "junk" genes Approximately 97% of the base sequences in human DNA have been considered "biologically meaningless" in that they do not encode proteins or RNAs or have any other known function. This so-called *junk DNA* may play other important roles that will emerge when the structure and function of the genome, chromosomes, nucleus, and nuclear proteins are fully understood. It has been suggested that RNA transcribed from junk DNA may directly influence the transcription of other sequences and participate in normal genome repair and regulation. When defective, it may lead to cancer. Some of the repetitive sequences of nontranscribed DNA form *telomeric DNA,* which is essential for correct formation and maintenance of chromosomes. Indeed, loss of telomeric DNA correlates with cell senescence and carcinogenesis. Therefore, sequences within junk DNA may influence the transcription or otherwise regulate the expression of numerous other genes.

Much of this wasteland of DNA is composed of highly repetitive sequences. Some of these sequences include *satellites, minisatellites, microsatellites, short interspersed elements (SINEs),* and *long interspersed elements (LINEs).* The most frequent of the repetitive DNAs is the 300 bp *Alu* sequence, named for the restriction enzyme used to identify it. The Alu sequence is a SINE that occurs 500,000 times in the human genome. Alu sequences are distributed through retroposition (Alu → Alu RNA → Alu cDNA → insertion) and may cause disease if one inserts within and disrupts a gene. This process accounts for one cause of type 1 neurofibromatosis. An important LINE is the L1 repeat sequence, composed of about 10,000 copies of 1–6 kb length each. The L1 repeat sequence has also been implicated as a cause of mutations.

Nucleus *The nuclear envelope.* The nucleus matrix is separated from the cytoplasm by an *envelope* consisting of two membranes fenestrated with openings called *nuclear pores.*

The nucleolus. The structure within the nucleus called the *nucleolus* contains the nucleolar organizer, which directs the synthesis of rRNA needed for translation.

Nuclear matrix. Each nucleus contains an estimated 50,000–100,000 genes plus all the biochemical enzymes and products needed for gene duplication, chromosomal formation, mitosis, meiosis, gene regulation, transcription, imprinting, X chromosomal inactivation, and intron removal from mRNA. The nuclear matrix also contains numerous DNA-binding proteins that are involved in controlling expression and affecting regulation.

Histones. Histones are basic proteins that bind with DNA to form the stable structures called *chromosomes.* The five major types of histones are H1, H2A, H2B, H3, and H4. The amino acid composition of histones is highly conserved, even among distantly related species. The weight of histones in the chromosomes is approximately equal to that of the DNA itself. Probably because large amounts of histones are required during the *S,* or *replication,* phase of cell division and the half-life of histone mRNA is only minutes, 20–50 copies of histone genes exist in the human genome.

Proteins have been discovered that remodel histones, allowing access to the DNA. It is unclear whether such unfolding plays a role in gene regulation or is merely a consequence of transcription. Some histones appear to target the transcription of certain genes while suppressing the transcription of others, a process called *gene silencing*. Transcriptional regulators may make direct contact with specific domains of histones.

Grunstein M. Histones as regulators of genes. *Sci Am.* 1992;267(4):68–74B.

Wolffe AP. Histone deacetylase: a regulator of transcription. *Science.* 1996;272: 371–372.

Nonhistone proteins. Nonhistone proteins appear to be involved in organizing long regions or domains of DNA that match the units of replication in mammalian chromosomes.

Chromatin. Chromatin is the state of DNA in which it is tightly coiled up with histones, which keep it from being transcribed into mRNAs. The DNA is wound onto the histones like thread around a spool. Segments of 146 bp of DNA are wrapped almost twice around a core of eight histones to form units called *nucleosomes,* which resemble beads on a string. One of the primary purposes of the nucleosomes appears to be compacting DNA by supercoiling it into 30 nm filaments that in turn can be arranged into thicker fibers composed of spirals or solenoid arrangements.

On histologic staining, chromatin within the nucleus appears as densely stained *heterochromatin,* which contains bound DNA not undergoing transcription, and more lightly stained *euchromatin,* which is the form of chromatin thought to be undergoing transcription for production of RNA. Centromeric and telomeric heterochromatin have been shown to be enriched for proteins that suppress gene transcription.

Chromosomal Structure

Chromosomes are complex aggregations of DNA, histones, and nonhistone nucleoproteins that form prior to cell division. The DNA exists in the nucleus in the chromatin state at other times, but during cell division the chromatin is further packaged into the chromosome structure specifically to allow duplicated sets of genes to segregate to each daughter cell. The complement of human chromosomes can be pictured as a set of 23 heavily packed suitcases that are created to allow compartmentalization of duplicated chromosomal material prior to cell division (*meiosis* and *mitosis*).

Chromosomes also provide the platform for maintaining variation of the species through crossover and exchange during meiosis. Considering the 22 autosomes and the two sex chromosomes, the different combinations of sets of whole chromosomes possible from the formation of *gametes* (reproductive cells) is 2^{23}, or 8,388,608. If recombination of whole chromosomes were the only determinant of variation, one out of every 8.4 million humans would be identical. However, with the added variation produced by meiotic crossover, the number becomes nearly infinite.

Chromatids The two individual strands of a duplicated chromosome as they exist during prophase and metaphase are often called *sister chromatids.* During anaphase, the chromatids separate, the long and short arm of each separating into the daughter cells.

Centromere The central constriction that divides the chromosome into two arms is called the *centromere*. Centromeric chromatin is a form of heterochromatin that encompasses many kilobases of DNA. The DNA that characterizes the centromere location on the chromosome is called *CEN*.

The role of the centromeres during mitosis and meiosis is complex. Centromeres appear to have at least four functions:

- Centromeres are the sites of formation of the *kinetochore*, which binds the spindle microtubules that are essential for the direction of chromosomal segregation during cell division.
- The centromere is the final site of attachment of sister chromatid pairing, and thus it must play a role in the process of releasing sister chromatids during the metaphase–anaphase transition.
- Centromeres are involved in cell-cycle checkpoint regulation during mitosis and, as the site of the kinetochore, draw the chromosomes to separate daughter cells during anaphase.
- The centromere acts as a station for chromosomal passenger proteins (mitosis-specific cytoskeletal proteins) that pass from chromosomes to the mitotic spindle.

Pluta AF, Mackay AM, Ainsztein AM, et al. The centromere: hub of chromosomal activities. *Science*. 1995;270:1591–1594.

Short (p) and long (q) arms The centromere divides the chromosome into a short arm designated *p* (for "petit") and a long arm designated *q*. The nomenclature of chromosomes is based on size and the position of the centromere.

Telomeres These structures at the ends of chromosomes are critical for the maintenance of chromosomal integrity. Telomeres protect chromosomes from DNA degradation, rearrangement, end-to-end fusion, and chromosomal loss. They are essential for avoidance of *end-replication problems*, or deletion during DNA duplication of terminal base pairs located at the ends of a linear chromosome. Telomeres in humans are composed of repetitive sequences of six nucleotides (TTAGGG) repeated a few hundred to several thousands of times, occupying up to several kilobases.

Telomeres are maintained by the enzyme *telomerase*, a ribonucleoprotein enzyme (containing both RNA and protein) that synthesizes the sequences by reverse transcriptase of a portion of its own RNA sequence. After synthesis, the new sequences are fused to the 3′ terminus of the chromosomes to replace the telomeric DNA loss that occurs normally with each round of DNA duplication.

The length of telomeres is thus a function of the number of duplications a cell has undergone and the activity of telomerase in the cell line. Loss of telomeres is a sign of cellular senescence, and it occurs to a greater extent in premature aging syndromes such as progeria.

Greider CW, Blackburn EH. Telomeres, telomerase and cancer. *Sci Am*. 1996;274(2): 92–97.

Zakian VA. Telomeres: beginning to understand the end. *Science*. 1995;270:1601–1607.

Gene Transcription (Expression)

Genes control cellular activity through two processes:

- *Transcription,* in which DNA molecules give rise to RNA molecules, followed by translation in most cases
- *Translation,* in which RNA directs the synthesis of proteins

Translation is discussed in the next section.

Transcription Factors and Regulation

Transcription factors contain DNA-binding domains that typically include a helical unit (α-helix) within or near positively charged amino acids. Four classes of structural protein motifs characterize 80% of transcription factors (Fig VIII-4):

- Helix-turn-helix (HTH)
- Zinc finger
- Leucine zipper
- Helix-loop-helix (HLH)

The *TATA box* is a highly conserved sequence of DNA in the 5′ untranslated region of a gene that is involved in the turning on, or *expression,* of genes. The first step in the assembly of the RNA polymerase transcription complex appears to be the binding of a transcription factor called the *TATA-box binding protein (TBP)* to the promoter region of DNA. Because of its importance in cellular function, the TBP is highly conserved over evolution, and its core domain of 180 amino-acid residues is 80% identical between yeast and humans. An architectural protein, TBP induces a 70° bend in the DNA that exposes sites for other proteins to bind. Several other gene-regulating proteins, of which about 50 are known, are then able to bind to this region (Fig VIII-5).

One of the several basal transcription factors involved in assembly of RNA polymerase onto promoters is *TFIIH,* which is also necessary for nucleotide excision DNA repair. The TBP then directs the start of transcription to a site at the 5′ end of the DNA sequence about 25 bp downstream from the TATA box. The TBP is the target of several activators, enhancers, and repressors and therefore participates in regulation as well.

Ophthalmology reveals many examples of disease resulting from transcription-factor mutations. *PAX2* mutations cause colobomas of the optic nerve and renal hypoplasia. *PAX3* mutations cause Waardenburg syndrome with dystopia canthorum (WS1 and WS3). *PAX6* mutations are the basis of virtually all cases of aniridia, occasional cases of Peter anomaly, and several other rarer phenotypes, specifically autosomal dominant keratitis and dominant foveal hypoplasia.

Farrer LA, Arnos KS, Asher JH Jr, et al. Locus heterogeneity for Waardenburg syndrome is predictive of clinical subtypes. *Am J Hum Genet.* 1994;55:728–737.

Glaser T, Walton DS, Cai J, et al. PAX6 mutations in aniridia. In: Wiggs JL, ed. *Molecular Genetics of Ocular Disease.* New York: Wiley-Liss; 1995:51–82.

Hanson IM, Fletcher JM, Jordan T, et al. Mutations at the PAX6 locus are found in heterogeneous anterior segment malformations including Peters' anomaly. *Nat Genet.* 1994;6:168–173.

Latchman DS. Transcription-factor mutations and disease. *N Engl J Med.* 1996;334:28–33.

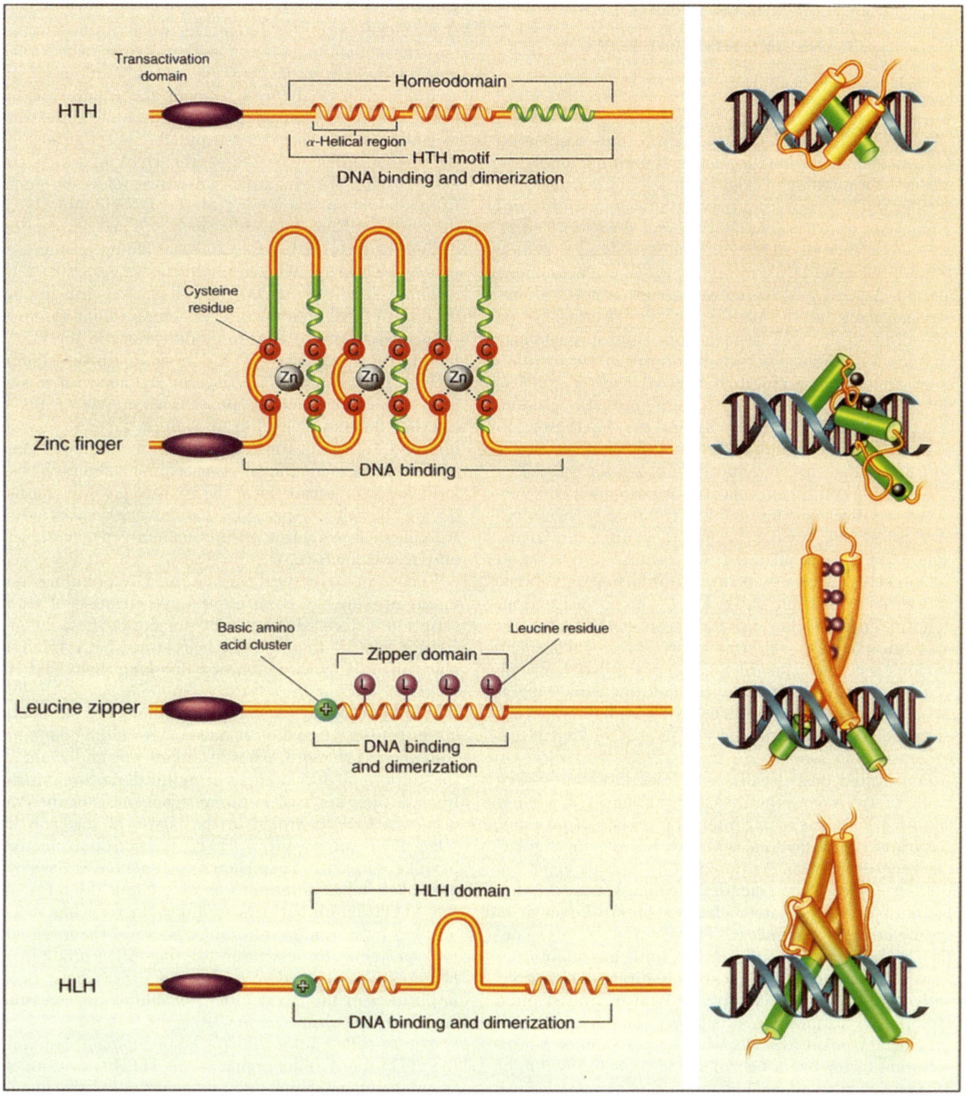

FIG VIII-4—The general protein structures of the four major classes of transcription factors are shown on the left. The structures include a transactivation domain linked to a DNA-binding domain and, in certain cases, a dimerization domain. The types of transcription factors take their names from the characteristic motifs involved in DNA binding and protein dimerization and are shown on the right, interacting with the DNA. The cylinders represent α-helical regions, and the areas that contact DNA directly are *green.* *HTH,* helix-turn-helix; *HLH,* helix-loop-helix; *Zn,* zinc; *C,* cysteine; *L,* leucine. The *plus sign* indicates a positive charge. (Reproduced with permission from Papavassiliou AG. Molecular medicine: transcription factors. *N Engl J Med.* 1995;332:46.)

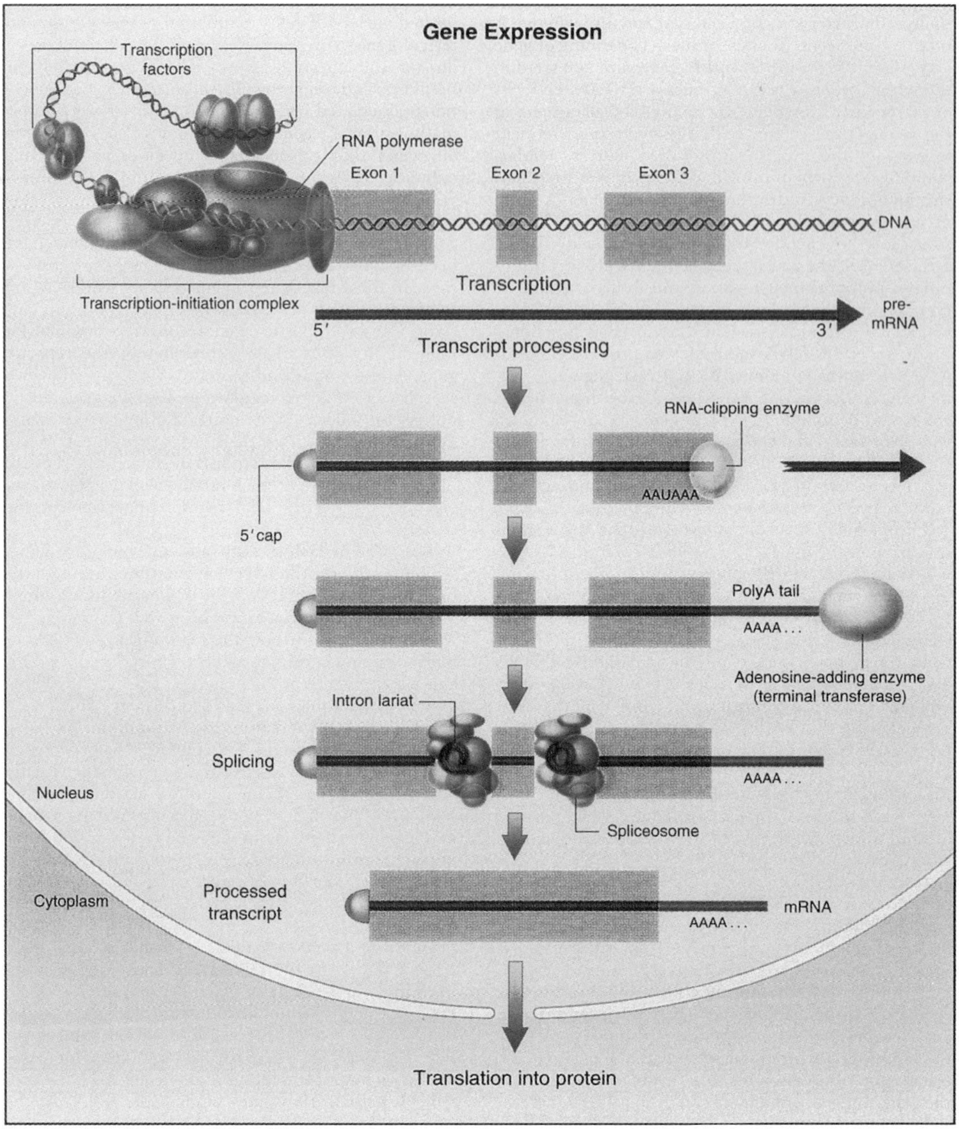

FIG VIII-5—Gene expression begins with the binding of multiple protein factors to enhancer sequences and promoter sequences. These factors help form the transcription-initiation complex, which includes the enzyme RNA polymerase and multiple polymerase-associated proteins. The primary transcript (pre-mRNA) includes both exon and intron sequences. Post-transcriptional processing begins with changes at both ends of the RNA transcript. At the 5′ end, enzymes add a special nucleotide cap; at the 3′ end, an enzyme clips the pre-mRNA about 30 bp after the AAUAAA sequence in the last exon. Another enzyme adds a polyA tail, which consists of up to 200 adenine nucleotides. Next, spliceosomes remove the introns by cutting the RNA at the boundaries between exons and introns. The process of excision forms lariats of the intron sequences. The spliced mRNA is now mature and can leave the nucleus for protein translation in the cytoplasm. (Reproduced with permission from Rosenthal N. Molecular medicine: regulation of gene expression. *N Engl J Med.* 1994;331:932.)

Mirzayans F, Pearce WG, MacDonald IM, et al. Mutation of the PAX6 gene in patients with autosomal dominant keratitis. *Am J Hum Genet.* 1995;57:539–548.

Papavassiliou AG. Molecular medicine: transcription factors. *N Engl J Med.* 1995;332: 45–47.

Rosenthal N. Regulation of gene expression. *N Engl J Med.* 1994;331:931–933.

Sanyanusin P, Schimmenti LA, McNoe LA, et al. Mutation of the PAX2 gene in a family with optic nerve colobomas, renal anomalies and vesicoureteral reflux. *Nat Genet.* 1995;9:358–364.

Tassabehji M, Read AP, Newton VE, et al. Waardenburg syndrome patients have mutations in the human homologue of the Pax-3 paired box gene. *Nature.* 1992;355: 635–636.

Creation of RNA from DNA

The three major classes of RNA are:

- Transfer RNA (tRNA)
- Ribosomal RNA (rRNA)
- Messenger RNA (mRNA)

The RNA polymerase transcription complex creates a complete complementary copy of the antisense strand of DNA, including all exons and introns, called *heteronuclear RNA,* or *hnRNA.* About 25% of the hnRNA is eventually modified by RNA splicing that removes the introns, leaving a mature product. This modified hnRNA is the mRNA that leaves the nucleus to be translated into protein. DNA can also be transcribed into RNA that never leaves the nucleus and is involved in functions that take place exclusively within it, including regulation of transcription and X-inactivation. RNAs, either alone or combined with protein, may also have enzymatic activities that are important for cellular function.

Processing of mRNA

After the DNA is transcribed to hnRNA, important modifications are made within the nucleus before the mature mRNA is transported to the cytoplasm. A cap that aids in the efficiency of ribosome binding is added to the 5′ end of the single-stranded RNA. Approximately 20 bp are removed from the 3′ untranslated end by a specific ribonuclease, and a stretch of adenine bases is attached, the *polyA* tail.

Intron Excision

The modified mRNA undergoes excision of the introns by a highly organized process called *splicing,* which leaves the mRNA composed of only exons, or coding segments. The exons can then undergo translation in the ribosomes. Splicing takes place in specialized structures composed of RNA and proteins called *spliceosomes.* The exact process of splicing is complex but involves intermediate steps that look like a lariat. Splicing must recognize precisely the beginning and end of each coding sequence, or exon, and errors of splicing can lead to genetic disease. Approximately 15% of point mutations that cause human disease do so by the generation of splicing errors that result in aberrations such as exon skipping, intron retention, or use of a cryptic splice site.

Alternative Splicing, Isoforms

Alternative splicing is the creation of multiple pre-mRNA sequences from the same gene by the action of different promoters. These promoters cause the transcription of the gene to skip certain exons. The protein products of alternative splicing are often called *isoforms*. The promoters are usually tissue specific, so different tissues express different *isoforms*. The gene for dystrophin is an example of alternative splicing: full-length dystrophin is the major isoform expressed in muscle; shorter isoforms predominate in the retina, peripheral nerve, and central nervous system.

Methylation

Evidence suggests a close correlation of methylation with gene inactivation. Regions of DNA that are undergoing transcription lack 5-methyl cytidine residues, which normally account for 1%–5% of total DNA. Regulation of DNA methylation may be responsible for the imprinting control.

X-Inactivation (Lyonization)

A major occurrence in early development is the normally random inactivation of one of the two X chromosomes in the female, resulting in the lack of expression of the great majority of genes on that chromosome. The time of X-inactivation is not precisely known but is thought to vary over a period of several cell divisions during the blastocyst–gastrula transition. X-inactivation is also known as *lyonization* after its discoverer, Mary Lyon.

The *XIST* (X inactivation–specific transcript) gene is the only gene that is exclusively expressed from the inactive X chromosome. Its site is at the region designated the X-inactivation center (XIC), located at Xq13.2. In some way, the XIC counts the number of X chromosomes in the cell and initiates the process of inactivation. An intact XIC is required in *cis* (on the same chromosome) for inactivation to occur. An X chromosome that is missing the XIC (through translocation or deletion) will not undergo inactivation. The *XIST* site on the active X chromosome is methylated and not expressed, whereas demethylation at the *XIST* site appears associated with expression of the gene and initiation of inactivation.

The *XIST* gene codes not for a protein but for an RNA 15–17 kb in length that remains intranuclear and directly binds to specific sites on the X chromosome that are to be inactivated. A dozen or more genes on the X chromosome, including the gene for choroideremia, can escape inactivation. Once established, X-inactivation is so stable that all daughter cells inactivate the same X chromosome. *XIST* RNA colocalizes with the Barr body during interphase. Certain DNA-binding nuclear proteins interact with the 5′ end of the *XIST* gene and may influence its action.

The proportion of paternally and maternally derived X chromosomes that are inactivated in women shows an approximately normal distribution (a mean of about 50:50). Thus, by chance alone, some women preferentially inactivate a greater proportion of one of the two X chromosomes. However, nonrandom X-inactivation skewing sometimes occurs and results in the inactivation of a greater proportion of the maternally or paternally derived X chromosome. X:autosome translocations are generally associated with preferential inactivation of the normal X chromosome, since the spreading of inactivation onto the autosome would result in monosomy for the genes on this segment.

Skewing of inactivation can also occur in single-gene mutations—for example, incontinentia pigmenti and focal dermal hypoplasia, wherein inactivation of the abnormal gene-containing X chromosome carries a selective survival advantage. X-inactivation skewing that favors the X chromosome with the abnormal gene can occur when the abnormal gene product results in increased proliferation, or when the normal function of the XIC is somehow disrupted. Skewed inactivation of the X chromosome has also been reported in monozygous twins for X-linked disorders, resulting in apparent discordance for expected phenotype.

Lee JT, Jaenisch R. The (epi)genetic control of mammalian X-chromosome inactivation. *Curr Opin Genet Dev.* 1997;7:274–280.

Lyon MF. The William Allan memorial award address: X-chromosome inactivation and the location and expression of X-linked genes. *Am J Hum Genet.* 1988;42:8–16.

Imprinting

Genetic imprinting, also called *allele-specific marking,* is a heritable yet reversible process by which a gene is modified, depending on which parent provides it. The mechanism is unclear but appears to operate at the chromatin organization level and involves heterochromatization and CpG methylation. Examples of genes that can be imprinted include the Wilms tumor–suppressor gene and the human *SNRPN* (small nuclear ribonucleoprotein polypeptide N) gene.

Prader-Willi and Angelman syndromes are examples of diseases resulting from abnormalities of imprinting. About 70%–80% of patients with Prader-Willi syndrome harbor a deletion of the paternally derived chromosome 15q11-q13, resulting in loss of the normal contribution of this region from the paternal line. About 70%–80% of patients with Angelman syndrome also have a deletion of 15q11-q13 but from the maternally derived chromosome, resulting in loss of the maternal contribution. Uniparental disomy, wherein both 15 chromosomes are inherited from the same parent, can also cause each syndrome. Again, the two chromosomes 15 in uniparental disomy are maternal in Prader-Willi and paternal in Angelman syndrome. The *SNRPN* gene maps to 15q11-q13 but appears to be expressed only from the paternally inherited allele.

Gene Translation

Protein Synthesis

Protein synthesis occurs in the cell cytoplasm, mediated by mRNA that has migrated from the nucleus. Synthesis is facilitated (1) by tRNAs that transport amino acids onto the sites of protein synthesis (the *ribosomes*) and (2) by rRNAs that function in connection with the ribosomes during the final stages of protein synthesis (Fig VIII-6).

Transfer RNA Specific tRNAs recognize each of the RNA triplets used in the genetic code. Because of the redundancy of the genetic code, more than one tRNA—as directed by its specific RNA triplet recognition site (or *anticodon*)—may insert the same amino acid into the growing polypeptide. Some diseases, such as the mitochondrial DNA disorder MELAS (*m*itochondrial *e*ncephalomyopathy, *l*actic *a*cidosis, and *s*trokelike episodes), result from a genetic defect of a specific tRNA, in this case

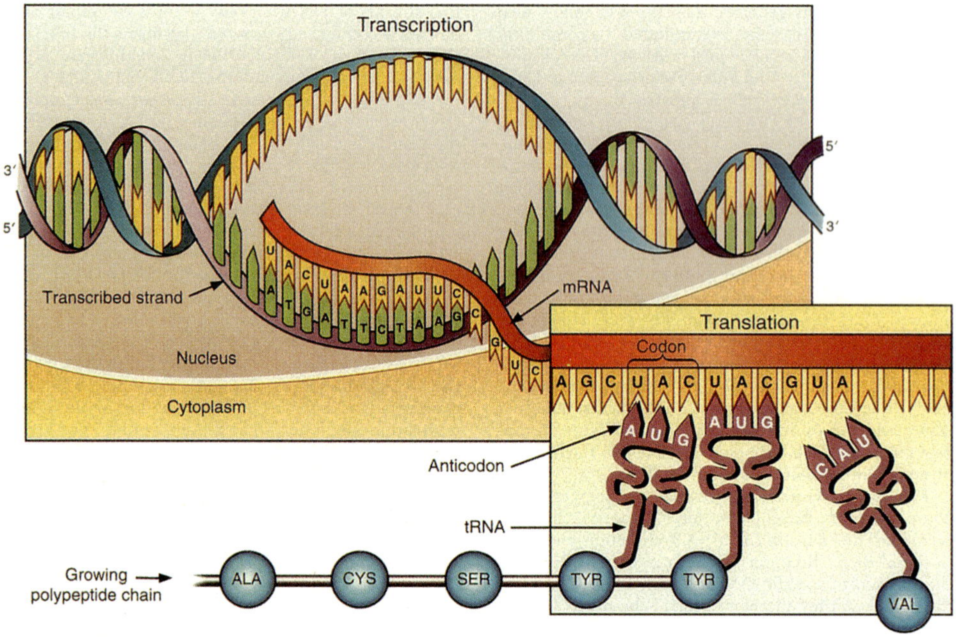

FIG VIII-6—Transcription in the nucleus creates a complementary nucleic acid copy (mRNA, in *red*) from one of the DNA strands in the double helix. The mRNA leaves the nucleus and associates with ribosomes in the cytoplasm, where it is translated into protein *(inset)*. Special transfer RNAs (tRNA, in *purple*) align the corresponding amino acids *(blue)* along the mRNA, using the three-base genetic code to transform the nucleic acid sequence into a protein sequence. (Reproduced with permission from Rosenthal N. Molecular medicine: DNA and the genetic code, *N Engl J Med.* 1994;331:40.)

one for leucine. In this disorder, a misreading of the genetic code is likely whenever the mitochondrial translation machinery requires one of the specified leucine tRNAs.

Transfer RNAs are small molecules, having only 70–80 nucleotides. A three-dimensional cloverleaf structure is formed, predominantly through G-to-C Watson-Crick base pairings of the RNA onto itself. One loop, the *anticodon loop*, contains the recognition site for the mRNA triplet for that specific tRNA.

Transfer RNAs must be charged by having their specific amino acid attached to a specific site in the molecule. This process is performed by 20 separate aminoacyl tRNA synthases; each attaches a specific amino acid onto the one or more tRNAs that code for that amino acid.

The tRNA–amino acid complex attaches to a site on the ribosome through the formation of hydrogen bonds between complementary base sequences on the tRNA and the rRNA. Protein synthesis begins at the N-terminal and proceeds through the carboxyl terminal. The N-terminal end of the final peptide therefore corresponds to the 5′ end of the open reading frame.

Ribosomal RNA Ribosomal RNAs are coded by multiple copies of genes scattered throughout the genome. Human cells contain approximately 200 copies of the largest 45S component rRNA genes spread out in small clusters over five chromosomes. Each genome contains about 2000 copies of the genes for the smaller 5S rRNA, which is about 120 bp in length; these are located in a smaller cluster far away from the other rRNA genes. Such redundancy is essential to maintaining protein synthesis at the levels required for higher organisms.

Post-Translational Modification

Glycosylation Following release from ribosomes, most newly synthesized proteins undergo glycosylation (glucosamine, mannose, and glucose) in the endoplasmic reticulum. This glycosylation occurs prior to transport through the Golgi apparatus, where further modifications occur, often with stripping of some of the sugars from the molecule.

Prenylation (protein lipidation) *Protein lipidation,* also called *prenylation,* is the process of post-translational insertion onto proteins of lipid side chains (eg, farnesyl or geranylgeranyl) to facilitate attachment to membranes and to mediate protein–protein interactions. Prenylation for the farnesyl moiety is called *farnesylation*; for the geranylgeranyl moiety it is called *geranylgeranylation*. This process of protein lipidation is essential for the important cellular processes of endocytosis, vesicular trafficking, and vesicular fusion (docking) with membranes.

Farnesylation of rhodopsin kinase is required for its function in phosphorylation of activated rhodopsin. Choroideremia is a disorder of geranylgeranylation caused by a defect in the gene *CHM* (also known as Rab Escort Protein-1, or *REP-1*), which is required for protein lipidation of a subset of G proteins essential for the retinal pigment epithelium and choriocapillaris.

Seabra MC, Ho YK, Anant JS. Deficient geranylgeranylation of Ram/Rab27 in choroideremia. *J Biol Chem.* 1995;270:24420–24427.

DNA Replication

Gene replication occurs with surprising fidelity. The final error rate for DNA replication is about 1×10^{-10} mutations per base pair per cell generation. Approximately 1 individual in every 100,000 harbors a new disease-producing mutation. The incidence of neutral variations, including some that may even be beneficial, is probably slightly greater. If the error rate for DNA replication were zero, genetic diversity and evolution would be impossible. If the rate were much higher, the likelihood of receiving a faithful copy from one's parents would decrease and mutations would accumulate with each cell division, eventually resulting in cell dysfunction, cell death, or the formation of cancers. Evidence suggests that the aging process itself involves loss of fidelity or efficiency of DNA repair processes.

DNA replication begins at specific sites on the chromosomes, each called an *origin of replication* (or *replication origin*), wherein the double strands of DNA separate and replication proceeds bidirectionally. A simple concept of DNA replication suggests that the genetic information residing within the nucleus is merely duplicated during the S phase prior to cell division. However, the actual events that must take place are complex. The logistics include copying the entire genome, which comprises billions of nucleotides; packaging the copies into duplicate chromosomes; and then directing the replicated chromosomes to segregate appropriately to the daughter cells through the intricate movements involved in the processes of mitosis and meiosis. Specific events must take place at the appropriate time over hundreds to thousands of sites. These events are coordinated by the joint actions of hundreds of gene products that govern the cell cycle, initiate replication, create chromosomes, and (in conjunction with the centromere) form the mitotic spindle. Many of these mechanisms are only now beginning to be understood.

DNA Repair

DNA Damage

DNA is constantly sustaining damage from mutagens such as ultraviolet light, chemicals, and spontaneous deamination. Each cell loses 10,000 bases per day from spontaneous DNA breakdown related to normal body temperature alone. This process may involve hydrolytic loss of purine bases or deamination of cytosine to uracil and, less frequently, adenine to hypoxanthine. Oxidation, alkylation, generation of free radicals, and other common metabolic reactions can also injure DNA. In the absence of repair, these mutations would accumulate and result in tumor formation. Damaged DNA is estimated to cause about 80%–90% of cancers in humans.

Damaged DNA sites are repaired chiefly by two mechanisms: *excision repair* and *mismatch repair*. The processes of replication, transcription, mismatch repair, excision repair, and gene expression are closely coordinated by cross-acting systems. Enzymes that cut or patch segments of DNA during crossing over at meiosis are also involved in DNA repair. Molecules that unwind double-stranded DNA (called *helicases*) are involved in replication, transcription, and DNA excision repair.

The *anti-oncogene p53* appears to play an extremely important role as the "guardian of the genome" by preventing cells from proliferating if their DNA is irreparably damaged. Levels of p53 increase after ultraviolet or ionizing radiation. p53 inhibits DNA replication directly and binds with one of the RNA polymerase transcription factors, TFIIH. If the degree of damage is slight, increased production of p53 induces reversible cell arrest until DNA repair can take place. If DNA damage is too great or irreversible, p53 production is massively increased and apoptosis occurs, probably through stimulation of the expression of the *BAX* gene, whose product promotes apoptosis. Loss of p53 causes cells to fail to arrest in response to DNA damage, and these cells do not enter apoptosis. Thus, mutations of p53 predispose to tumorigenesis.

The gene mutated in ataxia-telangiectasia, a protein kinase called *ATM*, also appears to be integrally involved in DNA repair, possibly by informing the cell of radiation damage. The ATM gene product associates with synaptonemal complexes, promotes chromosomal synapsis, and is required for meiosis. Persons with ataxia-telangiectasia have a threefold greater risk of cancer, which can involve breast, lung, pancreas, stomach, or biliary tree.

Latchman DS. Transcription-factor mutations and disease. *N Engl J Med*. 1996;334:28–33.

Levine AJ. p53, the cellular gatekeeper for growth and division. *Cell*. 1997;88:323–331.

Yu CE, Oshima J, Fu YH, et al. Positional cloning of the Werner syndrome gene. *Science*. 1996;272:258–262.

Excision Repair

Excision repair is essential for survival of the organism and indeed the species. Most DNA lesions block RNA transcription and DNA replication by interfering with normal polymerase function. Excision repair can operate on either single damaged bases *(base excision repair)* or larger, bulkier lesions in DNA *(nucleotide excision repair)*. Base excision repair is the most important mechanism for repair of oxidative and spontaneous DNA damage, whereas nucleotide excision repair is the more versatile system for excising more complex, if less frequent, lesions. Excision repair is a complex process involving the coordinated action of up to 30 different gene products.

The complementary nature of DNA strands sets the stage for excision repair. In essence, the information from the intact strand is used to repair the damaged base or nucleotide. In base excision repair, DNA glycosylase releases the damaged base and the enzyme AP(apurinic/apyrimidinic)-endonuclease excises the abasic sugar. (The nuclease repair process in excision repair is called *excinuclease* to distinguish it from activity usually associated with endonucleases.)

Unlike the mismatch repair system, which is discussed below, the excinuclease system is unable to differentiate which strand in a simple DNA mismatch is the correct one and which contains the mutation. Thus, if the excinuclease repairing a single mismatch chooses the normal strand to "repair," the effect will be fixation of the mutation. This may seem to be an excessive price to pay for a single damaged base or nucleotide, but excision repair appears to be the only system able to remove the nearly infinite spectrum of possible mutations in humans, especially the bulkier lesions. Fortunately, the mismatch repair system is much more efficient in making these smaller repairs, and such false repairs by the excinuclease system do not contribute significantly to the mutational load.

Three diseases in humans are associated with defects of genes involved in nucleotide excision repair:

- Xeroderma pigmentosum (XP)
- Cockayne syndrome (CS)
- Trichothiodystrophy (TTD)

All occur from defects of any of several excision repair genes. XP is characterized by defective replicational repair and associated with sun-induced photosensitivity, skin cancers, and neurologic abnormalities. CS is characterized by defective transcriptional repair and associated with mental retardation, retarded growth, cataracts, retinal degeneration, premature aging, and early death. Patients with TTD exhibit brittle hair, mental retardation, neurologic abnormalities, and skeletal abnormalities. Neither CS nor TTD patients are excessively susceptible to cancers during their life spans.

Mismatch Repair

Single base-pair mutations occur during DNA replication at the rate of 1 in 100,000 nucleotides copied. Such errors may occur because of insertion of the wrong base (eg, an A opposite a C) or from the addition of extra nucleotides, creating a loop of up to five or more unpaired bases in the helix during DNA replication. Proofreading functions of the replication complex appear to be able to correct all but 1 in 1000 of the misincorporation errors, reducing the order of magnitude of remaining errors to 10^{-8}. The process of repair of these remaining mistakes involves the mismatch repair system, which corrects all but 1 in 100 of the errors, resulting in the final overall replication error rate of 10^{-10}.

Amazingly, the mismatch repair system appears to know which is the correct copy and which is the mutant copy. The actual process of mismatch repair is targeted to the newly created strand because of the transient unmethylated state of adenines at GATC sequences that occurs characteristically in newly replicated DNA strands. The actual process of mismatch correction involves a complex mechanism involving ten separate activities. Four proteins—MutH, MutL, MutS, and MutU—are required for mismatch repair in *Escherichia coli,* the system in which the repair mechanism was first elucidated. Repair in *E coli* is initiated by binding of first MutS and then MutL, MutH, and MutU to the mismatch, activation of a latent GATC-directed endonuclease that eventually incises the sequence that includes the mismatch, and finally restoration of the incised segment by resynthesis and ligation. Human mismatch repair appears to occur similarly.

Mismatch repair is highly conserved in nature, and defects of this system would be expected to be associated with disease or tumor formation. A particular form of genetic cancer, hereditary nonpolyposis colorectal cancer (HNPCC), is associated with severe deficiency of strand-specific mismatch repair from defects of any of four genes required for the strand-specific mismatch repair process (*hMSH2, hMLH1, hPMS1,* and *hPMS2*). HNPCC is an autosomal dominant trait, but the tumor cells are always defective for both copies of the gene in question, as in retinoblastoma. Normal somatic cells repair mutations normally because of the existence of one good copy of the gene; however, cell lines derived from colorectal tumors collect mutations at a rate exceeding 100 times normal, suggesting that they have become homozygous for the mutant copy. The missing repair function leads to mutational events that defeat the normal control of cell proliferation and thereby result in cancer formation.

Mutations and Disease

Requirements for Identifying a Disease-Producing Mutation

The major characteristics required for a given DNA mutation to be verified as disease-producing are:

- It does not occur in the normal population (the variation cannot be more frequent than the disease).
- It produces a DNA sequence that alters protein function or expression.
- The presence of the variation cosegregates with disease in family members according to the inheritance type (the significance of cosegregation in a given family depends on the number of possible chances for noncosegregation).

Mutations

Mutations can involve a change in a single base pair; simple deletion or insertion of DNA material; or more complex rearrangements such as inversions, duplications, or translocations. Deletion, insertion, or duplication of any number of base pairs in other than groups of three creates frame shifts of the entire DNA sequence downstream, resulting in the eventual formation of a stop codon and truncation of the message.

Mutations that result in no active gene product being produced are called *null mutations.* Null mutations include missense or nonsense mutations that (1) produce either a stop mutation directly or a frame shift with creation of a premature stop codon downstream or (2) cause the loss or gain of a donor or acceptance splice junction site, resulting in the loss of exons or inappropriate incorporation of introns into the spliced mRNA.

Mutations can also lead to a gain of function that may be beneficial (leading to evolution) or detrimental (leading to disease). An example of a beneficial gain in function is the emergence of antibiotic resistance among bacteria. An example of a detrimental gain of function is a receptor protein that binds too tightly with its target protein, creating loss of normal physiologic function.

Single base-pair mutations may code for the same amino acid or a tolerable change in the amino acid sequence, leading to harmless polymorphisms or DNA variations that are in turn inherited. These are called *conserved base-pair mutations.*

Transitions and transversions *Transitions* are replacements of a purine with another purine or a pyrimidine with another pyrimidine (C to T, G to A, T to C, A to G). *Transversions* are replacements of a purine with a pyrimidine, or vice versa (C to A, C to G, G to T, T to G).

Private sequence variations A variation in the DNA sequence in association with a given disease does not by itself indicate that the variation causes the disease. The existence of tolerated mutations or mutations that do not significantly alter the gene product may, because of nonallelic disequilibrium, cosegregate with disease.

Polymorphisms A *polymorphism* is any variation in DNA sequence that occurs frequently in the normal population. By convention, variations that are present at a frequency of 1% or greater are called polymorphisms.

Cancer Genes

Cancer can result from any of a number of genetic mechanisms, including the activation of oncogenes and the loss of tumor-suppressor genes. The product of proto-oncogenes is often involved in signal transduction of external messages to the intracellular machinery that governs normal cell growth and differentiation (Fig VIII-7). As such, the DNA sequences of proto-oncogenes are highly conserved in nature between such different organisms as humans and yeast. Proto-oncogenes can be activated to oncogenes by loss or disruption of normal regulation.

Oncogenes Oncogenes were first detected in retroviruses that had acquired them from their host in order to take control over cell growth. Such oncogenes are often identified by names such as *ras* that refer to the viral source (*ra*t *s*arcoma virus). They are found to be activated not only in virus-induced malignancies but in common nonviral cancers in humans. Oncogenes behave as autosomal dominant traits, and only one mutant allele is needed for tumor formation, presumably by a dominant negative effect on regulation of signal transduction.

Some oncogenes result in congenital anomalies as well as tumor formation. The *RET* gene codes for a cell surface–membrane protein that appears to receive the messages from growth factors and relay the signal to intracellular pathways. When mutated at different nucleotide regions, *RET* can cause seemingly disparate phenotypes, including aganglionic megacolon (Hirschsprung disease), familial medullary

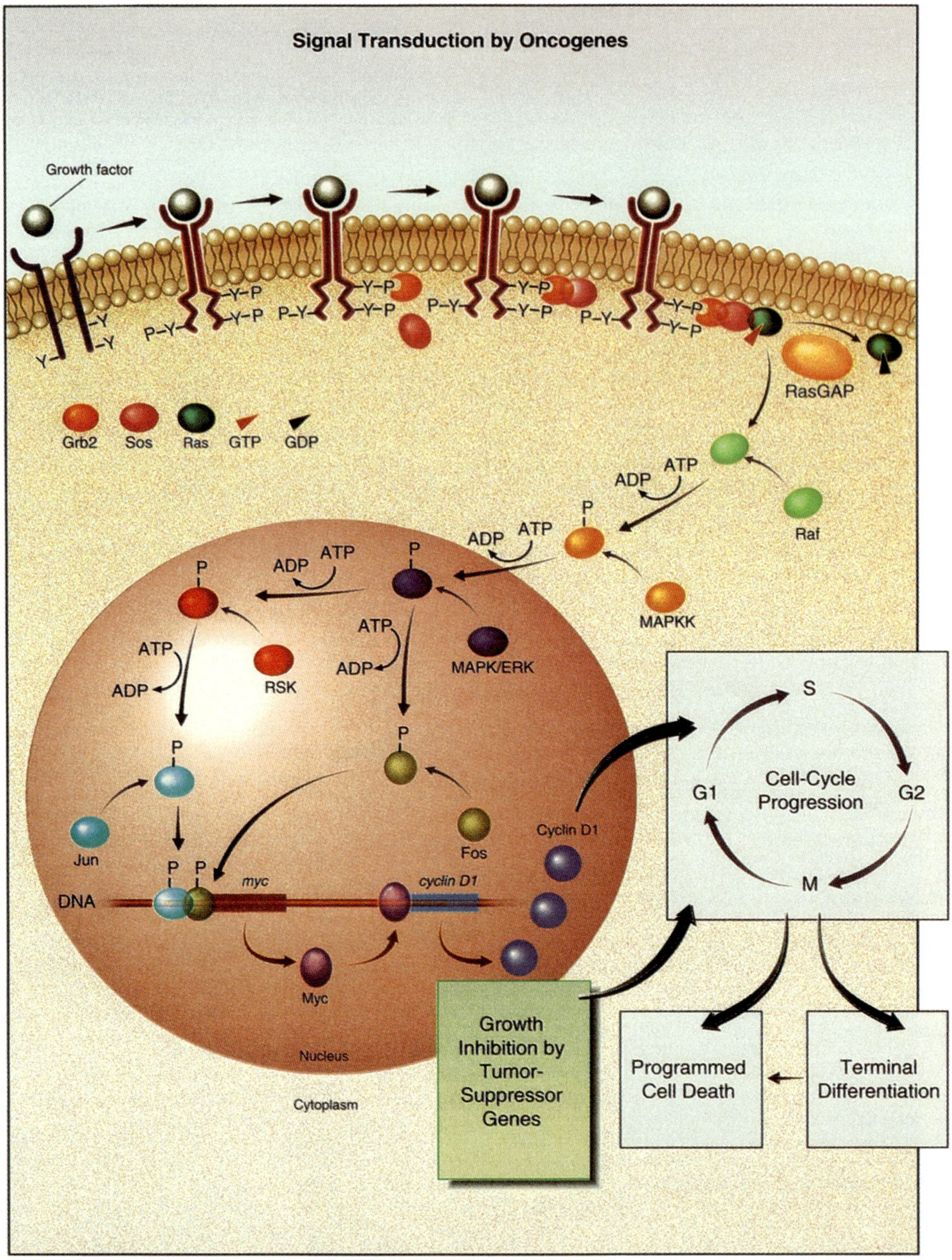

FIG VIII-7—In the cell cycle, the progression from DNA synthesis *(S)* to mitosis *(M)* includes phases before *(GI)* and after *(G2)* the replication of DNA. On receiving signals to differentiate, cells leave the cycle and enter the pathway of terminal differentiation. Under certain circumstances, cells may enter the pathway to programmed cell death (apoptosis). Signal transduction begins with the binding of a growth factor to its transmembrane receptor *(upper left)*. Usually, the next step is the dimerization of the receptor. The receptor subunits then phosphorylate one another on tyrosine residues *(Y)*. The phospho-tyrosines *(P)* create docking sites on the receptor for many proteins, some of which undergo phosphorylation; others recruit multicomponent complexes to the plasma membrane. One such interaction, shown here, is the activation of the Ras GTPase. In the cascade of phosphorylation initiated by the activation of Ras, the Raf kinase phosphorylates another kinase (mitogen-activated protein kinase kinase, or MAPKK), which in turn phosphorylates a third kinase, the mitogen-activated protein kinase, or MAPK. MAPK directly activates transcription factors and ribosomal S6 protein kinase *(RSK)*, which also phosphorylates transcription factors. MAPK probably represents at least two related proteins. Two transcription proteins, Fos and Jun, are shown. They join to form a fully active transcription factor. The phosphorylation of Fos by MAPK and of Jun by RSK causes them to bind to specific DNA sequences near the myc gene, thereby initiating transcription of the gene. The Myc protein itself is a transcription factor with several binding partners (not shown). The binding of Myc to its specific recognition sites on DNA activates another set of genes. Cyclin D1 initiates the progression cells through G1 to the S phase. (Reproduced with permission from Krontiris TG. Molecular medicine: oncogenes. *N Engl J Med.* 1995;333:304.)

thyroid carcinoma, and multiple endocrine neoplasia types IIA and IIB. *RET* is expressed in the developing central and peripheral nervous systems (sensory, autonomic, and enteric ganglia) and the excretory system. This wide range of expression is probably the basis of the multiplicity of systems involved.

Tumor-suppressor genes *Tumor-suppressor genes,* also called *anti-oncogenes,* are genes that must be present in one functional copy to prevent uncontrolled cell proliferation. Although some may represent genes whose products participate in checkpoints for the cell cycle, one characteristic of tumor-suppressor genes is the diversity of their normal functions. Examples of tumor-suppressor genes include the genes for retinoblastoma, Wilms tumor, neurofibromatosis types 1 and 2, tuberous sclerosis, ataxia-telangiectasia, von Hippel–Lindau disease, and HNPCC. All of these examples (except ataxia-telangiectasia) behave as autosomal dominant traits, but the mechanism of tumor formation is very different for tumor-suppressor genes compared to oncogenes. If one allele is already defective because of a hereditary mutation, the other allele must also be lost for tumor formation to occur. This loss of the second allele is termed *loss of heterozygosity,* and it can occur from a second mutation, gene deletion, chromosomal loss, or mitotic recombination.

Mutation-producing mutations Some mutations produce an increased frequency of chromosomal and locus-specific mutations, resulting in premature aging and, often, eventual cancers, lymphomas, or leukemias. Bloom syndrome, Werner syndrome, xeroderma pigmentosum, and Cockayne syndrome all result from genetic instability caused by mutations of DNA helicases involved in replication, transcription, or DNA repair. Hereditary nonpolyposis colorectal cancer is caused by a defect in DNA mismatch repair that increases the rate of mutation 100-fold, spawning mutations that finally lead to tumor formation.

Ellis NA. Mutation-causing mutations. *Nature.* 1996;381:110–111.

Krontiris TG. Oncogenes. *N Engl J Med.* 1995;333:303–306.

Mitochondrial Genome

Genetic Code

Mitochondrial DNA (mtDNA) constitutes 0.3% of the total DNA of a human cell but only 0.0005% of the human genome. The complete sequence, which in humans is 16,569 bp in length, is known for numerous species. Each small, circular chromosome contains 37 genes, including those for polypeptides of several respiratory complexes such as cytochrome oxidase as well as tRNAs and rRNAs. These extranuclear, or *cytoplasmic,* genes determine the production of 13 of the more than 69 peptides of the mitochondrial respiratory chain and adenosine triphosphate (ATP) synthase. The remainder of the peptides of the mitochondrial respiratory chain are encoded by nuclear DNA.

The genetic code for the translation apparatus for mtDNA differs from the "universal" code in the following respects:

- TGA, normally a stop codon, codes for tryptophan
- AGA and AGG, which code for arginine in nuclear DNA, are stop codons
- ATA and ATT, which code for isoleucine in nuclear DNA, code for methionine

Genetic transmission of mtDNA appears to be exclusively maternal. Each ovum contains about 100,000 copies of mtDNA, whereas somatic cells usually have 1000–10,000 copies. The midpiece of the mature sperm has about 50 copies and is degraded after the sperm enters the egg at conception.

Genomic Structure

Mitochondrial DNA exists as double-stranded circular DNA. Histones are not present. The DNA for the structural coding regions of mtDNA has no introns; instead, it is composed of an open reading frame similar to that of prokaryote DNA.

For DNA synthesis, mitochondria use a specific DNA polymerase (DNA polymerase γ), whereas the nucleus uses DNA polymerases α, β, δ, and ε. The mutation rate for mtDNA is 10–20 times higher than that for nuclear DNA. Contributing to this increased mutation rate are the following:

- Absence of the protective histones
- Absence of repair enzymes
- The high oxygen concentration in the mitochondria, leading to DNA damage by reactive oxygen species

Mitochondrial DNA codes for the two rRNAs found in mitochondrial ribosomes, the 22 tRNAs used to translate the mitochondrial mRNAs, and the structural genes for 13 polypeptides. Three of these polypeptides are subunits of the cytochrome-*c* oxidase complex, two are subunits of ATPase, seven are subunits of NADH-CoQ reductase complex (ND1, ND2, ND3, ND4L, ND4, ND5, ND6), and one is the cytochrome-*b* subunit of CoQ–cytochrome-*c* reductase.

Mitochondrial Disease

A significant number of disorders associated with the eye or visual system involve mitochondrial deletions and mutations. Mitochondrial diseases should be considered whenever the inheritance pattern of a trait suggests maternal transmission. The inheritance pattern might superficially resemble that of an X-linked trait. Maternal transmission differs from X-linked inheritance in that all of the offspring of affected females—both daughters and sons—can inherit the trait, but only the daughters can pass it on.

The phenotype and severity of mitochondrial disease appear to depend on the nature of the mutation, the presence or degree of heteroplasmy (coexistence of more than one species of mtDNA—ie, wild type and mutant), and the oxidative needs of the tissues involved. Spontaneous deletions and mutations of mtDNA accumulate with age, and the effect of this accumulation is to decrease the efficiency and function of the electron-transport system, reducing the availability of ATP. When energy production becomes insufficient to maintain the function of cells or tissue, disease occurs. There appears to be an important interaction between age and tissue threshold of oxidative phosphorylation need and the expression of inherited mutations of mtDNA.

With each cell division, the number of mutant mtDNA copies that are partitioned to a given daughter cell is random, obeying galtonian rather than mendelian inheritance characteristics. After a number of cell divisions, some cells, purely by chance, receive more normal or more mutant copies of mtDNA, resulting in a drift toward homoplasmy in subsequent cell lines. This process is called *replicative segregation*. With mtDNA deletions, preferential replication of the smaller deleted mol-

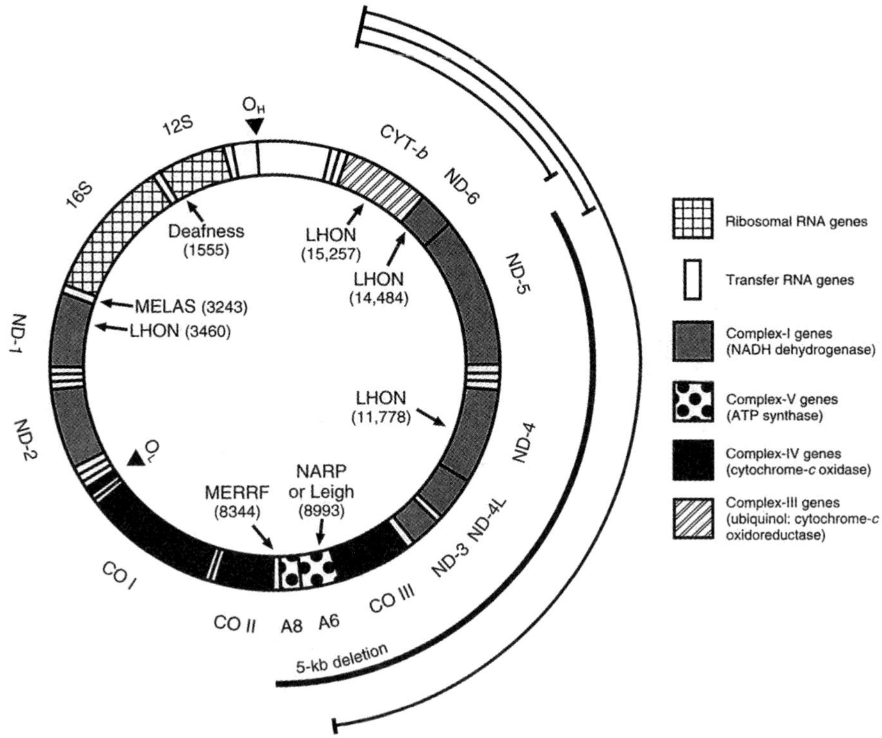

FIG VIII-8—Diagram of human mitochondrial DNA and the most common associated pathogenetic mutations. Point mutations in structural and protein-coding genes are shown inside the circle, with the clinical phenotype indicated and the nucleotide position of the mutation shown in parentheses. The position of the most common single deletion, which is 5 kb long, and the multiple deletions are indicated by the arcs outside the circle. *MERRF,* myoclonic epilepsy with ragged red fibers; *NARP,* neuropathy, ataxia, and retinitis pigmentosa; *Leigh,* maternally inherited Leigh disease; *LHON,* Leber hereditary optic neuropathy; *MELAS,* syndrome of mitochondrial encephalomyopathy, lactic acidosis, and strokelike episodes; O_H, origin of heavy-stranded DNA replication; O_L, origin of light-stranded DNA replication; *CYT-*b, apocytochrome-*b* subunit; *ND-1, ND-2, ND-3, ND-4L, ND-4, ND-5, ND-6,* NADH dehydrogenase subunits; *CO I, CO II, CO III,* cytochrome-*c* oxidase subunits; *12S and 16S,* ribosomal RNA subunits; *A6, A8,* ATPase subunits. The large open space at the top, which includes O_H, is the noncoding D (displacement) loop. (Reproduced with permission from Johns DR. Mitochondrial DNA and disease. *N Engl J Med.* 1995;333:641.)

ecules causes an increase of deleted copy over time. The trend toward homoplasmy helps explain the worsening of disease with age and new involvement of organ systems not previously involved in multisystem mitochondrial disease.

Mitochondrial diseases can be subdivided into these categories (Fig VIII-8):

- Disorders resulting from large rearrangements of mtDNA (deletions and insertions), such as chronic progressive external ophthalmoplegia (CPEO), Kearns-Sayre syndrome, and Pearson marrow-pancreas syndrome

- Mutations of mtDNA-encoded rRNA, such as maternally inherited sensorineural deafness and aminoglycoside-induced deafness
- Mutations of mtDNA-encoded tRNA, such as the syndromes of MELAS, myoclonic epilepsy with ragged red fibers (MERRF), adult-onset diabetes and deafness, and (in about 30% of cases) CPEO
- Missense and nonsense mutations such as Leber hereditary optic neuropathy (LHON) and neuropathy, ataxia, and retinitis pigmentosa (NARP)

Chronic progressive external ophthalmoplegia CPEO is a disorder involving progressive ptosis and paralysis of eye muscles associated with a ragged red myopathy, usually as a result of deletion of a portion of the mitochondrial genome. Patients with CPEO commonly have pigmentary retinopathy that does not create significant visual disability. Infrequently, they may have more marked retinal or other system involvement, the so-called *CPEO-plus syndromes*. In Kearns-Sayre syndrome, CPEO is associated with heart block and severe retinitis pigmentosa with marked visual impairment. Pearson marrow-pancreas syndrome results from a large deletion of mtDNA and presents in younger patients with an entirely different phenotype involving sideroblastic anemia and pancreatic exocrine dysfunction. However, in later years, Pearson marrow-pancreas syndrome can evolve into a phenotype resembling Kearns-Sayre syndrome.

Although roughly 50% of patients with CPEO have demonstrable mtDNA deletions, virtually all patients with Kearns-Sayre syndrome have large deletions. As many as 30% of patients with CPEO who do not harbor demonstrable mtDNA deletions may have a point mutation at nucleotide position 3243, the same mutation in the tRNA for leucine that in other persons is associated with MELAS syndrome. For all of the syndromes associated with deletions, such as Kearns-Sayre and CPEO, detection of the deletion usually requires study of muscle tissue.

Leber hereditary optic neuropathy The most important ophthalmologic disease of mitochondria is LHON, which is more prevalent in males than females but does not fit a classic X-linked pattern of transmission. The trait is not transmitted to the offspring of affected males; virtually every daughter and son of a female patient with LHON inherits the trait. In about 50% of cases, LHON development is correlated with a single base change (G to A at nucleotide position 11778 in the *ND-4* gene) in human mtDNA involved in the synthesis of NADH dehydrogenase. In addition to optic atrophy, patients can exhibit peripapillary microangiopathy and cardiac abnormalities, especially Wolff-Parkinson-White syndrome. LHON can also occur from other so-called primary mutations at nucleotide positions 3460 of *ND-1*, 14484 of *ND-6*, 14459 of *ND-6*, and (more controversially) 15257 of cytochrome-*b*. At least 12 secondary mutations have been associated with LHON, often when multiple mutations are present in an individual's mitochondria. Some authors think that these secondary mutations cause disease by additive detrimental effects on the electron-transport system of oxidative phosphorylation. Most of these secondary mutations appear in the general population. Debate persists on whether each mutation alone is truly pathogenic.

The likelihood of improvement with time in the recovery of visual acuity appears to differ among the separate mutations associated with LHON. Mutation at nucleotide position 11778 is associated with the least, and mutation at nucleotide position 14484 is associated with the greatest, likelihood of recovery. The mutation at 14459 of *ND-6* appears to be associated with two very different clinical phenotypes, one LHON and the other a severe, early-onset progressive dystonia with pseudobulbar syndrome, short stature, and reduced intelligence. The two phenotypes may reflect different proportions or distribution of mutant mtDNA.

Neuropathy, ataxia, and retinitis pigmentosa NARP is associated with a single base-pair mutation at nucleotide position 8993 in the ATPase-6 gene. The NARP phenotype occurs when the percentage of mutant mtDNA is lower than 80%, whereas the same mutation present at much higher proportions (greater than 95%) can cause Leigh syndrome, a severe neurodegenerative disease of infancy and early childhood. The 8993 mutation is demonstrable in fibroblasts and lymphoblasts.

Other mitochondrial diseases Aminoglycoside-induced deafness and streptomycin ototoxicity are instances wherein antibiotic administration, often only a modest dose, is associated with severe hearing loss. This susceptibility to ototoxicity is a maternally inherited trait. Aminoglycosides (kanamycin, gentamicin, tobramycin, and neomycin) "target" the evolutionarily related bacterial ribosome. The mechanism of action is thought to include interference with the production of ATP in the mitochondria of hair cells in the cochlea.

Finally, some mitochondrial diseases most assuredly involve defects of nuclear-encoded oxidative phosphorylation genes or interactions between products of mtDNA and nuclear DNA. Nuclear gene mutations may also be responsible for a phenotype similar to LHON or may contribute to the expression of LHON in at least some families.

Brown MD, Wallace DC. Molecular basis of mitochondrial DNA disease. *J Bioenerg Biomembr.* 1994;26:273–289.

Johns DR. Seminars in medicine of the Beth Israel Hospital, Boston. Mitochondrial DNA and disease. *N Engl J Med.* 1995;333:638–644.

Nikoskelainen EK, Savontaus ML, Wanne OP, et al. Leber's hereditary optic neuroretinopathy, a maternally inherited disease. A genealogic study in four pedigrees. *Arch Ophthalmol.* 1987;105:665–671.

Phillips CI, Gosden CM. Leber's hereditary optic neuropathy and Kearns-Sayre syndrome: mitochondrial DNA mutations. *Surv Ophthalmol.* 1991;35:463–472.

The Search for Genes in Specific Diseases

A variety of methods have been used to assign individual genes to specific chromosomes, to link individual genes to one another, and to link diseases to specific genes.

Synteny

The presence of genes on the same chromosome, even if the genes are too far apart to demonstrate linkage, is called *synteny*. Genes that are X linked, such as red-green color blindness, choroideremia, and hemophilia, are by definition syntenic. The term is also used to denote homologous chromosomal regions between species. For example, the mouse gene for rhodopsin is localized to the distal half of mouse chromosome 1, which is syntenic or homologous to human chromosome 3q.

Both the value and the limitations of the study of genetic disease in other animals can be seen in the following example: the use of the mouse in the discovery of the gene for one form of type I Usher syndrome (profound congenital deafness, vestibular dysfunction, and retinitis pigmentosa). A mouse mutant for deafness, shaker1, had been mapped to a conserved linkage region on mouse chromosome 7, and USH1B was linked to human chromosome 11q13, suggesting that the two might result from similar or homologous genes. The shaker1 gene was isolated in 1995 by positional cloning and the mutated gene identified as Myosin 7a. The human counterpart to this gene, *MYO7A*, was quickly identified as the gene mutated also in *USH1B*. The subsequent twist to the story is that the mouse mutant has deafness and vestibular dysfunction but not retinitis pigmentosa because of differences in tissue-specific expression. Thus, animal models may help to find genes that cause human disease, but the expression of mutations in the homologous genes may have important species differences.

El-Amraoui A, Sahly I, Picaud S, et al. Human Usher IB/mouse shaker-1: the retinal phenotype discrepancy explained by the presence/absence of myosin VIIA in the photoreceptor cells. *Hum Mol Genet.* 1996;5:1171–1178.

Gibson F, Walsh J, Mburu P, et al. A type V11 myosin encoded by the mouse deafness gene *shaker-1. Nature.* 1995;374:62–64.

Meisler MH. The role of the laboratory mouse in the human genome project. *Am J Hum Genet.* 1996;59:764–771.

Weil D, Blanchard S, Kaplan J, et al. Defective myosin VIIA gene responsible for Usher syndrome type 1B. *Nature.* 1995;374:60–61.

Cytogenetic Markers (Morphologically Variant Chromosomes)

If a specific chromosomal structure is abnormal or even normally variant, its transmission through a family with a hereditary disease, as mapped by a pedigree, may allow the assumption that the mutant gene and the variant chromosome are comigrating. Thus, the mutant gene is physically located on the variant chromosome.

Gene Dosage

If a portion of a chromosome containing a specific gene is physically deleted, the amount of the gene product will only be determined by the remaining homologue. For example, 50% of normal levels of esterase-D may be found in the serum of persons with an interstitial deletion of part of the long arm of chromosome 13. When several such persons were also found to have retinoblastoma, it was suggested that both the esterase and the retinoblastoma genes are located in the missing segment. By contrast, duplication mapping requires finding 150% of normal activity of a given gene product, together with either a chromosomal trisomy or triplication of a specific chromosomal segment.

Association

Certain combinations of traits may occur for reasons other than the physical relationship of genes. For example, blood group O and peptic ulcer are found together in the same person more often than would be expected from their individual frequencies in the population. This finding occurs not because the *ABO* gene and another gene for peptic ulcer are located on the same chromosome, but because persons with type O blood have a physiologic peculiarity that predisposes them to peptic ulcerations. In another example, retinal detachment occurs more frequently in patients with Marfan syndrome and homocystinuria than in the general population. Rather than resulting from the concurrent action of two linked genes, this association is the result of *pleiotropism,* the multiple effects of a single gene.

Goldberg MF, Renie WA, eds. *Genetic and Metabolic Eye Disease.* 2nd ed. Boston: Little Brown; 1986.

Linkage

Even if no information is known about the nature or function of a gene for a disease, linkage studies may be able to localize the gene to a given chromosome or specific marker.

The first linkage assignment In 1937, Bell and Haldane recognized the first linkage between two diseases on a human chromosome: congenital color deficiency and hemophilia on the X chromosome. Subsequent investigations have led to the chromosomal mapping of over 100 different human ocular diseases.

Other gene assignments As of January 2000, OMIM* *(Online Mendelian Inheritance in Man: www.ncbi.nlm.nih.gov/omim)* listed over 8,699 established human gene loci. Every chromosome has numerous defined genes. Human gene mapping has two major applications. The first is identification of the gene for a specific genetic disease by its linkage to a known marker. For example, suppose gene A causes a hereditary disease and gene B is a known enzyme or polymorphic marker closely linked to A. Even though no biochemical test exists for A, a tight linkage to B would allow a reasonable probability of identifying the disease for prenatal diagnosis and sometimes for carrier detection. The second impact of linkage is understanding the cause of the phenotypic malformations in specific chromosomal diseases. (For example, the phenotype of Down syndrome may result from triplication of only the distal long arm of chromosome 21 through a chromosome rearrangement rather than trisomy of the entire chromosome.)

Mets MB, Maumenee IH. The eye and the chromosome. *Surv Ophthalmol.* 1983;28: 20–32.

Markers: RFLPs Polymorphisms detectable by the presence or absence of a specific restriction endonuclease cleavage site are called *restriction fragment length polymorphisms (RFLPs).* The differences between two chromosomes in DNA genetic material fragment size between restriction endonuclease cleavage sites are inherited and become useful markers for following various genetically determined disorders.

Identification of RFLPs begins when DNA is isolated from peripheral blood lymphocytes. The DNA fragments are then produced when the DNA is cut with restriction endonucleases. Each restriction endonuclease recognizes a highly specific sequence of four to nine bases and cuts double-stranded DNA wherever this sequence occurs. Change of a single base within the recognition sequence results in the loss of that cleavage site and a change in the corresponding DNA fragment length. A single base-pair change elsewhere may create a new recognition cleavage

* *Online Mendelian Inheritance in Man (OMIM)* is the computer database based on *McKusick's Mendelian Inheritance in Man.* This catalog is now maintained by the National Center for Biotechnology Information (NCBI) of the National Library of Medicine.

site where none had existed. A variation in DNA sequence involving a single base pair occurs with a frequency of about 1 per 200–500 bp. The variable-length fragments, determined by the spacing of the restriction enzyme recognition sites, are then separated by agarose gel electrophoresis. The gel will contain millions of DNA fragments.

To identify a certain fragment that might be linked to a specific genetic locus (ie, a defective gene), a radioactively labeled DNA probe is hybridized with the DNA fragments generated by the restriction endonucleases. The various probes represent cloned DNA sequences, which are complementary (homologous in base-pair sequence) to a part of the DNA fragment containing the RFLPs. Any fragments containing part or all of the radioactively labeled sequence can be identified by radioactive or nonradioactive detection methods (Fig VIII-9).

RFLPs have been used to map the gene locus implicated in numerous diseases. In many cases, the gene in question has been identified by positional cloning or candidate screening, which are both discussed below.

It is possible to detect linkage by observing the frequency with which a polymorphic marker is inherited with a disease trait, provided that the disease locus is within 20–30 centimorgans (cM) of the *marker* site. The physical distance represented by 1 cM (0.01 recombination fraction) is about 1 million bp (1000 kb) and corresponds to a 1% chance that recombination will result from a single meiosis. When a genetic probe is sufficiently close to a disease gene, they are rarely separated by meiotic recombination. The frequency of separation by chromosomal exchange at meiosis is their *recombination frequency.* Linked markers should be no more than 20 cM apart. For perspective, the average chromosome contains about 150 cM; there are about 3300 cM in the entire human genome, which corresponds to 3×10^9 bp.

Botstein D, White RL, Skolnick M, et al. Construction of a genetic linkage map using restriction fragment length polymorphisms. *Am J Hum Genet.* 1980;32:314–331.

When determining linkage between a diseased gene and a marker, geneticists compare different models by calculating likelihood ratios. When the likelihood ratio is 1000:1 that the odds of one model are greater than those of another, the first is accepted over the second. The log of the likelihood ratio (*l*ogarithm of *od*ds score, or *LOD score*) is usually reported. An LOD score of 1–2 is of potential interest in terms of linkage; 2–3 is suggestive; and greater than 3 is generally considered proof of linkage. Although an LOD score of 3 gives a probability ratio of 1000:1 in favor of linkage versus independent assortment, this score does not indicate a Type 1 error as low as 0.001 but, in fact, it indicates an error that is close to 0.05, the standard significance level used in statistics. (BCSC Section 1, Chapter XVI, Epidemiology and Statistics, explains these concepts in depth.)

Markers: microsatellites, minisatellites, and satellites Within the genome exist variable lengths of repetitive DNA composed of multiple units that each may be 1–5 bp in length *(microsatellites),* 14–100 bp in length *(minisatellites),* or larger *(satellites).* One class of such repeats is also called *short tandem repeats* (STRs). These are tandemly repeated blocks of two to five nucleotides. The variability in the number of repeats produces polymorphisms that are useful for linkage studies (Fig VIII-10). STRs have moderate to high mutation rates. The instability of certain minisatellites can lead to cancer and acquired diseases, such as insulin-dependent diabetes mellitus.

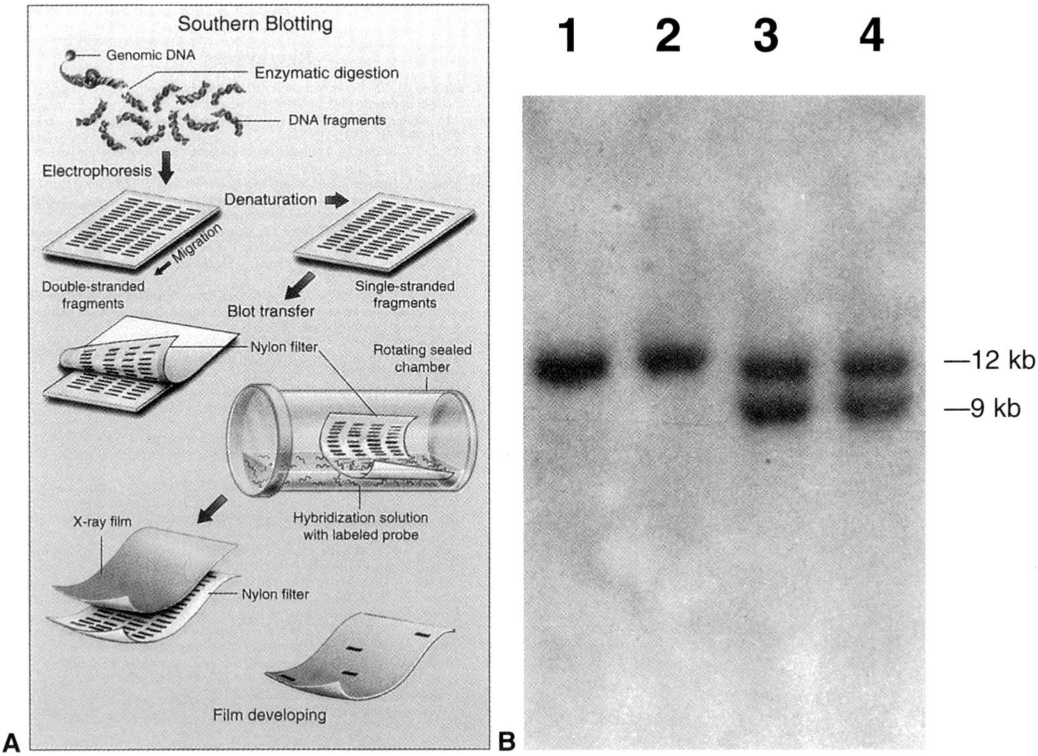

FIG VIII-9—**A**, Analysis of DNA by gel electrophoresis and Southern blotting. In Southern blotting, genomic DNA is cut with restriction enzymes into fragments before being separated according to size by gel electrophoresis. The four lanes on the gel represent the digestion of the DNA with four different restriction enzymes. After electrophoresis, the nucleic acids in the gel are transferred directly onto a charged nylon filter to which they are tightly bound. Thus, the filter contains a precise replica of the nucleic acid distribution in the gel. The filter is then hybridized in a rotating sealed chamber with a DNA or RNA probe specific for the target of interest (in this case, sequences in a microbial pathogen). Probes have traditionally been radioactively labeled with nucleotides containing phosphorus-32; however, use of nonradiolabeled probes is becoming more common. After the probe has hybridized to its target sequence, the nonhybridized probe is washed away and the filter is exposed to x-ray film. A DNA sequence complementary to the probe is seen as a dark band on the developed film. The position of the hybridized target sequence in each lane is unique to the restriction enzyme used to digest the DNA. (Reproduced with permission from Naber SP. Molecular pathology—diagnosis of infectious disease. *N Engl J Med.* 1994;331:1212.) **B**, Autoradiograph of a Southern blot with radiolabeled probe L1.28 after the DNA was cut with enzyme *Taq1*, separated by size on an agarose gel, and then transferred to a nylon filter. Four female carriers of X-linked retinitis pigmentosa are depicted. Note that two females, numbers 1 and 2, have only a 12 kb band, whereas carriers 3 and 4 have both 12 kb and 9 kb bands.

Housman D. Human DNA polymorphism. *N Engl J Med.* 1995;332:318–320.

Litt M, Luty JA. A hypervariable microsatellite revealed by in vitro amplification of a dinucleotide repeat within the cardiac muscle actin gene. *Am J Hum Genet.* 1989;44: 397–401.

Weber JL, May PE. Abundant class of human DNA polymorphisms, which can be typed using the polymerase chain reaction. *Am J Hum Genet.* 1989;44:388–396.

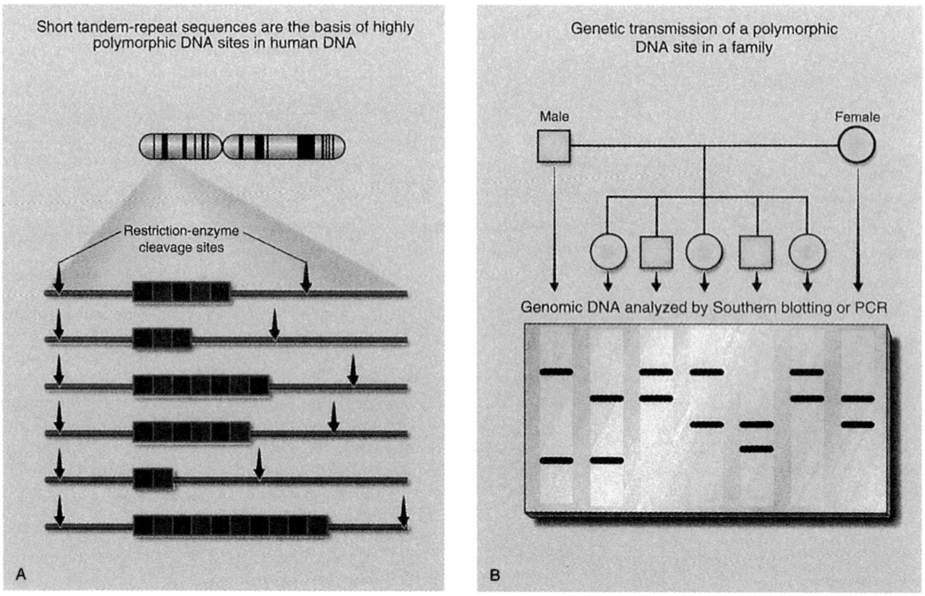

FIG VIII-10—**A**, Variable-length sequences in human DNA can be created by variations in the number of copies of a tandem-repeat DNA sequence. Each line in the figure represents a copy of a human DNA sequence. The copies are identical in sequence except for the tandemly repeated DNA sequence indicated by the boxes. The number of copies of the tandemly repeated DNA sequence is indicated by the number of boxes. The size of the DNA fragment that includes the tandem-repeat sequence is measured between two fixed points. In Southern blotting, the sites of restriction-enzyme digestion are the fixed points that determine the ends of the DNA fragment. **B**, A family in which a highly polymorphic marker is used for genetic analysis. The two copies of the DNA fragment from the offspring can be distinguished from the two copies of the fragment from the father, making the inheritance pattern from each parent clear for this chromosomal site. Detection may be carried out by Southern blotting or polymerase chain reaction, depending on the size of the tandem-repeat sequence. (Reproduced with permission from Housman D. Human DNA polymorphism. *N Engl J Med.* 1995;332:319.)

Positional Cloning

After linkage studies suggest that a gene resides in a specific chromosomal region, the gene may be isolated and cloned by one of several molecular genetic techniques that eventually distinguish the gene from surrounding genes and noncoding DNA. The process involves successive refinements of the linkage mapping by use of probes or markers generated for regions close to the gene. The region of interest is successively narrowed to a segment small enough to be isolated and introduced into a cloning vector to produce large quantities of DNA for subsequent molecular analysis. In this manner, a group of cloned nucleotide sequences that are contiguous, called a *contig,* is created over the entire region that spans the locus.

Genes of interest from such clones can be identified by a number of strategies:

- Screening of human candidate genes known to be in the region
- Screening for homologous sequences of known animal genes that are syntenic to the region

- Searching for microdeletions (common in X-linked diseases) by determination of the presence or absence of expressed sequence tags (ESTs) in the region
- Competitive hybridization mapping techniques (cDNA library screening using total yeast artificial chromosome [YAC] or cosmids)

The progress of the human genome sequencing project now makes it possible to identify genes located in particular chromosome regions directly from searches of human DNA sequence databases. One such database is maintained by the National Center for Biotechnology and can be searched through the BLAST protocol *(www.ncbi.nlm.nih.gov/BLAST)*.

Cremers FP, van de Pol DJ, van Kerkoff LP, et al. Cloning of a gene that is rearranged in patients with choroideraemia. *Nature.* 1990;347:674–677.

Glaser T, Walton DS, Cai J, et al. PAX6 mutations in aniridia. In: Wiggs JL, ed. *Molecular Genetics of Ocular Disease.* New York: Wiley-Liss; 1995:51–82.

Jordan T, Hanson I, Zaletayev D, et al. The human PAX6 gene is mutated in two patients with aniridia. *Nat Genet.* 1992;1:328–332.

Meindl A, Dry K, Herrmann K, et al. A gene (RPGR) with homology to the RCC1 guanine nucleotide exchange factor is mutated in X-linked retinitis pigmentosa (RP3). *Nat Genet.* 1996;13:35–42.

Ton CC, Hirvonen H, Miwa H, et al. Positional cloning and characterization of a paired box- and homeobox-containing gene from the aniridia region. *Cell.* 1991;67: 1059–1074.

Candidate Gene Approaches

Candidate gene screening This process involves the screening for mutations of genes that are abundantly expressed within a tissue and are either important for function or specifically expressed only in that tissue. Sometimes the candidate gene is one that causes an animal model similar to the human disease. Examples of candidate gene screening discoveries include the findings of mutations of peripherin/*RDS* in autosomal dominant retinitis pigmentosa and macular dystrophies and the finding of mutations of rod *cyclic guanosine monophosphate (cGMP)* β-subunit of rod phosphodiesterase and the cGMP-gated cation channel in autosomal recessive retinitis pigmentosa.

Dryja TP, Finn JIT, Peng Y-W, et al. Mutations in the gene encoding the α subunit of the rod cGMP-gated channel in autosomal recessive retinitis pigmentosa. *Proc Natl Acad Sci U S A.* 1995;92:10177–10181.

Kajiwara K, Sandberg MA, Berson EL, et al. A null mutation in the human peripherin/ RDS gene in a family with autosomal dominant retinitis punctata albescens. *Nat Genet.* 1993;3:208–212.

McLaughlin ME, Sandberg MA, Berson EL, et al. Recessive mutations in the gene encoding the β-subunit of rod phosphodiesterase in patients with retinitis pigmentosa. *Nat Genet.* 1993;4:130–133.

Nichols BE, Sheffield VC, Vandenburgh K, et al. Butterfly-shaped pigment dystrophy of the fovea caused by a point mutation in codon 167 of the RDS gene. *Nat Genet.* 1993;3:202–207.

Positional candidate gene screening Whenever linkage studies localize a gene to a given chromosomal region, genes already known to reside in the same region become candidate genes for that disease. Following are some examples of disease localization that resulted from linkage to a given region, which in turn led to finding the disease-causing gene by screening for mutations of genes in the region:

- Autosomal dominant retinitis pigmentosa from rhodopsin mutations (3q)
- Sorsby fundus dystrophy from *TIMP3* mutations (22q)
- Oguchi disease from point deletions within the arrestin gene (2q)

Dryja TP, McGee TL, Reichel E, et al. A point mutation of the rhodopsin gene in one form of retinitis pigmentosa. *Nature.* 1990;343:364–366.

Fuchs S, Nakazawa M, Maw M, et al. A homozygous 1-base pair deletion in the arrestin gene is a frequent cause of Oguchi disease in Japanese. *Nat Genet.* 1995;10: 360–362.

Weber BH, Vogt G, Pruett RC, et al. Mutations in the tissue inhibitor of metalloproteinases-3 (TIMP3) in patients with Sorsby fundus dystrophy. *Nat Genet.* 1994;8: 352–356.

Molecular Manipulation and Analysis of DNA

Recombinant Genetics

DNA libraries DNA libraries exist as a means of collecting and ordering genes of interest for future study. Libraries can be made from either genomic DNA or cDNA. *Genomic DNA libraries* are created by cleaving whole DNA from an organism, tissue, or cell type with restriction enzymes that produce fragments of DNA. These fragments can be cloned into vectors and plated onto media. The specific clones are identified with probes derived from the original sequence or gene of interest and then isolated and grown as needed (Fig VIII-11). *cDNA* libraries are created by using reverse transcriptase to generate complementary DNA from mRNA expressed by the cell or tissue to be studied. Recent technology allows many cDNA sequences or ESTs to be placed on a microscopic glass slide. Literally thousands of cDNAs can be screened using this microarray technique.

Rosenthal N. Stalking the gene—DNA libraries. *N Engl J Med.* 1994;331:599–600.

Recombinant DNA DNA that is coupled from different sources within a single DNA molecule is called *recombinant DNA*. One of the most important techniques of molecular genetics has been the creation of recombinant DNA by bacterial cloning. DNA is cleaved with a restriction enzyme to create fragments that can be inserted into vectors such as plasmids, cosmids, bacterial artificial chromosomes (BACs), or YACs. These vectors can then be used to produce nearly unlimited quantities of the inserted genetic sequences (Fig VIII-12, top).

Polymerase Chain Reaction (PCR)

The PCR is an in vitro method for the enzymatic synthesis of specific DNA sequences from a small amount of template DNA. An exponential increase in the quantity of specific fragments of DNA can be obtained by repetitive synthesis, starting with a minuscule amount of DNA as a template (Fig VIII-12, bottom). Automated PCR was made possible by the isolation of a thermostable DNA polymerase from the bacteria *Thermus aquaticus.*

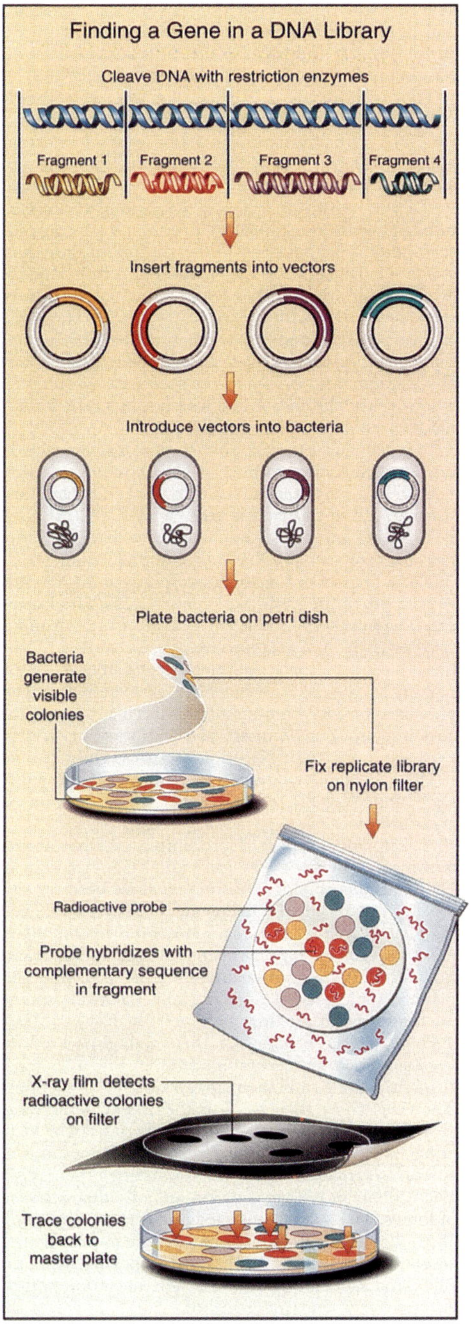

FIG VIII-11—The first step in making a library of DNA sequences is to cut DNA into fragments with restriction enzymes. These DNA fragments, when inserted into vectors, form recombinant molecules with the DNA of the vector (a plasmid vector is shown, but viral vectors are also used). Bacteria carrying the vectors can replicate on an agar-coated Petri dish, where they grow to form colonies. Each colony originates from a single bacterial cell and thus contains a single type of recombinant DNA fragment. A nylon filter put on the surface of the Petri dish picks up a portion of each colony. Chemical treatment of the filter lyses the bacterial cells, denaturing the DNA and fixing it in place. A radioactive probe for a known sequence of nucleotides can reveal the desired fragment on the filter. The filter, with the replicate library of the colonies on its surface, is incubated with a solution containing the radioactive DNA probe in a plastic bag (or glass dish); after the unbound probe is washed away, an x-ray film can locate the radioactive colonies *(black ovals)*. The position of the signals on the film serves as a map with which to locate the corresponding colonies on the original master plate. Once identified, these colonies can then be amplified in culture to produce large quantities of the desired recombinant DNA molecule. (Reproduced with permission from Rosenthal N. Molecular medicine: stalking the gene—DNA libraries. *N Engl J Med.* 1994;331:599.)

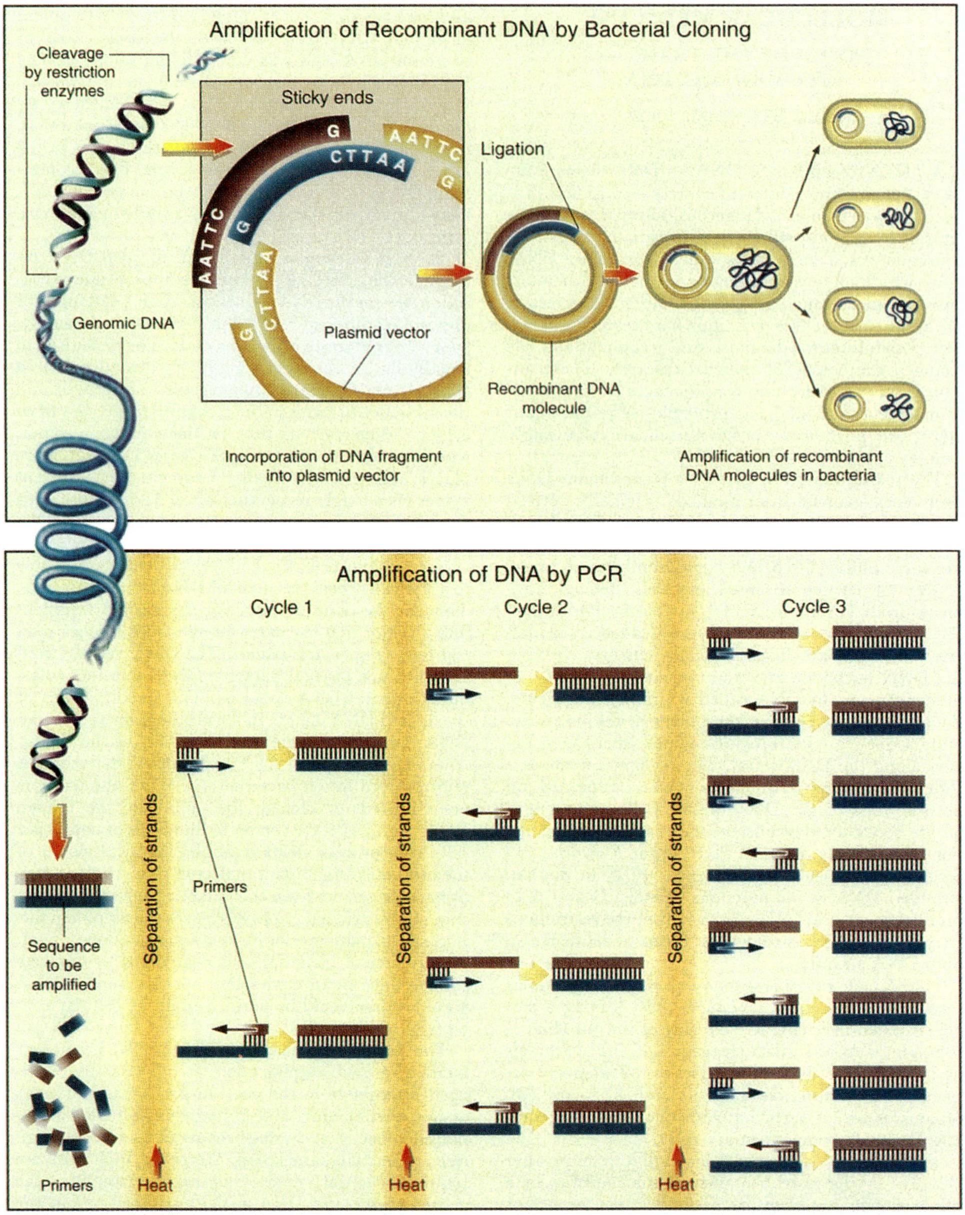

FIG VIII-12—**Top**, In the amplification of recombinant DNA, the DNA segment to be amplified is separated from surrounding genomic DNA by cleavage with a restriction enzyme. The enzymatic cuts often produce staggered or "sticky" ends. In the example shown here, the restriction enzyme *Eco*R1 recognizes the sequence GAATTC and cuts each strand between guanine *(G)* and adenine *(A)*; the two strands of the genomic DNA are shown as *blue* and *purple*. *C,* cytosine; *T,* thymine. The same restriction enzyme cuts the circular plasmid DNA *(tan)* at a single site, generating sticky ends that are complementary to the sticky ends of genomic DNA fragment. The cut genomic DNA and the remainder of the plasmid, when mixed together in the presence of a ligase enzyme, form smooth joints on each side of the plasmid-genomic DNA junction. This new molecule—recombinant DNA—is carried into bacteria, which replicate the plasmid as they grow in culture. **Bottom**, In the amplification of DNA by the polymerase chain reaction *(PCR)*, the DNA sequence to be amplified is selected by primers, which are short, synthetic oligonucleotides that correspond to sequences flanking the DNA to be amplified. After an excess of primers is added to the DNA, together with a heat-stable DNA polymerase, the strands of both the genomic DNA and the primers are separated by heating and allowed to cool. A heat-stable polymerase elongates the primers on either strand, thus generating two new, identical double-stranded DNA molecules and doubling the number of DNA fragments. Each cycle takes just a few minutes and doubles the number of copies of the original DNA fragment. (Reproduced with permission from Rosenthal N. Tools of the trade—recombinant DNA. *N Engl J Med.* 1994;331:316.)

PCR has transformed the means by which molecular biologists approach biological problems. The results of two relatively recent experiments demonstrated the power of PCR.

Pittler and colleagues amplified DNA recovered from a formalin-fixed, paraffin-embedded section of *rd* mouse retina and showed that the defective gene was identical to the one they were studying. Remarkably, the slides had been prepared almost 70 years before.

An even more remarkable feat was when Hunt and colleagues demonstrated the molecular basis of the color vision defect for John Dalton (1766–1844). John Dalton had written elegantly in 1794 about how his own color perceptions were different from those of others. Thinking that his color-vision defect must have been caused by an alteration of the color of the media of his eyes, he left his physician instructions to remove his eyes after death for examination. Dalton's physician sectioned one eye and, finding the media clear, preserved the other eye by air drying without fixation. Hunt and coworkers used PCR-extracted DNA from this eye to show that John Dalton was a deuteranope with a hybrid green pigment gene with predicted altered spectral sensitivity, rather than a protanope as many had assumed.

PCR requires the creation of two primers, one for the sense strand (the *sense primer*) and the other for the antisense strand (the *antisense primer*). For amplification of genomic DNA, the coding sequence of the flanking intronic sequences must be known so that each exon and adjacent splice sites can be selectively amplified.

Some large exons require overlapping subregions for amplification. Often a stretch of the 5′ UTR and the 3′ UTR will also be amplified to search for defects in the promoter or presumed regulatory regions of the gene.

Hunt DM, Dulai KS, Bowmaker JK, et al. The chemistry of John Dalton's color blindness. *Science.* 1995;267:984–988.

Pittler SJ, Keeler CE, Sidman RL, et al. PCR analysis of DNA from 70-year-old sections of *rodless* retina demonstrates identity with the *rd* defect. *Proc Natl Acad Sci U S A.* 1993;90:9616–9619.

Rosenthal N. Tools of the trade—recombinant DNA. *N Engl J Med.* 1994;331:315–317.

Southern, Northern, and Western blotting Genomic or complementary DNA that is partially digested by endonucleases can be separated on an electrophoretic gel. This process is called *Southern blotting,* named after the developer of the technique. *Northern blotting* is the analogous application of gel electrophoresis to separate sequences or fragments of mRNA. The nucleotide of interest for either Southern or Northern blotting can be identified by hybridization to cDNA probes. The separation of proteins on gel electrophoresis for identification by immunologic techniques is called *Western blotting.*

Mutation Screening

Single-stranded conformational polymorphism (SSCP) With this technique for mutation detection, single-stranded DNA is electrophoresed under nondenaturing conditions, allowing the molecules to fold on themselves according to their inherent similarity of sequences. Molecules of differing size and sequences fold differently and migrate at different rates of speed on the gel. Mutations that alter amino acid residues often change the way in which single-stranded DNA folds upon itself, creating different tertiary configurations that can be separated from the normal sequence by differences in mobility on gel electrophoresis.

Denaturing gradient gel electrophoresis (DGGE) With DGGE, double-stranded DNA samples are electrophoresed against a gradient of denaturing agent such as urea. Molecules of differing size and composition reach differing points on the gel before they become denatured. Mutations affect the point in the gel for which a given DNA molecule will denature and hence alter the migration patterns. The sensitivity of DGGE in detecting mutations is improved if a 40 bp sequence rich in GC is added to one primer before PCR (GC clamp). Identification of a polymorphism detected by DGGE can be determined by direct DNA sequencing of the PCR-amplified exonic product.

Direct sequencing One of the most important advances in molecular genetics has been the development of techniques for rapid sequencing of DNA. Currently, it is far cheaper to sequence a stretch of DNA than to sequence and characterize the amino acid peptide that it produces.

Although other mutation screening techniques exist, sequencing of DNA is the surest and most direct. Sequencing of cDNA derived from mRNA provides a quick look at the reading frames (exons) of the gene. Sequencing of genomic DNA is more time-consuming because of the presence of introns between the exons. The intron–exon boundaries must be known and multiple PCR assays set up to screen not only the exons and their splice-site junctions but also upstream and downstream regions that may be important to gene activation and regulation.

The two DNA sequencing techniques used today are the *enzymatic* (or *Sanger's*) *method,* which can be implemented manually or semiautomated; and *automated sequencing,* which (for high-volume laboratories) is faster and less prone to errors of reading. Figure VIII-13 illustrates these procedures.

Rosenthal N. Molecular medicine: fine structure of a gene—DNA sequencing. *N Engl J Med.* 1995;332:589–591.

Use of restriction endonucleases for mutation screening If a point mutation destroys or creates a restriction site, screening for this mutation can be quickly accomplished through the use of the particular enzyme that recognizes the changed restriction site. Figure VIII-14 illustrates the detection of mutations using restriction enzymes, oligonucleotide hybridization, PCR, and Southern blot analysis.

Allele-specific oligonucleotides (ASO) An allele-specific oligonucleotide is a synthetic probe made of a sequence of nucleotides. It is constructed to use hybridization to recognize a specific DNA sequence in order to detect a specific point mutation. Often, diagnostic testing is done with two separate ASOs, one that recognizes the specific base-pair change of a given genetic mutation and the other that recognizes the normal allelic sequence. ASOs are commonly used for diagnosing point mutations that occur frequently or for testing multiple members of a large family with a previously identified genetic disorder.

Transgenic and Knockout Animals

The fertilized egg of a *transgenic animal* has undergone introduction of new genetic material that becomes incorporated into the genome and is duplicated and transmitted to subsequent generations along with the normal genes (Fig VIII-15). The

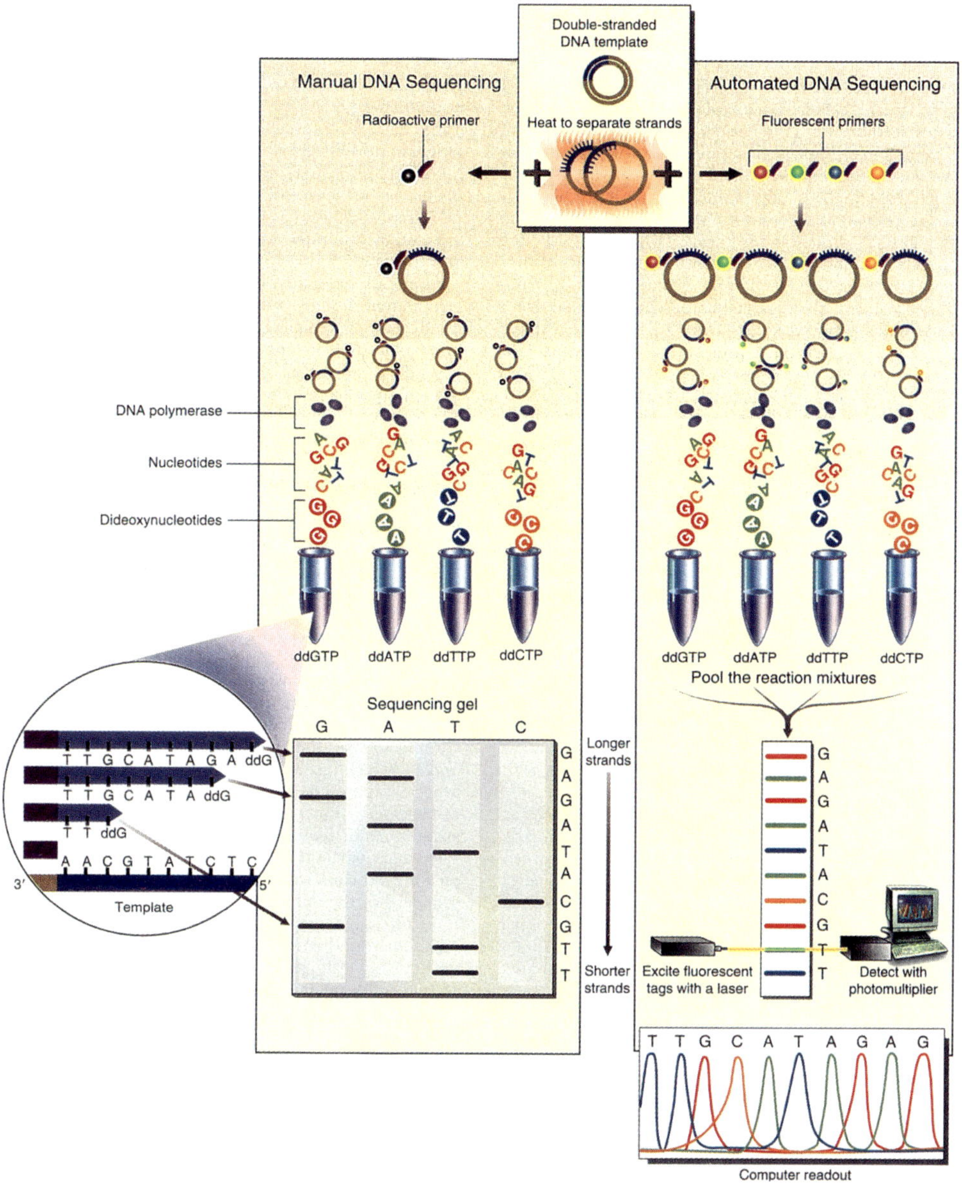

FIG VIII-13—**Left**, A double-stranded fragment of DNA whose sequence is unknown is cloned into a plasmid, which is then heated to separate the two strands. A primer with a sequence complementary to a short plasmid sequence near the junctional site is added and hybridizes to one of the two template strands. In manual sequencing, the DNA primer has a radioactive tag (alternatively, one of the nucleotide precursors is radioactive). The plasmid-primer hybrid is added to four tubes, each containing DNA polymerase, all four nucleotides, and a single dideoxynucleotide-dideoxyguanosine triphosphate *(ddGTP),* dideoxyadenosine triphosphate *(ddATP),* dideoxythymidine triphosphate *(ddTTP),* or dideoxycytidine triphosphate *(ddCTP).* DNA polymerase extends the DNA primer, which incorporates nucleotides (and occasionally dideoxynucleotides) into the growing DNA chains. The incorporation of dideoxynucleotides prevents further elongation of the chain (the *circular inset* shows the elongation products in the ddGTP sample). Electrophoresis of the four reactions on a very thin gel separates the radioactive fragments according to size. Autoradiography reveals these fragments as bands. Each band corresponds to a nucleotide in the DNA sequence. **Right**, Automated sequencing uses a DNA primer labeled with four different fluorescent tags. Each fluorescent primer hybridizes to the template DNA and undergoes the same synthesis reactions as in manual sequencing. After elongation of the primer, samples containing the four kinds of newly synthesized fragments are pooled and undergo electrophoresis in a single lane of a gel. A laser beam directed near the bottom of the gel excites each fragment, which emits a specific fluorescent signal as it passes through the beam. A photomultiplier senses the specific wavelength of each signal, which corresponds to the dideoxynucleotide incorporated in the elongation reaction. A computer stores and translates the signals as a nucleotide sequence. (Reproduced with permission from Rosenthal N. Molecular medicine: fine structure of a gene—DNA sequencing. *N Engl J Med.* 1995;332:590.)

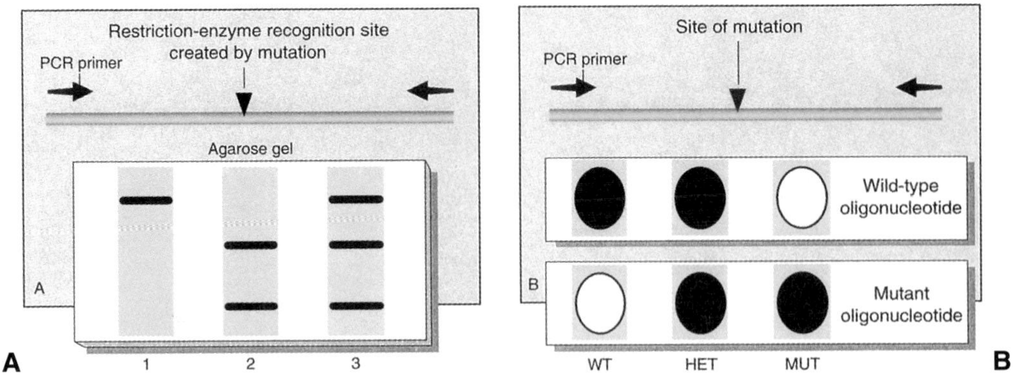

FIG VIII-14—**A**, The detection of a point mutation by digestion of DNA with a restriction enzyme. The mutation creates a new recognition site. The region surrounding the mutation is amplified by the polymerase chain reaction *(PCR)*, and the resulting PCR product is incubated with the restriction enzyme and then analyzed by agarose-gel electrophoresis. *Lane 1* shows DNA from a person without the mutation; only one band appears because the enzyme does not cut the DNA. *Lane 2* shows DNA from a person homozygous for the mutation; two bands represent the two fragments obtained after enzyme digestion. *Lane 3* shows DNA from a heterozygote; there is one uncut fragment and two cut fragments. **B**, A mutation detected by oligonucleotide hybridization. The segment of DNA is amplified by PCR, divided into aliquots, and spotted onto separate filter membranes, which are hybridized with a labeled oligonucleotide corresponding to the wild-type or mutant sequence. The amplified segment of DNA from a person with the wild-type sequence *(WT)* hybridizes only with the wild-type oligonucleotide, whereas the DNA from a person homozygous for the mutant sequence *(MUT)* hybridizes only with the mutant oligonucleotide. DNA from a heterozygote *(HET)* hybridizes with both oligonucleotides. **C**, The detection of a mutation by PCR. The mutant sequence differs from the wild-type sequence by the substitution of an A for a C. To search for the two kinds of sequences by PCR, two primers are necessary. A separate reaction is carried out with each, together with a common downstream primer. With a wild-type gene, the primer corresponding to the wild-type sequence yields a PCR product. Similarly, the mutant primer produces a product with the mutant sequence. However, with the wild-type primer and the mutant sequence, or the mutant primer and the wild-type sequence, there is no PCR product. The agarose-gel pattern shows that DNA from a person homozygous for the wild-type allele *(WT)* reacts only with the wild-type primer; DNA from a person homozygous for the mutant sequence *(MUT)* reacts only with the mutant primer; and DNA from a heterozygote *(HET)* yields PCR products with both primers. **D**, Detection of a triplet-repeat mutation by Southern blot analysis or PCR. In the fragile X syndrome, a CGG repeat occurs near the 5′ end of the gene. The number of repeats ranges from 5 to 50 in the general population and from approximately 50 to 200 in those with the fragile X syndrome. The abnormality is detected as follows: DNA is treated with a restriction enzyme that cuts at recognition sites flanking the CGG repeat. Hybridization on a Southern blot with labeled DNA from the region of the gene reveals a single band in a normal male subject (wild type). An asymptomatic male carrier will have a band of higher molecular weight, and a subject with a full mutation will have a very large, diffuse band because of the instability of the full-mutation allele. The normal and asymptomatic-carrier alleles can also be detected by PCR *(right)*. The full-mutation allele cannot be amplified by PCR because it is too large. (Reproduced with permission from Korf B. Molecular diagnosis. *N Engl J Med.* 1995;332:1500–1501.)

entire genetic material needed for transcription must be included in the transgene, including the following:

- Promoter
- 5′ transcription-initiation site
- 5′ untranslated region

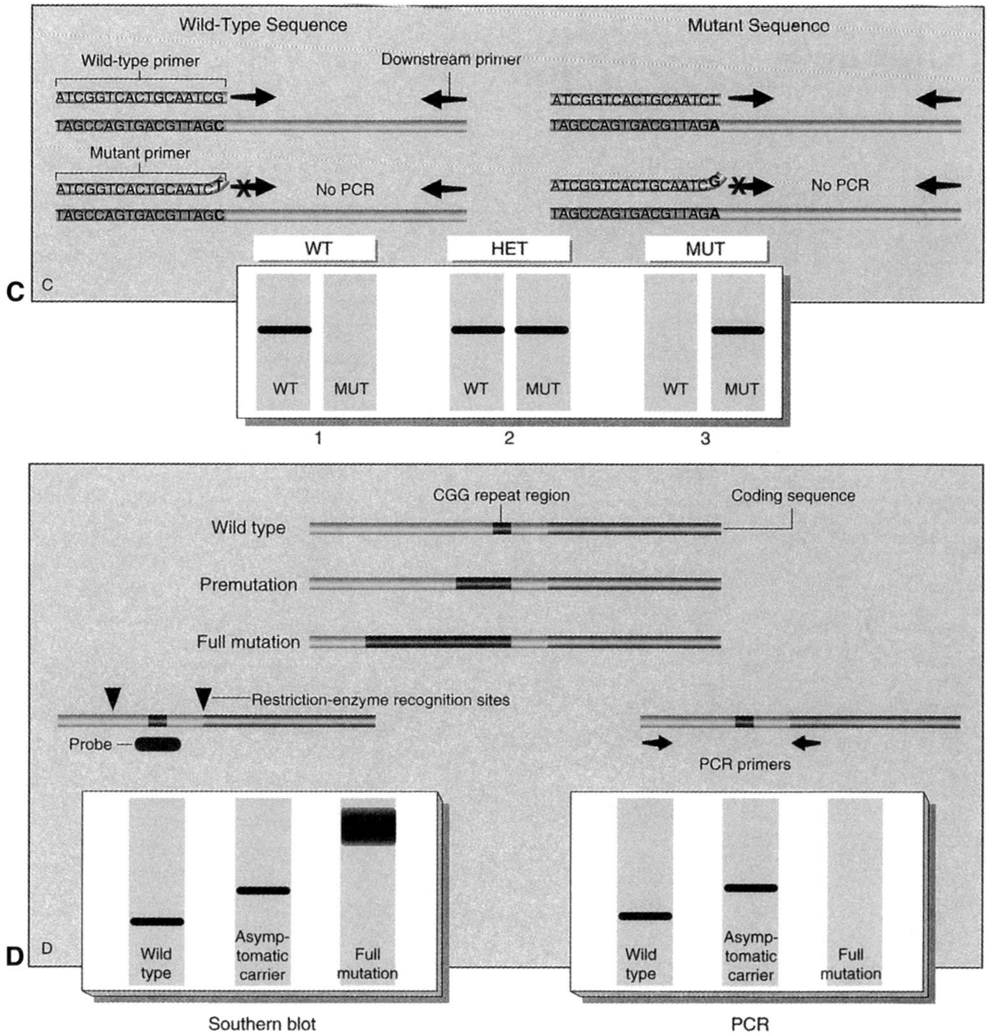

- □ Translation-initiation codon
- □ Coding section
- □ Stop codon
- □ 3′ untranslated region
- □ Polyadenylation site (in most instances)

The promoter is chosen to ensure expression within the tissue of interest.

Exactly where the transgene becomes inserted within the genome of the animal is usually unknown. Transgenic animals have been rescued or cured of autosomal recessive disease by insertion of a copy of the normal gene. Transgenic animals that

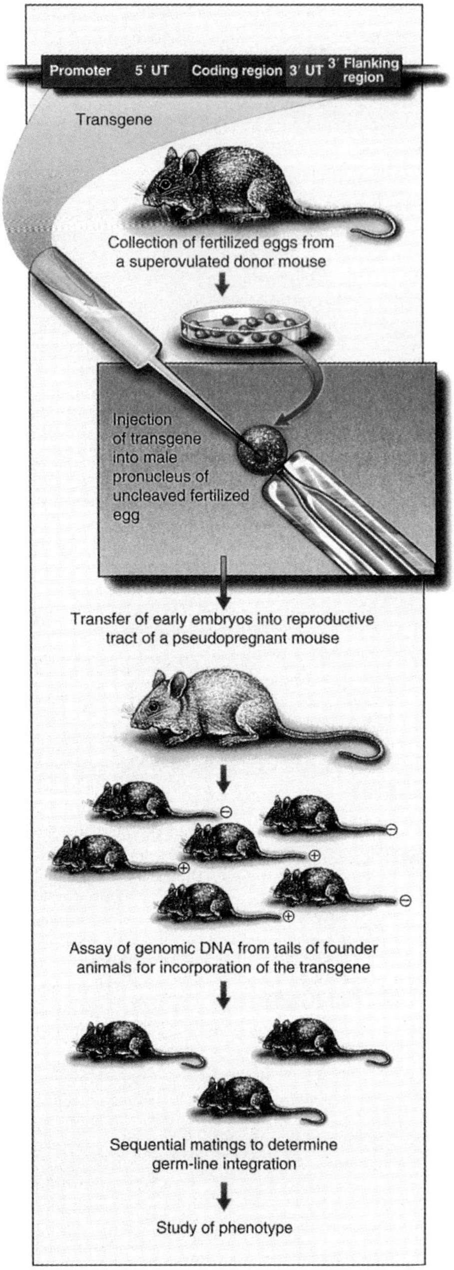

FIG VIII-15—The transgene containing the DNA sequences necessary for the expression of a functional protein is injected into the male (larger) pronucleus of uncleaved fertilized eggs through a micropipette. The early embryos are then transferred into the reproductive tract of a mouse rendered "pseudopregnant" by hormonal therapy. The resulting pups (founders) are tested for incorporation of the transgene by assaying genomic DNA from their tails. Founder animals that have incorporated the transgene *(+)* are mated with nontransgenic mice, and their offspring are mated with each other to confirm germ-line integration and to establish a line of homozygous transgenic mice. Several transgenic lines that have incorporated different numbers of transgenes at different integration sites (and thus express various amounts of the protein of interest) are usually studied. (Reproduced with permission from Shuldiner AR. Transgenic animals. *N Engl J Med.* 1996;334:654.)

carry copies of specific mutations of human genes (eg, mutant rhodopsins) that appear to cause dominant disease (retinitis pigmentosa) have been successfully created.

Shuldiner AR. Transgenic animals. *N Engl J Med.* 1996;334:653–655.

A *knockout mutation* in an animal strain is created by targeting a known gene to disrupt its function, usually by replacement in embryonic stem cells of a portion of the coding region with a sequence of DNA that destroys the original function. The desired gene is incorporated into the stem cell through homologous recombination with a specially constructed vector that contains the bacterial neomycin resistance gene sandwiched between two sequences that are homologous, or identical, to the wild strain (Fig VIII-16). The homologous regions undergo recombination; rarely, a double recombination occurs, resulting in the swapping of the new sequence for the targeted sequence.

The clone of mutant embryonic stem cells with the disrupted gene is introduced into a host embryo at the blastocyst stage, resulting in a chimeric animal. Some of the mutant cells are incorporated into the germ line, allowing for breeding of animals that are homozygous for the defective gene. The creation of "knockout animals" is useful for determining the phenotype of the lack of a normal gene product in the absence of a naturally occurring recessive animal model. The knockout technique is used to generate animal models for autosomal recessive disease in humans.

Capecchi MR. Targeted gene replacement. *Sci Am.* 1994;270(3):52–59.

Majzoub JA, Muglia LJ. Knockout mice. *N Engl J Med.* 1996;334:904–907.

Gene Therapy

Replacement of Absent Gene Product in X-Linked and Recessive Disease

For genetic diseases wherein the mutant allele produces either no message or an ineffective gene product (a so-called *null allele*), correction of the disorder may be possible by simple replacement of the gene in the deficient cells or tissues. It is theoretically possible to transfer normal genes into human cells that harbor either null or mutant genes that do not produce a stable, translated product. Vectors used to carry the genetic material into the cells include retroviruses, adenoviruses, and plasmid–liposome complexes. Early attempts have met with limited success because of problems with delivery, specificity of targeting, stability, and regulation.

Retroviral vectors carry the risk of inducing toxicity from overexpression or insertional mutagenesis if the inserted sequence disrupts a tumor-suppressor gene or activates an oncogene. The target cell cannot be a terminally differentiated cell, and it must proliferate in order to integrate the inserted DNA into its genome. Adenosine deaminase (ADA) has been transferred with expression in T cells, cord blood, and placental cells from children with ADA deficiency and has resulted in partial immune function. Low-density lipoprotein receptor cDNA has been transferred to autologous hepatocytes of patients with familial hypercholesterolemia. Cytokine cDNAs and p53 antisense DNA have been transferred to individuals and cells from patients with tumors.

Plasmid–liposome complexes may have advantages as vectors, because they can be used for nondividing cells and may be less likely to incite inflammation or immune responses. However, as vectors these complexes are inefficient and may not

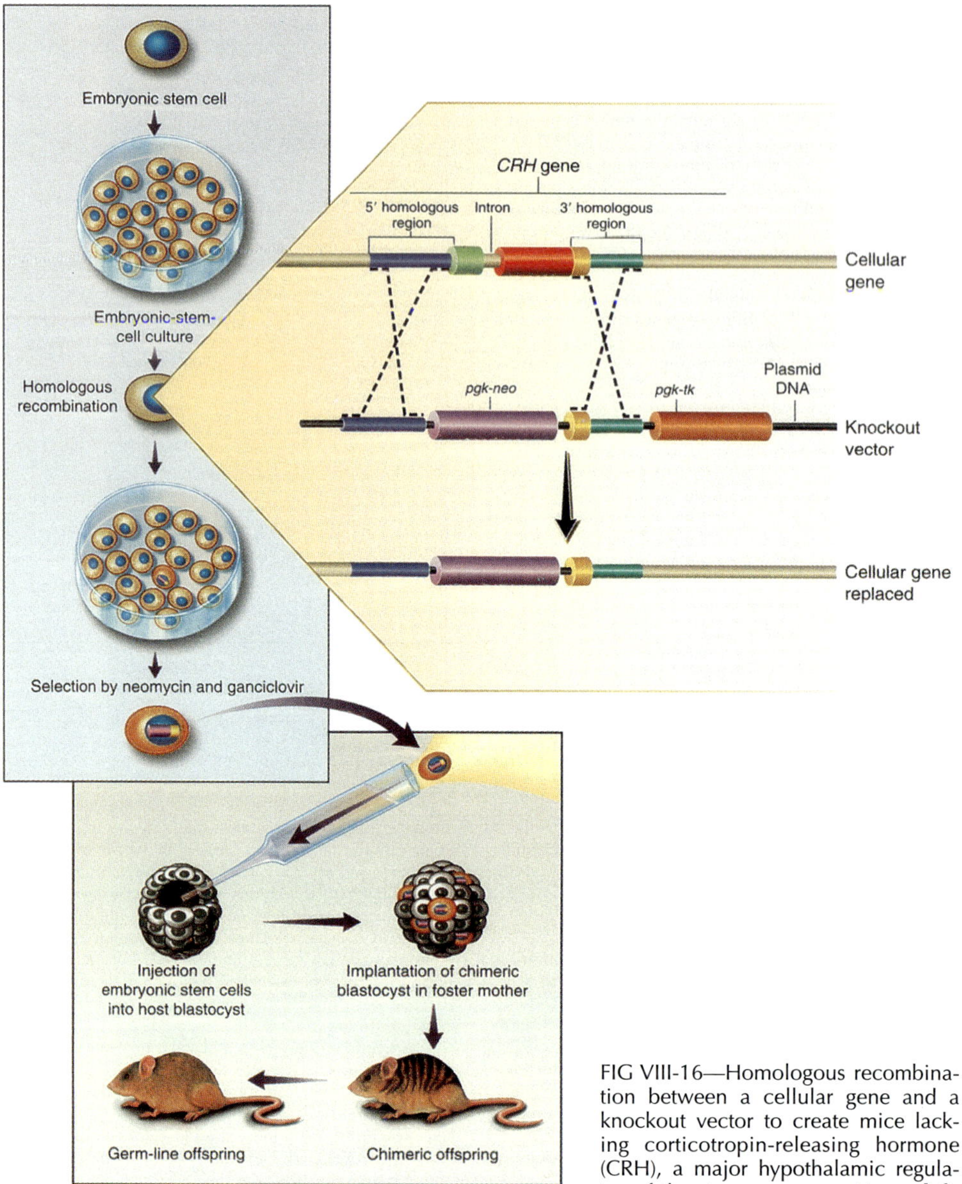

FIG VIII-16—Homologous recombination between a cellular gene and a knockout vector to create mice lacking corticotropin-releasing hormone (CRH), a major hypothalamic regulator of the stress response. **Upper left**, Embryonic stem cells contain the *CRH* cellular gene (**right**), which consists of exon 1 (*olive green,* a 5′ noncoding region), an intron, and exon 2 (*red,* a protein-coding region, and *yellow,* a 3′ noncoding region). A knockout vector, consisting of a collinear assembly of a DNA flanking segment 5′ to the cellular gene *(blue),* the phosphoglycerate kinase-bacterial neomycin gene *(pgk-neo, violet),* a 3′ segment of the cellular gene *(yellow),* a DNA flanking segment 3′ to the cellular gene *(green),* and the phosphoglycerate kinase-viral thymidine kinase gene *(pgk-tk, orange),* is created and introduced into the embryonic stem cell culture. Double recombination occurs between the cellular gene and the knockout vector in the 5′ homologous regions and the 3′ homologous regions *(dashed lines),* resulting in the incorporation of the inactive knockout vector, including *pgk-neo* but not *pgk-tk,* into the cellular genomic locus of the embryonic stem cell. The presence of *pgk-neo* and the absence of *pgk-tk* in these replaced genes will allow survival of these embryonic stem cells after positive-negative selection with neomycin and ganciclovir. **Bottom**, The clone of mutant embryonic stem cells is injected into a host blastocyst, which is implanted into a pseudopregnant foster mother and subsequently develops into chimeric offspring. The contribution of the embryonic stem cells to the germ cells of the chimeric mouse results in germ-line transmission of the embryonic stem cell genome to offspring that are heterozygous for the mutated *CRH* allele. The heterozygotes are mated to produce mutant mice homozygous for CRH deficiency, with impaired hormonal responses to multiple stressors. (Reproduced with permission from Majzoub JA, Muglia LJ. Molecular medicine: knockout mice. *N Engl J Med.* 1996;334:905.)

produce sufficient expression of the wanted product. Strategies are being developed to direct the new genetic material into the nucleus, with the possibility of subsequent incorporation into the genome, perpetuating expression of the new gene.

Blau HM, Springer ML. Molecular medicine: gene therapy—a novel form of drug delivery. *N Engl J Med.* 1995;333:1204–1207.

Crystal RG. Transfer of genes to humans: early lessons and obstacles to success. *Science.* 1995;270:404–410.

Hangai M, Kaneda Y, Tanihara H, et al. In vivo gene transfer into the retina mediated by a novel liposome system. *Invest Ophthalmol Vis Sci.* 1996;37:2678–2685.

Strategies for Dominant Diseases

Dominant diseases are caused by production of a gene product that is either insufficient *(haploid insufficiency)* or conducive to disease *(dominant-negative effect).* Haploid insufficiency theoretically should be treatable by gene replacement as outlined above for recessive or X-linked disease. (For dominant disorders produced by defective developmental genes, this correction would have to occur in early uterine development.)

Disorders resulting from a dominant-negative effect require a different approach. Thus, strategies for treatment of dominant disease differ, depending on whether a functional gene product is produced. Some genes code for RNA that can bind to mRNA from another gene and block its ability to make a protein. Greater understanding of these genes may allow for creation of either drugs or new gene-encoded RNAs that can block the translation of mRNA for defective alleles, thus allowing only the normal allele to be expressed.

Another approach is the use of oligonucleotides that are designed to bind with mRNA from mutant alleles, stopping the mRNA from being translated by ribosomes (Fig VIII-17). Although many problems need to be worked out for such therapy to be effective, this approach holds promise for autosomal dominant disorders wherein disease is caused by expression of the mutant gene product.

Askari FK, McDonnell WM. Molecular medicine: antisense-oligonucleotide therapy. *N Engl J Med.* 1996;334:316–318.

Della NG. Molecular biology in ophthalmology. A review of principles and recent advances. *Arch Ophthalmol.* 1996;114:457–463.

Musarella MA. Gene mapping of ocular diseases. *Surv Ophthalmol.* 1992;36:285–312.

Petrash JM. Applications of molecular biological techniques to the understanding of visual system disorders. *Am J Ophthalmol.* 1992;113:573–582.

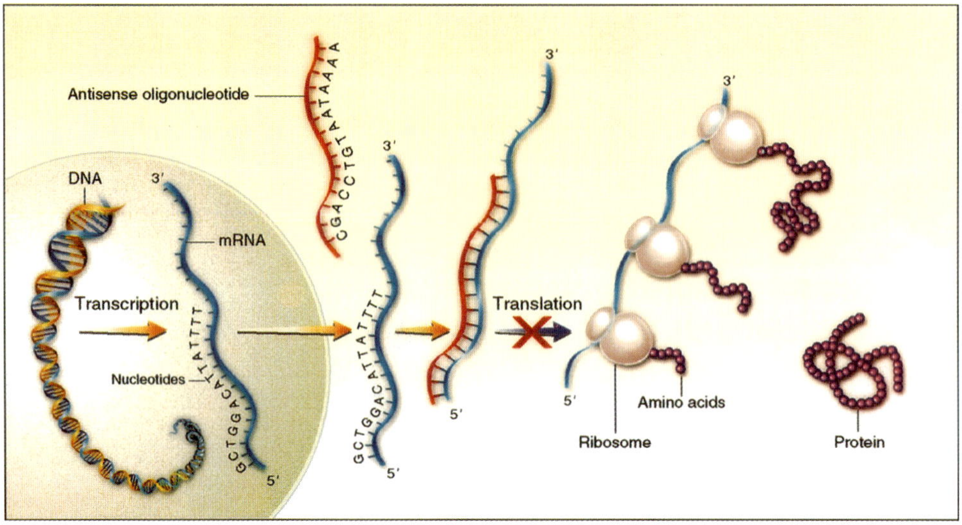

FIG VIII-17—Blockade of translation by antisense oligonucleotides. Normal gene transcription of DNA into mRNA is followed by translation of mRNA into protein. Antisense oligonucleotides complementary to a portion of mRNA bind mRNA, preventing translation—either by the steric effect of the binding process itself or (possibly) by inducing degradation of the mRNA by RNase. (Reproduced with permission from Askari FK, McDonnell WM. Molecular medicine: antisense-oligonucleotide therapy. *N Engl J Med.* 1996;334:316.)

CHAPTER IX

Clinical Genetics

Ophthalmology as a science of human medicine has played an important historical role in human genetics, providing the following breakthroughs:

- First modern description of a familial disease (daltonism, or dichromatic red-green color blindness, 1798)
- First textbook of human genetics (Waardenburg, 1932)
- Establishment of the first linkage of one human disease to another trait (color vision deficiency and hemophilia, 1937)
- First human disease linkage to an autosome (cataracta nucleus pulverulenta to the Duffy blood group on chromosome 1, 1963)
- First clear demonstration of mitotic recombination in human cell lines (in retinoblastoma, 1983)

It is important for a clinician not only to diagnose a disease state accurately and minimize its effects in the patient but to ensure that the siblings and parents are evaluated for milder or earlier forms of the disease. Any family with a genetic disorder should receive genetic counseling as a primary care responsibility. However, only the ophthalmologist will be sensitive to the wide variability of traits affecting the visual system and to the subtleties of carrier-state detection (both by direct evaluation and with indirect diagnostic technology). Furthermore, only the ophthalmologist can appropriately counsel the patient and family about the ocular effects of a given genetic disorder and its attendant risks and burdens.

Terminology: *Hereditary, Genetic, Familial, Congenital*

Not infrequently, uncertainty surrounds the accurate use of the terms *hereditary, genetic, familial,* and *congenital. Hereditary* indicates that a disease or trait under consideration results directly from an individual's particular genetic composition (or *genome*) and that it can be passed from one generation to another. *Genetic* denotes that the disorder is caused by a defect of genes, whether acquired or inherited. In some instances (eg, with large deletions of mitochondrial DNA [mtDNA] associated with chronic progressive ophthalmoplegia), the disease is clearly genetic but is not passed to subsequent generations and is therefore not hereditary. These deletions associated with ocular myopathies presumably arise in the oocyte or during early embryologic development. Thus, the terms *hereditary* and *genetic* are not exactly synonymous but are sometimes used to convey similar concepts. A hereditary or a genetic disorder may or may not be congenital.

A condition is *familial* if it occurs in more than one member of a family. It may, of course, be hereditary but need not be. A familial disorder can be caused by common exposure to infectious agents (eg, tuberculosis), traumatizing materials (eg, radiation), deficient or excess food intake (eg, vitamin deficiencies or obesity), or environmental agents such as asbestos or coal dust.

The term *congenital* refers to characteristics that are present at birth. These may be hereditary or familial, or they may occur as an isolated event, often as the result of an infection (eg, rubella, toxoplasmosis, or cytomegalic inclusion disease). The presence of such characteristics *at birth* is the defining factor. Findings suggesting but not establishing that a congenital anomaly may be genetic include a phenotype similar to that from known genetic disorders (eg, aniridia) or a tendency toward bilaterality (eg, bilateral colobomas) and symmetry. However, some nongenetic congenital disorders, such as cataracts from rubella, can be bilateral. Moreover, not all hereditary disorders are bilateral or symmetrical—for example, optic nerve coloboma in only one eye has been observed in multiple generations and presumably in this case is an autosomal dominant trait.

Although numerous hereditary disorders are expressed at the time of birth, the complex interrelationships between genomic expression and factors such as the environment may alter the time of onset and the extent to which a disorder is manifested. For example, the overt clinical onset of signs for diabetes mellitus—a condition with a heritable tendency mediated by multiple genes—in individual monozygotic twins can differ appreciably with each twin's level of carbohydrate intake or other factors. In addition, although the enzyme deficiency responsible for galactosemia is clearly inherited, the expression of systemic disease as well as cataract formation can be avoided by the removal of galactose from the diet. Von Hippel–Lindau disease and lattice dystrophy of the cornea are additional examples of genetic disorders in which signs of ocular disease may not be apparent at birth.

A condition known to be genetic and hereditary may appear in only one individual of a family (eg, retinitis pigmentosa). Such an individual is said to have a *simplex* presentation of a genetic disease. Some authors apply the confusing term *sporadic* to describe the occurrence of a trait in a single member of a kindred, often erroneously implying that this case is not genetic. A genetically determined trait may be isolated in the pedigree for several reasons:

- The pedigree is small
- The full expression of the disease has not been sought or has not manifested in other relatives
- The disorder represents a new genetic mutation
- The disorder is recessive and the investigation to determine whether the parents are carriers has been inadequate
- The disorder is caused by chromosomal changes

Clinically similar disorders may be inherited in several different ways—for example, retinitis pigmentosa can occur as an autosomal dominant, autosomal recessive, or X-linked trait or result from a mitochondrial mutation. These various genetic forms represent distinct gene defects with different alterations in gene structure and different biochemical pathogeneses, each of which has similar clinical phenotypic expressions. Clarification of genetic heterogeneity is important, since only with the proper diagnosis and inheritance pattern can appropriate genetic counseling and prognosis be offered.

Some genetic disorders originally thought to be single and unique are found on close scrutiny to be two or more fundamentally distinct entities. Further clarification of the inheritance pattern or biochemical analysis permits separation of initially similar disorders. Such has been the case for Marfan syndrome and homocystinuria. Although both disorders cause unusual body habitus and ectopia lentis, the presence of dominant inheritance, aortic aneurysms, and valvular heart disease in Marfan syndrome distinguishes it from the recessive pattern and thromboembolic disease of homocystinuria.

The term *heterogeneity* has been expanded and modified. *Genetic heterogeneity* is a general term that applies to the phenotypic similarity that may be produced by two or more fundamentally distinct genetic entities; this term implies that the genes are nonallelic. The term *locus heterogeneity* has been used when linkage studies have shown that different families with similar phenotypes map at different loci; hence, the phenotype can be caused by mutations of different genes. *Allelic heterogeneity* describes the situation in which different alleles at the same locus are capable of producing an abnormal phenotype. With *clinical heterogeneity,* different mutations at the same locus can produce different phenotypes.

Once the location on a chromosome is determined for a particular disease gene and the gene's molecular structure is identified, most examples of genetic heterogeneity cease to be a problem for diagnosis or classification. However, clinical, allelic, and locus heterogeneity can remain perplexing issues. For example, mutations of the same gene, *FGFR2,* which codes for fibroblast growth factor receptor 2, can cause Pfeiffer syndrome, Crouzon disease, and Apert syndrome. Identical mutations in the *FGFR2* gene cause both Pfeiffer syndrome and Crouzon disease phenotypes. Mutation of the Norrie disease gene, *NDP,* usually results in the typical phenotype of pseudoglioma from exudative retinal detachments, but some mutations of *NDP* have been associated with X-linked exudative vitreoretinopathy. Mutations of the proto-oncogene *RET* can give rise to medullary thyroid carcinoma, multiple endocrine neoplasia IIA and IIB, and Hirschsprung disease. Such examples give added meaning to the term *heterogeneity*. Ultimately, greater understanding of these and other disorders at the molecular and cellular level results in more reliable patient management and improved classification of genetic disease. However, considerably more information is needed about how seemingly similar mutations can produce such differing phenotypes.

Mulvihill JJ. Craniofacial syndromes: no such thing as a single gene disease. *Nat Genet.* 1995;9:101–103.

van Heyningen V. Genetics. One gene—four syndromes. *Nature.* 1994;367:319–320.

Wilkie AO, Slaney SE, Oldridge M, et al. Apert syndrome results from localized mutations of FGFR2 and is allelic with Crouzon syndrome. *Nat Genet.* 1995;9:165–172.

Genes and Chromosomes

In 1909, the Danish biologist Wilhelm Johannsen coined the word *genes,* from the Greek for "giving birth to," as a name for segments of the DNA molecule containing individual units of hereditary information. Genes are the basic units of inheritance, and they include the length of nucleotides that codes for a single trait or a single polypeptide chain and its associated regulatory regions. Human genes vary greatly in size from approximately 500 bp to more than 2 million bp. However, more than 98% range from less than 10 kb to 500 kb in size. Many are considerably larger than 50,000 bp. Whereas a single human cell contains enough DNA for 6 million genes, about 50,000–100,000 genes are found among the 23 pairs of known chromosomes. The function of the remaining 95% of the genetic material is unknown.

The relative sequence of the genes, which are arranged linearly along the chromosome, is called the *genetic map.* The physical position or region on the chromosome occupied by a single gene is known as a *locus.* The physical contiguity of various gene loci becomes the vehicle for close association of genes with one another *(linkage)* and their clustering in groups that characteristically move together or separately *(segregation)* from one generation to the next.

It was not until 1956 that Tjio and Levan (in Lund, Sweden) determined that each normal human somatic cell has 46 chromosomes composed of 23 homologous pairs, not 48 chromosomes as had been previously supposed. Each member of a homologous pair carries matched, although not necessarily identical, genes in the same sequence. One member of each chromosome pair is inherited from the father, the other from the mother. Each normal sperm or ovum contains 23 chromosomes, one representative from each pair; thus, each parent transmits half of his or her genetic information to each child. Of the 46 chromosomes, 44 are called *autosomes* because they provide information on somatic characteristics. (The X and Y chromosomes provide such information as well; eg, genes on the Y chromosome influence teeth and height.) At metaphase, the longest human autosome measures approximately 8.0–10.0 μm and the shortest is 1.2–1.5 μm.

In females, the pair of chromosomes that is not identical in the two sexes consists of two similar sex-determining chromosomes, *X chromosomes,* with a length of 4.0–5.0 μm each. Males have one X chromosome and another smaller sex chromosome, about 1.5 μm in length, called a *Y chromosome.* The Y chromosome is unique to phenotypically normal males. It determines development of the testes and other male secondary sexual characteristics.

The two female X chromosomes, although similar in length and position of the centromere, show differences during cell division: One is more condensed, is darker-staining, and replicates its DNA later. This mechanism of inactivation avoids an "overdose" of information from the two X chromosomes in females. Not all genes are unexpressed on the inactivated X chromosome. Certain genes, including those for steroid sulfatase *(STS)* and choroideremia *(CHM),* can escape inactivation. One gene that always escapes inactivation is the X-inactivation center on band q13. Expression of this gene from one X chromosome is directly involved in inactivation of the remainder of the genetic material for this chromosome.

In some female somatic cells, a mass of chromatin in the nucleus appears to represent the inactive X chromosome. This so-called *Barr body,* first identified by Barr and Bertram in cat nerve cells, is easiest to demonstrate in the epithelium of the buccal mucosa. However, the presence of the Barr body varies widely and depends on multiple factors, including the stage of the cell cycle and orientation of the nucleus on a microscopic slide; therefore, evaluation of smears of mucosal epithelium for Barr bodies has little usefulness in modern clinical genetics.

Bishop JE, Waldholz M. *Genome.* New York: Simon & Schuster; 1990.

Emery AEH, Rimoin DL, Connor JM, et al, eds. *Principles and Practice of Medical Genetics.* 4th ed. New York: Churchill Livingstone; 1999.

Goldberg MF, Renie WA, eds. *Genetic and Metabolic Eye Disease.* 2nd ed. Boston: Little, Brown & Co; 1986.

Schmickel RD. The genetic basis of ophthalmological disease. *Surv Ophthalmol.* 1980; 25:37–46.

Alleles

Alternative forms of a particular gene at the same locus on each of an identical pair of chromosomes are called *alleles* (Greek for "reciprocals"). If both members of a pair of alleles for a given autosomal locus are identical (ie, the DNA sequence is the same), the individual is *homozygous* (a *homozygote*); if the allelic genes are distinct from each other (ie, the DNA sequence differs), the individual is *heterozygous* (a *heterozygote*). Different gene defects can cause dramatically different phenotypes and still be allelic. For example, sickle cell disease (SS hemoglobinopathy) caused by homozygosity of one mutant gene is significantly different from the phenotypic expression of SC hemoglobinopathy, yet the Hb S gene and the Hb C gene are allelic. The term *polyallelism* refers to the many possible variants or mutations of a single gene.

As biochemical analysis has become more sophisticated, different alleles frequently have been shown to have slightly different biochemical properties. Among the mucopolysaccharidoses, for example, the enzyme alpha-L-iduronidase is defective in both Hurler disease and Scheie syndrome. Since these are mutations of the same gene, they are abnormalities of the same enzyme and thus allelic. However, the clinical severity of these two disorders (age of onset, age of detection, and severity of affliction of skeleton, liver, spleen, and cornea) is entirely different, presumably because the function of the mutant enzyme is less altered by the Scheie syndrome mutation. Since the enzyme is a protein composed of hundreds of amino acids, a mutation resulting in a base substitution within a certain codon might cause a change in one or more amino acids in a portion of the enzyme remote from its active site, reducing the effect on the enzyme's function. On the other hand, the substitution of one amino acid at a critical location in the active site of the enzyme might abolish most or all of its enzymatic activity. Several examples of allelic disorders appear among the mucopolysaccharidoses.

The phenotype of the usual heterozygote is determined by one mutant allele and one "normal" allele. However, the genotype of a compound heterozygote comprises two different mutant alleles, each at the same locus. The genetic Hurler-Scheie compound is biochemically proven and clinically manifests features intermediate between the homozygotes of the two alleles. Whenever detailed biochemical analysis is possible, the products of the two alleles manifest slightly different properties (such as rates of enzyme activity or electrophoretic migration). Among other autosomal recessive diseases, a spectrum of phenotypes can be caused by a diversity of mutant alleles occurring in various paired combinations. Current knowledge about most recessive diseases (as well as dominant disorders) does not permit more than speculation on this alternative in the clinical setting.

In contrast, close scrutiny may reveal some genetic disorders originally thought to be single and unique to be two or more fundamentally distinct entities. Occasionally, this genetic heterogeneity is seen with diseases that are inherited in the same manner, such as tyrosinase-negative and tyrosinase-positive oculocutaneous albinism. Since these two conditions are phenotypically similar, and each is inherited as an autosomal recessive trait, it was assumed for some time that they were allelic. When a tyrosinase-negative person bears children with a tyrosinase-positive person, the offspring appear clinically normal. This observation excludes the possibility that these two conditions are allelic: Each condition occurs only when an offspring is heterozygous for the gene causing the condition. Separate gene loci (the tyrosinase gene and the *P* gene) are now appreciated to cause oculocutaneous albinism. The offspring of such matings of individuals with phenotypically similar but genotypically different disorders are called *double heterozygotes* since they are heterozygous for each of the two loci.

Because a female has two X chromosomes, she may be either homozygous or heterozygous with respect to X-linked genes. A male is said to be *hemizygous* for X-linked genes since he has only a single X chromosome, and the Y chromosome has little comparable material. A person is also termed hemizygous for a given genetic locus when the second allele is missing, either through loss of an entire chromosome or through rearrangements resulting in deletion of any segment of one of a pair of chromosomes.

Mitosis

A cell may undergo two types of cell division—mitosis and meiosis. *Mitosis* gives rise to the multiple generations of genetically identical cells needed for the growth and maintenance of the organism. When mitosis is about to occur, the cell accurately duplicates all of its chromosomes. The replicated chromosomes then separate into two identical groups that migrate apart and eventually reach opposite sides of the cell. The cell and its contents then divide, forming two genetically identical daughter cells, each with the same diploid chromosome number and genetic information as the parent cell.

Meiosis

In contrast to mitosis, *meiosis* leads to the production of cells that have only one member of each chromosome pair (Fig IX-1). The specialized cells that arise from meiosis and participate in sexual reproduction are called *gametes.* The male gamete is a sperm, the female gamete an ovum. During meiosis, a modified sequence of divisions systematically reduces the number of chromosomes in each cell by one half to the *haploid* number. Consequently, each gamete contains 23 chromosomes, one representative of each pair. This assortment occurs randomly, except that one representative of each pair of chromosomes is incorporated into each sperm or egg.

At conception a sperm and an ovum unite, forming a single cell called a *zygote* that contains 46 chromosomes. Since both parents contribute equally to the genetic makeup of their offspring, new and often advantageous gene combinations may emerge. Meiosis also provides the essential mechanism for sex determination. Since females possess two X chromosomes, they always produce ova that contain 22 autosomes and one X chromosome. Males, in contrast, have one X and one Y chromosome. Half of their sperm carry 22 autosomes and one X chromosome; the other half carry 22 autosomes and one Y chromosome. When a sperm bearing an X chromosome fuses with an ovum, the result is a female. If, instead, the sperm contains a Y chromosome, a male is conceived.

Segregation

Two allelic genes, which occupy the same gene locus on two homologous chromosomes, separate with the division of the two chromosomes during meiosis, and each goes to a different gamete. Thus, the genes are said to *segregate,* a property limited to allelic genes, which cannot occur together in a single offspring of the bearer. For example, if a parent is a compound heterozygote for both hemoglobin S and hemoglobin C, which occupy the same genetic locus on homologous chromosomes, none of the offspring will inherit both hemoglobins from that parent; each will inherit either one or the other.

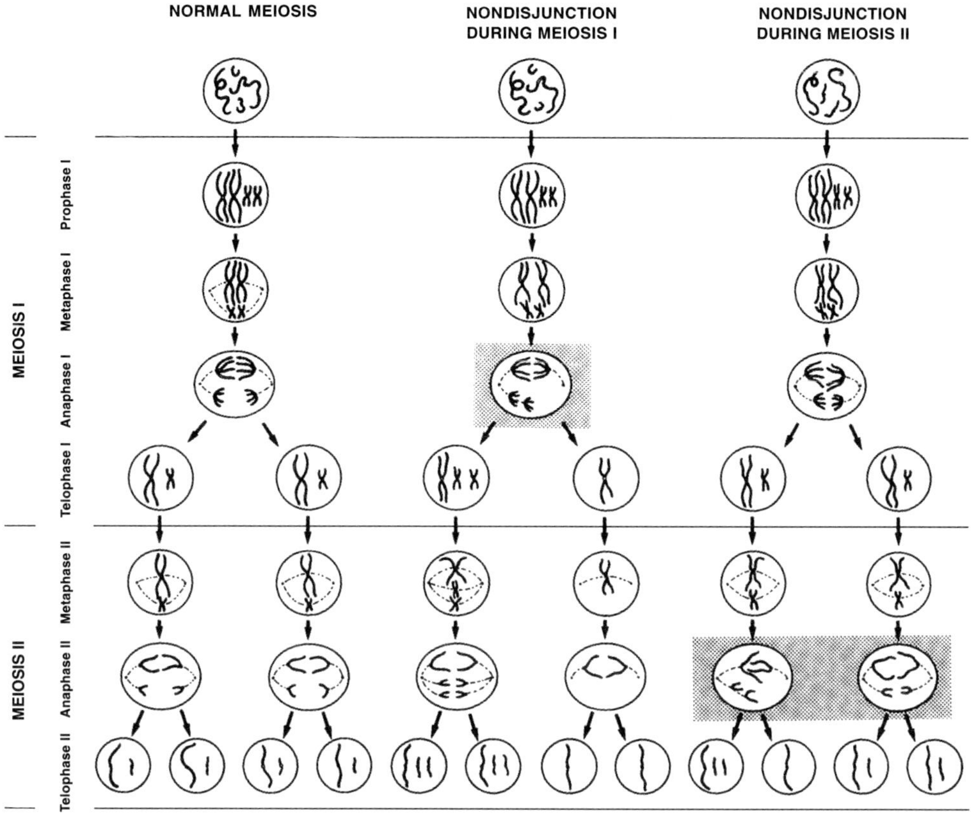

FIG IX-1—Normal meiosis and chromosomal nondisjunction occurring at different phases of meiosis. Nondisjunction is discussed in the text.

Independent Assortment

Genes on different *(nonhomologous)* chromosomes may or may not separate together during meiotic cell division. This random process is called *independent assortment,* based on Mendel's law of independent assortment, which states that nonallelic genes assort independently of one another. Since *crossing over* (exchange of chromosomal material between the members of a pair of homologous chromosomes) can occur in meiosis, two nonallelic genes originally on opposite members of the chromosomal pair may end up together on either of the two or remain separated, depending on their original positions and on the sites of genetic interchange. Thus, the gametes of an individual with two nonallelic dominant traits, or *syntenic traits,* located on the same chromosome could produce four possible offspring. A child may inherit:

- Both traits if the separate alleles remain on the same chromosome and the child inherits this chromosome

- Neither trait if the genes remain on one chromosome but the child inherits the opposite chromosome with neither allele
- Only one of the two alleles if crossing over occurred between the loci and the child received the chromosome with that particular allele

This scheme for nonallelic traits depends on the independent assortment of chromosomes in the first division of meiosis. About 50 crossovers (1–3 per chromosome) occur during an average meiotic division.

Linkage

Linkage is the major exception or modification to the law of independent assortment. Nonallelic genes located reasonably close together on the same chromosome tend to be transmitted together, from generation to generation, more frequently than chance alone would allow for; thus, they are said to be linked. The closer together the two loci, the less likely they are to be affected by crossovers. Linear physical proximity along a chromosome cannot be considered an automatic guarantor of linkage, however. In fact, certain sites on each chromosome may be more vulnerable to homologous crossing over than others.

Jorde LB, Carey JC, Bamshad MJ, et al. *Medical Genetics.* 2nd ed. St. Louis: Mosby; 1999:chap 8.

Chromosomal Analysis

Cytogenetics is a branch of genetics concerned with the study of chromosomes and their properties. Chromosomal defects are changes in the chromosome number or structure that damage sensitive genetic functions and lead to developmental or reproductive disorders. These defects usually result from (1) a disruption of the mechanisms controlling chromosome movement during cell division or (2) alterations of chromosome structure that lead to changes in the number or arrangement of genes or to abnormal chromosomal behavior.

Chromosomal abnormalities occur in about 1 of 200 term pregnancies and in 1%–2% of all pregnancies involving parents over the age of 35 years. About 7% of perinatal deaths and some 40%–50% of retrievable spontaneous abortuses have significant chromosomal aberrations. Virtually any change in chromosome number during early development profoundly affects the formation of tissues and organs and the viability of the entire organism. Most major chromosomal disorders are characterized by both developmental and mental retardation as well as a variety of somatic abnormalities.

Indications

The usual indications for chromosome analysis are listed in Table IX-1. Ophthalmologists should be aware of the value of constitutional and tumor karyotypes in infants with retinoblastoma, especially if the tumor represents a new genetic mutation. Chromosome analysis is also suggested in patients with isolated (nonfamilial) aniridia (which is often associated with Wilms tumor) and other systemic malformations and in patients who survive a neoplasia syndrome and experience a second neoplasm.

A chromosomally abnormal state in a child or an adult of reproductive age warrants consideration of amniocentesis or chorionic villus sampling for prenatal

TABLE IX-1

SOME INDICATIONS FOR CHROMOSOME ANALYSIS

Clinical diagnosis in newborns: multiple malformations, especially involving more than one organ system, with or without intrauterine growth retardation; perinatal death
Clinical diagnosis at any age: mental retardation with or without congenital malformations, in the absence of unequivocal identifiable cause; gonadal ambiguity; infertility, amenorrhea, or reproductive dysfunction; ocular malformations associated with any malformations of other organ systems with no known cause
Multiple miscarriages: spontaneous abortions or stillbirths without apparent cause, even in clinically normal parents
Studies of malignancy: constitutional and tumor karyotypes, especially leukemia, retinoblastoma, aniridia–Wilms tumor, and other embryonal malignancies; specific syndromes with high risks of malignancy–ataxia-telangiectasia, Bloom syndrome, Fanconi anemia; tumor karyotypes on all second tumors in neoplasia syndromes
Prenatal diagnosis: in advanced maternal age, known translocation carrier state, X-linked carrier state for sexing; previous child with chromosomal abnormalities; and as part of either amniocentesis or chorionic villus sampling

diagnosis in subsequent pregnancies to avoid the risk of recurrence. Amniocentesis can be undertaken at about the 16th week of pregnancy.

Preparation

Karyotype The systematic display of chromosomes from a single somatic cell is called a *karyotype.* Chromosome preparations can be made from any tissue whose cells will divide in culture. The tissue most commonly used is peripheral venous blood, although bone marrow, skin fibroblasts, and cells from amniotic fluid or chorionic villi are useful under specific circumstances. In special situations, chromosome analyses can be obtained directly from rapidly dividing neoplastic tissues, as has been done with fresh cells from retinoblastoma and Wilms tumor.

The blood of healthy, nonleukemic persons contains no dividing cells. Therefore, T lymphocytes in a small sample (often less than 5 ml) of fresh heparinized blood are stimulated to divide by adding phytohemagglutinin to a culture medium. After about 72 hours, as the dividing cells approach metaphase, a drug that has colchicine-like effects is added to prevent formation of the mitotic spindle apparatus.

Chromosomal banding After the cells are spread on slides, ruptured, and fixed, they can be stained directly with modified Giemsa techniques *(G-banding)* or quinacrine *(Q-banding),* or they can be prepared with a variety of other materials (Fig IX-2). Faint horizontal differences in staining density called *bands* are enhanced by these procedures. Before the era of chromosomal banding, the chromosomes were arranged in the karyotype into seven groups according to size. Within each group, the specific chromosomes were often impossible to differentiate. The development of chromosomal banding has allowed all 24 chromosomes (22 autosomes plus the X and Y chromosomes) to be uniquely identified, permitting each specific

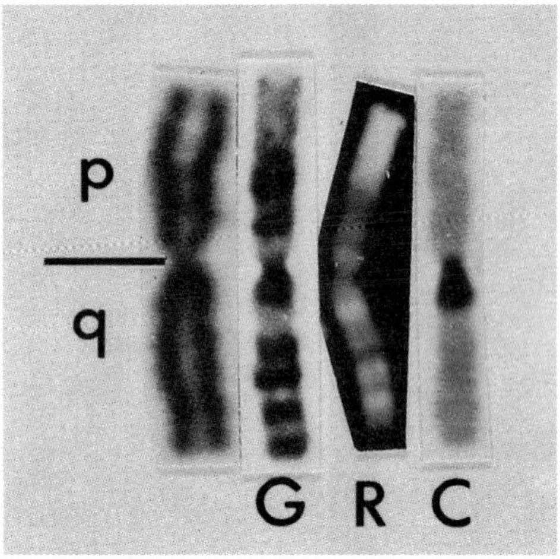

FIG IX-2—Human chromosome 1 prepared by different banding techniques. The unbanded chromosome of the far left was stained with Giemsa stain without any pretreatment. The line indicates the location of the primary constriction, or *centromere,* which divides the chromosome into a short *(p)* and a long *(q)* arm. The second chromosome was prepared by trypsin G-banding, the current standard technique in most clinical cytogenetics laboratories. The third chromosome was prepared with a fluorescent R-banding (reverse banding) technique, which produces a pattern essentially opposite that of G-banding. The chromosome on the far right was prepared by C-banding, which selectively stains regions containing constitutive heterochromatin (genetically inert regions of highly repeated DNA sequences).

autosome to be identified with the chromosome number alone. Arabic numerals written after the letters *p* or *q* designate a specific band region on the short or long arm of a chromosome (Fig IX-3).

The smallest human chromosome bands that can be recognized under the microscope contain 1–5 million bp of DNA and are, therefore, likely to contain many genes. Members of each of the autosomes plus the sex chromosomes can be distinguished by overall size, their morphology (with respect to the position of their centromere), and the pattern of banding (as determined by their staining properties). The longest chromosome is about 7.0–8.0 μm in length, and the shortest is about 1.5 μm.

The introduction of prophase rather than metaphase banding has provided even more detailed understanding of individual chromosomes and microscopic structural aberrations. Currently used banding techniques on prophase cells can distinguish up to 850 bands. Figure IX-4 shows the conventionally stained and arranged karyotype of a human male. Chromosome morphology is determined by the position of the centromere (or *central constriction*) that divides the chromosome into a short arm designated *p* (for *petit)* and a long arm designated *q*. Chromosomes are termed *metacentric* if the centromere lies in the center of the chromosome, *submetacentric* if the centromere is somewhat distant from the center, and *acrocentric* if the centromere lies near the end of the chromosome.

A karyotype is prepared from a photomicrograph of the chromosomes in one cell that are cut out, paired, ordered, and numbered according to their structural and

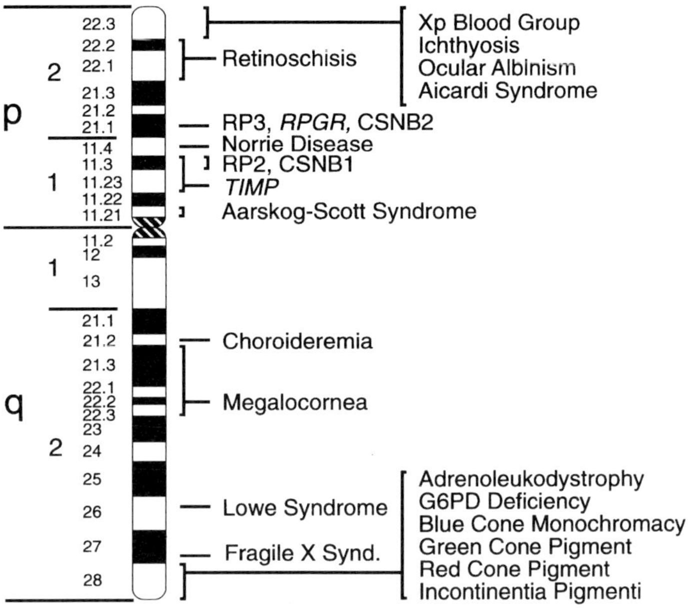

FIG IX-3—Approximate location on the X chromosome of genes for various X-linked disorders. (Courtesy of Richard G. Weleber, MD.)

banding features. Equipment has been developed to handle and process digital images of the chromosomes; this equipment promises greater sensitivity for detection of small chromosomal abnormalities and better understanding of complex chromosomal rearrangements in conjunction with the newer techniques for labeling of specific regions of a chromosome (discussed in the following pages).

Fluorescence In Situ Hybridization (FISH) and Chromosome Arm Painting (CAP)
With the FISH technique, DNA fragments from genes of interest are first tagged with a fluorescent compound and then annealed or hybridized to chromosomes. The regions of interest are stained to determine whether duplication, deletion, or rearrangement has occurred. Such molecular probes can detect and often quantify the presence of specific DNA sequences on a chromosome and can find microscopic abnormalities that would be indiscernible by conventional cytogenetic methods.

Probes have been developed from microdissections of chromosomal regions and FISH that label entire arms of chromosomes (CAP) and each of the individual chromosomes (multicolor spectral karyotyping and combinatorial multifluor FISH). With two-color FISH, both arms of each chromosome can be simultaneously labeled (Fig IX-5). These probes are valuable for detecting and understanding the mechanisms of complex chromosomal rearrangement (Fig IX-6), such as can occur in cancer. Chromosome painting is also being used to generate information on the three-dimensional organization of chromatin domains within the nucleus in interphase.

Guan X-Y, Zhang H, Bittner M, et al. Chromosomal arm painting probes. *Nat Genet.* 1996;12:10–11.

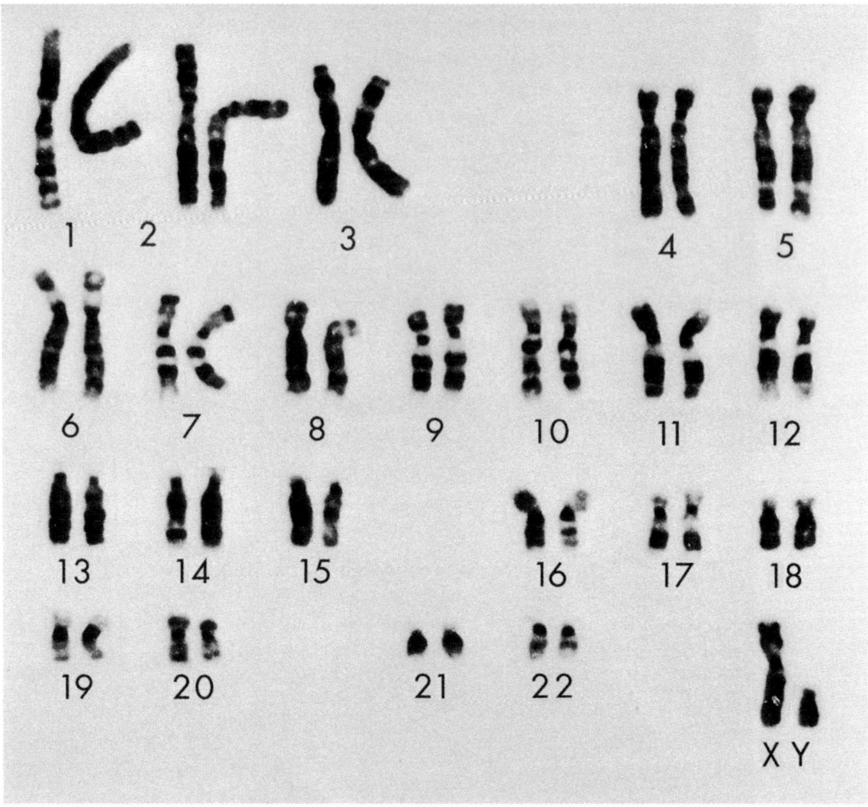

FIG IX-4—Photomicrograph of chromosomes arranged by size, shape, and banding pattern, according to the Paris classification. Note the two dissimilar chromosomes in the last row, called X and Y, which identify this subject as a male.

Schröck E, du Manoir S, Veldman T, et al. Multicolor spectral karyotyping of human chromosomes. *Science.* 1996;273:494–497.

Speicher MR, Ballard SG, Ward DC. Karyotyping human chromosomes by combinatorial multi-fluor FISH. *Nat Genet.* 1996;12:368–375.

Aneuploidy of Autosomes

Aneuploidy denotes an abnormal number of chromosomes in nongametic cells. The presence of three homologous chromosomes in a cell rather than the normal pair is termed *trisomy. Monosomy* is the presence of only one member of any pair of autosomes or only one sex chromosome. The absence of a single autosome is almost always lethal to the embryo; an extra autosome is often catastrophic to surviving embryos. Aneuploidy of sex chromosomes (such as X, XXX, XXY, and XYY) is less disastrous. Monosomies and trisomies are generally caused by mechanical accidents that increase or decrease the number of chromosomes in the gametes. The

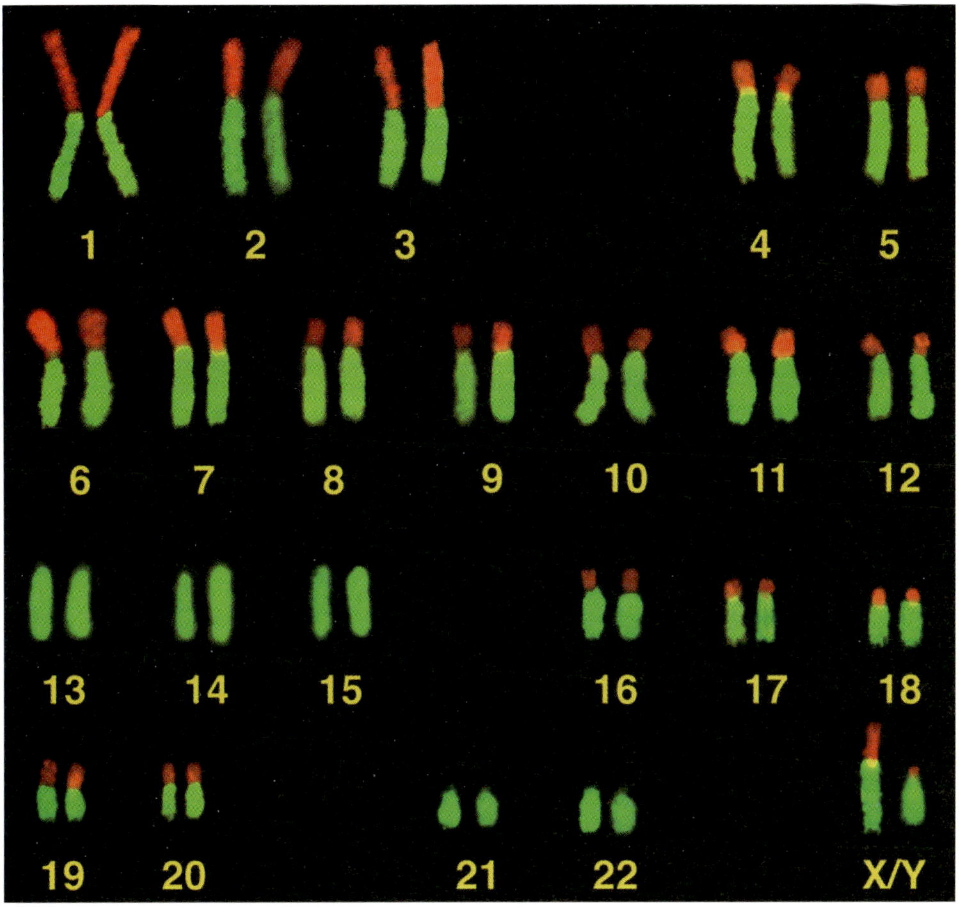

FIG IX-5—Composite karyotype of all human chromosomes hybridized with chromosome arm painting. Metaphase chromosomes were hybridized with corresponding short arm *(red)* and long arm *(green)* painting probes simultaneously, and a composite karyotype was generated. Short arm probes were not generated for the acrocentric chromosomes 13, 14, 15, 21, and 22. Minimal regions of overlap *(yellow)* between long and short arm probes were identified for chromosomes 2, 3, 11, 18, and X. The size distribution of polymerase chain reaction (PCR)-amplified microdissected DNA fragments of each individual arm were analyzed by running PCR products on 1% agarose gels. All PCR products showed a smear ranging from 200 to 600 bp with no apparent dominant bands. After PCR amplification, microdissected DNA fragments from all 19 short arms were labeled with SpectrumOrange fluorescent label for DNA probes. To avoid cross-hybridization between repetitive sequences localized on the short arms of the acrocentric chromosomes, the short arms of the acrocentrics were not dissected. Microdissected DNA fragments from all long arms were labeled with biotin and detected with fluorescein-conjugated avidin. (Reproduced with permission from Guan X-Y, Zhang H, Bittner M, et al. Chromosomal arm painting probes. *Nat Genet.* 1996;12:10.)

most common type of accident, meiotic *nondisjunction*, results from a disruption of chromosome movement during meiosis, as shown in Figure IX-1. *Polyploidy* describes a cell that contains an exact multiple of the normal diploid number: 3n(69), 4n(92), etc.

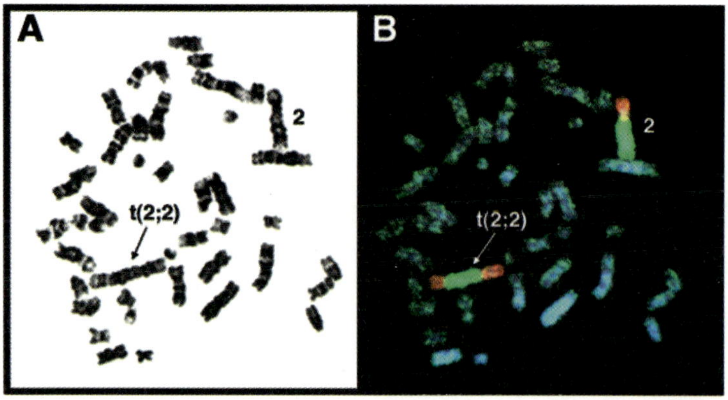

FIG IX-6—Application of chromosome arm painting (CAP) to detect complex chromosome rearrangements. A cell line with complex defined chromosome rearrangements was identified for hybridization [human lymphoma cell line SU-DHL-4]. **A**, G-banded metaphase from SU-DHL-4. **B**, The identical metaphase hybridized with CAPs 2p *(red)* and 2q *(green)*. A normal 2 and a rearranged chromosome 2[t(2;2)(q37;p13)] *(arrow)* was observed. (Reproduced with permission from Guan X-Y, Zhang H, Bittner M, et al. Chromosomal arm painting probes. *Nat Genet.* 1996;12:11.)

de Grouchy J, Turleau C. *Clinical Atlas of Human Chromosomes.* 2nd ed. New York: Wiley; 1984.

Jorde LB, Carey JC, Bamshad MJ, et al. *Medical Genetics.* 2nd ed. St. Louis: Mosby; 1999:chap 6.

Trisomy 13 syndrome (Patau syndrome) Trisomy for chromosome 13 occurs with a frequency of about 1:25,000 live births. About 50% of affected infants die within the first month, 75% die by the sixth month, and fewer than 5% survive more than 3 years. As in all trisomies, the risk of this syndrome increases with advancing maternal age. About 20% of cases are caused by unbalanced translocations.

The phenotype includes severe central nervous system (CNS) anomalies. The most characteristic CNS anomaly involves varying degrees of aberration in formation of the forebrain *(holoprosencephaly)*. Anomalies of the olfactory system *(arrhinencephalia)* and microcephaly are found, and 80% of patients have bilateral cleft lip and palate. Other manifestations may include:

- Binaural deafness
- Polydactyly (more frequently in the hands than the feet)
- A characteristic clenched fist
- Rocker-bottom feet
- Severe congenital heart defects, including atrial and ventricular septal defects, patent ductus arteriosus, and dextrocardia (80%)
- Polycystic kidney disease (30%)
- Hydronephrosis

TABLE IX-2

OCULAR FINDINGS IN TRISOMY 13 (PATAU) SYNDROME

Hypotelorism (sometimes hypertelorism)
Shallow supraorbital ridges
Absent eyebrows
Epicanthal folds
Cyclopia, rarely
Microphthalmos, sometimes clinical anophthalmos
Corneal "clouding"
Cataracts
Uveal colobomas, iridoschisis
Intraocular connective tissue, including cartilage
Persistent hyperplastic primary vitreous
Retinal dysplasia
Optic nerve hypoplasia, atrophy, or occasionally colobomas
Anterior "cleavage" syndrome

In 1960, Klaus Patau and coworkers correlated this complex of abnormalities with a specific chromosomal abnormality based on a trisomy of the 13–15 (D) chromosome group.

Many of the clinical features of this syndrome require trisomy—not of the entire chromosome 13, but only of the portion from 13q14 to q terminus. The features of the full clinical syndrome considerably overlap Meckel syndrome, an autosomal recessive entity with a recurrence risk of 25%. If the clinical signs suggest Patau syndrome, chromosome analysis is imperative to establish the diagnosis of trisomy and to identify translocations or other rearrangements. This chromosomal aberration is closely associated with severe intraocular abnormalities (Table IX-2).

de Grouchy J, Turleau C. *Clinical Atlas of Human Chromosomes.* 2nd ed. New York: Wiley; 1984:226–237.

Trisomy 18 syndrome (Edwards syndrome) *Trisomy 18* is the second most common chromosomal syndrome with multiple malformations in humans. The incidence is about 1:8000 live newborns. The incidence at conception is much higher, but early spontaneous abortion occurs in approximately 95% of cases. Half of the affected live newborns die in the nursery before 4 months of age, and 90% die before 1 year, although a few have lived as long as 15 years. There is a 3:1 preponderance of females affected, perhaps because of preferential survival. The outstanding features of trisomy 18 include the following:

- Low birth weight and growth failure
- Mental retardation
- A skull with a narrow bifrontal diameter and protruding occiput
- Micrognathia
- Misshapen ears

- Flexion deformities of the fingers, with the second finger overlapping the third, and the fifth finger overlapping the fourth
- Nail hypoplasia
- A low-arch dermal ridge pattern
- Atrial and ventricular septal defects
- Inguinal or umbilical hernias (and frequently a single umbilical artery)
- Cryptorchidism
- Rocker-bottom feet
- Myelomeningocele (occasionally)

A maternal age effect has been demonstrated in this trisomy also. The critical region for many of the phenotypic features appears to be near 18q12.2. The most frequent ocular abnormalities are listed in Table IX-3.

de Grouchy J, Turleau C. *Clinical Atlas of Human Chromosomes.* 2nd ed. New York: Wiley; 1984:290–302.

Trisomy 21 syndrome (Down syndrome) *Down syndrome* is the most common chromosomal syndrome in humans, with an overall incidence of 1:800 live births. It was the first chromosomal disease defined in humans. Clinical features of this syndrome have been well known since the British physician John Langdon Down originally described them in 1866.

The frequency of Down syndrome clearly increases with age of the mother, from about 1:1400 live births (to mothers aged 20–24 years) to about 1:40 live births (to mothers aged 44 years). Yet the frequency of Down syndrome is greater (1:1250) for mothers between 15 and 19 years of age than it is in the next-higher age range. Above age 50, the frequency is 1:11 live births. The eponym *Down syndrome* summarizes a clinical description of certain distinctive if variable phenotypic features, whereas the karyotype describes the chromosomal constitution of the cells and tissue studied.

TABLE IX-3

OCULAR FINDINGS IN TRISOMY 18 (EDWARDS) SYNDROME

Prominent epicanthal folds, ptosis
Blepharophimosis with unusually small or oblique palpebral fissures
Unusually thick lower eyelid
Hypertelorism (sometimes hypotelorism)
Hypoplastic supraorbital ridges
Congenital glaucoma
Corneal opacities
Microphthalmos
Optic disc anomalies
Iris and uveal colobomas

The chromosomal basis of this disorder was first demonstrated by French geneticist Jerome Lejeune and coworkers in 1959. Approximately 95% of children with this disorder have an extra chromosome 21 as a result of meiotic nondisjunction. Either parent may contribute the third chromosome 21, but the most important factor that influences the risk of having a child with Down syndrome is maternal age. When it has been possible to determine where the meiotic error took place, more than 80% occurred in the first meiosis, and more than 95% occurred with maternal rather than paternal meiosis.

Approximately 5% of patients with Down syndrome have a *translocation* resulting from the attachment of the long arm of chromosome 21 with the long arm of one of the other acrocentric chromosomes, usually 14 or 22. These translocations cause pairing problems during meiosis, and the translocated fragment of chromosome 21 appears in one of the daughter cells along with a normal 21. As in nondisjunction, the fragment becomes trisomic on fertilization. Trisomy of only the distal third of chromosome 21q is sufficient to cause the disorder. Genes that lie within the q22 band of chromosome 21 appear to be specifically responsible for the pathogenesis of Down syndrome.

The increased incidence of this disorder with maternal age is principally the result of a greater likelihood of nondisjunction, since translocation errors are not related to maternal age. The extra chromosome is of paternal origin in fewer than 5% of affected persons. A major positive correlation between advanced paternal age and an increased incidence of Down syndrome has not been established. The empirical recurrence risks to parents who have had one child with trisomy 21 are approximately 1%, although this number is higher among older women. Patients with Down syndrome exhibit the following features:

- Mental retardation
- Short stature
- Poor muscular control (hypotonia)
- Brachycephaly with a broad flat occiput
- Hypoplasia of the middle phalanx of the fifth finger
- Wide space between the first and second toes
- Small ears
- Various forms of congenital heart disease, including ventricular and atrial septal defects (40%) and, occasionally, duodenal atresia and tracheoesophageal fistulas
- Infertility
- Dental hypoplasia
- Characteristic dermatoglyphic findings

A single palmar crease occurs in about 50% of those with Down syndrome, but in only 1% of the general population. About half of the infants and young children with this disorder have unspecified hearing loss. The most common ocular findings of Down syndrome are presented in Table IX-4.

Additional medical complications in patients with Down syndrome include an increased susceptibility to infection and a 20- to 50-fold increase in the risk of leukemia. The shortened lifespan in patients affected by Down syndrome is partly secondary to these related medical problems. Studies of autopsy material from affected patients show that virtually all patients with Down syndrome over the age of 35 years develop abnormal microscopic senile plaques and neurofibrillary tangles in the brain, similar to those in Alzheimer disease. Down syndrome patients also

TABLE IX-4

OCULAR FINDINGS IN DOWN SYNDROME (TRISOMY 21)

Almond-shaped palpebral fissures
Upslanting (mongoloid) palpebral fissures
Prominent epicanthal folds
Blepharitis, usually chronic, with cicatricial ectropion
Strabismus, usually esotropic
Nystagmus (typically horizontal)
Aberrant retinal vessels (at disc)
Iris stromal hypoplasia
Brushfield's spots
Keratoconus
Cataract
Myopia
Optic atrophy

appear to be at significantly increased risk for the cognitive symptoms of Alzheimer disease. It has been shown that the amyloid beta precursor protein (a major component of the neurofibrillary plaques that accumulate in the brain of persons with Alzheimer disease) is identical to the protein that accumulates in apparently identical lesions in persons with Down syndrome who are older than 35 years. The relationship of the gene for amyloid precursor protein and the form of Alzheimer disease in Down patients is unknown.

Catalano RA. Down syndrome. *Surv Ophthalmol.* 1990;34:385–398.

de Grouchy J, Turleau C. *Clinical Atlas of Human Chromosomes.* 2nd ed. New York: Wiley; 1984:338–349.

Patterson D. The causes of Down syndrome. *Sci Am.* 1987;257:52–60.

Aneuploidy of Sex Chromosomes

Recognized characteristic eye findings do not appear in most sex chromosomal abnormalities, except possibly *Turner syndrome.* About 55%–60% of patients with Turner syndrome have a complete absence of one X chromosome in all cells (45,X), usually because of paternal X or Y chromosomal nondisjunction. The rest have a variety of structural changes of the X chromosome (such as isochromosomes for the long arm of X, deletions from the long or short arms, or a ring-X chromosome) or they exhibit mosaicism (see below). Affected persons always develop as females, except in X/XY mosaicism. They tend toward short stature, do not develop secondary sexual characteristics, and often have webbing of the neck with low-set nuchal hair, cubitus valgus, nonpitting edema of the hands and feet (at birth), and coarctation of the aorta. Their ovaries fail to develop properly (gonadal dysgenesis), resulting in sterility and amenorrhea.

Turner syndrome occurs in about 1 of every 5000 live female births. The incidence of X-linked color deficiency is the same in monosomy-X Turner syndrome as

TABLE IX-5

OCULAR FINDINGS IN TURNER SYNDROME

Pigmented areas on eyelids
Prominent epicanthal folds
Accentuated downslanting of palpebral fissures
Ptosis
Strabismus
Nystagmus
Blue sclera
Cornea with short horizontal axis
Pupillary heterotopia
Anterior axial embryonic cataract
Incidence of color deficiency same in 45,X as in normal male population

in normal males. Complete expression of other X-linked disorders in monosomy Turner syndrome (such as choroideremia, X-linked retinitis pigmentosa, or X-linked ocular albinism) could also be anticipated. Other ocular findings are summarized in Table IX-5.

Mosaicism

Occasionally, an individual or a tissue contains two or more cell lines with distinctly different chromosomal constitutions. Such persons or tissues are termed *mosaics*. Sometimes the peripheral blood, which is the usual source for chromosome analysis, contains populations of cells with completely different chromosomal constitutions. One population of cells may be so infrequent that a second tissue, such as skin fibroblasts, must be analyzed to demonstrate the mosaicism. The incidence of mosaicism in all tissues in humans is unknown.

Cytogenetic defects arise because of abnormal chromosomal distribution during the early stages of embryonic development. These embryos possess two or more chromosomally different cell populations. Mosaicism usually results from either *mitotic nondisjunction*, in which one replicated chromosome fails to separate in the dividing cell; or from *anaphase lag*, in which normal separation occurs, but one member of the replicated pair fails to migrate and is lost.

The clinical effects of mosaicism are difficult to predict because the distribution of abnormal cells in the embryo is determined by the timing of the error and other variables. If mitotic nondisjunction immediately follows conception, the zygote divides into two abnormal cells, one trisomic and one monosomic. The monosomic cells rarely survive and may decrease in number or even disappear entirely over time. Mitotic nondisjunction may occur when the embryo is composed of a small population of cells. Thus, three populations of cells are established, one normal and two abnormal, although some abnormal cell lines may be "discarded" or lost during development. If mitotic nondisjunction occurs at a more advanced stage of development, resulting abnormal populations constitute a minority of the embryo's cells, and mosaicism may have little or no measurable effect on development.

A small population of aneuploid mosaic cells may not have a direct effect on development. However, when cells of this type occur in the reproductive tissues of otherwise normal persons, some of the gametes may carry extra or missing chromosomes. Consequently, mosaic parents tend to be at high risk for chromosomally abnormal children.

The most common example of autosomal mosaicism is *trisomy 21 mosaicism.* Some patients with trisomy 21 mosaicism have the typical features of Down syndrome; others show no abnormalities in appearance or intelligence. The critical variable seems to be the frequency and the embryologic distribution of the trisomic cells during early development, which does not necessarily correlate with the percentage of trisomic cells in any one tissue, such as peripheral blood.

Several types of sex-chromosome mosaicism may occur. Again, the physical effects tend to vary, probably reflecting the quantity and distribution of the abnormal cells during development. For example, the cell population that lacks one of the X chromosomes can arise in a female embryo, leading to 45,X/46,XX mosaicism. In some cases, these patients develop normally; in other cases, some or all of the features of Turner syndrome appear. Similarly, the Y chromosome may be lost in some cells of a developing male embryo. This produces 45,X/46,XY mosaicism. X/XY mosaics may develop as normal males, as females with the features of Turner syndrome, or as persons with physical characteristics intermediate between the sexes (*intersexes,* or *pseudohermaphrodites*).

Mechanisms of Chromosomal Abnormalities

Several well-defined and common chromosomal syndromes result from the structural abnormality of a specific autosome. They present with major developmental aberrations of the visual system, which bring the patient to an ophthalmologist for recognition or management. These *structural aberrations* may be divided into several groups.

Fragility and breakage on specific chromosomes A break is a fracture in a segment of a chromosome when the fractured segment is retained. Breaks are found under normal physiologic conditions, but certain stresses (such as exposure to x-rays, various drugs, and viral infections) increase the number of breaks.

Deletion Loss of chromosomal material may arise in two fashions: as a *terminal deletion,* with one break resulting in loss of the extremity of a chromosome; or as an *interstitial deletion,* with two breaks along an arm, resulting in the loss of the portion between the breaks. A *ring chromosome* is a type of deletion in which both ends have been lost and the two broken ends have united to form a ring-shaped configuration. Sizable deletions of genetic material are often lethal.

Duplication The addition of genetic material along a chromosome results from unequal crossing over between two homologous chromosomes or between two sister chromatids.

Translocations The transfer of part of one chromosome to a nonhomologous chromosome is called *translocation. Robertsonian translocations* occur with breaks at the centromere when entire arms of the chromosomes are exchanged. A fragment without a centromere that has broken off a chromosome is not retained unless it becomes attached to another chromosome that has a raw end from having also been broken.

Reciprocal translocations occur when genetic material is exchanged between chromosomes without either addition or loss of chromosomal material. *Insertional translocations,* in which a broken piece of chromosome is inserted into a nonhomologous chromosome, require at least three chromosomal breaks to occur. *Isochromosomes* form when a duplicated chromosome divides through its centromere into two separate arms rather than into two sister chromatids. Thus, a duplicated chromosome divides through the centromere to form both an iso-short-arm chromosome and an iso-long-arm chromosome.

de Grouchy J, Turleau C. *Clinical Atlas of Human Chromosomes.* 2nd ed. New York: Wiley; 1984.

Etiology of Chromosomal Aberrations

Parental age The way in which maternal age is associated with nondisjunction is unknown. The significance of this mechanism has been discussed in the section on Down syndrome. Increased paternal age has been associated with de novo mutations of several autosomal dominant diseases, including the following:

- The craniosynostoses (Apert and Pfeiffer syndromes and Crouzon disease)
- Achondroplasia
- Oculodentodigital dysplasia
- Treacher Collins syndrome
- Marfan syndrome
- Waardenburg syndrome type I
- Multiple endocrine neoplasia type IIB (MEN2B)
- von Recklinghausen disease (neurofibromatosis type 1)

In some disorders, such as Apert syndrome and MEN2B, virtually all of the new mutations appear to have a paternal origin. The presumed reason for the paternal age effect is that spermatogenesis occurs over many years and the much larger number of cell divisions increases the likelihood of replication errors.

Carlson KM, Bracamontes J, Jackson CE, et al. Parent-of-origin effects in multiple endocrine neoplasia type 2B. *Am J Hum Genet.* 1994;55:1076–1082.

Moloney DM, Slaney SF, Oldridge M, et al. Exclusive paternal origin of new mutations in Apert syndrome. *Nat Genet.* 1996;13:48–53.

Chromosomal instability The prevalence of chromosomal breaks and rearrangements is high in both cultured cells and in vivo preparations derived from individuals with ataxia-telangiectasia, Fanconi syndrome, Bloom syndrome, and xeroderma pigmentosum. Patients with these conditions have a markedly increased risk of malignant neoplasias. It is likely that the increased likelihood of tumor formation is related to the chromosomal instability.

Environmental agents In some laboratory models, radiation, certain chemicals, and some viruses are responsible for a moderate frequency of chromosomal breaks and rearrangements. The specific mechanism by which each or any of these environmental agents induces genetic abnormalities is no more apparent than their roles in spontaneous birth defects or mutations.

Specific Chromosomal Abnormalities

The phenotypic features of three specific deletion syndromes are characteristic and important enough to warrant summarizing here.

Long arm 13 deletion (13q14) syndrome Retinoblastoma is one of several heritable childhood malignancies. Ocular tumors, which are usually noted before the age of 4 years, affect between 1 in 15,000 and 1 in 34,000 live births in the United States. The disease exhibits both hereditary occurrence (approximately 30%–40%), in which tumors tend to be bilateral and multicentric; and sporadic occurrence, in which unilateral and solitary tumors are the rule. Only about 10% of patients with hereditary retinoblastoma have a family history of the disease; the remaining 90% have a new mutation in their germ cells.

Retinoblastoma does not develop in approximately 10% of all obligate carriers of a germ-line mutation. In addition, a karyotypically visible deletion of part of the long arm of chromosome 13 occurs in 3%–7% of all cases of retinoblastoma. The larger this deletion, the more severe the phenotypic syndrome, which includes mental and developmental retardation, microcephaly, hand and foot anomalies, and ambiguous genitalia (Table IX-6).

TABLE IX-6

SYSTEMIC FINDINGS IN THE LONG ARM (13q14) DELETION SYNDROME

"Failure to thrive" growth retardation
Mental retardation
Microcephaly
Trigonencephaly; scalp defect
Micrognathia
Large, malformed, low-set ears
Cleft of highly arched palate
Facial asymmetry
Congenital heart disease
Pelvic girdle anomalies
Anal atresia
Cryptorchidism, bifid scrotum
Hypospadias or epispadias, underdeveloped labia
Hypoplastic thumbs
Short incurved fourth and/or fifth finger
Foot anomalies (clubfoot, short great toe, syndactyly of the fourth and fifth toes)
Associated esterase D deficiency

Although the hereditary pattern in familial retinoblastoma is that of an autosomal dominant mutation, the defect is recessive at the cellular level. The predisposition to retinoblastoma is caused by hemizygosity of the Rb locus within human chromosome band 13q14. The Rb locus is a member of a class of genes called *recessive tumor-suppressor genes*. The alleles normally present at these loci help to prevent tumor formation. At least one active normal allele is needed to prevent the cell from losing control of proliferation. Patients who inherit a defective allele from one parent are at greater risk for losing the other allele through a number of mechanisms. Thus, tumor formation in retinoblastoma is caused by the loss of function of both normal alleles. Homozygous deletions within the 13q14 region have been noted in retinoblastomas derived from enucleated eyes.

The first step in tumorigenesis is a recessive mutation of one of the homologous alleles at the retinoblastoma locus by inheritance, germinal mutation, or somatic mutation. Hereditary retinoblastomas arise from a single additional somatic event in a cell that carries an inherited mutation, whereas sporadic cases require two somatic events. In approximately 50% of tumors, homozygosity for such a recessive mutation results from the mitotic loss of a portion of chromosome 13, including the 13q14 band. The resulting homozygosity for recessive mutant alleles at this locus allows the genesis of the tumor. Retinoblastoma, therefore, seemingly represents a malignancy caused by defective gene regulation rather than the presence of a dominant mutant oncogene. Those who inherit a mutant allele at this locus have a high incidence of nonocular, second tumors thought to be caused by the same mutation. Almost half of these tumors are osteosarcomas.

The retinoblastoma gene was identified and cloned in 1986. Haplotype testing can determine the alleles for polymorphic marker loci immediately adjacent to and within the retinoblastoma gene in various family members and in the tumor. This testing allows the identification of the segment of DNA containing the mutant allele that has become homozygous or hemizygous (ie, underwent loss of heterozygosity) in the tumor. This information can be used to predict the risk of retinoblastoma in siblings or other family members. For optimal success, such testing requires availability of tumor tissue and the testing of parents, siblings, and others in the family who might carry the mutation. Even without the information gained from studying the tumor itself, haplotype testing can occasionally determine that a particular sibling did not inherit either of the haplotypes that are present in DNA from peripheral blood from the affected child. BCSC Section 4, *Ophthalmic Pathology and Intraocular Tumors*, discusses retinoblastoma in greater detail.

Cavenee WK, Dryja TP, Phillips RA, et al. Expression of recessive alleles by chromosomal mechanisms in retinoblastoma. *Nature.* 1983;305:779–784.

Dryja TP, Rapaport JM, Joyce JM, et al. Molecular detection of deletions involving band q14 of chromosome 13 in retinoblastomas. *Proc Natl Acad Sci U S A.* 1986;83:7391–7394.

Friend SH, Bernards R, Rogelj S, et al. A human DNA segment with properties of the gene that predisposes to retinoblastoma and osteosarcoma. *Nature.* 1986;323:643–646.

Friend SH, Dryja TP, Weinberg RA. Oncogenes and tumor-suppressing genes. *N Engl J Med.* 1988;318:618–622.

Godbout R, Dryja TP, Squire J, et al. Somatic inactivation of genes on chromosome 13 is a common event in retinoblastoma. *Nature.* 1983;304:451–453.

Lohmann DR, Brandt B, Oehlschläger U, et al. Molecular analysis and predictive testing in retinoblastoma. *Ophthalmic Genet.* 1995;16:135–142.

Wiggs J, Nordenskjold M, Yandell D, et al. Prediction of the risk of hereditary retinoblastoma, using DNA polymorphisms within the retinoblastoma gene. *N Engl J Med.* 1988;318:151–157.

Short arm 11 deletion (11p13) syndrome *Aniridia* (AN2) occurs from a defect of a gene that encodes a transcription factor needed for development of the eye. This developmental gene, *PAX6,* is located at 11p13. Aniridia is a panophthalmic disorder characterized by the following:

- Subnormal visual acuity
- Congenital nystagmus
- Strabismus
- Corneal pannus
- Cataracts
- Ectopia lentis
- Glaucoma
- Optic nerve hypoplasia
- Foveal or macular hypoplasia
- Iris absence or severe hypoplasia

Although almost all cases of aniridia result from *PAX6* mutations, a rare autosomal recessive disorder called *Gillespie syndrome* (MIM 206700) also produces partial aniridia, cerebellar ataxia, mental deficiency, and congenital cataracts.

Aniridia (often with cataract and glaucoma) can also occur sporadically in association with Wilms tumor, other genitourinary anomalies, and mental retardation, the so-called *WAGR syndrome.* This complex of findings is called a *contiguous gene-deletion syndrome* because it results from a deletion involving nearby genes in the region. Most of the affected patients have a karyotypically visible interstitial deletion of a segment of chromosome 11p13. This region also includes the gene for the enzyme catalase, and an adjacent locus (11p12) contains the gene *LDH-A* for lactic dehydrogenase. Patients with aniridia that is not clearly part of an autosomal dominant trait and those with coincident systemic malformations should undergo chromosomal analysis and observation for possible Wilms tumor.

The ophthalmologist confronted with a new aniridia patient should conduct a careful biomicroscopic examination of both parents for the variable expression of autosomal dominant aniridia. For female infants with isolated aniridia, a high-resolution banded chromosomal analysis is essential, as the genital variation caused by 11p deletion can be extraordinarily subtle. If a male infant with isolated aniridia has no genital aberrations, the chromosome analysis is desirable although probably not mandatory (because of the more severe expression of 11p deletions in males). In older children without other anomalies or developmental delay, a baseline intravenous pyelogram and periodic urinalysis (for microscopic hematuria) are recommended. Intravenous pyelography is probably a more sensitive procedure than either echography or computed tomography for the embryonal malignancy associated with this chromosomal deletion.

The PAX6 gene product is a transcription factor that is required for the normal development of the eye. Mutations of *PAX6* have also been reported in Peters anomaly, autosomal dominant keratitis, and dominant foveal hypoplasia. The mechanism for disruption of normal embryology and the degenerative disease in aniridia and other *PAX6* disorders appears to be *haploinsufficiency,* the inability of a single active allele to activate transduction of the developmental genes that are regulated by the PAX6 gene product. In this way, aniridia is different from retinoblastoma and Wilms tumor, which result from an absence of both functional alleles at each of homologous gene loci.

de Grouchy J, Turleau C. *Clinical Atlas of Human Chromosomes.* 2nd ed. New York: Wiley; 1984:208–209.

Fearon ER, Vogelstein B, Feinberg AP. Somatic deletion and duplication of genes on chromosome 11 in Wilms' tumours. *Nature.* 1984;309:176–178.

Littlefield JW. Genes, chromosomes, and cancer. *J Pediatr.* 1984;104:489–494.

Solomon E. Recessive mutation in aetiology of Wilms' tumor. *Nature.* 1984;309: 111–112.

Short arm 5 deletion A short arm deletion in chromosome 5 is found in the so-called *cat's cry,* or *cri du chat, syndrome.* The critical region for this syndrome is 5p15. Infants with cri du chat have a typical cry, which is present only in the first few months of life, that resembles the mewing of a cat. The abnormal cry is the result of hypoplasia of the larynx. Other systemic findings include:

- Severe mental retardation
- Microcephaly
- Hypotonia
- Micrognathia
- Low-set or poorly formed ears
- Low birth weight
- Slow growth
- Congenital heart disease

Ocular findings include:

- Hypertelorism
- Downward-slanting palpebral fissures
- Epicanthal folds
- Myopia
- Exotropia
- Iris coloboma
- Optic atrophy

Mutations

Change in the structure or sequence of a gene is called a *mutation.* A mutation can occur more or less randomly anywhere along the DNA sequence of a gene and may result when one nucleotide is substituted for another (sometimes called a *point mutation*). A mutation that occurs in a noncoding portion of the gene may or may not be of clinical consequence. Similarly, a mutation may structurally alter a protein but in a manner that does not notably compromise its function. A new mutation that compromises function appears in a given gene as the gene is transmitted from parent to offspring at a frequency of about 1 in a million. Mutations are more likely within certain genes than in others. Aniridia has a mutation rate (mutations/locus/generation) of $2.5–5.0 \times 10^{-6}$; retinoblastoma's rate is $5.0–12.0 \times 10^{-6}$. Two examples of disorders with even higher mutation rates are von Recklinghausen neurofibromatosis type 1 and Duchenne muscular dystrophy, each with an estimated mutation rate of approximately $0.4–1.0 \times 10^{-4}$.

A classic example of a simple point mutation is sickle cell anemia, which affects about 1 in 600 African Americans. This disorder results from a mutant gene that defines the sequence of amino acids in the β-polypeptide chain of adult hemoglobin. In sickle cell hemoglobin, the valine is substituted for glutamic acid at the sixth position in the β-polypeptide chain. This substitution is caused by an abnormal specific base, where adenine is substituted for thymine. This seemingly small alteration causes a profound reduction of solubility when hemoglobin is deoxygenated: red blood cells tend to become deformed into a characteristic sickle shape when the partial pressure of oxygen is low.

More gross mutations may involve deletion, translocation, insertion, or internal duplication of a portion of the DNA. Some mutations cause either destruction of the offspring or sterility. Others are less harmful or are potentially beneficial and become established in subsequent generations. Mutations can occur spontaneously for reasons that are not understood. They may also be produced by a variety of environmental agents called *mutagens,* such as radiation, viruses, and certain chemicals.

Mutations may arise in somatic as well as germinal cells, but these are not transmitted to subsequent generations. Somatic mutations in humans are difficult to identify, but some account for the inception of certain forms of neoplasia (eg, retinoblastoma).

Jorde LB, Carey JC, Bamshad MJ, et al. *Medical Genetics.* 2nd ed. St. Louis: Mosby; 1999:chap 3.

Polymorphisms

Many mutations have either little or no deleterious effect on the organism. A polymorphism is defined as the occurrence of two or more alleles at a specific locus with a frequency greater than 1% each. At least one third of all structural genes may exist in polymorphic forms. For example, at least 400 variants of hemoglobin are known, many with essentially no detectable phenotypic abnormalities. Similarly, several dozen functional and electrophoretic variants of glucose-6-phosphate dehydrogenase exist; again, many have no significant effect on the biochemical function of the affected individual. Finding additional polymorphisms will be important for the completion of gene mapping and for linkage to human diseases.

Genome, Genotype, Phenotype

The *genome* is the sum total of the genetic material within a cell or of an organism—thus, the total genetic endowment. By contrast, the *genotype* defines the genetic constitution, and thus biological capacity, with regard to a specific locus (eg, individual blood groups or a specific single enzyme). *Phenotype* indicates the total observable or manifest physical, physiologic, biochemical, or molecular characteristics of an individual, which are determined by the genotype but can be modified by the environment.

The phenotype of a disease is often known before its specific metabolic, genetic, or chromosomal basis. For example, several chromosomal aberrations lead to the phenotype of Down syndrome, and several different enzyme deficiencies may manifest themselves as a "Morquio syndrome phenotype." In these instances, only a tentative molecular diagnosis can be made from the clinical information. Specific information from a chromosomal analysis or an enzyme assay then establishes the genetic mechanism unique for the physical features: trisomy 21 (a chromosome 14 to 21 translocation) in the example of Down syndrome, or β-glucuronidase deficiency that causes a disease state often confused clinically with true Morquio syndrome.

A clinical picture produced entirely by environmental factors that nevertheless closely resembles, or is even identical with, a phenotype is known as a *phenocopy*. Thus, for example, the pigmentary retinopathy of congenital rubella has occasionally been confused with a hereditary dystrophic disorder of the retinal pigment epithelium. Similarly, chloroquine-induced changes in the corneal epithelium resemble those seen as cornea verticillata in the X-linked dystrophic disorder Fabry disease.

Single-Gene Disorders

About 4500 different diseases are known to be caused by a defect in a single gene. As a group, these disorders are called *monogenic*, or *mendelian, diseases*. They most often show one of three patterns of inheritance: autosomal dominant, autosomal recessive, or X linked. Disorders of mtDNA are inherited in a fourth manner termed *maternal inheritance*. These mtDNA disorders obey galtonian rather than mendelian inheritance characteristics.

Variability

Variability is an intrinsic property of human genetic disease that reflects the quantitative and qualitative differences in phenotype among individuals with the "same" mutant allele. Even within the homogeneous population of a single family with a genetic disease, every affected individual may not manifest the disease to the same degree, with the same features, or at the same age. Steinert myotonic dystrophy, for example, presents its features of motor myotonia, characteristic cataracts, gonadal atrophy, and presenile baldness with a wide variation in severity and age of detection. Even within a single family, the cataracts may begin to affect vision any time from the second to the seventh decade of life.

Such variability of clinical manifestation led to the concept of *anticipation*, the phenomenon of apparently earlier and more severe onset of a disease in successive generations within a family. Before 1990, most geneticists thought that anticipation was not a biological phenomenon but an artifact of ascertainment. With the relatively recent discovery of triplet or trinucleotide tandem repeat expansion diseases, anticipation has been shown to reflect the increased length of trinucleotide tandem repeats from one generation to the next. Myotonic dystrophy, fragile-X syndrome, Huntington disease, and one form of spinal and bulbar muscular atrophy called *Kennedy disease* are some of the diseases whose discovery contributed to the rejuvenation of the concept of anticipation.

Some human variability may result from the intrinsic differences in genetic background of every human being. Other recognizable or presumptive influences on the variable intra- or interfamilial phenotype of the same gene include the following:

- Sex influences or limitations
- Maternal factors such as intrauterine environment and even cytoplasmic (eg, mitochondrial) inheritance factors

- Modifying loci
- Genetic heterogeneity, including both isoalleles and genocopies
- Gene alterations induced either by position effects with other genes or by somatic mutations

Obviously, nongenetic factors extrinsic to a cell, tissue, or organism such as diet, temperature, and drugs may effect major changes in gene expression, either as phenocopies or through ecologic parameters.

Penetrance

The presence or absence of any effect of a gene is called *penetrance*. If a gene generates any evidence of phenotypic features, no matter how minimal, it is termed *penetrant*; if it is not expressed at any level of detection, it is termed *nonpenetrant*. Thus, penetrance is an all-or-nothing concept, statistically representing the fraction of individuals carrying a given gene that manifests any evidence of the specific trait. In families with an autosomal dominant mutant gene that has 100% penetrance of the phenotype, an average of 50% of the offspring will inherit the gene and show evidence of the disease.

Even though penetrance has an exact statistical definition, its clinical application is affected by the diagnostic sophistication and methodology applied in the examination. For example, many mild cases of Marfan syndrome would be missed without careful biomicroscopy of the fully dilated pupil and echocardiography of the heart valves and great vessels. Similarly, if the criteria for identification of the retinoblastoma gene include indirect ophthalmoscopy and scleral depression, some "nonpenetrant" parents or siblings in families with "dominantly inherited" retinoblastoma may be found to have a spontaneously involuted tumor, which clearly identifies them as bearers of the gene. In another example, some family members who have a gene for Best macular dystrophy will be identified not by clinical ophthalmoscopic examination but only by electro-oculographic testing. Therefore, in examining a potential bearer of a gene, the examiner must carefully search for any manifestations of the gene's effects in all susceptible tissues before dismissing someone as a "skipped generation."

Lack of penetrance does not prevent molecular diagnosis. Whenever the gene product or the presence of the gene defect can be sought directly at the molecular level, there is complete correlation with the expected genotype.

Expressivity

The presence of a defective gene does not necessarily imply a complete expression of every potential manifestation. The variety of ways and levels of severity in which a particular genetic trait manifests its presence among different affected individuals is called *expressivity*. In von Recklinghausen disease, for example, an affected child may have only café-au-lait spots. The affected parent may have Lisch nodules of the iris, extensive punctiform and pedunculated neurofibromas of the skin, a huge plexiform neurofibroma of one lower extremity, and a glioma of the anterior visual pathway. It is extremely rare that all affected members in the same family have uniform textbook presentations of the disorder.

Variation in the age of onset of manifestations is one way that expressivity commonly varies in dominant disorders. Using the von Recklinghausen example, the affected child may have only café-au-lait spots at birth, develop iris Lisch nodules that gradually increase in number and size at about age 5–10 years, develop punctiform neurofibromas of the skin in early adolescence, experience subareolar neurofibromas (in a postpubertal female), and experience visual impairment from the effect of an optic glioma in the late teens. Although all of these features are phenotypic components of the mutant gene, each has a characteristic age of onset and a natural history of growth and effect within the umbrella of the total disease.

Pleiotropism

Alteration within a single mutant gene may have consequences in various tissues in a given individual. The presentation of multiple phenotypic abnormalities produced by a single mutant gene is termed *pleiotropism*. For example, in Marfan syndrome, ectopia lentis is coupled with arachnodactyly, aortic aneurysms, and long extremities. Optic atrophy is found in association with juvenile diabetes mellitus, diabetes insipidus, and moderate perceptive hearing impairment in an autosomal recessive syndrome known as the *DIDMOAD* (*d*iabetes *i*nsipidus, *d*iabetes *m*ellitus, *o*ptic *a*trophy, and neural *d*eafness) *syndrome*. Neurosensory hearing loss can also be associated with hereditary hematuric nephritis, lenticular changes (anterior lenticonus, spherophakia, cataracts), arcus juvenilis, and whitish yellow retinal lesions in the dominantly inherited Alport syndrome. Similarly, the Bardet-Biedl syndrome comprises pigmentary retinopathy, obesity, genital hypoplasia, mental debility, and polydactyly. In each of these disorders, a single mutant gene is responsible for dysfunction in multiple systems.

Frequently, however, a disease is mistakenly termed *pleiotropic* when several different disorders with the same inheritance pattern and similar clinical manifestations are actually present. Thus, Leber congenital amaurosis has been attributed to a single pleiotropic gene. Based on the symmetry of the phenotype of affected siblings in individual families, it seems more likely that the clinical disease is heterogeneous and can be caused by several genes (not necessarily allelic), each of which is autosomal recessive.

Heterogeneity

Because of limited diagnostic discriminators, similar or apparently identical phenotypes may be caused by different genotypes. When multiple gene mutations can independently produce a single trait or traits that are difficult to resolve clinically, that trait is said to be genetically heterogeneous. *Allelic heterogeneity* refers to different mutations at the same genetic locus; *locus heterogeneity* refers to mutations at different loci. Some disorders, such as Ehlers-Danlos syndrome, show evidence of both allelic and locus heterogeneity among the different genotypes. An example of allelic heterogeneity is found in the human major histocompatibility complex *(HMC)* genes located on the short arm of chromosome 6.

The division of large groups of patients into smaller but more uniform subgroups by more careful analysis of phenotypes and investigations of the hereditary pattern has been essential for accurate clinical diagnosis and genetic counseling. For example, the phenotypes for Marfan syndrome and homocystinuria were formerly confused but can now be distinguished by biochemical testing, differences in ocular and cardiovascular involvement, and different inheritance patterns. Albinos also represent a heterogeneous group of genetic disorders. Some are tyrosinase-positive and others tyrosinase-negative. Differences in pigmentation also may be helpful in distinguishing different subgroups of albinos. Genetic heterogeneity clearly also exists in the group of disorders classified under the rubric of retinitis pigmentosa, in which the symptom of night blindness is coupled with a characteristic pigmentary retinopathy and abnormalities on psychophysical and electrophysiologic testing. These similar clinical manifestations of various genetically distinct disorders represent *genocopies.*

Racial and Ethnic Concentration of Genetic Disorders

Most genetic diseases occur without regard to the affected individual's racial or ethnic background. Some, however, are concentrated in certain population groups.

Tay-Sachs disease, with its characteristic macular cherry-red spot, occurs predominantly in persons of Eastern European Jewish (Ashkenazi) ancestry, especially those whose ancestors lived in northeastern Poland and southern Lithuania. An estimated rate of 1 in 30 for carriers of this disorder in the Jewish population of New York City compares with an estimated carrier rate of 1 in 300 in non-Jewish Americans. Although the reported incidence of this disorder among Ashkenazi Jewish newborns is 1 in 6000, the actual incidence among this population may be closer to 1 in 3600 births. About 50 new cases occur each year in the United States. In addition, familial dysautonomia *(Riley-Day syndrome)* with hypolacrima, corneal hypoesthesia, exodeviation, and methacholine-induced miosis also occurs more frequently in persons of Ashkenazi ancestry, as do *Gaucher disease* and *Niemann-Pick disease.*

An X-linked recessive gene that causes a *deficiency in the enzyme glucose-6-phosphate dehydrogenase (G6PD)* is found predominantly in people from southern Italy and surrounding areas on the Mediterranean and south into the interior of Africa. Somewhat less than 10% of African Americans show a reduced activity of the G6PD enzyme. Males with this trait develop hemolytic anemia after eating fava beans or ingesting antimalarial drugs such as primaquine or sulfanilamide. A variety of *achromatopsia* (complete color blindness) with *myopia* is common on the South Pacific island of Pingelap, affecting 5% of the Pingelapese population in the Caroline Islands of Micronesia. *Oguchi disease* is seen primarily, although not exclusively, in Japanese persons. Similarly, *sickle cell hemoglobinopathies* are inherited largely among blacks.

The prevalence of *oculocutaneous albinism* is high among the Cuna Indians in Panama. *Hermansky-Pudlak syndrome* occurs with a higher frequency in persons of Puerto Rican ancestry, especially those from the northwestern towns of Aguadilla and Arecibo. In this tyrosinase-positive phenotype of autosomal recessively inherited oculocutaneous albinism, findings include a history of easy bruisability and bleeding tendency, associated with a prolonged bleeding time and abnormal platelet aggregation.

Some specific human malformations occur with greater frequency in certain races than in others. For example, polydactyly is about 10 times more frequent in blacks than in whites, and preauricular sinus may be equally more frequent in blacks.

Goldberg MF. An introduction to basic genetic principles applied to ophthalmology. *Trans Am Acad Ophthalmol Otolaryngol.* 1972;76:1137–1159.

Palmer DJ, Miller MT, Rao S. Hermansky-Pudlak oculocutaneous albinism: clinical and genetic observations in six patients. *Ophthalmic Paediatr Genet.* 1983;3:147–156.

Patterns of Inheritance

Recessivity versus Dominance

The terms *dominant* and *recessive* were first used by Gregor Mendel, an Austrian monk who formulated the fundamental laws of heredity in 1865 while cultivating peas in a monastery garden in Brno. In classic genetics, a dominant gene is one that is always expressed with similar phenotype, whether the mutant gene is present in a homozygous or heterozygous state. Stated simply, a dominant gene is one that is expressed when present in only a single copy. A gene is called recessive when its expression is masked by a normal allele or, more precisely, when it is expressed only in the homozygote (or compound heterozygote) when both alleles at a specific locus are mutant.

A *trait* is the consequence of the gene's action. It is the trait, or phenotypic expression of the gene at a clinical level, rather than the gene itself that is dominant or recessive. A trait is recessive if its expression is suppressed by the presence of a normal gene (as in galactosemia) and dominant if it is apparently unaffected by a single copy of the normal allele (as in Marfan syndrome). If the alleles are different and yet they are both manifested in the phenotype, they are said to be *codominant*. Examples of codominant inheritance include the ABO and MN blood types, leukocyte antigens, and the hemoglobins.

As a result of transcription, a gene may have a greater or lesser effect on the individual or an organ, and therefore the trait may be more or less apparent. Thus, the designation of a trait as either dominant or recessive depends on the testing method used. For example, sickle cell hemoglobinopathy is recessive if the clinical disease is considered, dominant if the sickle preparation test is positive, and codominant if hemoglobin electrophoresis is used to look for the specific product of each allele.

Although, classically, a dominant gene is one that has the same phenotype when the mutant allele is present in either the heterozygous or the homozygous state, most dominant medical diseases stray from this strict definition. For many dominant disorders, individuals who are homozygous for a mutant allele or who harbor two mutant alleles (one on each homologous chromosome) will have more severe expression.

Benedict WF, Murphree AL, Banerjee A, et al. Patient with 13 chromosome deletion: evidence that the retinoblastoma gene is a recessive cancer gene. *Science.* 1983;219: 973–975.

Schmickel RD. The genetic basis of ophthalmological disease. *Surv Ophthalmol.* 1980; 25:37–46.

Experimentally, the biochemical mechanisms of "dominant" hereditary diseases appear different from those of "recessive" disorders. Recessive traits usually result from enzyme deficiency caused by structural mutations of the gene specifying the affected enzyme. The altered enzyme often can be shown to be structurally abnormal or unstable. Heterozygotes usually have approximately 50% of normal enzyme activity but are clinically unaffected, implying that half of the normal enzyme activity is compatible with near-normal function. If adequate biochemical testing can be performed and the specific enzyme isolated, the reduced enzyme activity can be quantified and the heterozygous genetic state inferred. Thus, clinically unaffected heterozygotes can be detected for such disorders as homocystinuria (decrease in cystathionine β-synthase), galactokinase deficiency (low blood galactokinase activity), classic galactosemia (galactose 1-phosphate uridyl transferase deficiency), gyrate atrophy of the choroid and retina (decreased ornithine-δ-aminotransferase), and Tay-Sachs disease (decreased hexosaminidase A). Table IX-7 outlines several disorders with ocular manifestations for which an enzyme defect is known.

Autosomal Recessive Inheritance

An autosomal recessive disease is expressed fully only in the presence of a mutant gene at the same locus on both homologous chromosomes (ie, homozygosity for a mutant gene) or two different mutant alleles at the same locus (compound heterozygosity). A single mutant allele is sufficient to cause a recessive disorder if the normal allele on the homologous chromosome is deleted. A recessive trait can remain latent through several generations until the chance mating of two heterozygotes for a mutant allele gives rise to an affected individual. The frequency of heterozygotes for a given disorder will always be considerably greater than that of homozygotes. It is estimated that all human beings inherit about six or seven mutations for different recessive disorders for which they are heterozygotes.

TABLE IX-7

KNOWN ENZYME DISORDERS AND CORRESPONDING OCULAR SIGNS

DISORDER	DEFECTIVE ENZYME	OCULAR SIGN
Storage diseases		
Fabry disease	Ceramide trihexosidase (alpha-galactosidase)	Corneal epithelial verticillate changes; aneurysmal dilation and tortuosity of retinal and conjunctival vessels
Krabbe leukodystrophy	Cerebroside beta-galactosidase	Macular cherry-red spot; optic atrophy
Mannosidosis	Alpha-mannosidase	Lenticular opacities
Metachromatic leukodystrophy	Arylsulfatase A	Retinal discoloration, degeneration
Hurler IH	Alpha-L-iduronidase	Corneal opacity; pigmentary retinal degeneration
Hunter II	Sulfoiduronate sulfatase	Corneal opacity (mild type); older age patients
Scheie IS	Alpha-L-iduronidase	Corneal opacity; pigmentary retinal degeneration
Sanfilippo III	Heparan sulfate sulfatase	Pigmentary retinal degeneration; optic atrophy
Tay-Sachs disease (GM_2 gangliosidosis, type I)	Hexosaminidase A	Macular cherry-red spot; optic atrophy
Sandhoff disease (GM_2 gangliosidosis, type II)	Hexosidase A and B	Macular cherry-red spot
GM_1 gangliosidosis, type I (generalized gangliosidosis)	Beta-galactosidase	Macular cherry-red spot; optic atrophy; corneal clouding (mild)

Table IX-7 (continued)

Known Enzyme Disorders and Corresponding Ocular Signs

DISORDER	DEFECTIVE ENZYME	OCULAR SIGN
Metabolic disorders		
Alkaptonuria	Homogentisic acid oxidase	Dark sclera
Albinism	Tyrosinase	Foveal hypoplasia; nystagmus; iris transillumination
Intermittent ataxia	Pyruvate dicarboxylase	Nystagmus
Crigler-Najjar syndrome	Glucuronide transferase	Extraocular movement
Ehlers-Danlos syndrome	VI lysyl hydroxylase	Microcornea; retinal detachment; ectopia lentis; blue scleras
Familial dysautonomia	Dopamine-beta-hydroxylase	Alacrima; corneal hypoesthesia; exodeviation; methacholine-induced miosis
Galactokinase deficiency	Galactokinase	Cataracts
Galactosemia	Galactose 1-phosphate uridyl transferase	Cataracts
Gyrate atrophy of the choroid and retina	Ornithine aminotransferase	Degeneration of the choroid and retina; cataracts; myopia
Homocystinuria	Cystathionine synthase	Dislocated lens
Hyperglycinemia	Glycine cell transport	Optic atrophy
Leigh necrotizing encephalopathy	Pyruvate carboxylase	Optic atrophy
Maple syrup urine disease	Branch chain decarboxylase	Ophthalmoplegia; nystagmus
Niemann-Pick disease	Sphingomyelinase	Macular cherry-red spot
Refsum syndrome	Phytanic acid oxidase	Retinal degeneration
Tyrosinosis	Tyrosine aminotransferase	Corneal dystrophy
Sulfite oxidase deficiency	Sulfite oxidase	Ectopia lentis
Tyrosinemia	Tyrosine aminotransferase	Lens opacity

For example, if the frequency of a mutant gene for a recessive disorder in a given population is 1 in 100, the chance that a given individual at random will carry the gene is 1 in 50 since there are two copies of each gene at a given locus in all cells. The likelihood of a carrier parent giving the abnormal gene to each child is 1/2 × 1/50, or 1/100. The chance that an individual at random will receive the gene from both parents is 1 in 10,000 (1/100 × 1/100). Therefore, the likelihood of being a heterozygote is 200 times greater than that of being a homozygote (10,000/50).

The frequency of a mutant allele for a recessive gene in a population and the chance of an individual's being a carrier can therefore be determined by calculating the square root of the frequency of homozygous or affected individuals. In the above example, if a recessive defect is observed in 1 out of every 10,000 persons, the gene frequency is 1 in 100 and the chance of an individual's being a carrier is 1 in 50.

Enzymatic defects Autosomal recessive diseases often result from defects in enzymatic proteins. Most of the so-called inborn errors of metabolism that result from enzymatic defects are autosomal recessive traits, although a few are X-linked recessive disorders (eg, Lesch-Nyhan syndrome). The defect in *alkaptonuria* involves homogentisic acid oxidase, an enzyme involved in the metabolism of homogentisic acid. Large amounts of homogentisic acid are excreted in the urine, which turns black when mixed with alkali or exposed to light or air. The black urine causes diaper stains, calling attention to the condition. In addition, aggregates of homogentisic acid accumulate in the body, becoming attached to the collagen of cartilage and other connective tissues. The cartilage of the ears and nose and the collagenous sclera are stained black or brownish blue. These manifestations are called *ochronosis.* The coloration in the sclera assumes a more or less triangular form, with a limbic base in the region of the palpebral tissue. In the joints, such as those of the spine, the accumulations lead to arthritis. Alkaptonuria is an example of a genetic enzyme block in which the phenotypic features are caused by the accumulation of excess substances just proximal to the block.

In some other disorders with genetic blocks in metabolism, the phenotypic consequences are related to the lack of a normal product distal to the block. An example is *albinism,* in which the metabolic block involves a step between the amino acid tyrosine and the formation of melanin. In still other inborn errors of metabolism, the phenotypic expression results from excessive production of a product through a normally alternative and minor metabolic pathway. *Phenylketonuria,* like alkaptonuria and albinism, is a genetic defect in aromatic amino acid metabolism. The defect is in the enzyme involved in the conversion of phenylalanine to tyrosine. In an affected person, hair and skin pigmentation is reduced. Severe mental retardation is one of the most prominent symptoms. Alternative metabolites of phenylalanine, especially phenylpyruvic acid, are excreted in the urine, providing one basis for diagnosis of the disorder.

The difference in phenotype of these three diseases—alkaptonuria, albinism, and phenylketonuria—is noteworthy, although they involve closely related metabolic pathways.

Carrier heterozygotes The heterozygous carrier of a mutant gene may show minimal evidence of the gene defect, particularly at a biochemical level. Thus, carrier heterozygotes have been detected by a variety of methods:

- Identification of abnormal metabolites by electrophoresis (eg, galactokinase deficiency)
- Liver biopsy (eg, phenylketonuria)
- Hair bulb assay (eg, oculocutaneous albinism and Fabry disease)
- Monitoring of enzyme activity in leukocytes (eg, galactose 1-phosphate uridyl transferase in galactosemia), fibroblasts from skin culture (eg, ornithine-δ-aminotransferase deficiency in gyrate atrophy of the retina and choroid), serum, and tears (eg, hexosaminidase A in Tay-Sachs disease)

In contrast to the transmission of dominant traits, most matings resulting in recessive disorders involve phenotypically normal heterozygous parents. Out of four offspring produced by carrier parents with the same gene for an autosomal recessive disease, usually one will be affected (homozygote), two will be carriers (heterozygotes), and one will be genetically and phenotypically normal. Thus, clinically normal heterozygous parents will produce offspring with a ratio of one clinically affected to three clinically normal. There is no predilection for either sex. In two-child families, the patient with a recessive disease is frequently the only affected family member. For instance, about 40%–50% of patients with retinitis pigmentosa have no family history of the disorder. However, their age of onset, rate of progression, and other phenotypic characteristics are similar to those with defined recessive inheritance patterns.

Once one child is born with a recessive disorder, the genetic risk for each subsequent child of the same parents is 25%. This concept has specific implications for genetic counseling. All offspring of an affected individual will be carriers; they are unlikely to be affected with the disorder unless their clinically unaffected parent is also by chance a carrier of the gene. However, since a specific method for identifying a carrier is lacking with most recessive diseases, the normal-appearing sibling of a child with a recessive disorder has a statistical risk of two chances in three of being a genetic carrier. This liability must be accounted for in any equation to predict the small risk that a normal-appearing sibling will have an affected child.

Consanguinity The mating of close relatives can increase the probability that their children will inherit a homozygous genotype for recessive traits, particularly for relatively rare ones. For example, the probability that the same allele is present in first cousins is 1 in 8. In the offspring of a first-cousin marriage, 1 of every 16 of the genes is commonly present in a homozygous state. It follows that each offspring from a first-cousin marriage has a 1 in 16 chance of manifesting an autosomal recessive trait within a given family. About 1% of all marriages may be consanguineous. A vigorous search for consanguinity between the parents should be made in any case of a rare recessive disease. Incest is one form of consanguinity that is not infrequent and often not acknowledged.

The expression of common recessive genes, by contrast, is less influenced by inbreeding, because most homozygous offspring are the progeny of unrelated parents. This is usually the case with such frequent disorders as sickle cell disease and cystic fibrosis. The characteristics of autosomal recessive inheritance are summarized in Table IX-8.

Francois J. *Heredity in Ophthalmology.* St Louis: Mosby; 1961:86–92.

Pseudodominance Occasionally, an affected homozygote mates with a heterozygote. Of their offspring, 50% will be carriers and 50% will be affected homozygotes. Because this segregation pattern mimics that of dominant inheritance, it is called *pseudodominance.* Fortunately, such matings are usually rare and are unlikely to affect more than two vertical generations.

Familial penetrance Penetrance of recessive disorders within families is rarely if ever incomplete. Expressivity of recessive disorders is characteristically more uniform among affected siblings within families, as each affected individual apparently has a double dose of the same gene. However, age of onset, severity, and rate of progression may vary appreciably among families with the same apparent genetic disease. These variations may reflect intrinsic constitutional differences among families or the modifying effects of unrelated, unknown genes in different families. Alternative (even nonallelic) genes, which cause distantly similar phenotypic diseases, may cause dissimilar expression, as might environmental modifiers.

Table IX-8

Criteria for Autosomal Recessive Inheritance

The mutant gene usually does not cause clinical disease (recessive) in the heterozygote.
Individuals inheriting both the genes (homozygote) of the defective type express the disorder.
Typically, the trait appears only in siblings, not in their parents or offspring or in other relatives.
The ratio of normal to affected in a sibship is 3:1. The larger the sibship, the more often will more than one child be affected.
The sexes are affected in equal proportions.
Parents of the affected person may be genetically related (consanguinity); this is increasingly likely, the rarer the trait.
Affected individuals have children who, although phenotypically normal, are carriers (heterozygotes) of the gene.

If the homozygote is defined as a specific base (pair) substitution in a codon, many "autosomal recessive" diseases result from genetically compound heterozygotes—that is, individuals who have two different (but both "defective") alleles at a given locus. Whenever detailed biochemical or molecular testing becomes possible, the products of different alleles will show slightly different properties or behaviors. Hemoglobin sickle cell disease and Hurler-Scheie syndrome are well-established compound heterozygote disorders.

Autosomal Dominant Inheritance

When an autosomal allele leads to a regular, clearly definable abnormality in the heterozygote, the trait is termed *dominant.* The first pedigree to be interpreted in terms of mendelian dominant inheritance was a family with brachydactyly (short fingers) reported by Farabee in 1903. Autosomal dominant traits often represent defects in structural nonenzymatic proteins, such as in fibrillin in Marfan syndrome or collagen in Stickler syndrome. In addition, a dominant mode of inheritance has been observed for some malignant neoplasia syndromes, such as retinoblastoma, von Hippel–Lindau disease, tuberous sclerosis, and Gardner syndrome. Although the neoplasias in these diseases are inherited as autosomal dominant traits, the tumors themselves result from loss of function of both alleles of autosomal recessive tumor-suppressor genes.

Almost all bearers of dominant disorders in the human population are heterozygotes. In dominant inheritance, the heterozygote is clinically affected and a single dose of the mutant gene interferes with normal function. Occasionally, depending on the frequency of the abnormal gene in the population and the phenotype, two bearers of the same abnormality marry and produce children. Any offspring of two heterozygous parents has a 25% risk of being an affected homozygote. This circumstance has been recorded in achondroplastic dwarfism. The homozygous achondroplastic dwarf has severe cranial and thoracic skeletal disorders and dies at an early age. Since the heterozygote has one normal allele and the homozygote has none, it is not surprising that the phenotype of the homozygote is more severely abnormal. Homozygotes (or double heterozygotes) for autosomal dominant retinitis pigmentosa also appear to have a much more severe form of retinal degeneration.

It has been suggested that dominant diseases are caused by mutations affecting structural proteins, such as cell receptor growth factors (eg, FGFR-2 in Crouzon disease), or by functional deficits generated by abnormal polypeptide subunits (eg, unstable hemoglobins). The dominant disorders aniridia and Waardenburg syndrome result from loss of one of the two alleles for the developmental transcription factors Pax-6 and Pax-3, respectively. However, it is not at all clear exactly how a single gene abnormality can produce the pleiotropic manifestations of such dominant diseases as von Recklinghausen neurofibromatosis or tuberous sclerosis.

In some instances, dominantly inherited traits are not clinically expressed. In other instances—such as with some families with autosomal dominant retinitis pigmentosa—pedigree analysis infrequently shows a defective gene in individuals who do not manifest any discernible clinical or functional impairment. This situation is called *incomplete penetrance*, or *skipped generation.*

Conclusive evidence of autosomal dominant inheritance requires demonstration of the disease in at least three successive generations. Transmission of the disorder from male to male, with both sexes showing the typical disease, must also occur. The criteria for autosomal dominant inheritance with complete (100%) penetrance are summarized in Table IX-9. In the usual clinical situation, any offspring of an affected heterozygote with a dominant disorder has one chance in two of inheriting the mutant gene and thereby demonstrating some effect, regardless of sex. The degree of variability in the expression of certain traits is usually more pronounced in autosomal dominantly inherited disorders than in other types of genetic disorders. Moreover, when a clinical disorder is inherited in more than one mendelian pattern, the dominantly inherited disorder is, in general, clinically less severe than the recessively inherited one.

TABLE IX-9

CHARACTERISTICS OF AUTOSOMAL DOMINANT INHERITANCE WITH COMPLETE PENETRANCE

Trait appears in multiple generations (vertical transmission).
Affected males and females are equally likely to transmit the trait to male and female offspring. Thus, male-to-male transmission occurs.
Each affected individual has an affected parent, unless the condition arose by new mutation in the given individual.
Males and females are affected in equal proportions.
Unaffected persons do not transmit the trait to their children.
The trait is expressed in the heterozygote but is more severe in the homozygote.
The age of fathers of isolated (new mutation) cases is usually advanced.
The more severely the trait interferes with survival and reproduction, the greater the proportion of isolated (new mutation) cases.
Variability in expression of the trait from generation to generation and between individuals in the same generation is expected.
Affected persons transmit the trait to 50% of their offspring on average.

Counseling for recurrence risk of autosomal dominant traits must involve thorough examination of not only the affected person (who may have the full syndrome) but also the parents. If one parent is even mildly affected, the risk of additional genetically affected siblings rises to 50%. It is unacceptable to miss variable expressivity when parents and other family members can be examined. In some ocular disorders, family members can inherit a gene for a dominant trait and not show clinically apparent manifestations; electrophysiologic testing must be used to detect the impairment. An example is Best vitelliform macular dystrophy, in which clinically normal family members can be diagnosed as having the gene for this disorder only by the presence of an abnormal electro-oculographic light:dark (peak:trough) ratio.

X-Linked Inheritance

A trait determined by genes on either of the sex chromosomes is properly termed *sex linked.* This genetic pattern became widely known with the occurrence of hemophilia in European and Russian royal families. The earliest known record of a correct analysis in an X-linked pattern of inheritance for a human trait appears to be in the Talmud, which decreed that if two boys in a family died from bleeding following circumcision, the later-born sons from the same mother or from her sisters need not be circumcised. Thus, even in antiquity, it was recognized that this trait affecting only males—subsequently identified as hemophilia—was transmitted through unaffected females.

The rules governing all modes of sex-linked inheritance can be derived logically by considering the chromosomal basis. Females have two X chromosomes, and one of these will go to each ovum. Males have both an X and a Y chromosome. The male parent contributes his only X chromosome to all his daughters and his only Y chromosome to all his sons. Traits determined by genes carried on the Y chromosome are called *holandric* and are transmitted from a father to 100% of his sons. Among these Y-chromosomal genes is the *testis-determining factor* (*TDF*—also called *sex-determining region Y,* or *SRY*). Genes controlling tooth size, stature, and spermatogenesis are also on the Y chromosome. Finally, a gene determining hairy pinnae (ie, hair on the outer rim of the ear) may also be located on the Y chromosome. All other sex-linked traits or diseases are thought to result from genes on the X chromosome and are properly termed *X linked.* Some X-linked conditions have considerable frequencies in human populations; the various protan and deutan color-vision defects were also among the first human traits assigned to a specific chromosome.

The distinctive feature of X-linked inheritance, both dominant and recessive, is the absence of father-to-son transmission. Since the male X chromosome passes only to daughters, all daughters of an affected male will inherit the mutant gene.

X-linked recessive inheritance A male has only one representative of any X-linked gene and therefore is said to be *hemizygous* for the gene, rather than homozygous or heterozygous. Since there is no normal gene to balance a mutant X-linked gene in the male, its resulting phenotype, whether dominant or recessive, will always be expressed. A female may be heterozygous or homozygous for a mutant X-linked gene. X-linked traits are commonly called recessive if they are caused by genes located on the X chromosome, which express themselves fully only in the absence of the normal allele. Thus, males (with their single X chromosome) are predominantly affected. All their phenotypically healthy but heterozygous daughters are carriers. By contrast, each son of a heterozygous woman has an equal chance of being normal or hemizygously affected.

A female will be affected with an X-linked recessive trait under a limited number of circumstances:

- She is homozygous for the mutant gene by inheritance (ie, from an affected father and a heterozygous mother)
- Her mother is heterozygous and her father contributes a new mutation
- She has Turner syndrome with only one X chromosome and therefore is effectively hemizygous
- She has a partial deletion of one X chromosome either by rearrangement or by formation of an isochromosome and is thereby effectively hemizygous
- She has a highly unusual skewing of inactivation of her normal X chromosome, as explained by the Lyon hypothesis (discussed later under Lyonization)
- Her disorder is actually an autosomal genocopy of the X-linked condition

Table IX-10 summarizes the criteria for X-linked recessive inheritance, which should be considered if all affected individuals in a family are males, especially if they are related through historically unaffected women (eg, uncle and nephew, or multiple affected half brothers with different fathers).

X-linked dominant inheritance X-linked dominant traits are caused by mutant genes expressed in a single dose and carried on the X chromosome. Thus, both heterozygous women and hemizygous men are clinically affected. Females are affected nearly twice as frequently as males. All daughters of affected males are affected. However, all sons of affected males are free of the trait unless their mothers are also affected. Since only children of affected males provide information in discriminating X-linked dominant from autosomal dominant disease, it may be impossible to distinguish these modes on genetic grounds when the pedigree is small or the available data are scarce. Some X-linked dominant disorders such as incontinentia pigmenti (Bloch-Sulzberger syndrome) may prove lethal to the hemizygous male. X-linked hypophosphatemic rickets (vitamin D–resistant rickets) is an example of an X-linked dominant disease. The characteristics of X-linked dominant inheritance are summarized in Table IX-11.

Jorde LB, Carey JC, Bamshad MJ, et al. *Medical Genetics.* 2nd ed. St. Louis: Mosby; 1999:chap 5.

X-linked disorders Females with X-linked diseases have milder symptoms than males. Occasionally, males may be affected severely enough that they die before the reproductive period, thus preventing transmission of the gene. Such is the case with Duchenne muscular dystrophy, in which most affected males die before their midteens. In other disorders, males are so severely affected that they die before birth, and only females survive. Families with such disorders would include only affected daughters, unaffected daughters, and normal sons at a ratio of 1:1:1. Incontinentia pigmenti is such a genetic lethal disorder. Perinatally, affected females develop an erythematous, vesicular skin eruption, which progresses to marbled, curvilinear pigmentation. The syndrome includes dental abnormalities, congenital or secondary cataracts, proliferative retinopathy and pseudogliomas, and tractional retinal detachment.

Among the most severe X-linked dominant disorders with lethality for the hemizygous males is Aicardi syndrome. No verified birth of males with this entity has ever been reported, although several XXY pseudomales have occurred. Females have

Table IX-10

Characteristics of X-Linked Recessive Inheritance

Usually only males are affected.
An affected male transmits the gene to all of his daughters (obligate carriers) and none of his sons.
All daughters, even phenotypically normal, of affected males are carriers.
Affected males in a family are either brothers or are related to one another through carrier females, eg, maternal uncles.
If an affected male has children with a carrier female, 50% of their daughters will be homozygous and affected and 50% will be heterozygous and carriers.
Heterozygous females may rarely be affected (manifesting heterozygotes) because of Lyonization.
Female carriers transmit the gene on average to 50% of their sons, who are affected, and to 50% of their daughters, who will in turn be carriers.

Table IX-11

Characteristics of X-Linked Dominant Inheritance

Both males and females are affected, but the incidence of the trait is approximately twice as great in females as in males (unless the trait is lethal in the male).
An affected male transmits the trait to all of his daughters and to none of his sons.
Heterozygous affected females transmit the trait to both sexes with equal frequency.
The heterozygous female tends to be less severely affected than the hemizygous male.

profound mental and developmental retardation; muscular hypotonia; blindness associated with a characteristic lacunar juxtapapillary chorioretinal dysplasia and optic disc anomalies; and central nervous system abnormalities, most characteristically agenesis of the corpus callosum. No recurrences have been reported among siblings, and parents can be reassured that the risk in subsequent children is minimal. All instances of the disease appear to arise from a new X-dominant lethal mutation, and females do not survive long enough to reproduce. The critical area appears to be on the distal end of the short arm of the X chromosome, since some patients with a deletion in this region have also been shown to have features of Aicardi syndrome.

Maternal Inheritance

When nearly all offspring of an affected woman appear to be at risk for inheriting and expressing the trait and the daughters are at risk for passing the trait on to the next generation, the pattern of inheritance is called *maternal inheritance*. The dis-

TABLE IX-12

OCULAR FINDINGS IN CARRIERS OF X-LINKED DISORDERS

DISORDER	OCULAR FINDINGS
X-linked retinitis pigmentosa	Regional fundus pigmentary changes, "gold-dust" tapetal-like reflex; ERG amplitude and implicit time abnormalities
Choroideremia	"Moth-eaten" fundus pigmentary changes with areas of hypopigmentation, mottling, and pigment clumping in a striated pattern near the equator
Ocular albinism	Chocolate brown clusters of pigment prominent in the midperipheral retina; mottling of macular pigment; iris transillumination
Congenital stationary night blindness with myopia	Reductions in ERG oscillatory potentials
Blue-cone monochromatism	Abnormalities in cone function on ERG, psychophysical thresholds, and color vision testing
Red-green color vision deficiencies (protan and deutan)	Abnormally wide or displaced color match on a Nagel anomaloscope; decrease in sensitivity to red light in protan carriers (Schmidt's sign)
Lowe syndrome	Scattered punctate lens opacities on slit-lamp examination
Fabry disease	Fingerprint or whorl-like (verticillate) changes within the corneal epithelium

ease stops with all-male offspring, whether affected or not. This form of inheritance is highly suggestive of a mitochondrial disorder. The structure and molecular aspects of the mitochondrial genome and a general discussion of mitochondrial disease are covered in Chapter VIII.

Lyonization (X-Chromosome Inactivation)

In classic human genetics, females with a gene for a recessive disease or trait on only one X chromosome should have no manifestations of the defect. However, ophthalmic examples of structural and functional abnormalities in females heterozygous for supposedly recessive X-linked traits abound. Such *carrier states,* usually mild but occasionally severe, have been described in carriers of choroideremia, X-linked Nettleship-Falls ocular albinism, X-linked retinitis pigmentosa, X-linked sutural cataracts, Lowe syndrome, Fabry disease, and color-vision defects of the protan and deutan types, among others (Table IX-12; Fig IX-7).

Detection of these carrier states of the X-linked traits has become clinically relevant, especially for sisters and maternal aunts of affected males. In 1961, Mary Lyon (a British geneticist) advanced an explanation for the unanticipated or partial expression of a trait by a heterozygous female. Briefly, the Lyon hypothesis stated that in every somatic cell of a female, only one X chromosome is actively functioning. The second X chromosome is inactive and forms a densely staining marginal nuclear structure demonstrated as a Barr body in a buccal smear or in "drumsticks,"

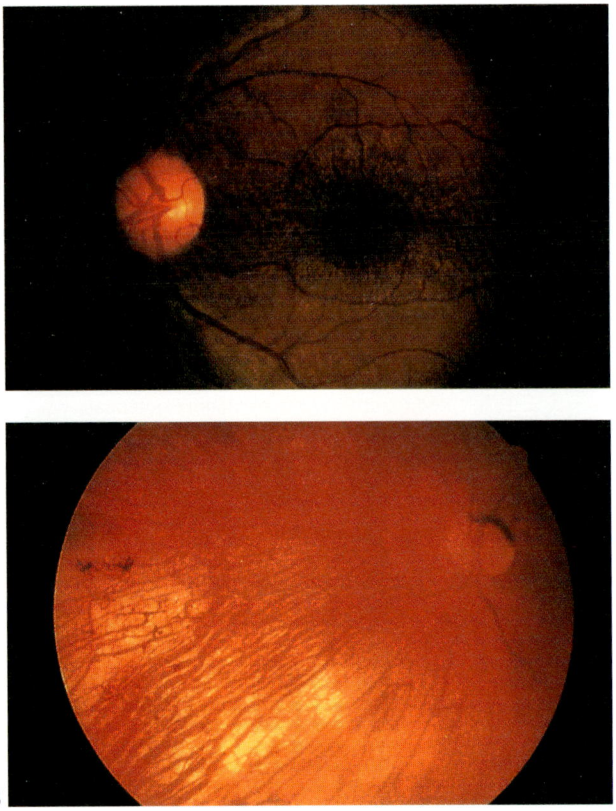

A

B

FIG IX-7—**A**, Yellow, "gold-dust" tapetal-like reflex in the left retina of a carrier for X-linked retinitis pigmentosa. **B**, Nasal midperipheral retina in the left eye of a carrier for X-linked retinitis pigmentosa, showing patchy bone spicule–like pigment clumping. **C**, Peripheral retina from the left eye of a carrier of choroideremia showing a "moth-eaten" fundus appearance from areas of hypopigmentation and hyperpigmentation. **D**, Characteristic iris transillumination from a carrier of X-linked ocular albinism. **E**, Midperipheral retina from the left eye of a carrier for ocular albinism showing a chocolate-brown pigmentation from areas of apparently enhanced pigmentation and clusters of hypopigmentation.

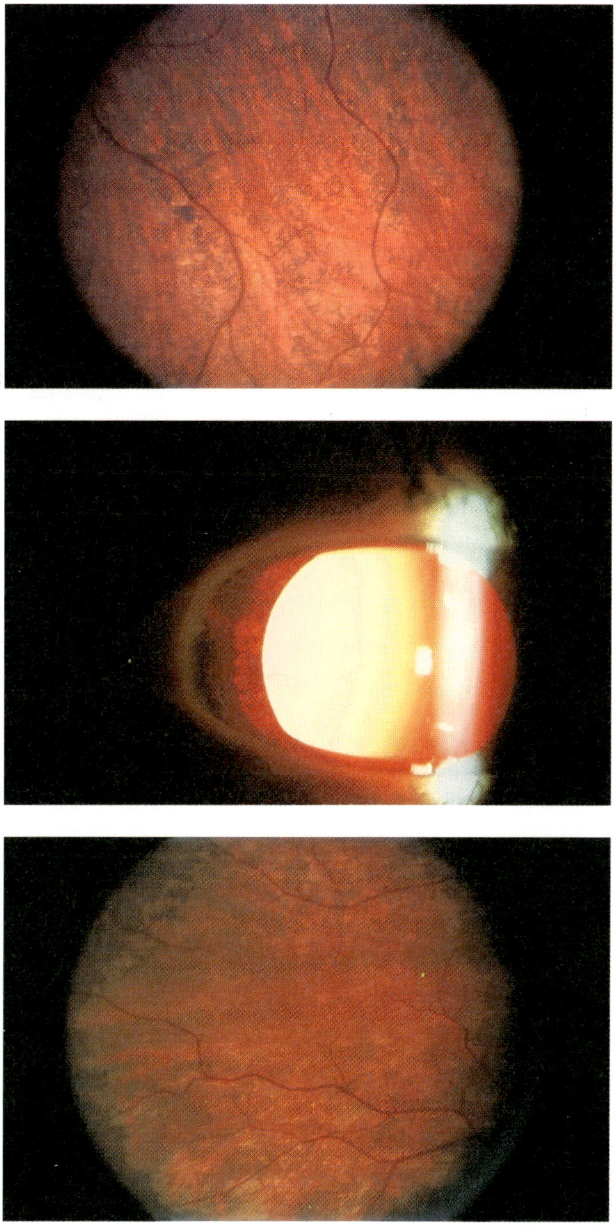

C

D

E

pedunculated lobules of the nucleus identified in about 5% of the leukocytes of the normal female. Warburg (1971) reasoned that X-chromosome inactivation occurs between approximately 6 and 11 days after fertilization, before the process of embryonic lateralization at 11–16 days of embryogenesis. The "decision" to inactivate one X chromosome is random, but once it is made, the same X chromosome will be irreversibly inactive in every daughter of each of these "committed" primordial cells. With only one X chromosome "functioning," the active gene is dominant at a cellular level. Thus, a heterozygous female for an X-linked disease will have two clonal cell populations (mosaic phenotype), one with normal activity for the gene in question and the other with mutant activity.

The proportion of mutant to normal X chromosomes inactivated usually follows a normal distribution, since presumably the inactivations in various cells are random events. Thus, an average of 50% of the paternal X chromosomes and 50% of maternal X chromosomes are inactivated. It is conceivable, however, that in some cases the mutant X is active in almost all cells; in other cases, the mutant X is inactivated in nearly all cells. By this mechanism, a female may express an X-linked disorder, and rare cases are known of women who have a classic color deficiency or X-linked ocular albinism, X-linked retinitis pigmentosa, or choroideremia.

Some possible clinical implications of X inactivation are the following:

- The abnormalities in carrier females from different families and even within the same family may vary greatly in degree because of random inactivation and the resultant tissue derivatives containing differing proportions of the active X cells.
- A large population sample should have as many severely affected as mildly affected heterozygous carriers.
- In some tissues, both normal and abnormal areas could be found if a known biochemical defect could be mapped in a carrier, especially if the gene product is nondiffusible. For example, each hair bulb is ultimately derived from a single primordial cell. Therefore, biochemical analysis of hair roots may demonstrate the mutant phenotype directly: In Fabry disease, X-linked Nettleship-Falls ocular albinism, and Lesch-Nyhan syndrome, the scalp is a mosaic of hairs that have either normal or defective enzyme activity but not an intermediate activity.
- In certain diseases, interactions between cells with the normal X active alter the ability of cells with the mutant X active to survive. The enzyme hypoxanthine phosphoribosyltransferase travels from "normal" skin fibroblasts through gap junctions to mutant cells, allowing them to survive. In other situations, normal cells survive by favorable growth characteristics. In bone marrow and white blood cells, there is progressive elimination of the abnormal X-active cells in the heterozygote and protective survival of cells with a normal X active.

Carriers of the X-linked variety of Nettleship-Falls ocular albinism may have a mottled mosaic fundus: In the pigmented retinal epithelial cells, the normal X chromosome is active; in the nonpigmented cells, the mutant X is active. However, these distinguishing features of the carrier state are not always present. The possibility that the patient is a carrier cannot be entirely eliminated if a given sign is not present. This is because a female might have undergone chance inactivation of the mutant X chromosome in most of her primordial cells, which evolved into the specific tissue observed and may appear phenotypically normal. This subtlety is even more important in evaluating family members with X-linked disease if the phenotypic carrier state is age dependent; thus, even in obligate carrier females for Lowe syndrome, lenticular cortical opacities are not necessarily seen before the third decade of life.

Krill AE. X-chromosome-linked diseases affecting the eye: status of the heterozygote female. *Trans Am Ophthalmol Soc.* 1969;67:535–608.

Warburg M. Random inactivation of the X chromosome in intermediate X-linked retinitis pigmentosa. Two hypotheses. *Trans Ophthalmol Soc U K.* 1971;91:553–560.

Polygenic and Multifactorial Inheritance

In chromosomal and mendelian (single-gene) disorders, genetic analysis of phenotypic, biochemical, or molecular parameters is imperative. However, a simple mode of inheritance cannot be assigned and a recurrence risk cannot be predicted for many common normal characteristics or disorders for which genetic variability clearly exists. Such traits as stature, facial features, refractive error, intraocular pressure, iris color, and intelligence are usually distributed as a continuous variation over a wide range without sharp distinction between normal and abnormal phenotypes. This distribution contrasts with the bimodal curve noted in conditions transmitted by a single gene. Common diseases are often superimposed on this substrate of normal variation, perhaps with a threshold level beyond which individuals may be regarded as abnormal. Consequently, the level of blood sugar in diabetes mellitus, the level of intraocular pressure for glaucoma, or the intermedial canthal distance for telecanthus is somewhat arbitrary. Such conditions are often termed *polygenic,* implying that they result from the operation of multiple collaborating genes, each with rather minor additive but individually indeterminate effects.

The term *multifactorial* denotes a combination of genetic and environmental factors in the etiology of disease without specifying the nature of the genetic influence. Examples in humans include intelligence, stature, blood pressure, atherosclerosis, and refractive index of the eye, among many others. The distinction between polygenic and multifactorial inheritance is not one of exclusion but rather of emphasis: most diseases can be thought of as constituting a spectrum of varying degrees of relative importance of genetic and nongenetic factors in their causation.

Counseling for recurrence may be difficult in this type of inheritance. Ideally, empirical data are summarized from exhaustive analyses of similarly affected families in the population. Regrettably, such empirical data are rarely available for ophthalmologic disorders. However, several general guidelines can be offered. If one offspring has the defect (such as cleft lip/palate) and the parents are normal, the chance that a subsequent child will inherit a similar set of genes and thus manifest the same type of malformation is considerably higher than the frequency of the defect in the general population but much lower than the risk of a mendelian defect. The usual estimate for such recurrence is 5% or less for common polygenic diseases. In addition, the more severe the abnormality in the index case, the higher the risk of recurrence of the trait in relatives, presumably because either a greater number of deleterious genes are at work or a fixed population of more harmful genes exists. The risk that an affected individual will have an affected offspring is also about 5%, similar to the recurrence risk in siblings. The risk of recurrence in future children is increased when more than one member of the family is affected, which is not true for mendelian disorders. Such observations have been offered for various forms of strabismus, glaucoma, and significant refractive errors.

Polygenic traits with a threshold (either present or absent) may be much more frequent in one sex if the threshold is sex-influenced. For example, isolated cleft lip is more common in males and isolated cleft palate is more common in females. The risk of recurrence should be higher among relatives of index cases of the less susceptible sex, who are genetically more highly predisposed or who carry more deleterious genes. For example, perinatal pyloric stenosis is much more common in males than females. Thus, if a female infant is affected, she presumably has either more deleterious genes or a higher personal liability to manifest the trait and, accordingly, a greater likelihood of having an affected sibling or relative.

Finally, if the malformation or disorder has occurred in both paternal and maternal relatives, the recurrence risk is distinctly higher because of the consanguineous sharing of multiple unspecifiable but potentially harmful genes in their offspring. Such empirical risks clearly increase the likelihood of diabetes mellitus in the offspring of two affected parents, even if neither parent has an antecedent family history of the disease.

Table IX-13 lists common polygenic disorders, and Table IX-14 summarizes the principles of recurrence risks for these disorders.

Jorde LB, Carey JC, Bamshad MJ, et al. *Medical Genetics.* 2nd ed. St. Louis: Mosby; 1999:chap 12.

Pedigree Analysis

Recording a family history for general medical and eye disease is an essential part of an ophthalmic consultation. Family data can be summarized in a pedigree chart, a shorthand method for recording data for visual reference. Although cumbersome for the novice, the chart should be incorporated into every medical record. The word *pedigree* is derived from the French expression *pied de grue,* or "crane's foot," from the branching pattern of the diagram.

The affected individual who brings a family to the attention of the physician is the *proband* (propositus or proposita). The person seeking counseling is most frequently identified as the *consultand.* The most commonly used symbols for drawing a pedigree are shown in Figure IX-8. In human pedigree charts, the usual practice is to place the male symbol first on the left; breeding records of other species generally list the female symbol first.

Accurate completion of the pedigree drawing is essential to its interpretation. The health history of family members may be as important as the ocular history. The interviewer must always inquire specifically about abortions, stillbirths, and deceased family members. Often, information on these persons is erroneously omitted, and prenatal or postnatal lethal disorders or relevant medical and genetic causes of death are overlooked. Ages at death may be useful in specific situations and can be recorded directly near the appropriate symbols. For example, a clinician evaluating a child with ectopia lentis and no family history of similar ocular disease can find the identification of a relative deceased from a dissecting thoracic aortic aneurysm in his fourth decade of life very informative, leading to a tentative consideration of Marfan syndrome in the differential diagnosis. The casual observation in a young adult of multiple patches of congenital hypertrophy of the retinal pigment epithelium in each eye may stimulate the recognition of a parent deceased at age 50 from metastatic adenocarcinoma of the colon and a sibling deceased from a brain tumor at 10 years of age, thus leading to a diagnosis of Gardner syndrome and referral to a gastroenterologist for further diagnostic evaluation.

The interviewer should always clarify whether brothers and sisters are half siblings or full siblings. This procedure may not only limit the possible patterns of inheritance but may also identify other persons at risk for the disorders under consideration. Occasionally, information about parentage must be pursued aggressively (but always privately and confidentially). In the United States in 1990, 28% of babies—3 of every 11—were born to parents who were not legally married. The frequency of

TABLE IX-13

COMMON POLYGENIC/MULTIFACTORIAL TRAITS

Anencephaly/spina bifida
Cleft lip/palate
Cleft palate
Clubfoot (pes equinovarus)
Common psychoses (schizophrenia, affective disorder)
Congenital heart disease (some forms)
Congenital scoliosis
Coronary heart disease (some forms)
Diabetes mellitus (some forms)
Hirschsprung disease
Hydrocephalus, nonspecific neural tube defects
Hypertension
Mental retardation
Open-angle glaucoma (some forms)
Pyloric stenosis
Refractive errors
Strabismus (some forms)
Urinary tract malformations

TABLE IX-14

RECURRENCE RISK FOR POLYGENIC DISORDERS

Increased risk above the general population for recurrence of the disorder among first-, second-, and third-degree relatives of the affected person
Baseline 3%–5% risk of recurrence for first-degree relatives of proband; about half that risk for second-degree relatives
Increasing risk with increasing numbers of affected genetic relatives, especially first- and second-degree
Increased risk of recurrence if affected person is of the sex usually less affected
Increased risk of recurrence with increased severity of the trait
Risk of recurrence greater for siblings in families with two affected than for families with one affected

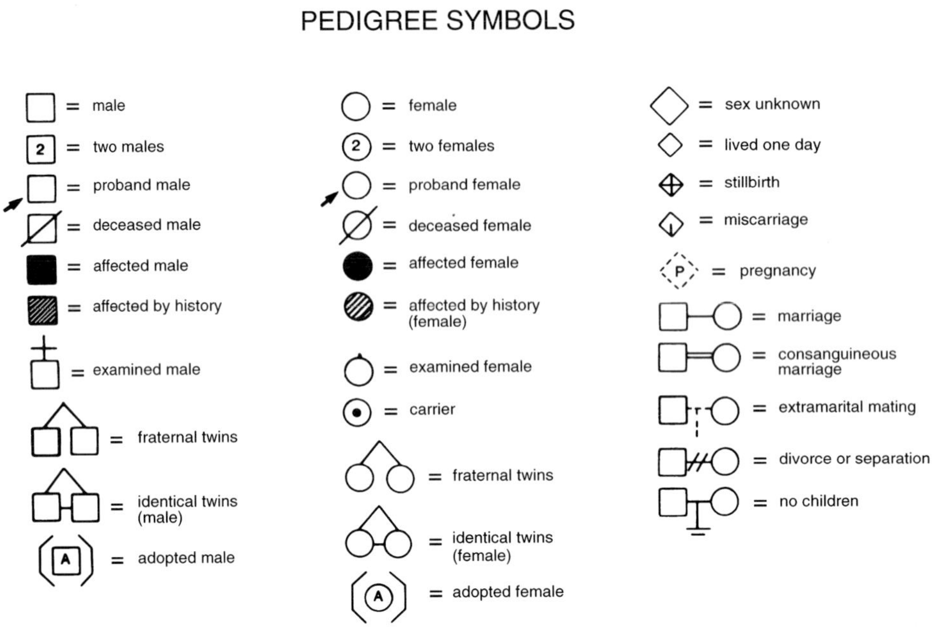

FIG IX-8—Symbols commonly used for pedigree analysis.

offspring born to teenage mothers outside of conventional marriages ranges from 30% to 80%. Both incest and nonpaternity are sensitive issues, but clearly neither is rare in our society. The national nonpaternity rate is presently estimated at 5%, but in some urban societies it may be as high as 15%. In considering rare autosomal recessive diseases, the interviewer must ask specifically about consanguinity in the following ways:

- Searching for common last names in the families of both parents
- Finding identical birthplaces for the parents
- Identifying similar parental backgrounds from known ethnic or religious isolates

Genetic Counseling

The ophthalmologist who understands the principles of human genetics has a foundation for counseling patients about their diseases. Genetic counseling imparts knowledge of human disease, including a genetic diagnosis and its ocular and systemic implications; information about the risk of occurrence or recurrence of the disorder within the family; and an open discussion of the options for reproduction. All genetic counseling is predicated on certain essential requirements.

- *Accurate diagnosis* The physician must be sufficiently aware of the range of human ocular pathology to derive an accurate and specific diagnosis. It is impossible to counsel or refer patients on the basis of "congenital nystagmus" or "color blindness" or "macular degeneration"; these are signs, not diagnoses.

- *Complete family history* A family history will narrow the choices of possible inheritance patterns, but it may not necessarily exclude new mutational events, isolated occurrences of recessive diseases, and chromosomal rearrangements in individual circumstances. The ophthalmologist must examine (or arrange to have examined) the parents, siblings, and other family members for mild manifestations of dominant diseases or characteristic carrier states in X-linked disorders. Identification of one young adult with the findings of Usher syndrome—prelingual deafness, night blindness, visual field constriction, and ultimately deterioration of central vision—obligates the ophthalmologist to evaluate a younger sibling who is congenitally deaf but "historically" has no eye problems. There is an overwhelming probability that the sibling has the same disease. Only an ophthalmologist will be cognizant of and attentive to the atypical findings of hereditary ocular disorders.
- *Understanding the genetic and clinical aspects of the disorder* The ophthalmologist should appreciate, perhaps more intimately than any other physician, how some clinically similar diseases inherited in the same pattern may be the result of different and even nonallelic defects. For example, the visual implications and prognosis of tyrosinase-positive and tyrosinase-negative oculocutaneous albinism are considerably different. Some entities that are clinically similar may be inherited differently and thus have a different impact on other family members. Pseudoxanthoma elasticum is often a late-onset disease in both its autosomal dominant and autosomal recessive modes that has serious implications for cardiovascular disease, stroke, and gastrointestinal bleeding. Informed counseling falls short if the ophthalmologist advises only about visual disability associated with angioid streaks without attention to the complete disease and risks to other family members.

Obstacles to Genetic Counseling

The ophthalmologist must remember that an individual affected by a heritable condition may represent a homozygous recessive trait; thus, the ophthalmologist should search for parental consanguinity or ambiguous parentage (nonpaternity, incest, and even occult adoption) or for a new mutation and should inquire about advanced paternal (or maternal grandparental) age. Heterogeneity may confuse the diagnosis. Somatic mutations also occur, as with segmental neurofibromatosis or unilateral unifocal retinoblastoma. Nonpenetrance or mild expressivity in other family members should be excluded by diligent examination. Chromosomal abnormalities and phenocopies caused by infections or drugs may account for the isolated affected person. Nonetheless, the ophthalmologist's obligation to explain the disorder begins with accurate diagnosis and establishment of the mode of heritability.

Counseling can be considered successful if the patient or family, having acquired the facts, makes a reproductive decision that is reasonable and appropriate. The counselor is an informer, not an advisor. Properly done, genetic counseling is nondirective. It is inappropriate, if not unethical, to tell the patient what to do (for instance, not to have any children).

In any circumstance, the counseling ophthalmologist should outline the options for family planning when it is necessary. The responsibility of the ophthalmic practitioner is to:

- Suspect and establish the diagnosis of inherited disease
- Inform the patient of the findings and their implications for health

- Provide accurate answers to direct and implied questions about risks of recurrence and burden of disease

Some persons may accept a high statistical risk and have children. This decision must be based on how they perceive the social and psychological burdens of the disorder. Attitude toward reproduction may be considerably different for a female carrier of protanopia than for a female carrier of X-linked retinitis pigmentosa or choroideremia, even though the statistical risk for an affected son is the same for each of them.

Some persons may elect to delay childbearing in hopes of medical advances in prenatal diagnosis or postnatal treatment of a disorder. Others may choose for a variety of personal and ethical considerations not to have natural offspring and may proceed with contraception, termination of pregnancy, sterilization, or adoption.

Artificial insemination by donor is a useful option in family planning if the father has a dominant disease or if both parents are carriers of a biochemically detectable recessive disorder. However, it is clearly not applicable if the mother is the carrier of an X-linked disorder or the individual affected by an autosomal dominant mutation. Finally, although its acceptance and legal implications may lag behind, embryo adoption (transplantation) and surrogate motherhood may soon become useful alternatives for some families.

Prenatal Diagnosis

Prenatal diagnosis with amniocentesis or chorionic villus sampling for biochemically identifiable disorders (eg, Tay-Sachs disease, many mucopolysaccharidoses, and about 100 other diseases) is also useful in the proper genetic settings. However, since most genes are expressed in a tissue-specific manner, biochemical diagnostic techniques are limited to diseases for which the gene products are expressed in amniocytes.

Other possible indications for amniocentesis include advanced maternal age, with its increased risk of chromosomal abnormalities; elevated maternal serum alpha-fetoprotein, suggesting a neural tube defect; and the presence of a familial disease detectable by DNA analysis.

Amniocentesis is usually performed at 15–16 weeks of gestation, when enough fluid and cells can be obtained for culture and the maternal risk of abortion is relatively low. The risk of spontaneous abortion or fetal morbidity from the procedure is about 0.5%. Earlier prenatal diagnosis of chromosome abnormalities, at about 10 weeks of gestation, is available through the use of chorionic villus sampling. In this procedure, tissue from the placenta is obtained under ultrasound visualization. It is then cultured and karyotyped in a manner similar to that used for amniocentesis. As a first-trimester procedure, chorionic villus sampling allows for an earlier diagnosis and a safer means of pregnancy termination. The rate of spontaneous abortion associated with this procedure is estimated at 1%–2%. Since the yield of DNA is greater than that from 20 ml of amniotic fluid withdrawn in amniocentesis, direct DNA analysis of cells can often be done without prior cell culture. Thus, information can be obtained considerably sooner than with amniocentesis.

New Developments

In the future, it is likely that certain aspects of genetic counseling will become more precise because of the coordinated international effort to study the entire human genome. A number of agencies around the world are orchestrating *the human genome project,* whose ultimate goal is to determine the precise nucleotide sequences not only of the human genome but also of several model organisms including yeast, nematodes, and fruit flies (the sequence of the entire *Drosophila* genome was published in March, 2000). In the United States, the National Center for Human Genome Research at the National Institutes of Health and the Department of Energy serve as coordinating agencies. It is anticipated that information obtained from this herculean effort will eventually aid in the identification and characterization of numerous additional genes that lead to human disease.

Pharmacogenetics

The study of heritable factors that determine how drugs are chemically metabolized in the body is called *pharmacogenetics.* This field addresses genetic differences among population segments that are responsible for variations in both therapeutic and adverse effects of drugs. Investigations in pharmacogenetics are important not only for more rational approaches to therapy but also because they facilitate a deeper understanding of drug pharmacology. Part 5 of this volume, Ocular Pharmacology, offers more detail.

Isoniazid is a drug that provides an example of pharmacogenetics. This antituberculosis drug is normally inactivated by the liver enzyme acetyltransferase. A large segment of the population, which varies by geographic distribution, has a reduced amount of this enzyme; these persons are termed *slow inactivators.* When they take isoniazid, the drug reaches higher-than-normal concentrations, thus causing a greater incidence of adverse effects. Family studies have shown this reduced level of acetyltransferase to be inherited as an autosomal recessive trait.

Several other well-documented examples of pharmacogenetics exist. One example occurs in 10% of the male African American population, a high percentage of male Sephardic Jews, and males from a number of other ethnic groups. They have an X-linked recessive trait that causes affected males to have glucose-6-phosphate dehydrogenase (G6PD) enzyme deficiency in their erythrocytes. As a consequence, a number of drugs (including sulfacetamide, vitamin K, acetylsalicylic acid, quinine, chloroquine, and probenecid) may produce acute hemolytic anemia in these persons. Pharmacogenetic causes have also been ascribed to variations in response to ophthalmic drugs, such as the increased intraocular pressure seen in a segment of the population after prolonged use of topical corticosteroids.

Several drugs have been shown to cause greater reaction in children with Down syndrome than in normal children. Some children with Down syndrome have died after systemic administration of atropine as a result of supersensitivity. This supersensitivity is also seen with the topical use of atropine: In these patients, it exerts a greater-than-normal effect on pupillary dilation. In several children with Down syndrome being treated for strabismus, hyperactivity has occurred several hours after local instillation of 0.125% echothiophate iodide.

One of the earliest examples of an inherited deficit in drug metabolism involved succinylcholine, a strong muscle relaxant that interferes with acetylcholinesterase, the enzyme that catabolizes acetylcholine at neuromuscular junctions. Normally, succinylcholine is rapidly destroyed by plasma cholinesterase (sometimes called *pseudocholinesterase*) so that its effect is short-lived—usually no more than a few minutes. Some persons are homozygous for a recessive gene that codes for a form of cholinesterase with a considerably lower substrate affinity. Consequently, at therapeutic doses of succinylcholine, almost no destruction occurs, and the drug continues to exert its inhibitory effect on acetylcholinesterase, resulting in prolonged periods of apnea.

Jorde LB, Carey JC, Bamshad MJ, et al. *Medical Genetics.* 2nd ed. St. Louis: Mosby; 1999:chap 7.

Clinical Management of Genetic Disease

Genetic disease may not be curable, but in most cases the patient benefits considerably from the physician's appropriate medical management. Such care should include all of the steps discussed below.

Accurate Diagnosis

Unfortunately, because health care providers may not be as knowledgeable about genetic diagnoses as they are about other areas of medicine, many cases are not precisely diagnosed or, worse yet, are diagnosed incorrectly. A case of deafness and pigmentary retinopathy may be called rubella syndrome when the patient really has Usher syndrome. Patients with retinitis pigmentosa may go unrecognized for a syndrome associated with their disease. For example, patients with retinitis pigmentosa and congenital hexadactyly (surgically corrected in infancy) may be not recognized as having Bardet-Biedl syndrome. The correct diagnosis in such cases is important to ensure that the patient's educational, career planning, and lifetime support needs are truly met.

Complete Explanation of the Disease

Patients are often very disturbed when they do not understand the nature of their disease. A careful explanation of the disorder, as currently understood, will often dispel myths that patients may have about how they became the way they are.

Virtually all genetic disorders confer burdens that may interfere with certain activities later in life. The appropriate time to discuss these burdens with patients and family members is often when they first ask about the consequences of a disease. Such explanations need to be tempered with compassion and sympathetic appreciation of the possible emotional and psychological effects of this information.

Treatment of the Disease Process

Although definitive cures—that is, reversing or correcting of underlying genetic defects—are yet to emerge for various heritable disorders, some conditions in which metabolic defects have been identified can often be managed through five fundamental approaches:

- Dietary control
- Chelation of excessive metabolites
- Enzyme or gene product replacements
- Vitamin and cofactor therapy
- Drug therapy to reduce accumulation of harmful products

Some genetic disorders affecting the eye that arise from an inborn error of metabolism can effectively be managed by dietary therapy. These include familial hyperlipoproteinemia, tyrosinosis, homocystinuria, Refsum disease, phenylketonuria, fructose or lactose intolerance, galactokinase deficiency, and galactosemia. Implementing a galactose-free diet can reverse such main features of galactosemia as hepatosplenomegaly, jaundice, and weight loss. Progression of cortical cataracts can be avoided, and less extensive lens opacities may even regress with a galactose-free diet. With time, galactosemic patients are able to metabolize galactose through alternative pathways, obviating the need for lifelong dietary restriction. In phenylketonuria, mental retardation can be prevented by early phenylalanine restriction.

Disorders that result from enzyme or transport protein deficiencies may lead to accumulation of a metabolite or metal that harms various tissues. For example, in Wilson disease, decreased levels of serum ceruloplasmin result in poor transport of free copper (Cu^{2+}) ions and storage of copper in tissues such as the brain, liver, and cornea. Resultant clinical signs can be reversed at least partially after the administration of D-penicillamine, a chelator of Cu^{2+} ions. Other copper chelators such as BAL (British antilewisite) can be employed along with copper-deficient diets to reverse clinical signs of Wilson disease.

In theory, replacement of deficient enzymes in disorders resulting from metabolic dysfunction by either plasma or leukocyte infusions could be beneficial in preventing the accumulation of toxic metabolites. Both in vitro and in vivo investigations on the transfusion of plasma or leukocytes in patients with Hunter or Hurler syndrome showed some preliminary promise for lessening the musculoskeletal signs of these disorders. Plasma infusions in patients with Fabry disease have succeeded in temporarily decreasing plasma levels of the accumulated substrate ceramide trihexoside, although clinical improvement was not detected. Further in-depth investigations are necessary before intervention with this type of therapy becomes meaningful, because circulating enzymes from infusions are rapidly degraded or excreted by the kidneys. Nevertheless, a slow-release depot preparation administered intramuscularly is at least feasible in providing short-term improvement in some metabolic disorders.

Organ transplantation can be considered as a form of regionalized enzyme replacement. In patients with cystinosis, cystine crystals accumulate in the kidney. If a normal kidney, with its rich source of enzymes, is transplanted into a patient with cystinosis, cystine does not accumulate in the cells of the renal tubules, and renal function tends to remain normal.

In addition to enzyme replacement, synthetic or recombinant gene product replacement can effectively manage a gene defect. Hemophilia, for example, can be treated through the administration of a missing clotting factor (VIII). The value of gene product replacement is demonstrated in the use of thyroid hormone for hypothyroidism, insulin for diabetes, erythropoietin for anemia, and growth hormone for pituitary dwarfism. However, caution is needed in light of the possible spread of AIDS caused by HIV if the gene product is extracted from pooled human tissues.

Vitamin therapy appears to be of benefit in two autosomal recessive disorders. In at least some patients with homocystinuria, vitamin B_6 (pyridoxine) administration has been shown to decrease homocystine accumulation in plasma and to reduce the severity of the disorder. Vitamin A and vitamin E therapy have been noted to benefit some patients with abetalipoproteinemia with regard to neurologic impairment; such therapy is also likely to slow or lessen the development and progression of retinal degeneration. More long-term therapeutic trials are necessary to better define the efficacy of vitamin therapy for these and perhaps other metabolic disorders.

Various genetically determined disorders can be managed by use of an appropriate drug. For example, excess accumulation of uric acid in primary gout can be prevented or reduced by (1) blocking the activity of the enzyme xanthine oxidase with the drug allopurinol or (2) increasing excretion of uric acid by the kidneys with the use of probenecid. In addition, a reduction in serum cholesterol found with familial hypercholesterolemia can often be achieved with the use of various cholesterol-lowering drugs or substances that bind bile acids in the gastrointestinal tract.

Goldberg MF, Renie WA, eds. *Genetic and Metabolic Eye Disease.* 2nd ed. Boston: Little Brown; 1986:569–578.

Sutton HE. *An Introduction to Human Genetics.* 2nd ed. New York: Holt, Rinehart & Winston; 1975:477–496.

Appropriate management of sequelae and complications Some of the sequelae of genetic diseases, such as glaucoma in Rieger syndrome or cataracts in patients with retinitis pigmentosa, can be successfully managed to preserve or partially restore vision. However, patients need to understand how the treatment of the sequelae or complications may differ in their situation, especially if treatment affects the expected outcome.

Genetic Counseling

The responsibility of the physician is to either provide genetic counseling or arrange for this service through referral to a geneticist. However, unless the genetic consultant is also an ophthalmologist, the referring ophthalmologist almost always has to provide the explanations of the ocular aspects of the disease and their ramifications for treatment and management. It must be emphasized that all genetic counseling should be nondirective. Physicians have no right to tell couples whether or not to have children but should instead strive to provide information about risks, burdens, and options so that the individuals can make informed decisions.

Referral to Providers of Support for Disabilities

Individuals and families often receive considerable benefit from referral to local, regional, or national agencies, support groups, or foundations that provide services for those with a particular disease. These organizations include local and state agencies for the blind or visually impaired, special school education programs, and appropriate consumer groups. Particularly when disability is chronic and progressive, these agencies or support groups can greatly aid the individual or family in adjusting to changing disabilities.A Directory of National Genetic Voluntary Organizations (list of support groups) is available from:

The Alliance of Genetic Support Groups
Internet: *http://ncbi.nlm.nih.gov/Omim* through the *Online Mendelian Inheritance in Man* home page website (under OMIM Allied Resources)

PART 4

BIOCHEMISTRY AND METABOLISM

CHAPTER X

Tear Film

The primary functions of the tear film are to provide a smooth optical surface at the air–eye interface, to serve as a medium for removal of debris, to protect the ocular surface, and to supply oxygen, growth factors, and other compounds to the corneal epithelium. The tear film carries tear constituents and debris to the puncta. In addition, it contains a vast number of antimicrobial agents, lubricates the cornea–eyelid interface, and prevents desiccation of the ocular surface. Human tears are distributed between:

- The marginal tear strip (or *tear meniscus*)
- The preocular film covering the exposed bulbar conjunctiva and cornea (precorneal tear film)
- The conjunctival sac (between the lids and bulbar conjunctiva)

The *precorneal tear film* is a trilaminar structure consisting of an anterior lipid layer, a middle aqueous phase, and a posterior mucin layer. Measurements of tear-film thickness have differed widely. Original measurements gave an average thickness for the precorneal tear film of about 8–9 μm, with the aqueous phase constituting nearly all the thickness (Fig X-1). A subsequent (but controversial) study, using confocal microscopy, found the mucin layer to be about 30 μm thick and the aqueous layer to be about 10 μm thick. In contrast, a recent study, using reflectometry, found the tear film to be only 3–4 μm thick. There is currently no consensus on the thickness of the tear film. Additionally, the separation between the mucin and aqueous layer may not be distinct since mucins absorb electrolytes and water. The steady-state volume of tears is 7.4 μl for the unanesthetized eye and 2.6 μl for the anesthetized eye; this volume decreases with age. Some properties of tears are given in Table X-1.

Lipid Layer

The anterior layer of the tear film (approximately 100 molecules thick) contains polar and nonpolar lipids secreted primarily by the *meibomian (tarsal) glands* (Fig X-2). These glands are located in the tarsal plate of the upper and lower eyelids, and they are supplied by parasympathetic nerves that are cholinesterase-positive and contain vasoactive intestinal polypeptide (VIP). Sympathetic and sensory nerves are present but sparsely distributed. Neuropeptide Y (NPY)-positive nerves are also abundant. There are approximately 30–40 meibomian glands in the upper eyelid and 20–30 smaller glands in the lower eyelid. Each gland orifice opens onto the skin of the eyelid margin, between the tarsal *gray line* and the mucocutaneous junction. The sebaceous glands of Zeis, located at lid margin in relation to the lash roots, also secrete lipid, which is incorporated into the tear film.

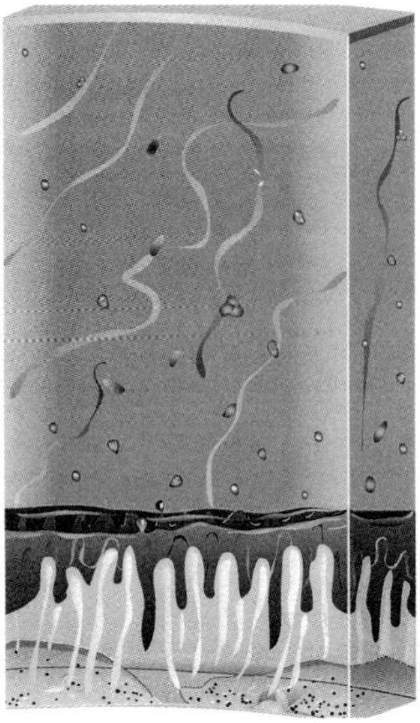

FIG X-1—Schematic drawing of the structure of the tear film showing the outer lipid layer, middle aqueous layer, inner mucus layer, and microvilli on the apical cells of the ocular surface epithelium. (From Marshall D. Tear layer mechanics. In: Bennett G, Weissman B, eds. *Clinical Contact Lens Practice.* Philadelphia: JB Lippincott; 1991:2.)

The four functions of the lipid layer are to:

- Retard evaporation
- Contribute to the optical properties of the tear film because of its position at the air–tear film interface
- Maintain a hydrophobic barrier *(lipid strip)* that prevents tear overflow by increasing surface tension
- Prevent damage to lid margin skin by tears

Because the polar lipids are charged compounds *(phospholipids),* they are located at the aqueous–lipid interface. The fatty acids of the phospholipids interact with the other hydrophobic lipids (cholesterol and wax esters, which make up the bulk of the lipid layer) through noncovalent, noncharged bonds. Tear lipids are not susceptible to lipid peroxidation because they contain extremely low levels of polyunsaturated fatty acids.

Aqueous Phase

The middle aqueous layer is secreted by the main and accessory lacrimal glands (see Fig X-2). It consists of electrolytes, water, and proteins. The main lacrimal gland is divided into two anatomical parts, the *orbital* and the *palpebral* portions, by the *levator aponeurosis.* The *glands of Krause,* which constitute two thirds of the accessory lacrimal glands, are located in the lateral part of the upper fornix. A number of

TABLE X-1

PROPERTIES OF HUMAN TEAR FILM

Composition	Water	98.2%
	Solid	1.8%
Thickness	Total	6.5–7.5 μm
	Lipid layer	0.1–0.2 μm
Volume	Unanesthesized	7.4 μl
	Anesthesized	2.6 μl
Secretory rate	Unanesthesized	
	Schirmer	3.8 μl/min
	Fluorophotometry	0.9 μl/min
	Anesthesized	
	Schirmer	1.8 μl/min
	Fluorophotometry	0.3 μl/min
Turnover rate	Normal	12–16%/min
	Stimulated	300%/min
Evaporation rate		0.06 μl/cm²/min
Osmolarity		296–308 mOsm
pH		6.5–7.6
Electrolytes (mmol/l)	Na^+	134–170
	K^+	26–42
	Ca^{2+}	0.5
	Mg^{2+}	0.3–0.6
	Cl^-	120–135
	HCO_3^-	26

Krause glands are also present in the lower fornix. The *glands of Wolfring* are variably located along the proximal margin of each tarsus. The accessory lacrimal glands are structurally like the main lacrimal gland.

The main lacrimal gland is richly innervated by parasympathetic nerves containing the neurotransmitters acetylcholine and VIP. The sympathetic innervation is less dense than the parasympathetic and contains norepinephrine and NPY as neurotransmitters. The sensory nerves are sparsely innervated with the neurotransmitters substance P and calcitonin gene–related peptide (CGRP). The accessory lacrimal glands are densely innervated, but the majority of nerves are unidentified. There is sparse innervation with nerves containing VIP, substance P, and CGRP. Corneal innervation is predominantly sensory, but there is also sympathetic and (to a lesser extent) parasympathetic innervation. The conjunctival epithelium is innervated by parasympathetic, sympathetic, and sensory nerves.

The aqueous layer of tears consists of electrolytes, water, protein, and a variety of other solutes secreted by the main and accessory lacrimal glands, as well as the corneal and conjunctival epithelia. In addition, with conjunctival inflammation, and in response to drugs such as histamine, the blood vessels of the conjunctiva can leak a plasma-like fluid into the aqueous layer of tears.

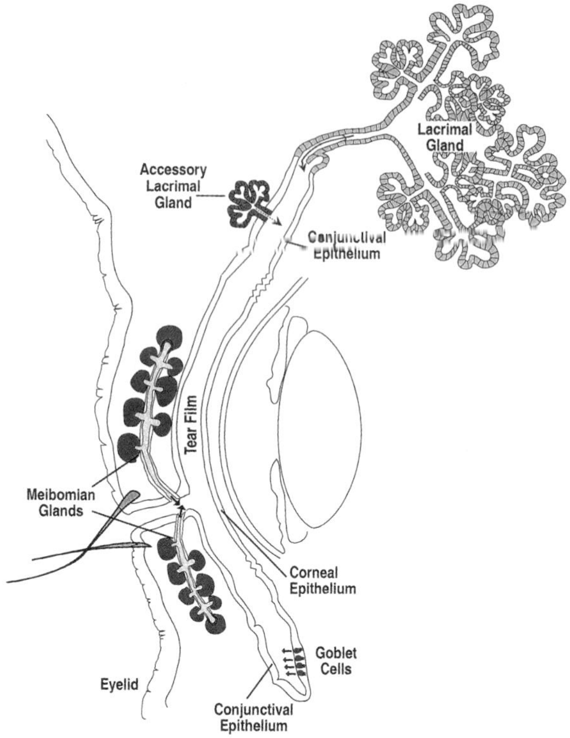

FIG X-2—Schematic drawing of the major tear glands and ocular surface epithelia that contribute to the tear film. Shown are the meibomian glands (secrete oily layer), main lacrimal gland, accessory lacrimal gland, conjunctival epithelium, corneal epithelium (secretes aqueous layer), and conjunctival goblet cells (secrete mucus layer). (From Dartt DA, Sullivan DA. Wetting of the ocular surface. In: Albert DM, Jakobiec FA, eds. *Principles and Practice of Ophthalmology.* Philadelphia: WB Saunders; 1994:967.)

Electrolytes and small molecules regulate the osmotic flow of fluids between the corneal epithelial cells and the tear film, buffer tear pH, and serve as enzyme cofactors in controlling membrane permeability. The Na^+ concentration of tears parallels that of serum; the concentration of K^+ is 5–7 times greater than that in serum. Na^+, K^+, and Cl^- regulate the osmotic flow of fluids from the cornea to the tear film. Bicarbonate regulates tear pH. Other tear electrolytes (Fe^{2+}, Cu^{2+}, Mg^{2+}, Ca^{2+}, PO_4^{3-}) are enzyme cofactors.

Tear film solutes include urea, glucose, lactate, citrate, ascorbate, and amino acids. All enter the tear film via the systemic circulation, and their concentrations parallel serum levels. Fasting tear glucose levels are 3.6–4.1 mg% in diabetic and nondiabetic persons. However, after a 100 mg oral glucose load, tear glucose levels exceed 11 mg% in 96% of diabetic persons tested.

Proteins in the tear film include immunoglobulin A (IgA) and secretory IgA (S-IgA). IgA is formed by plasma cells in interstitial tissues of the main and accessory lacrimal glands and the substantia propria of the conjunctiva. The secretory component is produced within lacrimal gland acini, with S-IgA being secreted into the lumen of the main and accessory lacrimal glands. IgA plays a role in local host-defense mechanisms of the external eye, as shown by increased levels of IgA and IgG in human tears associated with ocular inflammation. Other immunoglobulins in tears are IgM, IgD, and IgE. Vernal conjunctivitis causes elevated tear and serum levels of IgE, increased IgE-producing plasma cells in the giant papillae of the superior tarsal conjunctiva, and elevated histamine. Increased levels of tear histamine support the concept of conjunctival mast-cell degranulation triggered by IgE–antigen interaction.

Lysozyme, lactoferrin, group II phospholipase A_2, lipocalins, and defensins are important tear antimicrobial constituents. Also present in tears is interferon, which inhibits viral replication and may be efficacious in limiting the severity of ulcerative herpetic keratitis. Tears also contain a wide array of cytokines and growth factors, including tumor growth factor β, epidermal growth factor, β fibroblast growth factor, interleukin-1α and -1β, and tumor necrosis factor α. These may play a role in proliferation, migration, and differentiation of corneal and conjunctival epithelial cells. They may also regulate wound healing of the ocular surface.

The aqueous layer has the following functions:

- Supplying oxygen to the avascular corneal epithelium
- Maintaining a constant electrolyte composition over the ocular surface epithelium
- Providing an antibacterial and antiviral defense
- Smoothing minute irregularities of the anterior corneal surface
- Washing away debris
- Modulating corneal and conjunctival epithelial cell function

Mucin Layer

The mucin layer of the tear film coats the microplicae of the superficial corneal epithelial cells and forms a fine network over the conjunctival surface. It contains mucins, proteins, electrolytes, and water. Functions of the mucus layer include the following:

- The mucins convert the corneal epithelium from a hydrophobic to a hydrophilic layer, which is essential for the even and spontaneous distribution of the tear film.
- The mucins also interact with the tear lipid layer to lower surface tension, thereby stabilizing the tear film.
- The loose mucin network covering the bulbar conjunctiva traps exfoliated surface cells, foreign particles, and bacteria.
- The mucin layer provides lubrication for the eyelids as they pass over the globe.

Tear mucins are secreted principally by the conjunctival goblet cells, the stratified squamous cells of the conjunctival and corneal epithelia, and minimally by lacrimal glands of Henle and Manz (see Fig X-2). Goblet-cell mucin production is 2–3 μl/day, which contrasts with aqueous tear production of 2–3 ml/day. Both conjunctival and tear mucins are negatively charged, high-molecular-weight glycoproteins. Tear dysfunction may result when tear mucins are deficient in number (avitaminosis A, conjunctival destruction), excessive in number (hyperthyroidism; foreign-body stimulation; allergic, vernal, and giant papillary conjunctivitis), or biochemically altered (keratoconjunctivitis).

Tear Secretion

The lacrimal secretory system was once thought to have two components: *basic secretors* and *reflex secretors*. Basic secretion was ascribed to the accessory lacrimal glands of Krause and Wolfring, and reflex secretion to the main lacrimal gland. However, it is now thought that all lacrimal glands respond as a unit. In addition, the cornea and conjunctiva can also respond by secreting electrolytes, water, and mucins. Although the meibomian glands are innervated, it is not known whether nerves mediate lipid secretion from these glands. Reflex tear secretion is neurally mediated and induced in response to physical irritation (superficial corneal and conjunctival sensory stimulation by mechanical, thermal, or chemical means), by psychogenic factors, and by bright light via the optic nerve. Activation of sensory nerves by a local, neural reflex activates the parasympathetic and sympathetic nerves that innervate the tear glands and epithelia, causing secretion.

Parasympathetic and sympathetic nerves release their neurotransmitters, which interact with specific G protein–linked receptors in the lacrimal glands, cornea, and conjunctiva. These receptors then activate their respective signaling pathways. There are two main signaling pathways: Ca^{2+}/protein kinase C–dependent and cyclic adenosine monophosphate (cAMP)-dependent (Figs X-3 and X-4). In most tissues, the Ca^{2+}/protein kinase C–dependent pathway is activated by acetylcholine and, except in the main lacrimal gland, by norepinephrine. Acetylcholine, released from parasympathetic nerves, activates muscarinic receptors; norepinephrine, released from sympathetic nerves, activates α_1-adrenergic receptors. Stimulation of muscarinic and α_1-adrenergic receptors activates a guanine nucleotide–binding protein (G protein) of the $G\alpha_{q/11}$ subtype that then turns on phospholipase C. Phospholipase C breaks down a membrane lipid phosphatidylinositol-4,5,-bisphosphate into 1,4,5-inositol trisphosphate (IP_3) and diacylglycerol. IP_3 releases intracellular Ca^{2+}. The depletion of Ca^{2+} from intracellular stores causes the influx of extracellular Ca^{2+} to refill these stores. Ca^{2+} (either by itself or by activating Ca^{2+}-calmodulin–dependent protein kinases) stimulates protein and/or electrolyte and water secretion. The increase in diacylglycerol activates protein kinase C, a family of 11 isozymes that stimulate protein and/or electrolyte and water secretion.

The cAMP-dependent pathway is activated by VIP and norepinephrine. VIP, released from parasympathetic nerves, interacts with VIP receptors; norepinephrine, released from sympathetic nerves, activates β-adrenergic receptors. Stimulation of VIP or β-adrenergic receptors activates Gαs G protein, which in turn stimulates adenylyl cyclase. Activation of adenylyl cyclase produces cAMP from ATP. cAMP activates cAMP-dependent protein kinases to stimulate protein and/or electrolyte and water secretion. The action of cAMP is terminated when it is broken down by cAMP-dependent phosphodiesterases.

Another mechanism for stimulating tear secretion (in addition to nerves) is peptide and steroid hormones. Peptide hormones, including α-melanocyte-stimulating hormone and adrenocorticotropic hormone, stimulate protein secretion from the main lacrimal gland. These hormones activate the cAMP-dependent pathway described for VIP and β-adrenergic receptors. The steroid hormones, specifically the androgens, stimulate secretion of S-IgA from the main lacrimal gland and lipid secretion from the meibomian glands. Androgens diffuse into the nucleus and bind to receptors, which are members of the steroid/thyroid hormone/retinoic acid family of transcription factors. The monomeric activated androgen-receptor complex then associates with the response elements in the regulating region of the target gene (ie, for S-IgA secretion, the target would be the secretory component gene). This associ-

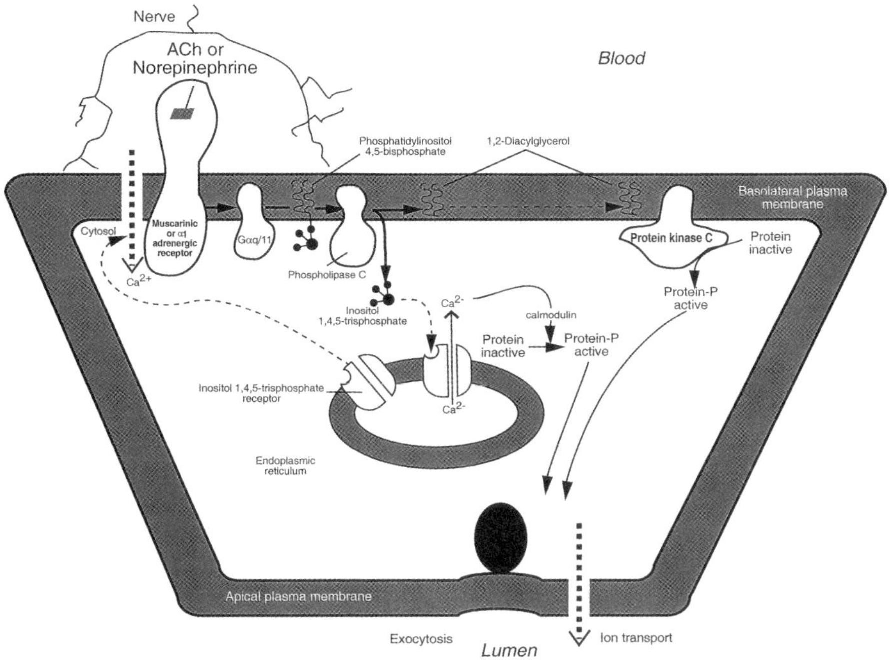

FIG X-3—Schematic of Ca^{2+}/protein kinase C–dependent signal transduction pathway activated by cholinergic and α_1-adrenergic agonists in epithelial cells to stimulate mucin, protein, or electrolyte and water secretion. *ACh,* acetylcholine; *Gαq/11,* q/11 subtype of guanine nucleotide binding protein; *Protein-P,* phosphorylated (activated) protein. (From Dartt DA. Regulation of tear secretion. *Adv Exp Med Biol.* 1994;350[1]:4.)

ation promotes dimerization of two androgen-receptor complexes that then activates gene transcription and eventually protein synthesis.

Eyelid movement is important in tear film renewal, distribution, turnover, and drainage. As the eyelids close in a complete blink, the superior and inferior fornices are compressed by the force of the preseptal muscles, and the eyelids move toward each other, with the upper eyelid moving over the longer distance and exerting force on the globe. This force clears the anterior surface of debris and any insoluble mucin and expresses secretions from meibomian glands. The lower eyelid moves horizontally in a nasal direction and pushes tear fluid and debris toward the superior and inferior puncta. When the eyelids are opened, the tear film is redistributed. The upper eyelid pulls the aqueous phase of the tear film by capillary action. The lipid layer spreads as fast as the lids move, so that no area of the tear film is left uncovered by lipid. The lipid layer increases tear film thickness and stabilizes the tear film. Polar lipids, present in the meibomian secretions, concentrate at the lipid–water interface and enhance the stability of the lipid layer.

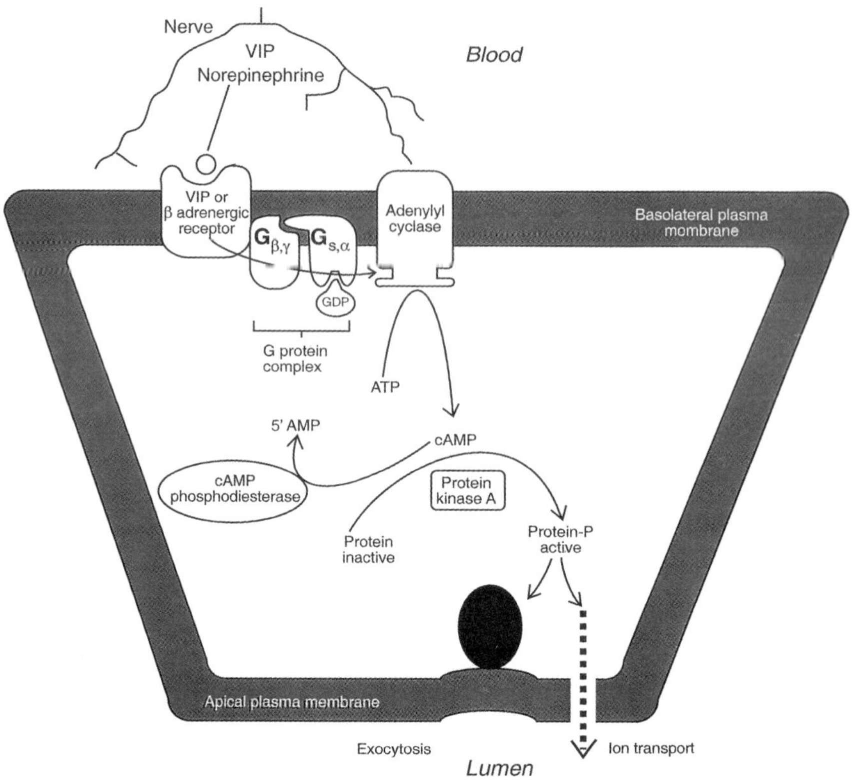

FIG X-4—Schematic of 3′,5′-cyclic adenosine monophosphate *(cAMP)*-dependent signaling pathway activated by vasoactive intestinal polypeptide *(VIP)* or norepinephrine to stimulate mucin, protein, or electrolyte and water secretion in epithelial cells. $G_{\beta,\gamma}$, β- and γ-subunits of guanine nucleotide binding protein; $G_{s,\alpha}$, stimulatory α-subunit of guanine nucleotide binding protein; *GDP*, guanosine 5′-diphosphate; *ATP*, adenosine 5′-triphosphate; *5′ AMP*, adenosine 5′-monophosphate; *Protein-P*, phosphorylated (activated) protein. (Modified from Dartt DA. Regulation of tear secretion. *Adv Exp Med Biol.* 1994;350[1]:5.)

Tear Dysfunction

A qualitative or quantitative abnormality of the tear film may occur as a result of:

- Change in the amount of tear film constituents
- Change in the composition of tear film
- Uneven dispersion of the tear film because of corneal-surface irregularities
- Ineffective distribution of the tear film caused by eyelid–globe incongruity

The amount or composition of the tear film can change because of aqueous deficiency, mucin deficiency or excess (with or without associated aqueous deficiency), and/or lipid abnormality (meibomian gland dysfunction). The preocular tear film is dispersed unevenly with an irregular corneal or limbal surface (inflammation, scarring, dystrophic changes) or poor contact lens fit. Eyelid–globe incongruity results from congenital, traumatic, or neurogenic eyelid dysfunction or absent or dysfunctional blink mechanism. Diagnostic tests for tear dysfunction include the tear breakup time, lissamine green staining, rose bengal staining, osmolarity tests, and Schirmer tests.

Tear dysfunction is managed by creating a more regular corneal or conjunctival contour or by facilitating eyelid–globe congruity. The tear-constituent imbalance can be corrected by decreasing evaporation of tears (through reduced room temperature or increased humidity) and changing contact lenses for regular glasses. Tear film instability (secondary to aqueous and/or mucous deficiency) can be improved by topical tear substitutes. Reduction of tear drainage by punctal occlusion prolongs the effect of artificial tears and preserves the natural tears. These treatments are all palliative, and no effective treatment of tear deficiency has yet been expanded to practice.

BCSC Section 7, *Orbit, Eyelids, and Lacrimal System,* discusses the lacrimal system in depth with numerous illustrations.

Tiffany JM. Tears and conjunctiva. In: Harding J, ed. *Biochemistry of the Eye.* London: Chapman & Hall; 1997:1–15.

Tsubota K. Tear dynamics and dry eye. *Prog Retin Eye Res.* 1998;17:565–596.

CHAPTER XI

Cornea

The cornea is a remarkable structure, with a high degree of transparency and excellent self-protective and reparative properties. The cornea is made up of the following histologic layers (Fig XI-1):

- Epithelium with basement membrane
- Bowman's layer
- Stroma (or *substantia propria*)
- Descemet's membrane
- Endothelium

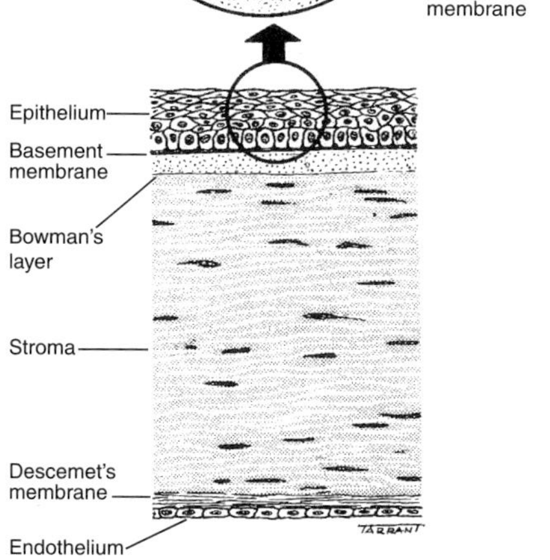

FIG XI-1—Diagram of different layers of the cornea. (Reproduced with permission from Kanski JJ. *Clinical Ophthalmology: A Systematic Approach.* 3rd ed. Oxford: Butterworth-Heinemann; 1994:100.)

The human cornea has a rich afferent innervation. The long posterior ciliary nerves (branches of the ophthalmic division of cranial nerve V_1) penetrate the cornea in three planes: scleral, episcleral, and conjunctival. Peripherally, approximately 70 to 80 branches of the long posterior ciliary nerves enter the cornea and lose their myelin sheath 1–2 mm from the limbus. A plexus posterior to Bowman's layer sends branches anteriorly into the epithelium.

Oxygen to the cornea is provided by the preocular tear film, lid vasculature, and aqueous humor. The primary metabolic substrate for the epithelial cells, stromal keratocytes, and endothelium is glucose. The stroma receives glucose primarily from the aqueous humor by carrier-mediated transport through the endothelium; the epithelium receives glucose by passive diffusion through the stroma. The preocular tear film and limbal vessels supply approximately 10% of the glucose used by the cornea. Glucose is metabolized in the cornea by all three metabolic pathways:

- Tricarboxylic acid (TCA) cycle
- Anaerobic glycolysis
- Hexose monophosphate (HMP) shunt (Fig XI-2)

In the epithelium and endothelium, the HMP pathway breaks down 35%–65% of the glucose, but the keratocytes of the stroma metabolize very little glucose via this pathway. The keratocytes appear to lack 6-phosphogluconate dehydrogenase, an important enzyme in the HMP pathway. The TCA cycle is much more active in the endothelium than in the epithelium. Pyruvic acid, the end product of glycolysis, is converted either to CO_2 and H_2O (via the TCA cycle under aerobic conditions) or to lactic acid (under anaerobic conditions). Production of lactic acid increases in conditions of oxygen deprivation, as in the presence of tight-fitting contact lenses of low oxygen permeability. Accumulation of lactic acid in the cornea has detrimental visual consequences, such as edema (due to an increase in an osmotic solute load) or stromal acidosis, which can change endothelial morphology and function.

Human corneas possess a remarkably high activity of aldehyde dehydrogenase and transketolase. Together, these two proteins constitute 40%–50% of the soluble proteins in corneal stroma. Like enzyme crystallins of the lens, both aldehyde dehydrogenase and transketolase are thought to contribute to the optical properties of the cornea. Both proteins are also thought to protect corneal cells against free radicals and oxidative damage by absorbing UV-B irradiation.

Epithelium

The *epithelium* is about 35 µm thick and constitutes 10% of total corneal thickness. It is composed of five to six layers, which include one to two layers of superficial squamous cells, two to three layers of broad wing cells, and the innermost layer of the columnar basal cells. Surface projections (microvilli and microplicae) are present on the apical surface of the most superficial cell layer of epithelium. These projections are coated with filamentous material known as *glycocalyx.* Mucin glycoproteins, the major constituents of glycocalyx, are thought to promote both stability of the tear film and wettability of the corneal surface. Plasma membrane proteins and lipids of corneal epithelial cells, like those of other cell types, are heavily glycosylated and play an important role in cell–cell adhesion as well as in the adhesion of the basal cells of corneal epithelium to the underlying basement membrane. The

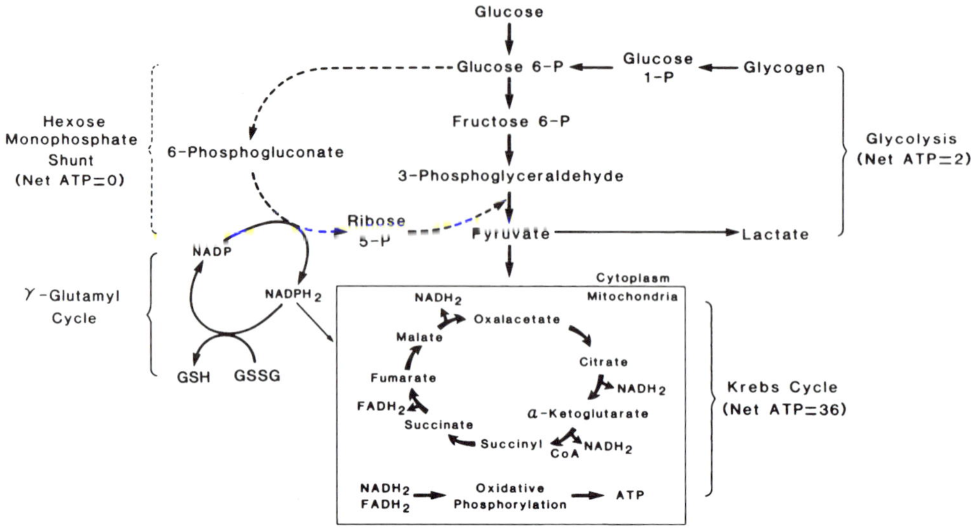

FIG XI-2—Schematic representation of the relationship among cytoplasmic glycolysis, the hexose monophosphate shunt, the γ-glutamyl cycle, and the mitochondrial Krebs cycle. *GSH,* reduced glutathione; *GSSG,* oxidized glutathione; *ATP,* adenosine triphosphate; *$FADH_2$,* reduced form of flavin adenine dinucleotide; *$NADH_2$,* reduced form of nicotinamide adenine dinucleotide. (Reproduced with permission from Anderson RE, ed. *Biochemistry of the Eye.* San Francisco: American Academy of Ophthalmology; 1983:25.)

sugar residues of the plasma membrane glycoproteins and glycolipids of corneal epithelium also play a role in:

- Wound-healing mechanisms, by mediating corneal epithelial sheet migration over the wound surface following ocular injury
- Pathogenesis of corneal infection, by serving as attachment sites for microbes

Hydrophilic molecules penetrate the epithelium poorly, but they may pass through intercellular tight junctions if the polar molecule is less than 500 daltons in apparent molecular mass. Knowing the ionic dissociation constant of a molecule is important for determining its permeability across the cornea. In order to diffuse across the epithelium, organic molecules should exist in an uncharged state. However, a charged molecule more readily penetrates the stroma. Therefore, to penetrate the cornea and enter the anterior chamber, an organic molecule should be able to dissociate at physiologic pH and temperature (ie, within the stroma).

Bowman's Layer

Bowman's layer is immediately posterior to the epithelial basal lamina. This layer is 12 µm thick and is composed of randomly packed type I and type V collagen fibers that are 30 nm in diameter. These fibers are enmeshed in a matrix consisting of proteoglycans and glycoproteins. Bowman's layer is secreted during embryogenesis by the anterior stromal keratocytes and epithelium. It is acellular, and it does not regen-

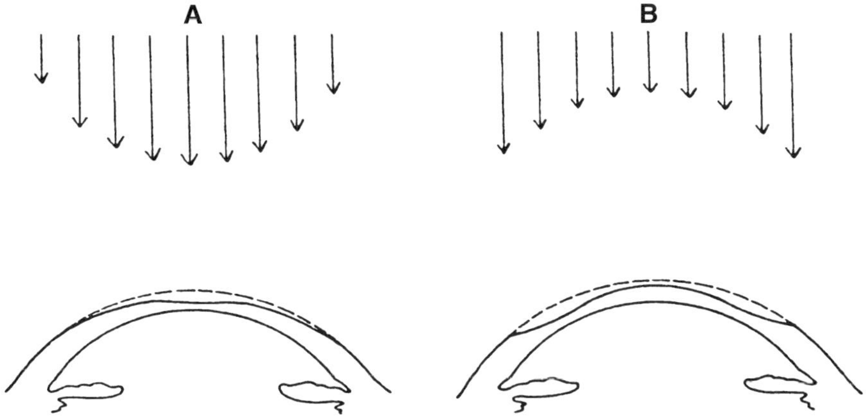

FIG XI-3—Re-profiling the corneal surface. **A**, If more photon energy is delivered to the center of the ablation, a concave erosion ensues. This will "flatten" the corneal surface and produce a myopic correction. **B**, If more energy is delivered to the periphery, a convex profile results. This produces steepening of the corneal surface and a hypermetropic correction. (From O'Brart DPS, Marshall J. Background of excimer laser refractive surgery. In: Wu HK, Thompson VM, Steinert RF, et al, eds. *Refractive Surgery.* New York: Thieme; 1999:221.)

erate when damaged. During excimer laser photorefractive keratectomy for correction of myopia, Bowman's layer (along with a small portion of anterior corneal stromal tissue) is removed from the center of the cornea so that the anterior dome of the cornea becomes flatter (Fig XI-3). Removal of about 10 µm of tissue corrects 1.0 D of myopia. For hypermetropic correction, the anterior stromal tissue is removed from the periphery so that the anterior dome of the cornea becomes steeper (see Fig XI-3).

Stroma

The *stroma* makes up 90% of the corneal thickness. Stromal cells are known as *keratocytes.* Depending on age, keratocytes constitute 10%–40% of corneal volume. Usually, these cells reside between the collagen lamellae. The stroma is made up of roughly 200 layers of lamellae, which are 1.5–2.5 µm in thickness and are constituted of collagen fibrils enmeshed in a matrix consisting of proteoglycans, proteins, and glycoproteins. The stromal fibrils within each lamella are narrow and uniform in diameter. In humans, the average fibril diameter is 30 nm.

Collagen fibrils within each lamella run parallel to each other from limbus to limbus. In adjacent lamellae, the fibrils make a large angle with each other, less than 90° in anterior stroma but almost 90° in the posterior stroma. Therefore, alternate arrays of fibrils cut nearly perpendicular and tangentially are observed in the electron micrographs of cross sections of corneal stroma. Also, collagen fibrils in each lamella are regularly spaced, with a center-to-center distance of 55–60 nm. The narrow and uniform diameter of collagen fibrils and their regular arrangement are characteristic of collagen of the corneal stroma and are necessary for the transparency of this tissue. Type I is the major collagen component of the corneal stroma.

It constitutes approximately 70% of the total stromal dry weight. Immunohistochemical and biochemical studies have demonstrated that normal adult corneal stroma also contains collagens type V, VI, VII, XII, and XIV. Type III collagen production is associated uniquely with stromal wound healing.

After collagen, proteoglycans are the second most abundant biological constituents of the cornea. They constitute about 10% of the dry weight of the cornea. It is the proteoglycans that confer hydrophilic properties to the stroma. Proteoglycans are glycosylated proteins with at least one glycosaminoglycan (GAG) chain covalently bound to the protein core. GAGs are composed of repeating disaccharides. The GAGs found in corneal stroma include:

- Keratan sulfate
- Chondroitin sulfate
- Dermatan sulfate

Two major proteoglycan populations have been identified in corneal stroma, one containing keratan sulfate chains, the other containing both dermatan sulfate and chondroitin sulfate chains. Regulation of spacing between the stromal collagen fibrils is thought to result from highly specific interactions between the proteoglycans and collagen fibrils; when these interactions are disturbed, the ability of cornea to remain transparent is profoundly affected.

Matrix metalloproteinases (MMPs) are a family of Zn^{2+}-dependent enzymes responsible for degradation of the components of the extracellular matrix (including proteoglycans and various types of collagens) during normal development as well as in disease processes. Of more than a dozen known metalloproteinases, only MMP2 proenzyme has been found in the normal healthy cornea. However, after corneal injury, additional MMPs (including MMP1, MMP3, and MMP9) are synthesized. The proteinase inhibitors of the cornea play a key role in corneal protection by restricting damage during corneal inflammation, ulceration, and wound healing. The following proteinase inhibitors have thus far been identified in the cornea:

- α_1-Proteinase inhibitor
- α_1-Antichymotrypsin
- α_2-Macroglobulin
- Plasminogen activator inhibitors 1 and 2
- Tissue inhibitors of metalloproteinases

Many of these inhibitors are synthesized by resident cells of the cornea; some are derived from tears, aqueous humor, and limbal blood vessels.

Descemet's Membrane and Endothelium

Descemet's membrane is a 10 µm thick specialized basement membrane present between the endothelium and the posterior stroma. It is secreted by endothelium and is constituted of an anterior banded portion and a posterior nonbanded portion. Type IV is the most abundant collagen in Descemet's membrane. The corneal endothelium is a single layer posterior to Descemet's membrane and is composed of polygonal cells 20 µm in diameter. In young adults, the normal endothelial cell count is approximately 3000/mm^2. The number of endothelial cells decreases with aging, with a concomitant spreading and thinning of the remaining cells. A group of tight junctions forms the apical junctional complex between cells that occludes the lateral extracellular spaces from the aqueous humor. Approximately 20–30 short microvilli per cell extend from the apical plasma membrane into the aqueous humor. The endothelium functions as a permeability barrier between the aqueous humor and the corneal stroma and as a pump to maintain the cornea in a partially dehydrated state. Water is pumped across the endothelial apical plasma membrane into the aqueous humor in an energy-dependent process. In vivo, the endothelium derives sufficient oxygen from the aqueous humor to maintain normal pump function.

After injury to the endothelium, healing occurs mainly via cell migration, rearrangement, and enlargement of the residual cells. Substantial cell loss or damage results in irreversible edema because human corneal endothelial cells have limited ability to divide after birth. Infiltration of polymorphonuclear leukocytes in response to severe corneal injury induces endothelial cells to become fibroblastic and to synthesize *retrocorneal fibrous membrane* (RCFM). The RCFM forms between Descemet's membrane and corneal endothelium and causes significant loss of visual acuity. Unlike normal corneal endothelial cells, which accumulate little type I collagen protein, the fibroblastic cells isolated from the RCFM predominantly express type I collagen.

Panjwani N. Cornea and sclera. In: Harding JJ, ed. *Biochemistry of the Eye.* London: Chapman & Hall; 1997:16–51.

CHAPTER XII

Iris and Ciliary Body

The uveal tract consists of three structures: iris, ciliary body, and choroid. The smooth muscles of the iris and ciliary body, unlike smooth muscle elsewhere in the body, are derived from the neuroectoderm.

The iris is a highly pigmented tissue that functions as a delicate and movable diaphragm between the anterior and posterior chambers of the eye, regulating the amount of light reaching the retina. It is a dynamic structure, capable of precise and rapid changes in pupillary diameter in response to both light and specific pharmacologic stimuli.

The ciliary body regulates the composition and production of aqueous humor and affects the ionic environment and metabolism of the lens, cornea, and trabecular meshwork. These functions require adaptation within the ciliary body to accommodate both for the rapid changes in surface area from constriction to dilatation and for the movement of ions. The ciliary body is the main pharmacologic target in the treatment of glaucoma. Many treatments employed to lower intraocular pressure (IOP) in glaucoma, such as adrenergic and cholinergic drugs and prostaglandins (PGs), work through receptors and their respective signal transduction pathways. In recent years, considerable progress has been made in these areas, especially as it relates to receptors and second messengers and their functions in the iris.

The iris–ciliary body is enriched in many types of receptors that bind to various agonists and antagonists, including adrenergic, muscarinic cholinergic, and peptidergic; PG; serotonin; platelet-activating factor; and growth factors. Several reports have appeared on the biochemistry of the smooth muscles of the iris, including:

- Contraction-relaxation
- Receptor characterization
- Second-messenger formation and regulation
- Protein phosphorylation
- Phospholipid metabolism
- Arachidonic acid (AA) release and eicosanoid biosynthesis

The following discussion concerns biochemical aspects of the iris–ciliary body such as aqueous humor formation, eicosanoids, and membrane signal transduction and second-messenger systems. Chapter III of Part 1 of this volume, Anatomy, discusses and illustrates all of the various structures mentioned in this chapter.

Aqueous Humor Dynamics

The aqueous humor, the transparent fluid that fills the anterior and posterior chambers, is formed from blood plasma and secreted by the nonpigmented ciliary epithelium. It is the nutrient source for the avascular lens and cornea and provides a route for the removal of waste products. Aqueous humor is essentially protein-free, which allows for optical clarity. The protein level in rabbit and human aqueous humor is very low, only about 1/200 and 1/500 of the plasma protein, respectively. Albumin accounts for about one half of the total protein. Other components include growth factors; several enzymes such as carbonic anhydrase, lysozyme, diamine oxidase, plasminogen activator, dopamine β-hydroxylase, and phospholipase A_2; and PGs, cyclic adenosine monophosphate (cAMP), catecholamines, steroid hormones, and hyaluronic acid.

Ocular fluids are separated from blood by barriers formed by tight junctions between epithelial and endothelial cells. These barriers are called either *blood–aqueous* or *blood–retina,* depending on their location in the eye. Because of these barriers, the composition and amounts of all materials entering and leaving the eye can be carefully controlled, except for exit through Schlemm's canal. Perturbations of these blood–ocular barriers result in mixing of blood constituents with the ocular fluids and may be the cause of plasmoid aqueous, retinal exudates, or retinal edema.

Aqueous humor enters the posterior chamber from the ciliary processes by means of four physiologic mechanisms:

- Diffusion
- Ultrafiltration
- Carbonic anhydrase II (CA II) activity
- Active secretion

Diffusion and ultrafiltration are passive, whereas both CA II and secretion are active energy-requiring processes. Diffusion involves movement of ions like sodium across a membrane toward the side with the most negative potential. Ultrafiltration is that nonenzymatic component of aqueous formation that is dependent on IOP, on the blood pressure, and on the blood osmotic pressure in the ciliary body. CA II in humans is present in pigmented and nonpigmented epithelium. Its inhibitors cause a reduction in the rate of entry of sodium and bicarbonate in to the aqueous, leading to a reduction in aqueous flow. The formation of aqueous humor is largely a product of *active secretion* by the inner, nonpigmented ciliary epithelium involving membrane-associated Na^+,K^+-ATPase.

The following observations support the involvement of active-transport mechanisms in secretion:

- Aqueous concentrations of Na^+, K^+, Cl^-, *myo*-inositol, certain amino acids, and glucose are maintained by specific active-transport systems located in the ciliary epithelium.
- The high level of ascorbic acid in the aqueous humor suggests that there is an active pump mechanism for its secretion into the aqueous.
- Both the iris and ciliary body accumulate ρ-aminohippuric acid from the aqueous against a concentration gradient (the accumulation shows saturation kinetics; is inhibited at 0°C; and is depressed by cyanide, dinitrophenol, ouabain, iodopyracet, and probenecid).
- The low protein concentration of aqueous humor relative to serum results from exclusion of large molecules by the blood–aqueous barrier. The presence of such a barrier requires active-transport systems either for moving substances across the cellular layer into the eye or for their removal from the aqueous humor.
- The rate of aqueous humor formation differs between species, being about 2 μl/min in humans and 3–4 μl/min in rabbits. IOP is maintained by steady aqueous

formation and drainage, which allows the surrounding tissues to remove waste products of metabolism. Inhibitors of enzymatic processes decrease aqueous humor inflow by different amounts and thus provide additional evidence for active secretory processes. CA inhibitors and beta blockers (discussed below) are used systemically and topically in the treatment of glaucoma to reduce the rate of aqueous humor formation. See Part 5 (Ocular Pharmacology), Chapter XXI of this volume and BCSC Section 10, *Glaucoma,* for more detail.

Eicosanoids

Eicosanoids, which include PGs, prostacyclin (PGI_2), thromboxanes (TXA_2), and leukotrienes, are an important family of compounds with hormonal activity. They are synthesized as a result of phospholipase A_2 (PLA_2) stimulation that causes the release of AA (20:4w-6) from membrane glycerolipids (Fig XII-1). These agents affect both the male and the female reproductive systems, the gastrointestinal system, the cardiovascular and renal systems, the nervous system, and the eye. PGI_2 and TXA_2 are natural biological antagonists. PGI_2, which is synthesized mainly in endothelial cells of vascular tissues, is a potent vasodilator, a potent platelet-antiaggregating agent, and a stimulator of adenylate cyclase. In contrast, TXA_2, which is synthesized mainly by platelets, is a potent vasoconstrictor and a platelet-aggregating agent.

PGs have profound effects on inflammation in the eye, aqueous humor dynamics, and blood–ocular barrier functions. PGs of the E and F types and AA, when administered intracamerally or topically at high concentrations, cause miosis, an elevation of IOP, an increase in aqueous protein content, and the entry of white cells into the aqueous and tear fluid. Evidence indicates that some antiglaucoma drugs such as epinephrine may affect IOP by influencing the production of PGs. More recently, PGs and their derivatives have been found to be useful as antiglaucoma agents. For example, latanoprost, a $PGF_{2\alpha}$ receptor agonist currently available, has been shown to decrease IOP by 27%–35%. Unlike β-blockers, CA inhibitors, and α_2-agonists, latanoprost acts on outflow rather than formation of aqueous humor.

The release of AA from plasma-membrane phospholipids can be brought about by a wide variety of stimuli: inflammatory, immunologic, neural, chemical, or simple mechanical agitation. Free AA reacts either with cyclo-oxygenase (also known as PG synthetase), the first enzyme of the PG biosynthetic sequence, or with lipoxygenase to generate hydroperoxy fatty acids (see Fig XII-1). In the cyclo-oxygenase reaction, the released arachidonate is converted into endoperoxides (PGG_2 and PGH_2) by the membrane-bound cyclo-oxygenase. The endoperoxides are then converted to TXA_2 by thromboxane synthetase or to various PGs by isomerase or reductase enzymes.

PG biosynthesis from AA, via COX-1, can be blocked by most nonsteroidal anti-inflammatory drugs (NSAIDs) such as indomethacin and aspirin. The NSAIDs bind irreversibly to the cyclo-oxygenase enzyme. Topical NSAIDs have been used in the treatment of anterior segment inflammation, aphakic and pseudophakic cystoid macular edema, allergic conjunctivitis, and pain after refractive surgery. Flurbiprofen 0.03% (Ocufen) does not dilate the normal pupil, and suprofen 1% (Profenal) drops are used preoperatively for the prevention of PG-mediated pupillary miosis during ocular surgery. Diclofenac 0.1% (Voltaren) has been approved for the treatment of postoperative inflammation following cataract extraction. Ketorolac tromethamine 0.05% (Acular) is indicated for the relief of itching from allergic conjunctivitis.

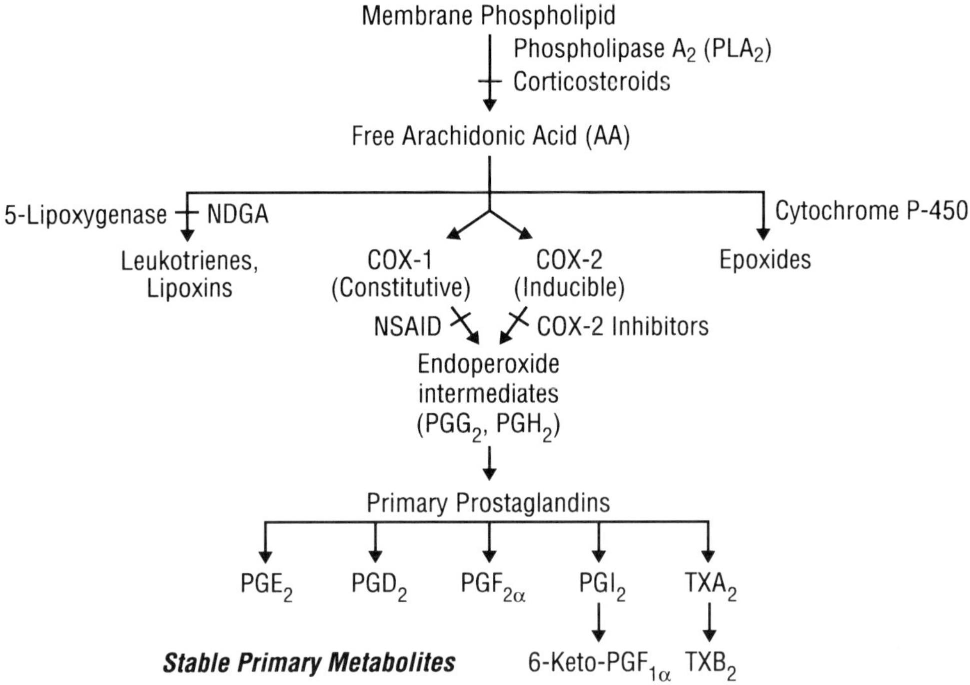

FIG XII-1—An outline of the synthesis of prostaglandins *(PGs)* and leukotrienes from arachidonic acid. In response to stimulation of a target cell with a relevant stimulus (eg, cytokine, neurotransmitter, various pharmacologic agents), phospholipase A_2 is activated and arachidonic acid is released from the *sn*-2 position of membrane phospholipids. Arachidonic acid is then converted by cyclo-oxygenase 1 *(COX-1)* or cyclo-oxygenase 2 *(COX-2)* to PGH_2, and then PGH_2 is isomerized to biologically active prostanoid products. Arachidonic acid can also be metabolized through the 5-lipoxygenase and cytochrome P-450 pathways to generate leukotrienes and epoxides, respectively. Phospholipase A_2 can be inhibited by corticosteroids such as dexamethasone; COX-1 by nonsteroidal anti-inflammatory drugs *(NSAIDs)* such as indomethacin and aspirin; COX-2 by DUP697, SC58125, L-745-337, and NS398; and the 5-lipoxygenase pathway by nordihydroguaiaretic acid *(NDGA)*. *TXA_2*, thromboxane A_2.

PG biosynthesis, via COX-2, can be blocked by the recently developed COX-2 inhibitors, DUP697, SC58125, L-745-337, and N5398. All currently available NSAIDs inhibit both COX-1 and COX-2 and compete with arachidonate for binding to the cyclo-oxygenase–active site. These compounds are effective anti-inflammatory agents, but they are also all quite ulcerogenic when given systemically. Many pharmaceutical firms have developed new cyclo-oxygenase inhibitors that selectively inhibit COX-2. These efforts were initially driven by two notions, which subsequently proved to be correct:

- COX-2 is the relevant enzyme in inflammation.
- COX-1 but not COX-2 is present in the stomach.

Indeed, COX-2 inhibitors have been reported to be anti-inflammatory and analgesic and to lack gastrointestinal toxicity. All of the COX-2–selective agents are time-dependent, reversible inhibitors of COX-2.

The leukotrienes are another group of compounds that are formed from AA by a variety of tissues. Their formation in ocular tissues through the lipoxygenase pathway has not yet been thoroughly investigated. Leukotrienes C_4 and D_4 have been identified as the major active constituents of the slow-reacting substance of anaphylaxis (SRS-A). These compounds are powerful smooth-muscle contractors that alter muscle permeability. They are much more active than histamine. NSAIDs do not inhibit lipoxygenase, but nordihydroguaiaretic acid (NDGA) does.

There is a large family of G-protein–coupled, seven-transmembrane PG receptors. Complex specificity and regulatory functions arise because:

- Individual receptors have differing and overlapping specificities for individual PGs.
- Individual receptors are specifically distributed among different cells and tissues.
- There are different coupling mechanisms in different cells. With the variety of PGs, this leads to very complicated and incompletely understood pathophysiologic functions for most PGs. In general, PGs play key roles in regulation of smooth-muscle contractility, in mediation of pain and fever, in regulation of blood pressure and platelet aggregation, and in other physiologic defense mechanisms, including immune and inflammatory responses. Inhibition of cyclo-oxygenases is the mechanism for much of the analgesic, anti-inflammatory, antipyretic, and antithrombotic effects of NSAIDs.

Abdel-Latif AA. Release and effects of prostaglandins in ocular tissues. *Prostaglandins Leukot Essent Fatty Acids.* 1991;41:71–82.

Garavito RM, DeWitt DL. The cyclooxygenase isoforms: structural insights into the conversion of arachidonic acid to prostaglandins. *Biochim Biophys Acta.* 1999;1441: 278–287.

Jampol LM. Nonsteroidal anti-inflammatory drugs: 1997 update. In: *Focal Points: Clinical Modules for Ophthalmologists.* San Francisco: American Academy of Ophthalmology; 1997;15:6.

Neurotransmitters, Receptors, and Signal Transduction Pathways

In the iris–ciliary body the sphincter and ciliary muscles are innervated by the third cranial (oculomotor) nerve (parasympathetic), and cholinergic impulses are transmitted to the muscle by acetylcholine (ACh). In the ciliary processes, nonmedullated nerve fibers, many of them adrenergic, surround the blood vessels. The dilator muscle fibers of the iris are innervated by the sympathetic nerves from the superior cervical ganglion, and the adrenergic nerve impulses are transmitted to the muscle cells by norepinephrine (NE). The neurons that synthesize, store, and release ACh are called *cholinergic neurons;* those that synthesize, store, and release NE are called *adrenergic neurons.*

There are also two major types of autonomic receptors: cholinergic receptors receive input from cholinergic neurons, and adrenergic receptors from adrenergic neurons. These receptors are further divided as shown in Table XII-1. The receptors of both the iris sphincter and the ciliary muscle are of the cholinergic muscarinic type, and the iris dilator is mainly of the α-adrenergic receptor type. In addition, the iris muscles contain sensory nerves. The sensory neurotransmitters include substance P and calcitonin gene-related peptide (CGRP), which may be involved in trophic effects, inflammatory reactions, and direct or indirect regulation of iris muscle tone in many species, including humans.

Table XII-1

Cholinergic and Adrenergic Receptors

RECEPTORS	AGONISTS	BLOCKING AGENTS
Cholinergic (sphincter)	Acetylcholine	
Muscarinic	Muscarine	Atropine
Nicotinic	Nicotine	d-Tubocurarine
Adrenergic (dilator)	Norepinephrine	
Alpha*	Phenylephrine	Phentolamine and phenoxybenzamine
α_1	Phenylephrine	Prazosin, thymoxamine, dapiprazole
α_2	Apraclonidine	Yohimbine
Beta	Isoproterenol	Propranolol and timolol
β_1	Tazolol	Betaxolol
β_2	Albuterol	Butoxamine

The cholinergic agonists and the adrenergic blockers listed cause miosis; the adrenergic agonists and the cholinergic blockers listed cause dilation.

*The prefixes α_1 and α_2 have been proposed for post- and presynaptic α-adrenoceptors, respectively. According to the present view, the classification into α_1 and α_2 subtypes is based exclusively upon the relative potencies and affinities of agonists and antagonists, regardless of their function and localization.

Miotics

Miotics act either by stimulating the sphincter (cholinergic agonists) or by blocking the dilator (adrenergic blockers).

Agonists Sphincter stimulators produce responses similar to acetylcholine. They contract the sphincter (resulting in an increase in pupillary contraction) and the ciliary muscles (resulting in accommodation). Sphincter activity can be affected by two mechanisms:

- Direct action on the muscles through ACh, carbachol, and pilocarpine
- Inhibition of acetylcholinesterase (AChE) either reversibly with physostigmine (Eserine) or neostigmine (Prostigmin), or irreversibly with diisopropyl fluorophosphate (DFP) or echothiophate iodide (Phospholine iodide)

The AChE inhibitors allow ACh to accumulate at the parasympathetic third-nerve endings and to produce accommodative spasm.

Antagonists Two mechanisms for the action of dilator blockers are available:

- Inhibition of the release of norepinephrine at the myoneural junction by guanethidine (Ismelin), which acts by depleting NE stores at the nerve terminals (once these stores are depleted, miosis ensues)
- Blockage of the α-adrenergic receptors of the dilator muscle by thymoxamine, dapiprazole, phenoxybenzamine, dibenamine, or phentolamine, which prevent contraction and produce miosis (miosis is due to the unopposed action of the iris sphincter, which is tonically innervated)

The pupillary dilator muscle has predominantly α-adrenergic receptors, whereas the ciliary muscle is under parasympathetic control. Thymoxamine and dapiprazole are selective α_1-adrenergic blocking agents and can cause miosis through dilator-muscle paralysis without affecting the ciliary muscle–controlled facility of outflow, IOP, or amplitude of accommodation. Thymoxamine has been advocated for many uses but is not commercially available in the United States. Dapiprazole is commercially available for reversal of pupillary dilation after pharmacologic mydriasis. With thymoxamine and dapiprazole, the pupil returns to its predilated state in 30 minutes (versus 3–4 hours without these agents). Angle-closure glaucoma will not be prevented with these agents, but they move the pupil through the more dangerous mid-dilated state more quickly.

Mydriatics

Mydriatics act by stimulating the dilator (adrenergic agonists) or by blocking the sphincter (cholinergic blockers).

Agonists Dilator stimulators increase dilator activity in three ways:

- By increasing NE release, as with hydroxyamphetamine, which causes NE to be released rapidly, thus resulting in mydriasis
- By interfering with NE reuptake, as with cocaine, which in addition to acting as a local anesthetic prevents NE reuptake (inactivation) and thus prolongs or potentiates the action of released NE
- By directly stimulating the α_1 receptors of the dilator, as with phenylephrine (Neo-Synephrine)

Blockers An adrenergic agent is only mydriatic, but an anticholinergic is mydriatic and cycloplegic. An effective anticholinergic (sphincter-blocking) agent that produces both mydriasis and cycloplegia is atropine, which blocks the action of ACh and pilocarpine at the muscle. Its effects are long lasting. A drug that produces rapid cycloplegia but has a relatively short action is tropicamide. Cyclopentolate is a blocker of intermediate action and duration.

Calcium Channels and Channel Blockers

Calcium channels are membrane-bound protein receptors that contain multiple subunits, one of which is the active site for binding of calcium-channel blockers. Calcium plays a major role in influencing cellular function, and the most common pathway for entry in cells is through the calcium channels. Six subclasses of calcium channel blockers have been identified: L, T, N, P, Q, and R. The L-type blocker is predominant in skeletal, cardiac, and vascular smooth muscle. Calcium-channel blockers bind to membrane-bound calcium channels and inhibit the influx of extracellular calcium in vascular smooth muscle, thereby causing direct arteriolar vasodilation and depression of myocardial contractility. They are widely used for treatment of hypertension and do not affect glucose tolerance, lipoproteins, uric acid, or serum electrolytes in therapeutic doses. Topically applied calcium channel blockers lower IOP. Since normal-tension glaucoma may be associated with vasospastic disease, calcium-channel blockers may play a role in the treatment of this poorly understood condition. See also BCSC Section 10, *Glaucoma.*

Netland PA, Erickson KA. Calcium channel blockers in glaucoma management. *Ophthalmol Clin North Am.* 1995;8:327–334.

Signal-Transduction Pathways in the Iris and Ciliary Body

Signal Transduction

The concept that most drugs, hormones, and neurotransmitters produce their biological effects by interacting with receptors has been substantiated by the isolation and identification of various receptors and their subtypes. Receptors for neurotransmitters and peptide hormones are located on the surface of the cell, whereas receptors for steroid hormones are intracellular. The iris and ciliary body both contain adrenergic, muscarinic cholinergic, peptidergic, PG, serotonin, and purinergic receptors. Membrane receptors have several characteristics:

- They are responsible for transmitting a signal across the membrane by recognizing molecules such as hormones, peptides, drugs, and neurotransmitters, which are the first messengers, or *ligands.*
- They are specific for agonists or hormones and can cause positive and negative responses (different cells can have different responses to the same messenger).
- The number of receptors occupied by ligands determines the response of the cell.

Receptor–Effector Coupling

Signal transduction across the plasma membrane occurs by several basic mechanisms. They include:

- Ligand-gated ion channels: a ligand (eg, neurotransmitters such as glycine, glutamate, γ-aminobutyric acid) binds to a receptor that functions as an ion channel and causes it to open, allowing cations to pass into the cells
- Receptors that have integral enzyme activity such as tyrosine kinase
- G protein–coupled receptors activate effector proteins, which include ion channels and enzymes

G protein–coupled receptors involve a cascade of three steps: A stimulus excites a receptor protein to activate a G protein, which in turns activates an effector protein (E). The effector proteins are usually enzymes such as adenylyl cyclase, phospholipase C (PLC), phospholipase A_2, or phosphodiesterase (PDE). They produce second messengers like cyclic nucleotides (cAMP, cGMP) and lipid-derived molecules (PGs, discussed above, and IP_3 and DAG, discussed below). Rhodopsin, one of the most studied G protein–coupled receptors, is discussed in detail in Chapter XVI.

Cyclic AMP and Polyphosphoinositide Turnover

Activated receptors, such as β-adrenergic and CGRP receptors, stimulate the adenylate cyclase system via the stimulatory G proteins (Gsα); others, such as M_2-muscarinic and α_2-adrenergic receptors, inhibit this effector enzyme via an inhibitory G protein (Giα) (Fig XII-2). Activation of Ca^{2+}-mobilizing receptors, such as M_3-muscarinic and α_1-adrenergic receptors, causes the release of Ca^{2+} from the sarcoplasmic reticulum (SR) via the second messenger IP_3 (Fig XII-3). PLC exists as several isoforms, one of which (PLC-β) is coupled to these receptors via the G protein, Gq. The primary control for many intracellular events is the concentration of free Ca^{2+} in the cytosol.

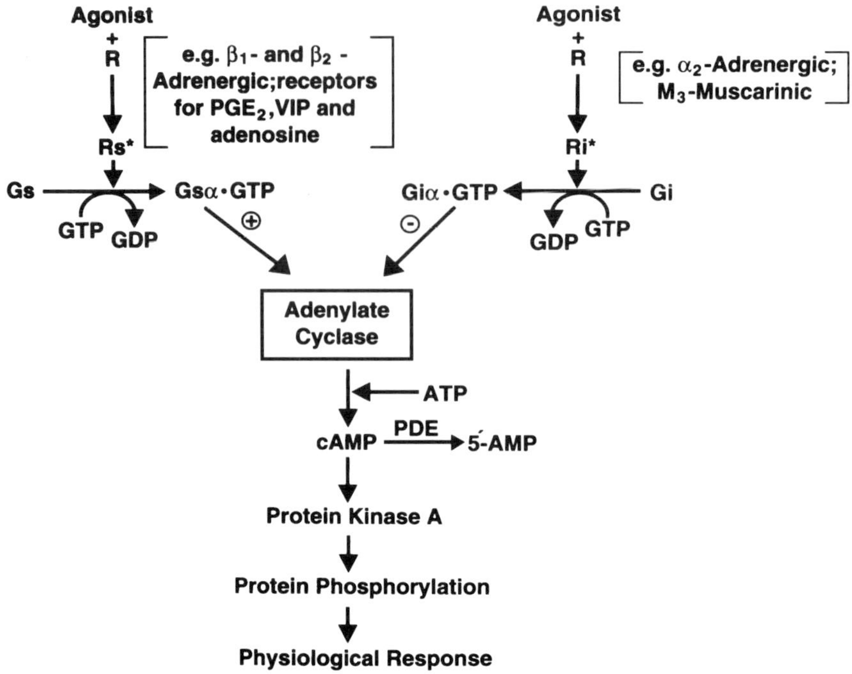

FIG XII-2—Scheme showing the activation and inhibition of the adenylate cyclase second-messenger system. *R**, activated receptor complex; *Gs* and *Gi,* stimulatory and inhibitory G proteins, respectively. Adenosine diphosphate (ADP) ribosylation of $G_{s\alpha}$ by cholera toxin maintains $G_{s\alpha}$ in active state, whereas ADP-ribosylation of $G_{i\alpha}$ by pertussis toxin results in inactivation of $G_{i\alpha}$. *ATP,* adenosine triphosphate; *5′-AMP,* adenosine 5′-monophosphate; *cAMP,* 3′,5′-cyclic adenosine monophosphate; *GDP,* guanosine diphosphate; *GTP,* guanosine triphosphate; *PDE,* phosphodiesterase; *PGE_2,* prostaglandin E_2; *VIP,* vasoactive intestinal polypeptide.

Many Ca^{2+}-mobilizing hormones and neurotransmitters, such as ACh, NE, and $PGF_{2\alpha}$, act by changing the intracellular Ca^{2+} concentration $[Ca^{2+}]_i$.

Recently, much attention has been focused on phosphatidyl-inositol-4,5-bisphosphate (PIP_2) turnover and its role in the regulation of $[Ca^{2+}]_i$. PIP_2 is hydrolyzed by phospholipase C into inositol-1,4,5-triphosphate (IP_3) and diacylglycerol (DAG) (see Fig XII-3), which have a physiologic role as second messengers. IP_3 opens Ca^{2+} channels, triggering smooth-muscle contraction, glycogen breakdown, and exocytosis. DAG activates protein kinase C, which phosphorylates serine and threonine residues in many proteins (see Fig XII-3). For example, phosphorylation of glycogen synthetase by protein kinase C inhibits synthesis of glycogen. Thus, protein kinase C complements IP_3-induced glycogen breakdown mediated by increase in cytosolic Ca^{2+}.

In the iris sphincter, as well as in other types of smooth muscle, activation of Ca^{2+}-mobilizing (eg, muscarinic) receptors leads to IP_3 production, Ca^{2+} mobilization, and muscle contraction. In contrast, activation of β-adrenergic receptors leads to cAMP formation, reduction in intracellular Ca^{2+} concentration, and muscle relax-

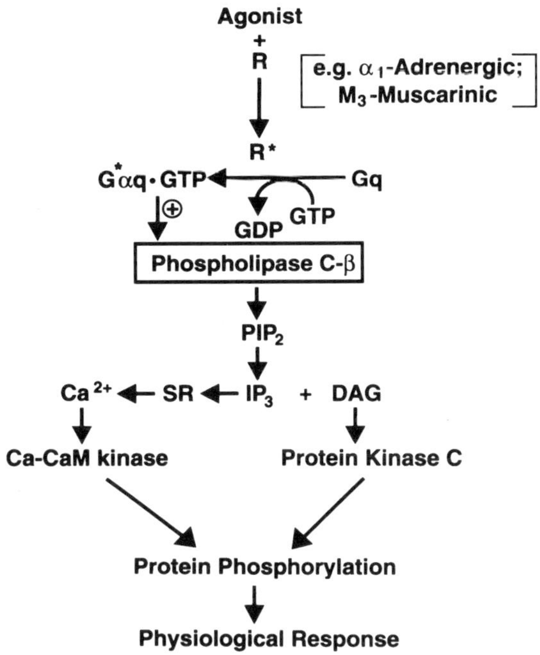

FIG XII-3—Scheme showing agonist-stimulated breakdown of phosphatidyl-inositol-4,5-bisphosphate *(PIP_2)* into inositol 1,4,5-triphosphate *(IP_3)* and diacylglycerol *(DAG)*. The agonists bind to the Ca^{2+}-mobilizing receptor to activate the G protein $G_{\alpha q}$ and stimulate phospholipase C-β_1 to hydrolyze PIP_2 into IP_3 and DAG. IP_3 diffuses through the cytosol and binds to the IP_3 receptor on the sarcoplasmic reticulum *(SR)* to release Ca^{2+} from a specific Ca^{2+} pool. Ca^{2+} binds to calmodulin (CaM); Ca^{2+}-calmodulin then activates a protein kinase to phosphorylate a protein substrate and provoke a cellular response. IP_3 is then rapidly metabolized by a series of dephosphorylation and phosphorylation pathways. DAG acts by binding to protein kinase C. Once activated, protein kinase C translocates to the plasma membrane and there phosphorylates a number of proteins on serine and threonine residues. Phorbol esters mimic the action of DAG on protein kinase C; in fact, the enzyme is presumed to be the receptor for phorbol esters. *GDP,* guanosine diphosphate; *GTP,* guanosine triphosphate.

ation. The IP_3-DAG-Ca^{2+} signaling system contains a number of potential target sites for the regulation of smooth-muscle tension responses. Contraction–relaxation responses are regulated by increases in the level of cAMP. In the iris sphincter, PGs can trigger either IP_3 production and muscle contraction or cAMP formation and muscle relaxation, depending on the type of PG, species, cell type, and tissue. Such interactions between the two signaling systems could constitute the biochemical basis for the multiple functions observed for PGs in a wide variety of systems. Ample evidence now suggests that the sympathetic nervous system, through alterations in cAMP concentrations, can regulate or inhibit the cholinergic stimulation of IP_3 production and muscle contraction.

TABLE XII-2

OCULAR RECEPTOR SUBTYPES AND THEIR SIGNAL TRANSDUCTION MECHANISMS

RECEPTOR	RECEPTOR SUBTYPES AND THEIR SIGNAL TRANSDUCTION MECHANISMS			
Beta adrenoceptor	β_1 cAMP↑	β_2 cAMP↑	β_3 cAMP↑	
α_1 adrenoceptor	α_1A IP_3/Ca^{2+}/DAG	$\alpha_1\beta$ IP_3/Ca^{2+}/DAG	α_1D IP_3/Ca^{2+}/DAG	
α_2 adrenoceptor	α_2A cAMP↓	$\alpha_2\beta$ cAMP↓	α_2C cAMP↓	α_2D cAMP↓
Muscarinic cholinergic	M_1 IP_3/Ca^{2+}/DAG	M_2 cAMP↓	M_3 IP_3/Ca^{2+}/DAG	M_4 cAMP↓
Calcitonin gene-related peptide	$CGRP_1$ cAMP↑	$CGRP_2$ cAMP↑		
Endothelin	ET_A IP_3/Ca^{2+}/DAG, cAMP↓	ET_B IP_3/Ca^{2+}/DAG		
Prostaglandin	DP, EP (EP_1, EP_2, EP_3, EP_4) FP, IP, TP Increase in cAMP or IP_3 depending on the receptor subtype			

TABLE XII-3

MODE OF ACTION OF ANTIGLAUCOMA AGENTS THAT ACT THROUGH RECEPTORS

PRIMARY MECHANISM OF ACTION	DRUG CLASS	EXAMPLES
1. Decrease aqueous humor production	a. β-adrenergic antagonists	Timolol, betaxolol, carteolol, levobunolol
	b. α_2-adrenergic agonists	Apraclonidine
2. Increase trabecular outflow	a. Miotics	Pilocarpine
	b. Adrenergic agonists	Epinephrine, dipivalyl epinephrine
3. Increase uveoscleral outflow	a. Prostaglandins	Latanoprost

Receptors

Examples of ocular receptors, their subtypes, and their signal transduction mechanisms are given in Table XII-2. In recent years, a number of reports have appeared on the biochemical and pharmacologic characterization of these receptors. Understanding ocular receptors and their transduction mechanisms will help us design better therapeutic agents for the treatment of glaucoma. This can be appreciated from the summary on the mode of action of antiglaucoma agents given in Table XII-3. The therapeutic actions of these drugs are mediated through specific receptors.

Abdel-Latif AA. Iris–ciliary body, aqueous humor and trabecular meshwork. In: Harding JJ, ed. *Biochemistry of the Eye.* London: Chapman and Hall; 1997:52–93.

CHAPTER XIII

Aqueous Humor

The aqueous humor is important in the physiology of the mammalian eye. It provides nutrients (eg, glucose and amino acids) to support the function of tissues of the anterior segment, such as the avascular lens, cornea, and trabecular meshwork; it also removes the metabolic wastes from these tissues (eg, lactic acid, pyruvic acid). Aqueous humor helps to maintain appropriate intraocular pressure. These functional properties are essential to the eye's structural integrity. In addition, since the aqueous is devoid of blood cells and of more than 99% of the plasma proteins, aqueous humor provides an optically clear medium for transmission of light along the visual path.

The aqueous humor is secreted by the ciliary epithelium at a flow rate of 2–3 μl/min. The ciliary epithelium is a bilayer of polarized epithelial cells lining the surface of the ciliary body and is composed of two cell layers:

- Nonpigmented ciliary epithelium (NPE), which faces the aqueous humor through the cells' basal plasma membrane
- Pigmented ciliary epithelium (PE), which faces the stroma, also through the cells' basal plasma membrane

Therefore, the apical plasma membranes of both NPE and PE cells appose each other, establishing cell-to-cell communication through numerous gap junctions. Of the two cell layers that constitute the ciliary epithelium, the NPE cells establish a blood–aqueous barrier by the presence of tight junctions proximal to the apical plasma membrane, thus preventing the free passage of plasma proteins and other macromolecules from the stroma to the posterior chamber through the paracellular space of NPE cells. In contrast, the PE cell layer is considered a leaky epithelium, and it allows the movement of solutes through the intercellular space between the PE cells.

Composition of the Aqueous Humor

Table XIII-1 summarizes the composition of the aqueous humor. It is clear from the values given in the table that the fluid and electrolyte composition is similar to that of plasma. However, the secretion is not an ultrafiltrate of plasma (as was once speculated), because it is produced by energy-dependent processes in the epithelial layer of the ciliary body. This production reflects the necessity of maintaining precise control over the composition of fluid bathing structures essential for normal vision.

Macknight AD, MacLaughlin CW, Peart D, et al. Formation of the aqueous humor. *Clin Exp Pharmacol Physiol.* 2000;27;100–106.

The ionic composition of aqueous humor is determined by selective active-transport systems (eg, Na^+,K^+-$2Cl^-$ symport, Cl^--HCO_3^- and Na^+,H^+ antiports, cation channels, water channels, Na^+,K^+-ATPase, K^+ channels, Cl^- channels, H^+-ATPase) that participate in secretion of aqueous humor by the ciliary epithelium. Their activ-

Table XIII-1

Composition of Aqueous Humor

COMPONENTS (mmol/kg H_2O)	PLASMA	AQUEOUS POSTERIOR	AQUEOUS ANTERIOR	VITREOUS
Rabbit				
Na	143	159	138	134
K	4.6	4.7	4.3	4.6
Cl	108	97	101	105
HCO_3	25	34	30	26
Ascorbate	0.04	1.4	1.1	0.46
Lactate	10.3	9.9	9.3	12
Glucose	6	6	6	4.6
Human				
Na	146	163		144
Cl	109	134		114
HCO_3	28	20		20–30
Ascorbate	0.04	1.06		2.21
Glucose	6	3		3.4

From MacKnight AD, MacLaughlin CW, Peart D, et al. Formation of the aqueous humor. *Clin Exp Pharmacol Physiol.* 2000;27:100–106.

ity and cellular distribution along the cell membranes of PE and NPE cells determine unidirectional net secretion from the stroma to the posterior chamber, which involves three steps:

- Uptake of solute and water at the stromal surface by PE cells
- Transfer from PE to NPE cells through gap junctions
- Transfer of solute and water by NPE cells into the posterior chamber

By the same token, it is thought that there is a mechanism for transporting solute and water from the posterior chamber back into the stroma. In this unidirectional reabsorption, another set of transporters may be involved in extruding Na^+,K^+ and Cl^- back into the stroma.

Krupin T, Civan MM. The physiologic basis of aqueous humor formation. In Ritch R, Shields MB, Krupin T, eds. *The Glaucomas.* 2nd ed. St. Louis: Mosby; 1996:251–280.

Molecular studies have shown that the secretory properties of the ciliary epithelium are not limited to ions and electrolytes, but extend to a wide range of molecules of different molecular mass. A common feature for many of these molecules is that they are synthesized locally in the ciliary epithelium and then secreted by the

NPE cells through the regulatory pathway into the aqueous humor. Among the proteins identified whose mRNA expression has been demonstrated are:

- Plasma proteins (eg, complement component C4, α_2-macroglobulin, selenoprotein P, apolipoprotein D, plasma glutathione peroxidases, angiotensinogen)
- Proteinases (eg, cathepsin D, cathepsin O)
- A component of the visual cycle (eg, cellular retinaldehyde-binding protein, or *CRALB*)
- A neurotrophic factor (eg, pigment epithelium–derived factor)
- Neuropeptide-processing enzymes (eg, carboxypeptidase E, peptidyl-glycine-α-amidating monoxygenase)
- Neuroendocrine peptides (eg, secretogranin II, neurotensin, galanin)
- Bioactive peptides and hormones (eg, atrial natriuretic peptide, brain natriuretic peptide)

These new findings support the view that the ciliary epithelium exhibits neuroendocrine properties that are directly related to the makeup of the aqueous humor and its regulation. The aqueous humor composition is in dynamic equilibrium determined both by its rate of production and outflow and by continuous exchanges with the tissues of the anterior segment. It contains:

- Inorganic ions and organic anions
- Carbohydrates
- Glutathione and urea
- Proteins
- Growth-modulatory factors
- Oxygen and carbon dioxide

Inorganic Ions

The concentrations of sodium, potassium, and magnesium in the aqueous are similar to those in plasma, but the level of calcium is only half that of plasma. The two major anions are chloride and bicarbonate. In some species, their concentrations in the aqueous are as much as 20%–30% above and below (in inverse ratios) those in the plasma. Phosphate is also present in the aqueous (aqueous:plasma ratio, ≈0.5 or less), but its concentration is too low to have any significant buffering capacity. Iron, copper, and zinc are all found in the aqueous humor at concentrations of approximately 1 mg/ml—essentially the same as the levels in plasma.

Organic Anions

Lactate is the most abundant of organic anions in the aqueous, and its concentration is always higher than that in plasma. Plasma and aqueous levels of lactate are directly related, and the contribution resulting from the glycolytic metabolism of intraocular tissues is significant. Ascorbic acid (vitamin C) is perhaps the most unique constituent of the aqueous humor. In most mammalian species, its concentration ranges from 0.6 to 1.5 mM, levels that are some 10–50 times higher than in the plasma. Furthermore, levels of ascorbic acid in the aqueous humor in the diurnal mammal may be 20 times higher than in nocturnal mammals. Ascorbic acid is an important antioxidant, both in the aqueous humor and in tissues of the anterior segment. Ascorbic acid is actively transported by the ciliary epithelial cells from the stroma side into the posterior chamber by a Na^+-dependent L-ascorbic acid transporter.

Carbohydrates

Glucose concentration in the aqueous is roughly 70% of that in plasma. The rate of entry of glucose into the posterior chamber is much more rapid than would be expected from its size and lipid solubility, suggesting that its passage across the ciliary epithelium occurs by facilitated diffusion. Saturation studies have demonstrated that a specific carrier is involved, but no evidence of active transport has been detected, nor has insulin been found to affect the entry of glucose into the aqueous. Aqueous glucose levels are increased in diabetic persons, leading to higher concentrations in the lens. Inositol, important for phospholipid synthesis in the anterior segment, occurs at a concentration about 10 times that in plasma.

Glutathione and Urea

Glutathione, an important tripeptide with a reactive sulfhydryl group, is found in the aqueous humor. Its concentration in primates ranges from 1 to 10 μM. Blood contains a high concentration of glutathione, but virtually all glutathione resides within the erythrocytes, and plasma has only a low concentration of 5 μM or less. Although aqueous glutathione may be derived by diffusion from the blood or by an active-transport system in the ciliary epithelium analogous to that of the lens, it probably also arises by loss from the lens and cornea. Glutathione acts as a stabilizer of the redox state of the aqueous by reconverting ascorbate to its functional form after oxidation as well as by removing excess hydrogen peroxide. Glutathione also serves as a substrate in the enzymatic conjugation by a group of cytosolic enzymes involved in the cellular detoxification of electrophilic compounds. These enzymes (glutathione S-transferases) are important in protecting tissues from oxidative damage and oxidative stress and are highly expressed in the ocular ciliary epithelium.

The concentration of urea in the aqueous is between 80% and 90% of that in the plasma. This compound is distributed passively across nearly all biological membrane systems, and its high aqueous:plasma ratio indicates that this small molecule (molecular weight of 60) crosses the epithelial barrier quite readily. Urea is effective in the hyperosmotic infusion treatment for glaucoma. However, mannitol (with a molecular weight of 182) is preferred to urea because it crosses the barrier.

Proteins

Because the nonpigmented ciliary epithelial cell layer establishes a blood–aqueous barrier that prevents the diffusion of plasma proteins from the stroma into the posterior chamber, one possible route for the plasma proteins to enter into the aqueous humor is through the root of the iris. In humans, the normal aqueous contains about 0.02 g of protein per 100 ml, as compared with the typical plasma level of 7 g/100 ml. The most abundant plasma proteins identified in aqueous humor are albumin and transferrin, which may account for 50% of all the protein content.

However, there is now compelling evidence that proteins that make up the aqueous humor might actually have been synthesized within the ciliary body and secreted directly into the aqueous humor. Molecular techniques (such as the screening of cDNA libraries constructed from the intact human and bovine ciliary bodies) has resulted in isolation and identification of genes encoding proteins, which are also present in the plasma. These studies therefore challenge the long-held view that plasma proteins in aqueous humor are transported into the aqueous humor from outside of the eye. Among cDNAs isolated from the ciliary body encoding plasma proteins are:

- Complement component C4, which participates in immune-mediated inflammation responses
- α_2-Macroglobulin, a multifunctional protein (functions include proteinase inhibition, clearance, and targeting; carrier protein; and processing of foreign peptides)
- Apolipoprotein D, which binds and transports hydrophobic substances, including cholesterol, cholesteryl esters, and arachidonic acid
- Selenoprotein P, which has antioxidant properties

In addition, a number of proteinases and proteinase inhibitors have also been identified in the aqueous humor. These proteinases include cathepsin D and cathepsin O, which are synthesized and secreted by the ciliary epithelial cells. Cathepsin D is involved in degradation of neuropeptides and peptide hormones and has been found in high levels in the cerebrospinal fluid of patients with Alzheimer disease. Less is known about cathepsin O; it may be involved in normal cellular protein degradation and turnover. Of the proteinase inhibitors, α_2-macroglobulin and α_1-antitrypsin are perhaps the most extensively studied. An imbalance in equilibrium between proteinases and proteinase inhibitors could lead to an alteration in aqueous humor composition, which may result in disease.

Activators, proenzymes, and fibrinolytic enzymes are present in the aqueous, and these enzymes could play a role in the regulation of outflow resistance. Plasminogen and plasminogen activator are both found in human and monkey aqueous, but only traces of plasmin have been reported. In addition to the proteins mentioned above, other enzymes have been reported in the aqueous humor. Several of these are of interest chiefly because of their increased levels in certain pathologic conditions such as retinoblastoma, wherein tissue damage results in the release of intracellular enzymes. The frequent absence of coenzymes or substrates of such enzymes from the aqueous (as in the case of nicotinamide adenine dinucleotide for lactic dehydrogenase or of oxaloacetate for glutamic-oxaloacetic transaminase) leads to the conclusion that these enzymes have no significant catalytic role in the normal aqueous. Three enzymes in the aqueous appear to be exceptions to this nonreactive role:

- *Hyaluronidase* may be important in the normal regulation of the resistance to outflow through the trabecular meshwork, since injection of this enzyme into the anterior chamber has been shown to increase outflow facility in some species.
- *Carbonic anhydrase* is present in trace amounts. Because it is an enzyme with an extremely high turnover, even a low concentration may be significant in the catalysis of the equilibrium between bicarbonate and CO_2 plus water.
- *Lysozyme* is found in the aqueous of the rabbit at 1 μg/ml—about 25% of the plasma level—which is a high aqueous:plasma ratio for a protein, even though its molecular mass is only 15 kD. It provides significant antibacterial protection. Most probably, the origin of lysozyme is only in part from the blood; in cases of ocular inflammation, the intraocular level may be raised tenfold or more, reaching concentrations well above those in plasma.

Recent studies have identified the neurotrophic factor secreted in the aqueous humor by the ciliary nonpigmented cells. This factor, originally identified as a secretory protein in cultured retinal pigment epithelial cells, is called *pigment epithelium–derived factor.* This protein is a new member of the serine proteinase inhibitor (serpin) gene superfamily, and it has been determined that it behaves more like a substrate than as an inhibitor, exhibiting neuroprotective activities on neurons.

Finally, the neuroendocrine properties of the ciliary epithelium could also determine the composition of aqueous humor. It is known that neuropeptide-processing enzymes and bioactive peptides are synthesized and secreted by the ciliary epithelial cells into the aqueous. Among the neuropeptide-processing enzymes are prohormone convertases, carboxypeptidase E, and peptidylglycine α-amidating monooxygenase; among neuropeptides and peptide hormones are secretogranin II, galanin, neurotensin, and natriuretic peptides. The regulatory peptides and hormones are synthesized as large precursors, which are then subjected to endoproteolytic cleavage, generating bioactive peptides. These peptides are then released from the ciliary epithelial cells into the aqueous via a regulated secretory pathway in a calcium-dependent manner. Although the function of these bioactive peptides in the aqueous humor secretion is not clearly understood, they may be involved in the regulation of aqueous outflow and intraocular pressure.

Coca-Prados M, Escribano J, Ortego J. Differential gene expression in the human ciliary epithelium. *Prog Retinal Eye Res.* 1999;18:403–429.

Cowdrey G, Firth M, Moss R, et al. The analysis of aqueous humor constituents using capillary zone electrophoresis. *Exp Eye Res.* 1998;67:449–455.

Growth Modulatory Factors

The physical and chemical properties of the aqueous humor have a substantial role in modulating the proliferation, differentiation, functional viability, and wound healing of ocular tissues in health and disease. These properties are largely influenced by a number of growth-promoting and differentiation factors that have been identified or quantified in aqueous humor. They include:

- Transforming growth factor 1 and 2 (TGF-1 and -2)
- Acidic and basic fibroblast growth factor (aFGF and bFGF)
- Insulin-like growth factor 1 (IGF-1)
- Insulin-like growth factor binding proteins (IGFBPs)
- Vascular endothelial growth factor (VEGF)
- Transferrin

The growth factors in the aqueous humor perform diverse, synergistic, or sometimes opposite biological activities. Normally, the lack of significant mitosis of the corneal endothelium and trabecular meshwork in vivo is probably controlled by the complex coordination of effects and interactions among the different growth-modulatory substances present in the aqueous humor (see Part 5, Ocular Pharmacology). Disruption in the balance among various growth factors that occurs with the production of plasmoid aqueous humor may explain the abnormal hyperplastic response of the lens epithelium and corneal endothelium observed in chronic inflammatory conditions and traumatic insults to the eye. Ultimately, however, the effect of a given growth factor in the aqueous humor is determined primarily by its bioavailability. Bioavailability depends on many factors, including the expression of receptors on target tissues, interactive effects with components of the extracellular matrix, and the levels of circulating and matrix-bound proteases.

The role of several growth factors has been studied in diabetic patients. IGFBPs are elevated fivefold in human diabetic patients without retinopathy, and IGF-1 levels are elevated in patients with diabetic retinopathy. This elevation suggests that the increase in vitreal IGFBPs is not the result of a preexisting endstage retinopathy but is rather an early ocular event in the diabetic process.

VEGF levels are increased not only in patients with active ocular neovascularization from proliferative diabetic retinopathy, but also after occlusion of the central retinal vein and with iris neovascularization. Actually, VEGF mediates the retinal vasoproliferative response to hypoxic stimuli observed in a variety of ocular ischemic diseases. Its expression is increased by hypoxia in retinal endothelial cells, retinal pericytes, Müller cells, and retinal pigment epithelium cells.

Takagi H, King GL, Ferrara N, et al. Hypoxia regulates vascular endothelial growth factor receptor *KDR/Flk* gene expression through adenosine A2 receptors in retinal capillary endothelial cells. *Invest Ophthalmol Vis Sci.* 1996;37:1311–1321.

Tripathi RC, Borisuth NSC, Tripathi BJ, et al. Growth factors in the aqueous humor and their therapeutic implications in glaucoma and anterior segment disorders of the eye. *Drug Dev Res.* 1991;22:1–23.

Waldbillig RJ, Jones BE, Schoen TJ, et al. Vitreal insulin-like growth factor binding proteins (IGFBPs) are increased in human and animal diabetes. *Curr Eye Res.* 1994; 13:539–546.

Oxygen and Carbon Dioxide

Oxygen is present in the aqueous humor at a partial pressure of about 55 mm Hg, roughly one third of its concentration in the atmosphere. It is derived from the blood supply to the ciliary body and iris, for there is no net flux of oxygen from the atmosphere across the cornea. Indeed, the corneal endothelium is critically dependent on the aqueous oxygen supply for the active fluid-transport mechanism that maintains corneal transparency. The lens and the endothelial lining of the trabecular meshwork also derive their oxygen supply from the aqueous humor.

Carbon dioxide content of the aqueous humor is in the range of 40–60 mm Hg, contributing approximately 3% of the total bicarbonate. The relative proportions of CO_2 and HCO_3^- determine the pH of the aqueous, which in most species has been found to be in the range of 7.5–7.6. CO_2 is continuously lost from the aqueous by diffusion across the cornea into the tear film and atmosphere. The Na^+,K^+-Cl^- cotransporter is also important in the trabecular meshwork and with control of aqueous outflow.

Gong H, Tripathi RC, Tripathi BJ, et al. Morphology of the aqueous outflow pathway. *Microsc Res Tech.* 1996;33:336–367.

Spector A, Ma W, Wang RR. The aqueous humor is capable of generating and degrading H_2O_2. *Invest Ophthalmol Vis Sci.* 1998;39:1188–1197.

Clinical Implications of Breakdown of the Blood–Aqueous Barrier

With compromise of the blood–aqueous barrier in conditions such as ocular insult (trauma or intraocular surgery), uveitis, and other inflammatory disorders, the protein content of aqueous humor may increase 10–100 times, especially in the high-molecular-weight polypeptides. The levels of inflammatory mediators, immunoglobulins, fibrin, and proteases rise, and the balance among the various growth factors is disrupted (see Chapter XXI). The clinical sequelae include fibrinous exudate and clot (with or without a macrophage reaction and formation of cyclitic membranes) and synechiae formation (peripheral and posterior), as well as an abnormal neovascular response, which further exacerbates breakdown of the barrier. Chronic disruption of the blood–aqueous barrier is implicated in the abnormal hyperplastic response of the lens epithelium, corneal endothelium, trabecular meshwork, and iris and in the formation of complicated cataracts. Degenerative and proliferative changes may occur in various ocular structures as well. The use of anti-inflammatory steroidal and nonsteroidal drugs, cycloplegics, protease activators or inhibitors, growth and antigrowth factor agents, and even surgical intervention may be necessary to combat these events.

CHAPTER XIV

Lens

The lens is a transparent, avascular body that, in concert with the cornea, focuses incident light onto the sensory elements of the retina. To perform this function, the lens must be transparent and must have an index of refraction higher than the surrounding fluids. Maintenance of transparency depends on precise organization of the cellular structure of the lens and a high degree of short-range order of the protein matrix of the lens-fiber cytoplasm. Transparency must be maintained as the lens changes shape during accommodation. The high refractive index is due to the presence of a very high concentration of protein in the lens cells, especially of the soluble proteins called *crystallins.* Further, since there is little if any turnover of protein in the central region of the lens (where the oldest, denucleated cells are found), the proteins of the human lens must be extremely stable to remain functionally viable for a lifetime. Given the mode of growth of the lens and the stresses to which it is chronically exposed, it is remarkable that in most people, lenses retain good transparency. Nearly all humans develop opacities by the seventh or eighth decade of life.

This chapter discusses the structure and composition of the lens as well as aspects of membrane function, metabolism, and regulatory processes within the lens. Further information on the lens and on cataractogenesis is provided in BCSC Section 11, *Lens and Cataract.*

Structure of the Lens

Capsule

The lens is enclosed in an elastic basement membrane called the *lens capsule* (Fig XIV-1). The capsule is noncellular and is composed primarily of type IV collagen with smaller amounts of other collagens and extracellular matrix components (including glycosaminoglycans, laminin, fibronectin, and heparan sulfate proteoglycan). The capsule is a very thick basement membrane, particularly on the anterior side of the lens, where the epithelial cells continue to secrete capsular material throughout life. On the posterior side of the lens, where there is no epithelium, the posterior fiber cells have limited capacity to secrete such material. The zonular fibers, from which the lens is suspended, insert into the capsule near the equator on both the anterior and posterior sides. The capsule is not a barrier to diffusion of water, ions, other small molecules, or proteins up to the size of serum albumin (molecular weight, 68,000).

Epithelium

A single layer of epithelial cells covers the anterior surface of the lens. These cells have full metabolic capacity and play the primary role in regulating the water and ion balance of the entire lens. Although the cells of the central epithelium are not

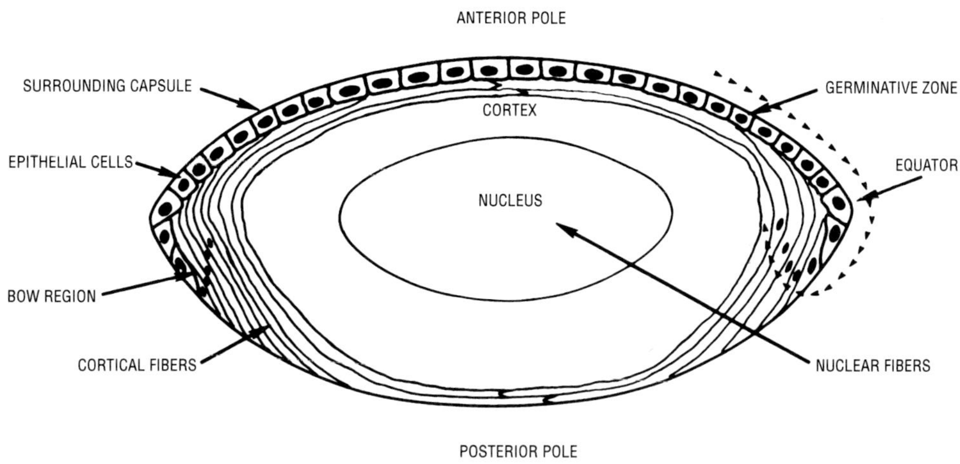

FIG XIV-1—Schematic representation of the mammalian lens in cross section. *Arrowheads* indicate direction of cell migration from the epithelium to the cortex. (Reproduced with permission from Anderson RE, ed. *Biochemistry of the Eye.* San Francisco: American Academy of Ophthalmology; 1983:112.)

mitotically active, a germinative zone exists as a ring anterior to the equator where the epithelial cells divide. The new cells migrate toward the equator and begin to differentiate into lens fibers.

Cortex and Nucleus

Aside from the single layer of epithelial cells on its anterior surface, the rest of the lens is composed of lens fibers, very long ribbon-like cells. All fibers are formed from epithelial cells at the lens equator; hence, younger fibers are always exterior to older ones. The lens structure can be equated with the growth rings of a tree in that the oldest cells are in the center with progressively younger layers, or shells, of fiber cells toward the periphery. Unlike the case with many tissues, no cells are sloughed from the lens and cells produced before birth remain at the center of the lens throughout life. A single lens thus provides the opportunity to study the process of aging at the molecular level by comparing cells, and cellular components, of ages spanning the entire life of the individual.

As new fiber cells are elongating and differentiating into mature fibers, their cell nuclei form the *bow zone* at the lens equator. Elongating fibers greatly increase their volume and their surface area and express large amounts of lens crystallins and of a lens fiber–specific membrane protein called the *major intrinsic protein (MIP).* As the fibers become fully elongated and make sutures at each end with fibers that have elongated from the opposite side of the lens, they become mature, terminally differentiated fiber cells. The cell nuclei disintegrate, as do mitochondria and other organelles. This process happens quite abruptly by mechanisms that remain obscure. What is understood is that elimination of cellular organelles is necessary in the central portion of the lens because such bodies are sufficiently large to scatter light and thereby degrade visual acuity. It should also be noted that with the loss of cell nuclei, the mature fibers lose the machinery required for synthesis of proteins. The fiber mass of the adult lens can be divided into the *cortex* (ie, the outer fibers laid down after the age of about 20 years) and the *internal nucleus* (composed of cells produced from embryogenesis through adolescence).

Chemical Composition of the Lens

Membranes

The chemical composition of the lens-fiber plasma membranes suggests that they are both very stable and very rigid. A high content of saturated fatty acids, high cholesterol:phospholipid ratio, and high concentration of sphingomyelin all contribute to the tight packing and low fluidity of the membrane. Although lipids make up only about 1% of the total lens mass, they constitute about 55% of the plasma membrane's dry weight, with cholesterol being the major neutral lipid. As the lens ages, the protein:lipid and cholesterol:phospholipid ratios increase as a result of phospholipid loss, especially in the nucleus.

Lens Proteins

The lens probably has the highest protein content of any tissue. In some species, more than 50% of lens wet weight is protein. Lens crystallins, a diverse group of proteins expressed in high abundance in the lens fiber cells and thought to play critical roles in providing the transparency and refractile properties essential to lens function, constitute 90%–95% of total lens protein. In addition to the crystallins, the lens also has a full complement of enzymes and regulatory proteins that are present primarily in the epithelium and immature fiber cells, where most metabolic activity occurs.

Crystallins Crystallins are water-soluble proteins named because of their high abundance in the "crystalline lens." Until the 1980s, crystallins were considered lens-specific proteins, lacking biological activity, which were highly evolved structural elements forming the transparent protein matrix of the lens. It is now clear that most (if not all) crystallins are expressed in other tissues and have specific biological functions distinct from their roles in the lens as refractile elements (Table XIV-1). All crystallins appear to be "borrowed" proteins, recruited by the lens for a function completely distinct from their biological functions in other tissues. Although the specific criteria that make a protein suitable to function as a crystallin are not well understood, crystallins must have two obvious attributes:

- Since the proteins of the lens are probably the longest-lived proteins in the body, crystallins must be very stable structures.
- Crystallins must remain soluble under conditions of high protein concentration without forming large aggregates, which would be light-scattering centers within the lens.

Crystallins can be divided into two groups. One group includes α-crystallin and the β,γ-crystallin family, which appear to be present in all vertebrate lenses. The second group is the taxon-specific crystallins, which are each present at crystallin levels only in phylogenetically restricted groups of species.

α-Crystallin. α-Crystallin is the largest of the crystallins, having a native molecular mass in the 600 to 800 kD range. It is composed of two subunits, αA and αB,

TABLE XIV-1

CRYSTALLINS OF VERTEBRATES

	RELATED PROTEINS	COFACTOR	DISTRIBUTION
Ubiquitous			
α	Small heat-shock protein	None	All vertebrates
β,γ-family	Microbial stress protein	None	All vertebrates
Taxon-specific			
δ	Argininosuccinate lyase*	None	Most birds, reptiles
ε	Lactate dehydrogenase B*	NADH	Birds, crocodiles
ζ	NADPH: quinone oxidoreductase*	NADPH	Hystrichomorph rodents, camellids
η	Aldehyde dehydrogenase*	NAD^+	Elephant shrews
λ	Hydroxyacyl-CoA dehydrogenase	NADH	Rabbits, hares
μ	Glutamyl-tRNA reductase	NADPH	Some marsupials
π	Glyceraldehyde-PO_4 dehydrogenase*	NAD^+	Geckos *(Phelsuma)*
ρ	Aldo-keto reductases	NADPH	Frogs *(Rana)*
ι	Retinol binding protein	Vitamin A_2	Geckos *(Lygodactylus)*
τ	α-Enolase*	None	Various species

*Related proteins are identical to the crystallin form. In the other cases, the crystallin is related to the listed protein or protein group.

which are approximately 20 kD in mass and which are nearly 55% identical in sequence. Native α-crystallin is polydisperse, apparently because it is a dynamic structure wherein the number of subunits varies somewhat and subunit exchange occurs between the native multimers. Probably because of its polydispersity, α-crystallin has resisted all attempts at crystallization, so a definitive three-dimensional structure for the molecule has not been determined.

α-Crystallin is a member of the small heat-shock protein family. αB, which is expressed in many nonlenticular tissues, is inducible by heat and other stresses in some cell types. αA, which is apparently not heat-inducible, is also expressed outside the lens but in only a few tissues. Both αA and αB have a chaperone-like activity whereby they bind proteins that are beginning to denature and prevent further denaturation and aggregation. Since protein aggregates in the lens will scatter light and cause loss of transparency, the antiaggregative function of α-crystallin is probably crucial to the long-term maintenance of transparency in the fibers of the lens nucleus, where synthesis of new protein is impossible and where protein molecules must exist for decades.

β-, γ-Crystallins. Until the primary sequences of the various β- and γ-crystallins were determined, they were thought to be two unrelated families of proteins. However, it is now known that the two groups are related members of the same protein superfamily. The β-crystallins are a complex group of oligomers composed of polypeptides ranging from 23–32 kD. The native oligomers range from dimers to units composed of up to eight subunits in various combinations. The bovine lens contains six distinct β-crystallin genes, which produce a total of seven polypeptides. In many species, the β-crystallins constitute the predominant crystallin group in the lens.

The γ-crystallins are monomeric proteins with molecular masses near 20 kD. There are six highly homologous and clustered γ-crystallin genes coding for polypeptides called *γA–F*. In rats, all six genes are active; in the human lens, two of them are pseudogenes and only two of the others (*γC* and *γD*) are expressed at significant levels. Most expression of γ-crystallins occurs early in development, and thus they tend to be most concentrated in the nuclear region of the lens. Bovine γB was the first crystallin to be crystallized; the structure was determined by x-ray diffraction. It is a compact and symmetrical structure, which can pack very densely; γ-crystallin tends to be highly concentrated in hard lenses, which have little to no accommodative ability. A final member of this group is γS-crystallin, which is also monomeric but less closely related to the other γ-crystallins. Unlike the other γ-crystallins, γS is expressed primarily later in life and thus is found primarily in the adult lens cortex.

Although no specific biological functions have been identified for the β- and γ-crystallins, at least some of them are expressed outside the lens, suggesting that such functions do exist. Members of the superfamily have also been identified in microorganisms, where they are expressed during spore or cyst formation (suggesting a possible role in stress response).

Taxon-Specific Crystallins. In addition to the α- and β,γ-crystallins found in all vertebrate lenses, other proteins are expressed in large quantities in various phylogenetic groups (see Table XIV-1). The first such taxon-specific crystallin was δ-crystallin, which occurs in many birds and some reptiles. Like most γ-crystallins, it is largely an embryonic protein and constitutes up to 80% of the total protein in chick lens. During the 1980s, several additional taxon-specific crystallins were reported, but it wasn't until 1987 that it was discovered that these proteins were actually enzymes, being expressed in the lens at greatly increased levels.

ε-Crystallin, which constitutes over 20% of total lens protein in some birds, was found to actually be lactate dehydrogenase and to be catalytically active. In rapid succession, it was demonstrated that all the known taxon-specific crystallins were identical or related to known enzymes or enzyme families. δ-Crystallin was shown to be the urea cycle enzyme argininosuccinate lyase and to exist in two forms resulting from gene duplication; however, with all other crystallins that have been investigated, only a single gene seems to express the protein at catalytic quantities in various tissues and at the level of a crystallin in the lens. This appears to be accomplished through selective up-regulation of transcription in the lens via development of lens-specific alternative promoters or enhancers in the gene. This situation—in which a single gene product apparently serves a completely different function in the lens from its function elsewhere—has been called *gene sharing*. An interesting question is why the lens recruited these particular proteins as crystallins. Two possibilities have been put forward:

- To replace γ-crystallin and make lenses softer
- To serve a protective role

Most taxon-specific crystallins are oxidoreductases, which bind pyridine nucleotides, and their presence in the lens tremendously increases the concentration of the bound nucleotide. Reduced nucleotides absorb UV light and could serve to protect the retina from UV-induced oxidation; on the other hand, these strongly reducing compounds could be part of the antioxidant defenses of the lens. It is noteworthy that taxon-specific crystallins generally occur only in strongly diurnal species that are chronically exposed to increased oxidative stress.

Cytoskeletal and membrane proteins Although the great majority of proteins in the normal lens are water-soluble, a number of important structural proteins can be solubilized only in the presence of chaotropic agents or detergents. These include the cytoskeletal elements *actin* (actin filaments), *vimentin* (intermediated filaments), and *tubulin* (microtubules), as well as two additional proteins called *filensin* and *phakinin,* which have been found only in lens-fiber cells and which compose a cytoskeletal structure unique to the lens called the *beaded filament.* The filamentous structures of the cytoskeleton provide structural support to the cells and also play critical roles in such processes as differentiation, motility and shape change, and organization of the cytoplasm.

Lens-fiber membranes have one quantitatively dominant protein, which has received a great deal of attention, called MIP (as discussed earlier). MIP is only expressed in lens fiber cells and was once thought to be a gap junction protein; in fact, it is not a connexin but rather an aquaporin. Current data suggest that MIP functions as a water channel.

Post-translational modifications to lens proteins As described above, lens-fiber cells are never sloughed and lose the ability to synthesize new protein as they terminally differentiate. The proteins of the lens thus are probably the longest-lived in the body, with the oldest ones (in the center of the lens nucleus) having been synthesized before birth. As would be expected, these proteins become structurally modified in a variety of ways: oxidation of sulfur and aromatic residue side chains, inter- and intra-polypeptide cross-links, glycation, racemization, phosphorylation, deamidation, and carbamylation. Until recently, these modifications were thought to be accumulated as a function of age and to reflect a slow degeneration of the structure of the proteins. It has been found that many of these modifications occur quite early in life and are probably part of a programmed modification of the crystallins that is required for their long-term stability and functionality. There is evidence that certain of these processes (phosphorylation, thiol oxidation) are reversible and may serve a regulatory function, although this remains to be proven.

What is known is that with increasing age (particularly in some cataracts), certain oxidative modifications do accumulate, contributing to the cross-linking of crystallin polypeptides, alterations in fluorescent properties, and increase in protein-associated pigmentation. In particular, the formation of disulfide cross-links in the proteins of the lens nuclear region is associated with formation of protein aggregates, light scattering, and cataract.

Physiologic Aspects

Because of its avascularity and its mode of growth, the lens faces some unusual problems. All nutrients must be obtained from the surrounding fluids; likewise, all waste products must be released into those fluids. Most of the cells of the adult lens have reduced metabolic activity and lack the membrane machinery to independently regulate ionic homeostasis. Elucidating how the lens maintains ionic balance and how solutes move from cell to cell throughout the lens is critical to understanding the normal biology of the organ as well as the process of cataractogenesis.

In the normal lens, sodium is low (~10 mM) and potassium is high (~120 mM); in the aqueous humor, sodium is about 150 mM and potassium about 5 mM. When normal regulatory mechanisms are abrogated, potassium leaks out of the lens and sodium floods in, followed by chloride. Water then enters in response to the osmotic gradient, causing loss of transparency by disrupting the normally smooth gradient of refractive index. The ionic balance in the lens is maintained primarily by the Na^+,K^+ pump, an intrinsic membrane protein that hydrolyzes adenosine triphosphate (ATP) to transport sodium out of and potassium into the lens. Functional Na^+,K^+ pumps are found primarily at the anterior surface of the lens, in the epithelium and outer, immature fibers. Studies using ouabain, a specific inhibitor of the pump, have established the pump's role as the primary determinant of the normal ionic state of the lens. Lens cells also contain membrane channels that pass ions; in particular, K^+-selective channels have been studied by patch-clamp techniques and found to be present primarily in the epithelial cells.

Communication between lens cells is provided by gap junctions, which are thought to account for most ion and small-molecule movement between cells. True gap junctions occur in the lens and are composed of members of the connexin family. Junctions between epithelial cells are composed of connexin 43; fiber–fiber gap junctions contain connexins 46 and 50. MIP also forms junction-like structures, although they apparently do not provide direct communication from cell to cell. As noted above, MIP is an aquaporin, not a connexin, and is now thought to function in water transport. Within the fiber mass, gap-junctional coupling is greatest in the outer layers of the lens, where the junctions can be uncoupled by lowering the pH. In the lens nuclear region, such uncoupling does not occur. This may support the recently proposed idea that the lens nucleus may be a syncytium, perhaps as a result of membrane fusion between adjacent fibers.

Some years ago, the vibrating probe technique was used to demonstrate that currents of ions flow around and through the normal lens. As depicted in Figure XIV-2, the currents flow in at the anterior and posterior poles and out at the equatorial region of the lens. The source of these currents is not completely understood, but it is apparent that the various ion pumps and channels near the lens surface and the communicating pathways between cells play important roles in establishing and maintaining the currents. It has been suggested that fluid flow would follow the flow of ions, and that this would in effect provide the lens with a circulatory system that could do the work done by blood vessels in other tissues. That is, nutrients could be carried deep into the lens and waste products moved toward the surface. This model implies that the homeostasis of the cells throughout the lens depends on these ionic currents; if that is true, the maintenance and regulation of the currents would be crucial to the functional integrity of the lens.

Lens Metabolism and Formation of Sugar Cataracts

Energy Production

Energy, in the form of ATP, is produced in the lens primarily through anaerobic glycolysis. This is necessitated by the fact that the oxygen tension in the lens is much lower than in other tissues, since oxygen reaches the avascular lens only via diffusion from the aqueous humor. Furthermore, since mitochondria degenerate as lens fibers mature, most of the lens lacks the machinery of oxidative phosphorylation. It has long been known that lenses placed in organ culture function quite well in the total absence of oxygen as long as glucose is provided. Removing glucose from the

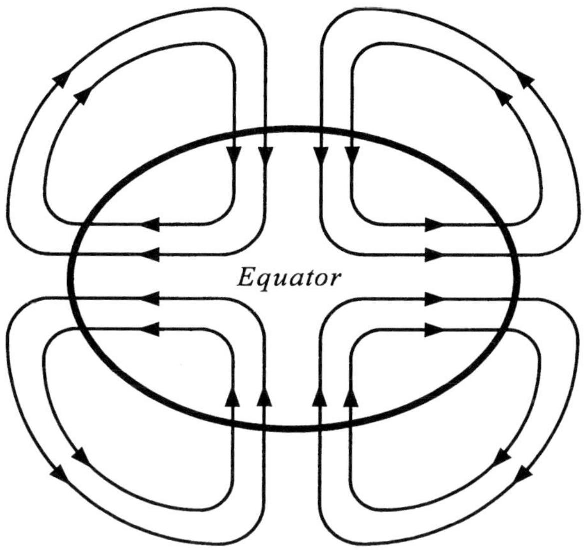

FIG XIV-2—Schematic representation of ionic current flow around and through the lens. The flow is inward at the anterior and posterior poles and outward at the equatorial area. (From Mathias RT, Rae JL, Baldo GJ. Physiological properties of the normal lens. *Physiol Rev.* 1997; 77:21–50.)

medium, however, causes rapid impairment of lens metabolic activity and loss of transparency in the presence or absence of oxygen.

Most of the glucose entering the lens is phosphorylated to glucose-6-phosphate by hexokinase, the rate-limiting enzyme of the glycolytic pathway. Under normal conditions, most glucose-6-phosphate passes through glycolysis, wherein two molecules of ATP are formed per original molecule of glucose. A small proportion of glucose-6-phosphate is metabolized through the pentose phosphate pathway (hexose monophosphate shunt). This pathway is activated under conditions of oxidative stress since it is responsible for replenishing the supply of NADPH that becomes oxidized through the increased activity of glutathione reductase under such conditions.

Sugar Cataracts

Much of the research activity on lens carbohydrate metabolism has been stimulated by interest in "sugar cataracts" associated with diabetes and galactosemia. True diabetic cataract is a rapidly developing bilateral "snowflake" cataract occurring in the lens cortex of type I diabetics with poorly controlled disease. Persons with adult-onset diabetes do not develop this type of cataract, but do have a higher prevalence of age-related cataract with a slightly earlier onset. It is likely that for such persons, diabetes is simply one additional factor contributing to the development of age-related cataracts.

Defects in galactose metabolism also cause sugar cataracts. Classic galactosemia is caused by deficiency of galactose-1-phosphate uridyltransferase. Infants with this inborn error of metabolism develop bilateral cataracts within a few weeks of birth unless milk (lactose) is removed from the diet. Cataracts are also associated with deficiency of galactokinase. Under certain conditions wherein sugar levels are elevated significantly, some glucose (or galactose) is metabolized through the polyol pathway (Fig XIV-3). Aldose reductase is the key enzyme for the pathway and converts the sugars into the corresponding sugar alcohols. Since aldose reductase has a very high K_m for glucose (or galactose), under normal conditions little or no activity occurs through this pathway; however, under conditions of hyperglycemia, aldose reductase competes with hexokinase for glucose (or galactose).

Studies using animal models have established the importance of the polyol pathway in experimental sugar cataracts. Animals that are diabetic (either natural or induced diabetes) develop cataracts that are associated with the presence of sorbitol in the lens and with influx of water. The osmotic hypothesis was proposed to account for these facts. It invokes the activity of aldose reductase as central to the pathology by markedly increasing the sorbitol content of the lens. Sorbitol is largely unable to penetrate cell membranes and is thus trapped inside the cells. Since its further conversion to fructose by polyol dehydrogenase is slow, sorbitol builds up in lens cells under conditions of hyperglycemia such that it creates an osmotic pressure that draws water into the lens, swelling the cells, damaging membranes, and causing cataract.

Confirmation of the theory has come from a variety of sources. First, ease of induction of cataract by hyperglycemia varies from species to species, depending on the level of aldose reductase activity in the lens. Rats, the most commonly used animal, have high activity and develop cataracts readily; mice with almost no activity in the lens do not form sugar cataracts. Second, rats fed diets high in galactose develop cataract more rapidly and severely than do diabetic rats. This correlates with the fact that dulcitol levels in their lenses reach higher levels than does sorbitol in diabetic rats because:

- Galactose is a better substrate than glucose for aldose reductase
- Dulcitol is not further metabolized since it is not a substrate for polyol dehydrogenase

Most convincingly, a number of potent inhibitors of aldose reductase have been developed that can completely prevent cataract in the diabetic or galactose-fed animals.

Unfortunately, although it is certain that aldose reductase activity is the critical factor in the cascade leading to sugar cataract in these animal models, the situation is not at all clear with respect to human diabetic cataracts. The levels of aldose reductase are much lower in human lens than in the lenses of animals used in sugar cataract models but the levels of polyol dehydrogenase are much higher. There is controversy as to whether sorbitol can accumulate in the lens of human diabetics to levels capable of causing significant osmotic influx of water. Oxidative stress and/or glycation of proteins in the lens may also be involved in human cataracts associated with diabetes.

Garland DL, Duglas-Tabor Y, Jimenez-Asensio J, et al. The nucleus of the human lens: Demonstration of a highly characteristic pattern by two-dimensional electrophoresis and introduction of a new method of lens dissection. *Exp Eye Res.* 1996;62:285–291.

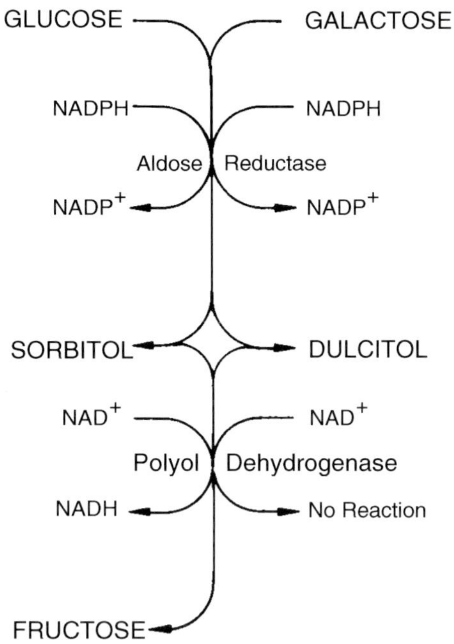

FIG XIV-3—Sorbitol pathway for glucose and galactose metabolism. The reduction of glucose and galactose to sorbitol and dulcitol, respectively, is catalyzed by aldose reductase using the reduced form of nicotinamide-adenine dinucleotide phosphate *(NADPH)* as cofactor. The second step is the oxidation of sorbitol (dulcitol is not a substrate) to fructose, catalyzed by polyol dehydrogenase using NAD^+ as cofactor. This step is reversible in the human lens, as indicated by the *arrows.*

Jaffe NS, Horwitz J. In: Podos SM, Yanoff M, eds. *Textbook of Ophthalmology, Volume 3: Lens and Cataract.* New York: Gower; 1992.

Mathias RT, Rae JL, Baldo GJ. Physiological properties of the normal lens. *Physiol Rev.* 1997;77:21–50.

Piatigorsky J. Gene sharing in lens and cornea: Facts and implications. *Prog Retin Eye Res.* 1998;17(2):145–174.

Quinlan RA, Sandilands A, Proctor JE, et al. The eye lens cytoskeleton. *Eye.* 1999;13:409–416.

Slingsby C, Clout NJ. Structure of the crystallins. *Eye.* 1999;13:395–402.

Wistow G. *Molecular Biology and Evolution of Crystallins: Gene Recruitment and Multifunctional Proteins in the Eye Lens.* Georgetown, TX: RG Landes; 1995.

Zigler JS Jr. Lens proteins. In: Albert DM, Jakobiec FA, eds. *Principles and Practice of Ophthalmology.* Philadelphia: Saunders; 1994.

CHAPTER XV

Vitreous

The vitreous body is a specialized connective tissue that has two basic functions:

- To serve as a transparent gel occupying the major volume of the globe
- To serve as a route of nutrition to the lens

The basic physical structure of the vitreous is a gel composed of a collagen framework interspersed with hydrated hyaluronan molecules. The hyaluronan contributes to the viscosity of the vitreous humor and is thought to help stabilize the collagen network, although most of the hyaluronan can be removed enzymatically without collapse of the gel.

Composition

The vitreous contains about 98% water and 0.15% macromolecules, including collagen, hyaluronan, and soluble proteins. The remainder of the solid matter consists of ions and low-molecular-weight solutes. The two major structural components are collagen and hyaluronan. However, several noncollagenous structural proteins and glycoproteins have been identified in the vitreous, including versican, link protein, fibulin-1, nidogen-1, fibronectin, and two novel proteins called opticin and VIT1.

The latter proteins were initially identified after extraction of a pellet of collagen fibrils obtained from the bovine vitreous after centrifugation. In addition, the human vitreous contains hyaluronidase and at least one matrix metalloproteinase (called *MMP-2* or *gelatinase*), suggesting that turnover of vitreous structural macromolecules can occur.

Collagen

At present, 19 types of collagen are known (Table XV-1), and the genes for several more have been identified. Vitreous collagen fibrils are composed of three different collagen types:

- *Type II,* which forms the major component of the fibrils
- *Type IX,* which is located on the surface of the fibril
- *Type V/XI,* which may be located such that its amino terminus projects from the surface of the fibril (Fig XV-1)

Type V/XI collagen is unique to the vitreous in that native triple-helical molecules can be isolated that contain the α_1 (XI) and α_2 (V) chains (see Table XV-1). The vitreous collagens are closely related to the collagens of hyaline cartilage and differ from the type I, type III, type XII, and XIV collagens commonly found in tissues such as dermis, scar tissue, cornea, and sclera.

The precise mechanism by which the diameter of the vitreous collagen fibrils is controlled remains poorly understood, although type V/XI collagen is thought to play

TABLE XV-1

COLLAGEN TYPES AND TISSUE DISTRIBUTION

TYPE	CHAINS	MOLECULAR ORGANIZATION	OCCURRENCE
Fibril-forming (interstitial)			
I	α_1(I) α_2(I)	$[\alpha_1(I)]_2\alpha_2(I)$ $[\alpha_1(I)]_3$ (rare)	Most connective tissues
II	α_1(II)	$[\alpha_1(II)]_3$	Hyaline cartilage, vitreous
III	α_1(III)	$[\alpha_1(III)]_3$	Blood vessels, kidney, lung, liver, cornea, sclera
V/XI	α_1(V) α_2(V) α_3(V) α_1(XI) α_2(XI) α_3(XI)	$[\alpha_1(V)]_2\alpha_2(V)$ $\alpha_1(V)\alpha_2(V)\alpha_3(V)$ $\alpha_1(XI)\alpha_2(XI)\alpha_3(XI)$ $[\alpha_1(XI)]_2\alpha_2(V)$	Most tissues Placenta Cartilage Vitreous
Fibril-associated			
IX	α_1(IX) α_2(IX) α_3(IX)	$\alpha_1(IX)\alpha_2(IX)\alpha_3(IX)$	Cartilage, vitreous
XII	α_1(XII)	$[\alpha_1(XII)]_3$	Widespread
XIV	α_1(XIV)	$[\alpha_1(XIV)]_3$	Widespread
XVI	α_1(XVI)		Fibroblasts, keratinocytes
XIX	α_1(XIX)		
Basement membrane			
IV	α_1(IV) α_2(IV) α_3(IV) α_4(IV) α_5(IV) α_6(IV)	$[\alpha_1(IV)]_2\alpha_2(IV)$ $\alpha_3(IV)\alpha_4(IV)\alpha_5(IV)$	All basement membranes Glomerular basement membrane
Short chain			
VIII	α_1(VIII) α_2(VIII)	$[\alpha_1(VIII)]_3$ $[\alpha_2(VIII)]_3$	Descemet's membrane, blood vessel endothelium
X	α_1(X)	$[\alpha_1(X)]_3$	Hypertrophic cartilage
Microfibril			
VI	α_1(VI) α_2(VI) α_3(VI)	$\alpha_1(VI)\alpha_2(VI)\alpha_3(VI)$ (other combinations?)	Most tissues
Epithelium-associated			
VII	α_1(VII)	$[\alpha_1(VII)]_3$	Anchoring fibrils
XVII	α_1(XVII)	$[\alpha_1(XVII)]_3$	Hemidesmosome-associated
Multiplexin			
XV	α_1(XV)		Widespread
XVIII	α_1(XVIII)		Liver, kidney, placenta
Transmembrane			
XIII	α_1(XIII)		Widespread

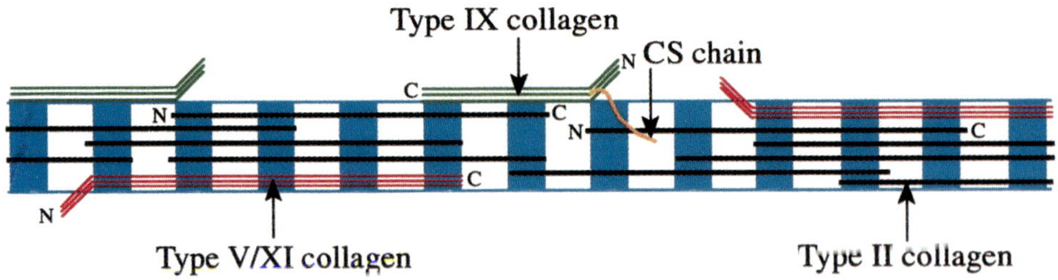

FIG XV-1—Model for the structure of a collagen fibril from the vitreous. Note that three different collagen types (types II, IX, and XI) are assembled to form the fibril. Type II collagen forms the major structure of the vitreous with stagger to form overlap and gap regions. Type IX is located on the surface of the fibril such that part of the molecule projects from the surface. It is in an antiparallel direction compared with the type II collagen molecules. Type IX also has a single chondroitin sulfate chain *(CS)* that may project from the surface of the fibril. Type V/XI collagen is also located close to the surface of the fibril with a part of the molecule projecting from the gap region. However, the location of this molecule is controversial; other models suggest it is located in the center of the fibril, where it may form a microfibril. *N,* amino terminus; *C,* carboxyl terminus. (Adapted from Olsen BR. New insights into the function of collagens from genetic analysis. *Curr Opin Cell Biol.* 1995;7:720–727.)

a critical role. In most connective tissues, the collagen fibrils (as observed in the electron microscope) show an axial periodicity of 67 nm, but the vitreous collagen fibrils rarely show this periodicity with direct staining, although the periodicity can be observed with negative staining. The collagen fibrils of the vitreous are only loosely attached to the inner limiting membrane (ILM) of the retina; however, at the vitreous base, the fibrils are firmly anchored to the peripheral retina and pars plana together with the margins of the optic disc.

The origin of vitreous collagen in mammals is not well established. It has been shown in the chicken that during early development, the cells of the neural retina synthesize and probably secrete type II collagen. However, other cells (hyalocytes?) present within the vitreous cavity may also contribute. In the developing chicken eye, in situ hybridization of the retina initially yields positive results for type II collagen, whereas transcripts for type IX collagen are present only at the region of the ciliary body. Later in development, type II collagen mRNA also becomes localized only to the presumptive ciliary region. The origin of type V/XI collagen within the vitreous is unknown.

Hyaluronan

Hyaluronan is a polysaccharide (glycosaminoglycan) that has a repeating unit of glucuronic acid and N-acetyl-glucosamine linked with a $\beta(1 \rightarrow 3)$ glycosidic bond (Fig XV-2). The repeating units are further linked with $\beta(1 \rightarrow 4)$ glycosidic bonds to form a long, linear (unbranched) molecular chain. At physiologic pH, hyaluronan is a weak polyanion because of the ionization of the carboxyl groups present in each glucuronic acid residue. In free solution, hyaluronan occupies an extremely large volume relative to its weight and probably occupies all of the space in the vitreous except for the collagen fibrils. Hyaluronan molecules of the vitreous may

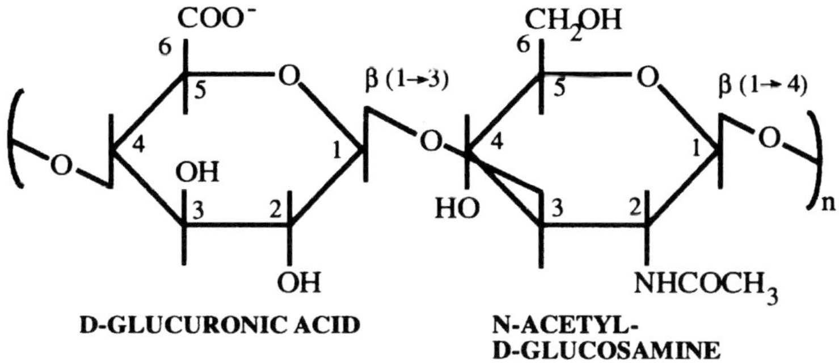

FIG XV-2—Hyaluronan consists of a repeating disaccharide (glucuronic acid and N-acetylglucosamine). It is a polyanion at physiologic pH with a net negative charge resulting from ionization of the carboxyl groups of the glucuronic acid residues.

undergo lateral interactions with each other, and such interactions may be stabilized by noncollagenous proteins. Link protein, which binds to hyaluronan in cartilage, is known to be present in the vitreous in small amounts. Hyaluronan is present in nearly all vertebrate connective tissues and is nontoxic, noninflammatory, and nonimmunogenic.

Both the concentration and the molecular weight of hyaluronan in the vitreous vary, depending on species, location in the vitreous body, and type of analysis. Hyaluronan in human vitreous can achieve a molecular weight of greater than 1×10^6, whereas in bovine vitreous the average molecular weight is 1.7×10^5.

The source of hyaluronan is also poorly understood. Hyaluronan may arise from the neural retina, although another possible source is the hyalocytes (macrophage-like cells) of the vitreous. Three forms of hyaluronan synthetase are known, but which isoform is responsible for synthesizing the hyaluronan of the vitreous has not been determined. In all animal species that have been analyzed, the hyaluronan concentration is highest in the posterior cortical layer near the retina and lowest in the anterior portion behind the lens. The physical structure of the vitreous is governed by the presence of collagen fibrils and hyaluronan, with the relative amount of collagen apparently determining the existence of a vitreous liquid or a gel. When only a low concentration of collagen fibrils is present, as in the owl monkey (25 μg/ml), a liquid vitreous predominates; in humans, in whom the concentration is much greater (286 μg/ml), the vitreous is a gel. The rigidity of the gel is greatest in regions of highest collagen concentration. The collagen fibrils are responsible for forming the gel, with hyaluronan interspersed between the fibrils. The collagen fibrils supply a resistance to tensile forces and give plasticity to the vitreous; the hyaluronan resists compression and confers viscoelastic properties. Regeneration of the vitreous after a vitrectomy occurs very slowly if at all.

Laurent TC, ed. *The Chemistry, Biology and Medical Applications of Hyaluronan and Its Derivatives.* Miami: Portland; 1998.

Soluble and Fibril-Associated Proteins

Many proteins remain in solution after the collagen fibrils and other insoluble elements present in the vitreous gel are removed by filtration or centrifugation. Serum albumin is the major soluble vitreous protein, followed by transferrin. Poorly defined glycoproteins make up 20% of the total vitreous protein and are thought to originate from surrounding tissues and not from blood. Other proteins include neutrophil elastase inhibitor (which may play a role in resisting neovascularization) and tissue plasminogen activator (which may have a fibrinolytic role in the event of vitreous hemorrhage). Serum albumin is thought to originate from the plasma, whereas transferrin at least partially originates from the region of the ciliary body. The concentration of soluble proteins estimated from a number of species is approximately 1.0 mg/ml. However, the concentration of serum proteins in the vitreous gel depends on the integrity of the retinal vasculature and the degree of any intraocular inflammation that may be present. Consequently, the concentration of soluble proteins within the vitreous cavity can rise dramatically if the blood–retinal barrier is compromised.

Some structural proteins are specifically associated with the collagen fibrils and are isolated by extraction of collagen fibrils after centrifugation of the vitreous. These include a novel leucine-rich-repeat glycoprotein called *opticin* and another novel glycoprotein called *VIT1*. The latter contains two von Willebrand A domains and is closely related to a protein of the cochlea called *COCH* or *cochlin*. Both opticin and VIT1 are thought to play key roles in the structure of the collagen fibril, and VIT1 may also interact with hyaluronan.

Zonular Fibers, Lipids, and Low-Molecular-Weight Solutes

Some zonular fibers are present in the anterior vitreous and can be observed by electron microscopy. However, most of these fibers form the zonular apparatus, which is the structural connection between the lens and the ciliary body. The major structural protein of these fibers is a large linear protein called *fibrillin,* which possesses an unusually high cysteine content.

Lipids account for about 7% w/w of the pellet obtained after centrifugation of the vitreous. The major fatty acids in human vitreous include palmitate (25%), stearate (18%), oleate (23%), and arachidonate (17%). Little variation occurs with age.

Ions and organic solutes originate from adjacent ocular tissues and blood plasma. The barriers that control entry into the vitreous include the vascular endothelium of retinal vessels, the retinal pigment epithelium, and the inner layer of the ciliary epithelium. The concentrations of Na^+ and Cl^- are similar to those in plasma, but K^+ is somewhat higher than in plasma.

Bishop PN. Structural macromolecules and supramolecular organization of the vitreous gel. *Prog Retin Eye Res.* 2000;19:323–344.

Mayne R, Brewton RG, Ren Z-X. Vitreous body and zonular apparatus. In: Harding JJ, ed. *Biochemistry of the Eye.* London: Chapman & Hall; 1997:135–143.

Sebag J. *The Vitreous: Structure, Function and Pathobiology.* New York: Springer Verlag; 1989.

Biochemical Changes With Aging and Disease

Vitreous Liquefaction and Posterior Vitreous Detachment

The human vitreous gel undergoes progressive liquefaction with age, so that typically by the age of 80–90 years, more than half of the vitreous is liquid. Myopia is associated with more rapidly progressive vitreous liquefaction leading to early posterior vitreous detachment (PVD). Liquefaction is associated with the aggregation of the thin (12–15 nm) vitreous collagen fibrils into thick fibers visible by low-powered slit-lamp microscopy and called *fibrillar opacities.* As liquefaction proceeds, the collagen fibrils become condensed into the residual gel phase and are absent from (or in low concentration in) the liquid phase. In terms of hyaluronan concentration or molecular weight, there are no differences between the gel and liquid phases. With increasing age, there is a weakening of adhesion at the vitreoretinal interface between the cortical vitreous gel and the inner limiting lamina. These combined processes eventually result in rhegmatogenous PVD in approximately 50% of the population.

PVD is a separation of the cortical vitreous gel from the ILM as far anteriorly as the posterior border of the vitreous base; the separation does not extend into the vitreous base owing to the unbreakable adhesion between the vitreous and retina in that zone. PVD is often a sudden event during which liquefied vitreous from the center of the vitreous body passes through a hole in the posterior vitreous cortex and then dissects the residual cortical gel away from the inner limiting lamina. As the residual vitreous gel collapses anteriorly within the vitreous cavity, retinal tears sometimes occur, which subsequently can result in rhegmatogenous retinal detachment. A PVD can protect against proliferative diabetic retinopathy by denying a scaffold for fibrovascular proliferation emanating from the disc and the retina.

A PVD can be achieved surgically during macular hole surgery. However, it is now clinically recognized that in many eyes thought to have a PVD, collagen fibrils are still extensively attached to the ILM. Even after production of an acute PVD during vitreous surgery, some collagen fibrils typically remain adherent to the ILM. Removal of the ILM itself is now more frequently the goal in limiting the extent of traction maculopathy.

Myopia

When the axial length of the globe is greater than 26 mm, both collagen and hyaluronan concentrations are approximately 20%–30% lower than in emmetropic eyes. This finding suggests a correlation between the macromolecular composition and the physical state of the vitreous.

Vitreous as an Inhibitor of Angiogenesis

Numerous studies have shown that the normal vitreous is an inhibitor of angiogenesis. This inhibitory activity is decreased during diabetic vitreoretinopathy. However, the molecular basis of the phenomenon remains poorly understood. Known inhibitors of angiogenesis, such as thrombospondin I and pigment epithelium–derived factor, are present within the mammalian vitreous and may inhibit angiogenesis in normal eyes. In contrast, vascular endothelial growth factor, a promoter of angiogenesis, is markedly elevated in the vitreous of patients suffering from proliferative diabetic vitreoretinopathy.

Injury with Hemorrhage and Inflammation

If blood penetrates the vitreous cortex, platelets come in contact with vitreous collagen, aggregate, and initiate clot formation. The clot in turn stimulates a phagocytic inflammatory reaction, and the vitreous becomes liquefied in the area of a hemorrhage. In severe cases, hemoglobin-laden macrophages may cause secondary glaucoma by blocking the trabecular outflow channels. Rigid, degenerated blood cells, called *ghost cells,* can also cause secondary glaucoma (see BCSC Section 10, *Glaucoma*).

If the vitreous is largely liquefied (as in myopic, aphakic, or senile eyes) the clot that is formed is loosely aggregated and early resolution is more likely. Hemorrhage into vitreous gel is less freely dispersed, and a more compact clot is formed. The subsequent inflammatory reaction varies for unknown reasons.

Streeten BAW, Wilson DJ. Disorders of the vitreous. In: Garner A, Klintworth GK, eds. *Pathobiology of Ocular Disease.* 2nd ed, Part A. New York: Dekker; 1994:701–742.

Involvement of Vitreous in Macular Hole Formation

With the development of optical coherence tomography, it is possible to directly visualize the vitreous cortex and the retina in the region of the macula. The results suggest that macular holes sometimes originate from traction generated by attachment of the vitreous specifically to the fovea, with the subsequent generation of additional tangential tractional force along the ILM, causing hole enlargement.

Madreperla SA, McCuen BW. *Macular Hole: Pathogenesis, Diagnosis, and Treatment.* Boston: Butterworth-Heinemann; 1999.

Genetic Disease Involving the Vitreous

In Stickler syndrome or Marshall syndrome, the vitreous collapses prematurely, often inducing retinal detachment. Recently, mutations in both the α_1 (II) and α_1 (XI) collagen chains have been shown to be responsible for this condition. However, other families have also been identified in which the genetic basis is still not understood.

Snead MP, Yates JR. Clinical and molecular genetics of Stickler syndrome. *J Med Genet.* 1999;36:353–359.

Enzymatic Vitreolysis

Considerable interest exists in enzyme preparations that may aid in the clearing of blood from the vitreous and in potentially performing noninvasive vitrectomy or producing a PVD in young eyes. Enzymes that have been proposed for injection into the vitreous cavity include hyaluronidase, plasmin, dispase, and chondroitinase.

Sebag J. Pharmacologic vitreolysis. *Retina.* 1998;18:1–3.

BCSC Section 12, *Retina and Vitreous.*

CHAPTER XVI

Retina

The retina is composed of two laminar structures, an outer retinal pigment epithelium (RPE) and an inner neural retina. (This chapter discusses the neurosensory retina; the RPE is discussed in its own chapter, Chapter XVII.) These laminar structures arise from an invagination of the embryonic optic cup that folds an ectodermal layer into apex-to-apex contact with itself. The two layers form a hemispheric shell on which the visual image is focused by the anterior segment of the eye. The neural retinal cell types are as follows:

- Photoreceptors—rods and three types of cones
- Bipolar cells—rod on-bipolars and cone on- and off-bipolars
- Interneurons—horizontal and amacrine cells
- Ganglion cells and their axons, forming the optic nerve
- Astroglia, oligodendroglia, Schwann's cells, microglia, and vascular endothelium and pericytes

Neural Retina—The Photoreceptors

Rod Phototransduction

Catching light and converting its minute amount of energy into a neural response distinguishes the retina from all other neural structures, which it otherwise resembles. This process occurs within a specialized organelle of the photoreceptor cell, the *outer segment.* Most of our knowledge of phototransduction comes from rods, which are sensitive nocturnal light detectors. Much more biochemical material can be obtained from rods than from cones because they are much more numerous in most retinas and contain much more membrane than cones. More membrane contributes to its higher sensitivity.

The outer segment of a rod is composed almost only of plasma-membrane material organized in an unusual way. Most of the membrane is in the form of membrane sacs flattened along the long axis of the outer segment. There are about 1000 sacs within a rod outer segment and about a million rhodopsin molecules in each sac. The sacs float within the cytoplasm of the outer segment like a stack of coins disconnected from the plasma membrane. The sacs contain the protein machinery to capture and amplify light energy. This abundance of outer-segment membrane increases the number of rhodopsin molecules, which can absorb light. Some deep-sea fishes, which need great sensitivity to detect the little light available, have much longer rods than humans.

Light is absorbed by rhodopsin concentrated in outer-segment membrane of rods. Rhodopsin is a freely diffusible membrane protein similar to α- and β-adrenergic receptors. It has seven helical loops embedded in the lipid membrane (Fig XVI-1). Phosphorylation sites exist on the cytoplasmic side of the protein, where rhodopsin

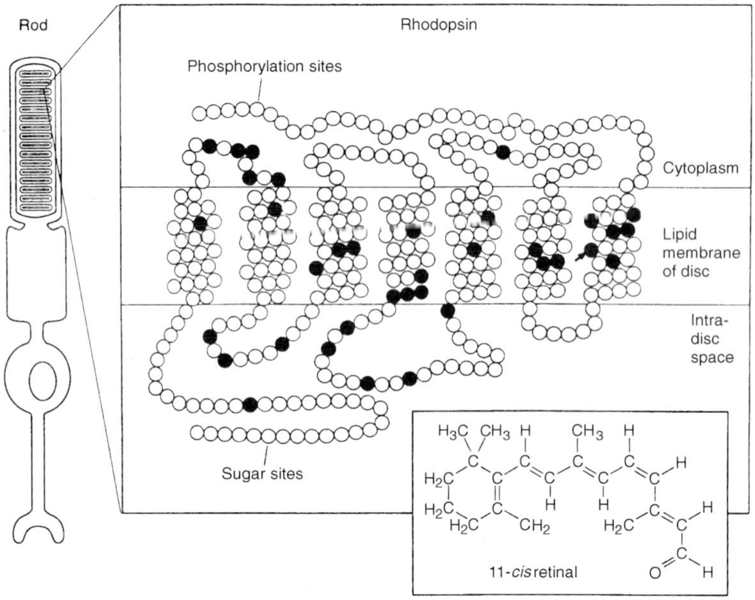

FIG XVI-1—The rhodopsin molecule is embedded in the lipid membrane of the outer segment with seven helical loops. Each circle is an amino acid with the highly conserved ones in *black*. An *arrow* shows the lysine to which the vitamin A chromophore is linked. Phosphorylation sites occur on the cytoplasmic and sugar attachment sites on the intradiscal (extracellular) ends of the rhodopsin molecule. (Courtesy of Peter Gouras, MD.)

is inactivated and sugar is attached on the intradiscal side. The 11-*cis* retinal chromophore is bound to a lysine at amino acid 296 on the seventh membrane loop by a protonated Schiff base linkage. Each molecule responds to a single quantum of light. Rhodopsin absorbs green light best at wavelengths of about 510 nm. It absorbs blue and yellow lights less well and is insensitive to longer wavelengths (red light). This tuning of rhodopsin to this part of the electromagnetic spectrum is due to the amino-acid sequence of the protein and the binding of the 11-*cis* isomer of retinaldehyde, creating a molecular antenna.

Once rhodopsin absorbs a quantum of light, the 11-*cis* double bond of retinal is broken and the opsin molecule undergoes a series of rapid configurational changes, leading to an activated state, *meta-rhodopsin II*. Activated rhodopsin starts a reaction that controls the inflow of cations into the rod outer segment (Fig XVI-2). The target of this reaction is a cyclic guanosine monophosphate (cGMP) gated cationic channel located on the outer membrane of the outer segment. This channel controls the flow of Na and Ca ions into the rod. In the dark, Na and Ca ions flow in through this channel kept open by cGMP. Ionic balance is maintained by a Na,K-ATPase pump in the inner segment and a Na/K-Ca exchanger in the outer-segment membrane, both of which require metabolic energy. Depolarization of the rod causes the transmitter glutamate to be released from its synaptic terminal, starting the neural signals for vision.

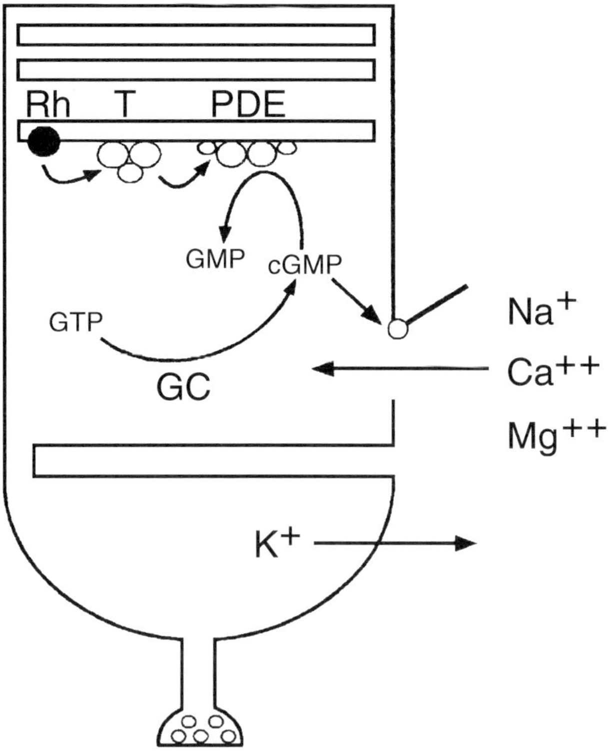

FIG XVI-2—The phototransduction cascade in the rod begins with light-activating rhodopsin *(Rh),* which then activates transducin *(T),* which activates phosphodiesterase *(PDE).* The latter hydrolyzes cyclic guanosine monophosphate *(cGMP),* which leads to the closure of the gated channel on the outer membrane, which stops the entrance of Na, Ca, and Mg. The reduction of intracellular Ca stimulates guanylate cyclase *(GC)* to synthesize more cGMP, which starts to reverse the process. K leaves the rod passively following the voltage gradient. (Courtesy of Marie Burns, Stanford University.)

Light-activated rhodopsin drives a second molecule, *transducin,* by causing an exchange of guanosine diphosphate (GDP) for guanosine triphosphate (GTP). One rhodopsin molecule can activate a hundred transducin molecules, amplifying the reaction. Activated transducins excite a third protein, *rod phosphodiesterase (rod PDE),* which hydrolyzes cGMP to 5′-noncyclic GMP. The decrease in cGMP closes the gated channels, which stops Na and Ca entry and hyperpolarizes the rod. Hyperpolarization stops glutamate's release from the synaptic terminal.

When the light goes off, the rod returns to its dark state as the reaction cascade turns off. Rhodopsin is inactivated by phosphorylation at its C-terminal by rhodopsin kinase, assisted by the binding of arrestin. Transducin is inactivated by the hydrolysis of GTP to GDP by transducin's intrinsic GTPase activity, which inactivates PDE. Guanylate cyclase, the enzyme that synthesizes cGMP from GTP, is activated by the decrease in intracellular Ca caused by the channel closure; the enzyme's action is assisted by guanylate cyclase–assisting proteins (GCAPs). As cGMP levels increase, the gated channels close and the rod is re-depolarized. The corresponding rise in intracellular Ca restores guanylate cyclase activity to its dark level. Calcium feedback may also regulate rhodopsin phosphorylation by recoverin as well as the sensitivity of the gated channel.

"Rim" Proteins

The rim of each rod sac has a unique collection of proteins (Table XVI-1). Two are peripherin and ROM-1, which play a role in the development and maintenance of the sac's curvature. Peripherin and ROM-1 are also found in cone outer segments. A third protein is a member of a superfamily of ATP binding cassette (ABC) transporters. These include the cystic fibrosis transmembrane regulator (CFTR), P-glycoprotein (involved in multidrug resistance), TAP 1 and 2 (which transport peptides in lymphocytes), prokaryotic permeases, and others. The ABC protein is unique to rod sacs and is not found in cones. Rod sacs differ from those of cones because they are disconnected from the outer plasma membrane. This ABC protein may transport retinal into the sac membrane, but this is unproven.

Outer-Segment Energy Metabolism

ATP is necessary to drive the reactions that control the ionic current generators as well as the transporters in the outer segment. Since only the inner and not the outer segment contains mitochondria, oxidative metabolism is confined to the former. The outer segment has glycolysis, including the hexose monophosphate pathway and the phosphocreatine shuttle that produces ATP and GTP and modulates NADPH. The latter is used for the reduction of retinal to retinol before it is returned to the RPE for isomerization and for glutathione reduction, which protects against oxidative stress (see Table XVI-1).

Cone Phototransduction

Qualitatively, cone phototransduction resembles that of rods. Light-activated cone opsins start an enzymatic cascade that hydrolyzes cGMP and closes cone-specific cGMP gated cation channels on the outer-segment membrane. Cone phototransduction is comparatively insensitive but fast and capable of adapting enormously to the ambient levels of illumination. The greater the ambient light levels, the faster and more temporally accurate is the response of a cone. Speed and temporal fidelity are important for all aspects of cone vision. This is one reason acuity improves progressively with increased illumination.

Several factors contribute to light adaptation. For example, higher levels of illumination bleach away photopigments, making the outer segment less sensitive to light. Increasing light levels also increases the noise, which reduces sensitivity. Biochemical and neural feedback speed up the cone response. This feedback must be increased as the light intensity increases and the cone absorbs more and more light. The biochemistry responsible for this speed up is not yet deciphered. All the processes that turn the rod response off are probably stronger in cones. The life span of activated cone opsin, its turn off, and the turn-off of cone transducin (a G protein) must be faster than they are in rods. In addition, increasing light must enhance the turn-off mechanisms in cones.

TABLE XVI-1

PROTEINS ASSOCIATED WITH ROD OUTER SEGMENTS

PROTEIN	LOCATION	FUNCTION
Rhodopsin	Sac, membrane	Absorbs light, drives transducin
Transducin	Sac	G-protein, activates PDE
Phosphodiesterase (PDE)	Sac	Hydrolyzes cGMP
Rhodopsin kinase	Sac	Inactivates rhodopsin
Arrestin	Sac	Inactivates rhodopsin
Recoverin	Sac	Ca-dependent kinase control
Guanylate cyclase	Sac	Forms cGMP
Dehydrogenase	Sac	Reduces AT-retinal
Peripherin	Rim of sac	Structure
ROM-1	Rim of sac	Structure
ABC transporter	Rim of sac	Transports lipid retinal
cGMP gated channel	Membrane	Controls entry of Na and Ca
Glucose transporter	Membrane	Entry of glucose
Na/K, Ca exchanger	Membrane	Ca removal
Ca-dependent GCAP	Sac, cytoplasm	Modulates cyclase
Calmodulin	Sac, cytoplasm	Modulates cGMP channel
Na, K-ATPase	Inner segment	Removes Na

There is also neurally mediated negative feedback on cones. Horizontal cells of the inner nuclear layer synapse antagonistically back on to cones, releasing gamma amino-butyric acid (GABA), an inhibitory transmitter. When light hyperpolarizes a cone, the cone hyperpolarizes neighboring horizontal cells. This inhibits the horizontal cells, stopping the release of GABA, which depolarizes (disinhibits) the cone by a recurrent synapse. This depolarization antagonizes the hyperpolarization produced by light. It tries to put the cone back in the dark. It occurs with a synaptic delay so that its main effect is on the later response of the cone. Horizontal cell feedback occurs with strong stimuli, undoubtedly preventing the cone from being overloaded. This also turns the cone response off more quickly, allowing it to respond more rapidly to a new stimulus. This increases the flicker fusion frequency, which is much higher in cones (about 100 Hz) than in rods (about 30 Hz).

Because of their ability to adapt, cones are indispensable to good vision. Without cones, one is legally blind, losing the ability to read and see colors. Lost rod function is a minor handicap. Molecular genetics has now demonstrated that cones evolved before rods, undoubtedly in the presence of bright ambient light. Rods must have evolved as some organisms fled to dimly illuminated niches, where high sensitivity became important. It also seems logical that short-wave cones also evolved earlier than rods to provide diurnal organisms with color before nocturnal vision.

Trivariant Color Vision (Three Cone Opsins)

In order to see colors, it is necessary to have at least two different spectral classes of cones. Most normal humans have three types of cones and consequently a three-variable color-vision system. Most mammals have divariant color vision with a middle-wavelength-sensitive (M) cone detecting high-resolution achromatic (black and white) contrast and short-wavelength-sensitive (S) cones used only for color by comparing its signals with those of the M-cones. This creates blue/yellow color vision. Because the S-cones only contribute to color, they are much less numerous than M-cones.

In primates, a third cone mechanism evolved to enhance color vision by splitting the high-resolution M-cones into long (L) and middle (M) wavelength cones (Fig XVI-3). This creates red/green color vision. L- and M-cones both contribute to achromatic and chromatic contrast. Therefore, both L- and M-cones are more numerous than S-cones in the human retina.

Most color-vision defects involve red/green discrimination and involve the genes coding for the L- and M-cone opsins. These genes are in tandem on the X-chromosome. There is one copy of the L-cone opsin gene at the centromeric end of the X-chromosome and one to six copies of the M-cone gene arranged in a head-to-tail tandem array. Normally, only the most proximal of these two genes are expressed. Most color-vision abnormalities are due to unequal crossing over between the L- and M-cone opsin genes (see Fig XVI-3B). This creates hybrid opsins that have different spectral absorption functions, usually less ideal. Some males have a serine-to-alanine substitution at amino acid 108 on the cone opsin gene, allowing more sensitivity to red light. Therefore, female subjects with both the serine-containing and alanine-containing opsins could have tetravariant color vision.

Genetic Approaches to Photoreceptor Function and Dysfunction

Gene technology is revolutionizing our understanding of photoreceptor function and retinal degenerations. Amino acid sequencing of rhodopsin led to the cloning of the rhodopsin gene on chromosome 3 and the cone opsin genes. The L- and M-cone opsins were cloned from the X chromosome, and the S-cone opsin from chromosome 7. The first genetic defect linked to retinal degeneration was in the rhodopsin gene in a dominant form of retinitis pigmentosa. Now, more than 70 different mutations in this rhodopsin gene have been found responsible for autosomal dominant retinitis pigmentosa (ADRP).

Figure XVI-4 shows 10 different rod-specific proteins that cause dysfunction or degeneration. The genes coding for these proteins can have different mutations. Some mutations produce severe or mild forms of degeneration or stationary forms of retinal dysfunction, and sometimes quite different phenotypes. The same gene can cause macular degeneration or retinitis pigmentosa (depending on the mutation). This is called *allelic heterogeneity*. Different genes can cause similar phenotypes. Mutations in the rhodopsin or peripherin or ABC-transporter genes can cause retinitis pigmentosa. This is called *genetic heterogeneity*. Therefore, genotyping is more precise than phenotyping.

Figure XVI-5 shows six different RPE-specific gene defects that cause photoreceptor dysfunction or degeneration. It is important to distinguish RPE from photoreceptor-specific defects because such distinctions can influence future therapies. For example, a vector can be programmed to express a new gene in the RPE, the photoreceptor, or a rod rather than a cone.

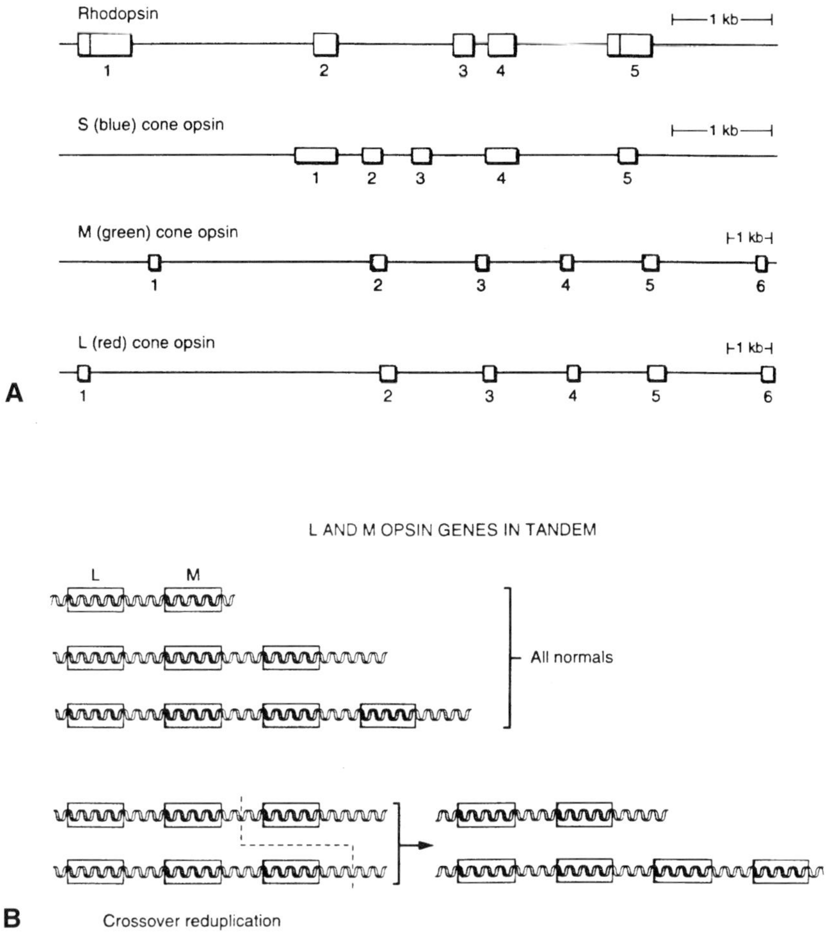

FIG XVI-3—**A**, The genes for rhodopsin and the cone opsins are similar. The rhodopsin gene and the S cone opsin gene have five exons. The L and M cone opsin genes have six exons and longer introns. **B**, The tandem arrangement of the L and M cone opsin genes on the X chromosome is shown with the additional copies of the M cone opsin gene. Crossover of the similar genes has been responsible for the duplication of the M cone opsin gene. (Courtesy of Peter Gouras, MD.)

Rod-Specific Gene Defects

Rhodopsin More than 70 different mutations cause ADRP. Mutations occur in different ways; they can alter transduction, protein folding, or localization of the protein. Deletion of amino acids 68–71 causes failure of binding to 11-*cis* retinal. Amino acids 58, 135, and 137 affect transducin activation. Rhodopsins, like P23H (responsible for 10% of RP in America), do not fold properly and accumulate in the rough endoplasmic reticulum. In the Q344ter mutation, rhodopsin folds properly but is not inserted into the rod outer segment. Generally, mutations affecting the

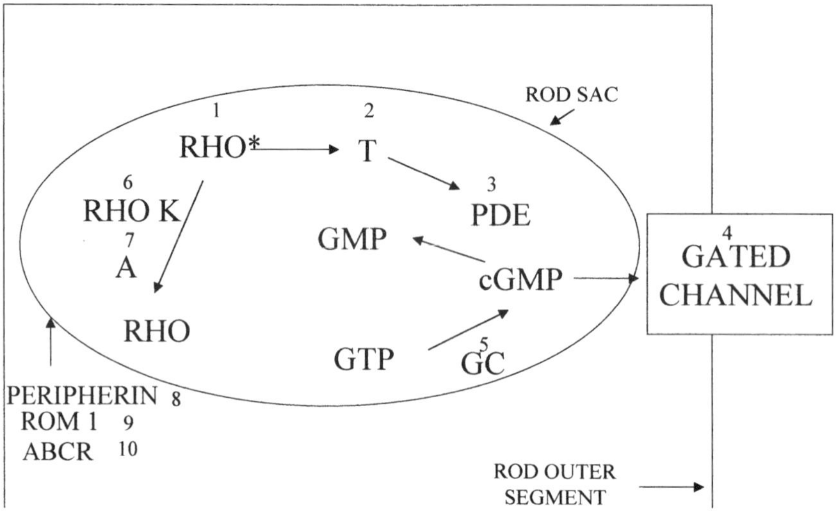

FIG XVI-4—*Rod-specific gene defects.* Gene defects that occur exclusively in rods occur in ten different proteins that cause either degeneration or vision dysfunction. A, adenosine; ABCR, *ABCR* gene; cGMP, cyclic guanosine monophosphate; GC, guanylate cyclase; GMP, guanosine monophosphate; GTP, guanosine triphosphate; PDE, phosphodiesterase; RHO, rhodopsin; ROM 1, *ROM1* gene; T, transducin. (Courtesy of Peter Gouras, MD.)

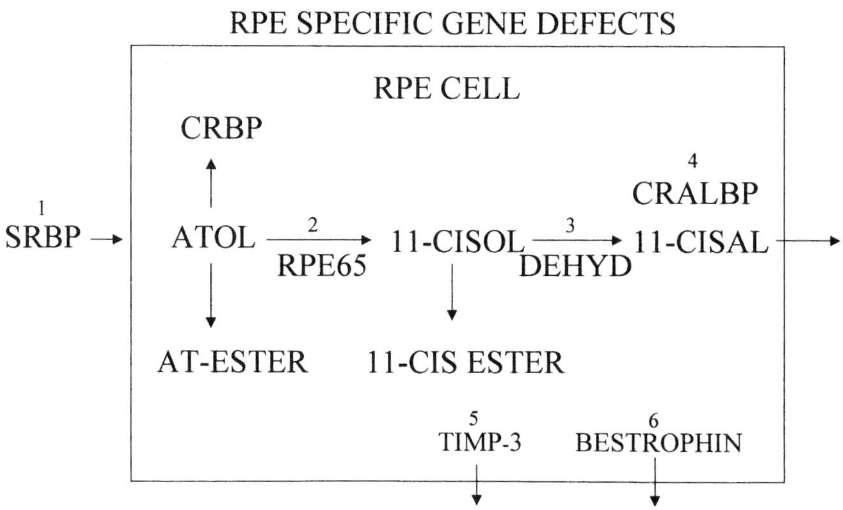

FIG XVI-5—Gene defects that occur exclusively in retinal pigmented epithelium *(RPE)* occur in six different proteins that cause either degeneration or vision dysfunction. (Courtesy of Peter Gouras, MD.)

intradiscal area and amino terminal of rhodopsin are less severe than in the cytoplasmic region and the carboxyl tail. Alterations in the middle of the gene, coding for the transmembrane regions, are intermediate in severity.

A homozygous null mutation at codon 249 in the rhodopsin gene produces autosomal recessive retinitis pigmentosa (ARRP). This opsin misses its sixth and seventh transmembrane domains including the 11-*cis*-retinal attachment site.

A heterozygous missense mutation, A292E, causes autosomal dominant stationary nyctalopia. This constitutively activates transducin without the need of rhodopsin. It light-adapts and desensitizes the rod. There is no degeneration, which contradicts the "equivalent light" hypothesis for rod cell death. A heterozygous G90D mutation also causes a stationary form of nyctalopia.

Rod transducin A dominant G38D mutation produces Nougaret disease, the oldest known form of autosomal dominant stationary nyctalopia. Transducin becomes continuously activated, another example of constitutively active rods that do not degenerate.

Rod cGMP phosphodiesterase (rod PDE) Defects in either the alpha (PDEA) or beta (PDEB) subunits cause ARRP. These are nonsense mutations, which truncate the catalytic domain of the protein. A H258D mutation in PDEB also causes dominant stationary nyctalopia. This mutation is near the binding site of the gamma subunit of PDE and may lead to constitutively active PDE.

Rod cGMP gated channel Null mutations of the beta subunit cause ARRP. No degeneration from the alpha or gamma units has been reported.

Arrestin A homozygous defect in codon 309 causes Oguchi disease, a form of stationary nyctalopia. It produces a frame shift and truncation of arrestin. There is genetic heterogeneity because rhodopsin kinase gene defects also cause Oguchi disease.

Rhodopsin kinase Null mutations cause Oguchi disease. This also retards activated rhodopsin's turn-off.

Guanylate cyclase Null mutations cause Leber amaurosis, a childhood autosomal recessive form of RP. Leber amaurosis shows genetic heterogeneity.

Rod ABC transporter Recessive defects cause Stargardt disease. There is allelic heterogeneity, which reflects the severity of the gene defect. Mild defects cause macular degeneration; intermediate ones cause cone–rod dystrophy and severe ones cause RP. Heterozygous defects are also found in 4% of age-related macular degeneration.

L-type calcium channel This gene codes for an alpha subunit. Defects cause X-linked stationary nyctalopia. The protein seems to determine transmitter release from the rod synaptic terminal and may also affect cones.

Cone- and Rod-Specific Gene Defects

Peripherin (RDS) There is great allelic heterogeneity in this gene. Defects cause several dominantly inherited retinal degenerations that range from ADRP to macular degeneration, pattern macular dystrophy, vitelliform macular dystrophy, butterfly macular dystrophy, and fundus flavimaculatus. A null mutation of the homologous murine gene causes a semidominant form of degeneration with failure of rod outer-segment development and slow degeneration.

ROM 1 Double-heterozygotic mutations in both the *ROM 1* and the peripherin gene cause "digenic" RP. A *ROM 1* gene defect alone has been reported in a patient with a vitelliform macular dystrophy, but this gene is not responsible for Best macular dystrophy (see below).

Myosin VIIA A heterozygous null mutation in a form of myosin, VIIA, causes Usher syndrome type I. Affected subjects have deafness and vestibular ataxia at birth and develop ARRP.

Oxygen-regulated protein Homozygous defects cause autosomal dominant RP. Expression of this unusual gene is modulated by oxygen.

Cone-Specific Gene Defects

Cone cGMP gated channel A homozygous defect in the alpha subunit causes achromatopsia, loss of all cone function.

L- and M-cone opsins Two genetic steps lead to "blue-cone achromatopsia." One reduces the tandem array of these genes to one gene, and the second eliminates the residual gene. Another defect in a sequence upstream from this tandem array can also cause blue-cone achromatopsia. These occur in males because of the gene's location on the X chromosome.

L- or M-cone opsins Defects in one or the other of these X-linked genes cause red/green color deficiencies, again almost exclusively in males.

RPE-Specific Gene Defects

RPE 65 Homozygous defects cause Leber amaurosis, a generalized loss of photoreceptor function. Mice with this defect retain some cone function. The protein influences the formation of 11-*cis* retinol. Cones may have access to another pool of this isomer.

Bestrophin Heterozygous missense mutations produce Best disease, a dominantly inherited form of macular degeneration, which involves the entire RPE layer but only causes damage in the macula. This protein is membrane-bound with four transmembrane regions.

TIMP3 Heterozygous point mutations produce Sorsby macular dystrophy. This protein is an inhibitor of a metalloproteinase that regulates extracellular matrix.

CRALBP (cytoplasmic retinal binding protein) Homozygous defects cause retinitis punctata albescens. This protein facilitates 11-*cis* retinal formation and transport; it is also found in Müller's cells.

11-cis retinol dehydrogenase Homozygous defects cause fundus albipunctus, a form of stationary nyctalopia. This enzyme forms 11-*cis* retinal from 11-*cis* retinol and may also exist in Müller's cells.

EFEMP1 (EGF-containing fibrillin-like extracellular matrix protein) A single heterozygous nonconservative mutation causes malattia leventinese (Doyne honeycomb retinal dystrophy), a dominant form of macular degeneration. It is uncertain whether the protein is unique to RPE.

Ubiquitously Expressed Genes Causing Retinal Degenerations

REP 1 (Rab escort protein 1) X-linked gene that causes choroideremia. This protein is involved in prenylating Rab proteins, which facilitates their binding to cytoplasmic membranes and promoting vesicle fusion. Photoreceptors, RPE, and/or choroid must be uniquely vulnerable.

OAT (ornithine amino transferase) Homozygous defects cause gyrate atrophy. The enzyme breaks down ornithine, which seems toxic to the RPE in high concentrations.

MTP (microsomal triglyceride transfer protein) Homozygous defects cause abetalipoproteinemia with faulty fat absorption and ARRP. It is treatable with fat-soluble vitamins.

PEX1 (infantile Refsum disease) Homozygous defects cause Refsum disease with RP, retardation, and hearing deficits. This is the least severe of a constellation of *PEX* genes, which encode for peroxins, proteins needed for peroxisome biogenesis.

PAHX (Refsum disease) Homozygous defects cause Refsum disease with RP, cerebellar ataxia, and peripheral polyneuropathy. This enzyme degrades phytanic acid and is located in peroxisomes. It is treatable with a phytanic acid–restricted diet.

Inner Nuclear Layer

This layer has three classes of neurons (bipolar, horizontal, and amacrine cells) and a glial cell (the Müller cell). There are separate bipolar cells for cones and rods. There are at least two distinctly different types of cone bipolars, *on-bipolars* and *off-bipolars* (Fig XVI-6). The former are inhibited, the latter excited by the glutamate transmitter released by cones. Thus, when light hyperpolarizes the cones, the on-bipolar is excited (turned on) and the off-bipolar inhibited (turned off). When a shadow depolarizes the cones, the reverse occurs.

Some cone bipolars synapse only with L-cones and others only with M-cones, which is necessary for color vision. In the fovea, some cone bipolars synapse with a single L- or M-cone (Fig XVI-7), which provides the highest spatial acuity. This cone selectivity is preserved through the ganglion cell layer. This selectivity for L- or M-cone inputs is transmitted by a tonic responding system of small ganglion cells. Separate L- and M-cone on-bipolars and off-bipolars transmit a faster, phasic signal to a parallel system of larger ganglion cells. Rods and probably S-cones have only on-bipolar cells. Neither rods nor S-cones are involved in high spatial resolution. The S-cones are involved in color vision and the rods in twilight vision.

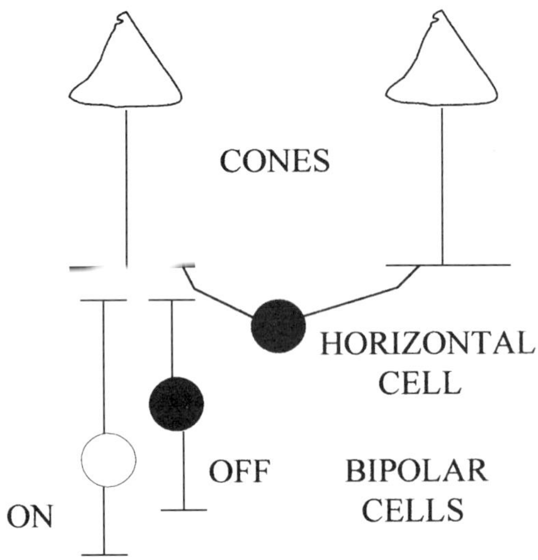

FIG XVI-6—Basic circuitry of the cones. A separate "on" and "off" bipolar cell contacts each cone. In the fovea, a cone has "midget" bipolar cells contacting only a single cone for high spatial acuity. Horizontal cells are antagonistic neurons between cones; when the cone absorbs light, it is hyperpolarized; this hyperpolarizes the horizontal cell. It therefore resembles an "off" bipolar cell. (Courtesy of Peter Gouras, MD.)

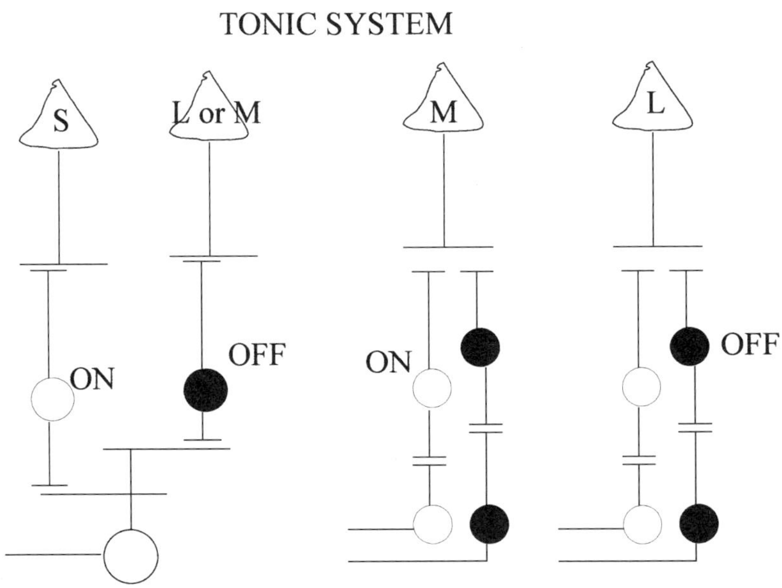

FIG XVI-7—The tonic system transmits signals from the cones that are relatively maintained for the duration of the light or dark stimulus. This system provides the brain with information about a separate cone system, necessary for color vision. It preserves the polarity of the signal for each cone type. This is best shown here in the S cone channel, which receives a signal of opposite polarity (off) from the L and M cones.

The horizontal cells are antagonistic interneurons that inhibit photoreceptors (see Fig XVI-6) by releasing gamma amino butyric acid (GABA) when depolarized. The dendrites of horizontal cells go to cones. One class of horizontal cells goes to L- and M-cones. Another class goes mainly to S-cones. A thin axon terminal that emanates from the cell body of horizontal cells sends dendrites to rods. The dendrites of horizontal cells receive glutamate from cones and rods and release GABA back on to cones and rods. This provides negative feedback. When light causes the cone to hyperpolarize and stop its transmitter release, the horizontal cell is also hyperpolarized (turned off). This stops the release of GABA from the horizontal cell on to the cone. This depolarizes the cone.

Cone amacrine cells mediate antagonistic interactions among on-bipolars and off-bipolars and ganglion cells. The rods have an unusual amacrine cell that receives the inputs of rod bipolars and delivers signals to on- and off-bipolar ganglion cells. Thus, rod signals undergo additional synaptic delays before they reach the ganglion cell output.

The retinal ganglion cells can be classified into two main types: *on* (center cells excited) and *off* (center cells inhibited by light in the center of their receptive field). A shadow does the converse to these two cell types. There are three main subgroups. Tonic cells driven by either L- or M-cones include small cells concentrated in the fovea (responsible for high acuity) and others located extrafoveally (see Fig XVI-8). They project to the parvo cellular layers of the lateral geniculate nucleus (the main relay station to visual cortex) and mediate both high spatial resolution and color vision. Tonic cells driven by S-cones have a unique physiology designed for detecting successive color contrast, blue/yellow or gray/brown borders. These ganglion cells are excited by short waves entering and long waves leaving their receptive fields (see Fig XVI-7). The phasic cells are larger, less concentrated in the fovea, and faster-conducting (Fig XVI-8). They project to the magnocellular layers of the lateral geniculate nucleus and may be more important in movement detection.

Müller's cells are the least understood retinal cells. Non-neural, they play a supportive role to the neural tissue extending from the inner segments of the photoreceptors to the inner limiting membrane, which their end feet form. They buffer the ionic concentrations in the extracellular space, seal off the subretinal space by forming the external limiting membrane, and may play a role in the vitamin A metabolism of cones.

The other non-neural cells of the retina are *macroglia* (astrocytes, oligodendroglia, and Schwann's cells) and *microglia*. These cells provide physical support, respond to injury, regulate the ionic and chemical composition of the extracellular milieu, participate in the blood–retinal barrier, form the myelination of the optic nerve, guide neuronal migration during development, and exchange metabolites with neurons. Neuroglia have high-affinity transmitter-uptake systems and voltage-dependent and transmitter-gated ion channels; they can release transmitters, but their role in signaling, as in many other functions, is unclear.

In addition, the neural retina contains blood vessels with endothelial cells and pericytes. Pericytes play a role in the autoregulation of retinal blood vessels. Pericytes are an early target in diabetes. The pathogenesis of diabetic retinopathy seems to be due to defects in the polyol pathway. Aldose reductase, the rate-limiting first enzyme in the conversion of many sugars to their alcohols, has a known role in the formation of diabetic cataracts. Accumulation of sorbitol and its aldose reductase–mediated metabolite, fructose, causes repeated osmotic insults and cataractogenesis. Similar conditions may lead to thickening of the basement membrane and pericyte loss in the blood vessels of the retina.

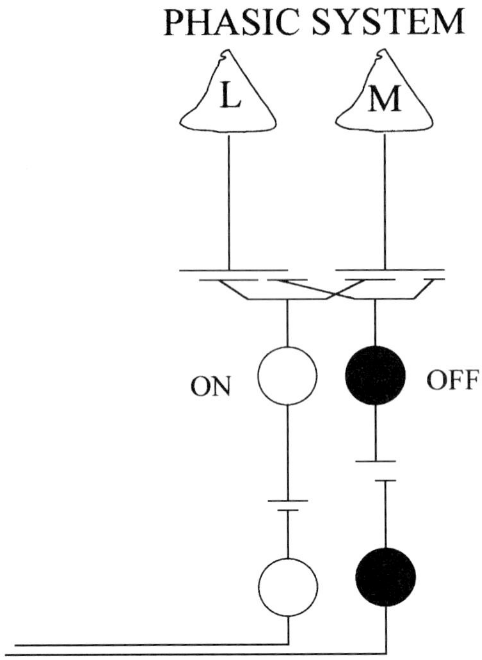

FIG XVI-8—The phasic system transmits signals at the beginning or end of a light stimulus. This produces a brief or transient response. L and M cone signals of the same polarity mix in driving the phasic system.

Retinal Electrophysiology

Changes in the light flux on the retina produces electrical changes in all of the retinal cells, including the RPE and Müller cells as well as neurons. These changes result from ionic currents that flow when ion-specific channels are opened or closed. These currents reach the vitreous and the cornea, where they can be detected noninvasively in the form of an electroretinogram. The initiating process is the ionic response started in the rods and cones that influences the ionic current both *directly* by changes in Na and K fluxes and *indirectly* by synaptically modifying second-order retinal neurons.

The changes in the electrical potentials of the rods and cones are shown in Figure XVI-9, which depicts the responses of rods and cones to a pulse of light or a pulse of darkness. Light hyperpolarizes cones and rods. The cone response is rapid, turns off while the light is still on, and overshoots the dark potential. The rod response is more prolonged and turns off very slowly. Dark depolarizes the cone and has little influence on the rod, which is saturated at high light levels and too slow to respond to the "shadow." The ionic changes are due to the shift in the photoreceptors' conductivity to Na and K ions. The concentration gradients for these ions are reversed across the membrane of the photoreceptors so that changes in the conductivity to these ions move currents in the opposite directions.

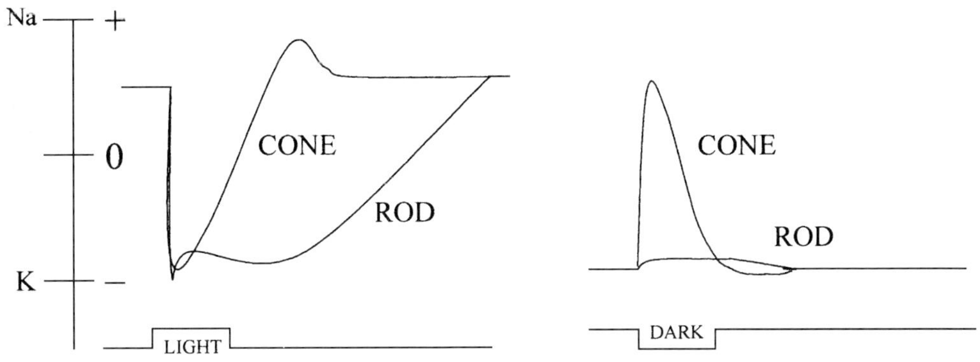

FIG XVI-9—The response of a rod and cone to a pulse of light and a pulse of darkness. The light pulse hyperpolarizes both photoreceptors. The rod responses are prolonged. The cone responses turn off quickly even while the pulse of light is on. Darkness depolarizes the cone rapidly but has only a small effect on the slower rod response.

Baylor D. Photoreceptor signals and vision. *Invest Ophthalmol Vis Sci.* 1987;28:34–49.

Liou GI, Fei Y, Peachey NS, et al. Early onset of photoreceptor abnormalities induced by targeted disruption of the interphotoreceptor retinoid-binding protein gene. *The J Neurosci.* 1998;18:4511–4520.

Molday RS. Photoreceptor membrane proteins, phototransduction, and retinal degenerative diseases. The Friedenwald Lecture. *Invest Ophthalmol Vis Sci.* 1998;39: 2491–2513.

CHAPTER XVII

Retinal Pigment Epithelium

The retinal pigment epithelium (RPE) is a single layer of cuboidal epithelial cells that constitutes the outermost layer of the retina. The RPE is located between the highly vascular choriocapillaris and the outer segments of photoreceptor cells. In humans, there are approximately 4–6 million RPE cells per eye. The ratio of photoreceptor cells to RPE cells is roughly 45:1. The RPE is derived embryologically from the same neural anlage as the sensory retina but differentiates into a secretory epithelium. Although it has no photoreceptive or neural function, the RPE is essential to the support and viability of photoreceptor cells.

Anatomical Description

RPE cells are polarized epithelial cells. They have long, microvillous processes on their apical surfaces that interdigitate with outer segments of photoreceptor cells. Their basal surface, which is adjacent to Bruch's membrane (an extracellular matrix between the RPE and choriocapillaris), has many infoldings. RPE cells are joined near their apical side by tight junctions that block the passage of water and ions. As such, the RPE contributes to the blood–retinal barrier. In addition to the organelles found in most cells (eg, the nucleus, golgi apparatus, smooth and rough endoplasmic reticulum, and mitochondria), the RPE has melanin granules and phagosomes that reflect two of it important roles (discussed shortly). The RPE is particularly rich in microperoxisomes, suggesting that it is quite active in detoxifying the large number of free radicals and oxidized lipids that are generated in this highly oxidative and light-rich environment.

Biochemical Composition

Biochemically, the RPE is a dynamic and complex cell. It must meet demands for its own active metabolism, its extraordinary phagocytic function, and its role as a biological filter for the neurosensory retina. These processes impose a very high energy requirement by the RPE; not surprisingly, the cells contain all the enzymes of the three major biochemical pathways: glycolysis, Krebs cycle, and the pentose phosphate pathway. Glucose is the primary carbon source used for energy metabolism and conversion to protein. Although the RPE does make a minor contribution to the glycosaminoglycan- and proteoglycan-containing interphotoreceptor matrix, glucose is not converted to glycogen in the RPE. Glucosamine, fucose, galactose, and mannose are all metabolized to some extent in the RPE, although mannose seems to be passed on almost directly to the photoreceptors.

Regarding the chemical composition of RPE, over 80% of wet weight is contributed by water. Proteins, lipids, and nucleic acids contribute most of the remaining weight.

Proteins

Nearly 850 proteins have been identified in RPE. Up to 200 acidic proteins are present, and approximately 180 plasma-membrane proteins have been identified. Many proteins that are found in other cells are present also in RPE. For example, hydrolytic enzymes such as glutathione, peroxidase, catalase, and superoxide dismutase, which are important for detoxification, are present in RPE. The cytoskeletal proteins actin, myosin, α-actinin, fodrin, and vinculin are present also in RPE.

Other proteins are present in RPE but are localized differently than in other cells. A well-known example of this is the Na^+ pump that uses energy derived from adenosine triphosphate (ATP) hydrolysis to transport Na^+ and K^+ against their electrochemical gradients. Na^+,K^+-ATPase has a unique location in RPE cells. Whereas most polarized epithelial cells localize this protein to their basolateral surface, the RPE places it on the apical surface. It is thought that Na^+,K^+-ATPase is apically located to maintain the balance of Na^+ and K^+ in the subretinal space. Additional proteins have been shown to have a reversed polarity in RPE cells compared with other polarized epithelial cells, including N-CAM-140 and folate receptor α. In addition to proteins with a unique location in RPE, other proteins are expressed only in the RPE. One such protein, RPE65, was described in the early 1990s and has been shown to be an obligate component of the isomerization of vitamin A that is required for regeneration of visual pigment (described below).

Lipids

Lipids account for approximately 3% of the wet weight of the RPE. About half of these lipids are phospholipids. Phosphatidylcholine and phosphatidylethanolamine make up more than 80% of the total phospholipid content. In general, levels of saturated fatty acids in the RPE are higher than in the adjacent outer segments. The saturated fatty acids palmitic acid and stearic acid are used for retinol esterification and for energy metabolism by the RPE mitochondria. The level of polyunsaturated fatty acids, such as docosahexaenoic acid (22:6ω3), is much lower in the RPE than in the outer segments, although the level of arachidonic acid is relatively high. A number of studies have suggested that the retina may be spared the effects of essential fatty acid deficiency because RPE efficiently sequesters fatty acids from the blood. The RPE actively conserves and efficiently reuses the fatty acids, thus preventing their loss as waste products.

Nucleic Acids

About 1% of the wet weight of the RPE is contributed by RNA. RNA is synthesized continually by RPE owing to production of many enzymes needed for cell metabolism, phagocytosis of shed disks, and maintenance of the retinoid pathway and transport functions.

Major Physiologic Roles of RPE

The RPE has a number of physiologic roles. Critical among these are:

- Visual pigment regeneration
- Phagocytosis of shed photoreceptor outer-segment disks
- Transporting necessary nutrients and ions to photoreceptor cells and removing waste products from photoreceptors

- Absorption of scattered and out-of-focus light via pigmentation
- Adhesion of the retina

These five functions are discussed briefly below. Other important functions subserved by the RPE include its role in synthesis and remodeling of the interphotoreceptor matrix, formation of the blood–retinal barrier, and elaboration of humoral and growth factors.

Visual Pigment Regeneration

The process of regeneration of the visual pigment rhodopsin has been studied extensively and involves both photoreceptors and the RPE. The RPE plays a major role in the uptake, storage, and mobilization of vitamin A for use in the visual cycle. Indeed, RPE is second only to liver in its concentration of vitamin A.

The basic function of the RPE cell in the visual process is to generate 11-*cis*-retinaldehyde (used in the formation of rhodopsin). As described in detail elsewhere in this section, the photoreceptor cell synthesizes opsin, which uses 11-*cis*-retinaldehyde in the regeneration of rhodopsin. In the photoreceptor cell, rhodopsin is photolyzed and undergoes a *cis*-to-*trans* isomerization. All-*trans* retinaldehyde is released and converted to all-*trans*-retinol by retinoldehydrogenase. The retinol is returned to the RPE in the presence of interphotoreceptor retinoid binding protein (IRBP). In RPE, retinol is converted to retinyl ester in the presence of the enzyme lecithin retinol acyltransferase. When needed for regeneration of rhodopsin, the retinyl ester is converted by an isomerohydrolase (isomerase) to 11-*cis*-retinol and is subsequently converted to 11-*cis*-retinal by a dehydrogenase. The 11-*cis*-retinal is returned to the photoreceptor cell along with IRBP. The recently described protein RPE65 is thought to play a role in the isomerization step because RPE65 knockout mice cannot regenerate rhodopsin.

The RPE acquires vitamin A in three ways:

- Release during bleaching of rhodopsin and return via the regeneration process of the visual cycle (Fig XVII-1)
- From circulation, presumably through a receptor-mediated mechanism
- Via phagocytosis of shed photoreceptor outer-segment disks

The aldehyde and alcohol forms of vitamin A are membranolytic; hence, several retinoid-binding proteins mediate both vitamin A metabolism within the RPE and its exchange with adjacent outer segments. A number of retinoid-binding proteins have been isolated and characterized in RPE, in the subretinal space, and in photoreceptors (Table XVII-1). RPE esterifies retinol with available fatty acids (predominantly palmitic acid and, to a lesser extent, stearic and oleic acids) and stores retinol as a retinyl ester, a form no longer lytic to cell membranes. In conditions of hypervitaminosis A, toxicity to RPE is minimal because of vitamin A storage as the ester.

Phagocytosis of Shed Photoreceptor Outer-Segment Disks

RPE plays a crucial role in turnover of the photosensitive membrane of rod and cone photoreceptors. In the mid-1960s, autoradiography was used to establish that proteins were synthesized in the inner segments of the photoreceptor cells and were transported to the outer segment and incorporated into new disks forming at the base of the outer segment. The band of radioactive protein was displaced toward the apex

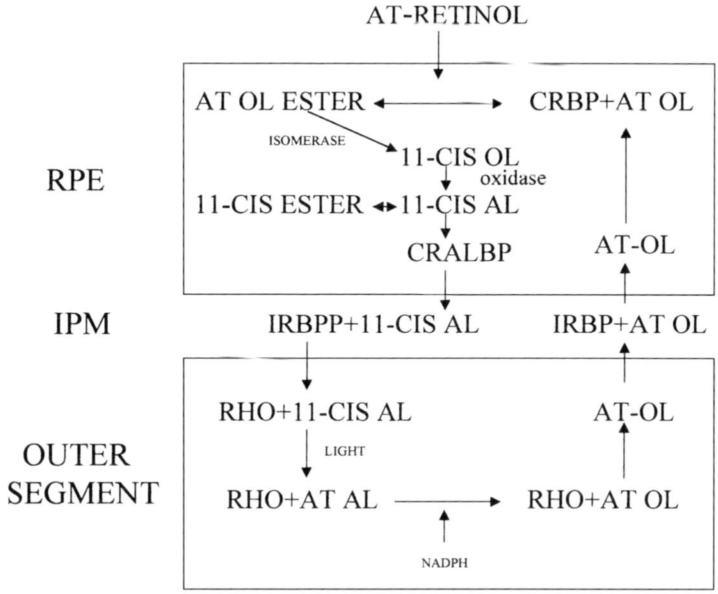

FIG XVII-1—The vitamin A cycle of vision (sometimes called the Wald cycle) involves the retinal pigment epithelium *(RPE)* cell and the rod, shown here. The cycle begins with all trans-retinol entering from the serum and being isomerized, oxidized, and transferred to the rhodopsin in the rod outer segment. Light re-isomerizes retinal to the trans-form. It is then reduced and re-enters the RPE to complete the cycle. *AT,* all trans; *OL,* retinol; *AL,* retinaldehyde (retinal); *RHO,* rhodopsin; *IPM,* interphotoreceptor matrix; *CRBP,* cellular retinol binding protein; *CRALBP,* cellular retinaldehyde-binding protein; *IRBP,* interphotoreceptor retinol binding protein. (Courtesy of Sylvia Smith, PhD.)

TABLE XVII-1

LOCATION OF RETINOID-BINDING PROTEINS RELATED TO THE VISUAL CYCLE

RETINOID-BINDING PROTEINS	COMPARTMENTS
RBP (retinol-binding protein)	Blood
CRALBP (cellular retinaldehyde-binding protein)	RPE and Müller cells
IRBP (Interphotoreceptor retinoid-binding protein)	IPM (interphotoreceptor matrix)
CRBP (cellular retinol-binding protein)	Müller cells
CRABP (cellular retinoic acid-binding protein)	Amacrine cells and Müller cells

of the cell over a period of about 9–11 days. The vital role of RPE in phagocytosis of these disks was demonstrated when the radioactive distal tip of the outer segment arrived in the RPE cell and was subsequently phagocytosed.

The shed outer-segment disks are encapsulated in phagosomes, which in turn fuse with lysosomes and are digested. During degradation of the disks, building blocks are recycled to photoreceptors for use in synthesis and assembly of new disks. The lipofuscin characteristic of RPE is derived from photosensitive membranes. Each photoreceptor cell sheds approximately 100 outer-segment disks per day. Since many photoreceptors interdigitate with a single RPE cell, each RPE cell ingests/digests more than 4000 disks daily! The shedding event follows a circadian rhythm: In rods, shedding is most vigorous within 2 hours of light onset; in cones, shedding occurs more vigorously at onset of darkness. Recent evidence suggests that the neurotransmitter dopamine acts within the photoreceptor–pigment epithelial complex to control disk shedding. Defects in the phagocytic function of the RPE are seen in the Royal College of Surgeons rat; these defects lead to degeneration of the photoreceptor cells.

Transport

The health and integrity of retinal neurons depends on a well-regulated extracellular environment. A critical function of the RPE that contributes to this regulation is the control of volume and composition of fluid in the subretinal space through transport of ions, fluid, and metabolites. The distribution of transport proteins that reside in the apical and basolateral membrane domains of the cell is clearly asymmetrical, and this difference is what allows the epithelium to carry out vectorial transport. The membrane proteins remain in their proper location because of tight junction proteins. Intercellularly, asymmetry or polarity of the cell is maintained because of the intracellular molecular machinery that synthesizes new proteins and delivers them preferentially to the apical or basolateral cell membranes. Cytoskeletal proteins are fundamental in determining cell polarity and regulating transport.

The aqueous environment of the subretinal space is actively maintained by the ion-transport systems of RPE. The active transport of a variety of ions (K^+, Ca^{2+}, Na^+, Cl^-, and HCO_3^-) across the RPE has been well documented. This transport is vectorial in most cases; for example, Na is actively transported from the choriocapillaris toward the subretinal space, whereas K is transported in the opposite direction. The apical membrane of RPE appears to be the major locus of this transport. As mentioned above, the ouabain-sensitive Na^+,K^+-ATPase is present at the apical, but not the basal, side. Similarly, an active bicarbonate-transport system appears to be located in this portion of the RPE membrane. High carbonic anhydrase activity seems to be associated with both the apical and basal sides of the cell.

Net ionic fluxes in RPE are responsible for the transepithelial electrical potential that can be measured across the RPE apical membrane, a potential rapidly modified in the presence of a variety of metabolic inhibitors (eg, ouabain and dinitrophenol). In addition, the RPE apical membrane must be responsive to the changing conditions of phototransduction. For example, light evokes a decrease in K^+ ion concentration in the subretinal space, thus hyperpolarizing the RPE. Since the activity of Na^+,K^+-ATPase is controlled in part by K^+ ion concentration, light can affect the ionic composition of the subretinal space and the transport functions of RPE. Active vectorial transport systems for other retinal metabolites (eg, taurine, methionine, and folate) have also been demonstrated. Therefore, it appears that RPE is important in maintaining the ionic environment of the subretinal space, which in turn is responsible for maintaining the integrity of the RPE–photoreceptor interface. The trans-RPE potential is the basis for the electro-oculogram, which is the most common electrophysiologic test for evaluating the RPE.

Pigmentation

A characteristic feature of RPE is the presence of melanin pigment. Pigment granules are abundant in the cytoplasm of adult RPE cells, predominantly in the apical and mid-portions of the cell. During development, activation of the tyrosinase promoter triggers the onset of melanogenesis in this cell and marks the commitment of the neuroectoderm to become RPE. Although most melanogenesis occurs before birth, melanin production in RPE does occur throughout life, albeit at a slow rate. As humans age, the melanin granules fuse with lysosomes; thus, an elderly fundus is less pigmented than that of a young person.

The exact role of melanin inside cells remains speculative. One universally recognized role is that melanin acts as a neutral-density filter in scattering light. In so doing, melanin may have a protective function. In spite of the minimization of light scatter, visual acuity in the minimally pigmented fundus can be normal. Visual problems in albinos are attributable to foveal aplasia, not optical scatter. Genetic ocular disorders associated with melanin include varying forms of albinism. *OCA* refers to oculocutaneous albinism. *OCA1* and *OCA2* are due to defects in the tyrosinase gene and the pink-eyed dilution gene, respectively. When melanin levels are below a critical level, there is aberrant neuronal migration in the visual pathway, lack of foveal development, low vision, nystagmus, and strabismus (ocular albinism is characterized by a lack of pigment in the eye, but relatively normal pigmentation of skin and hair). Melanin is thought to play a role in retinal development because albino mammals have underdeveloped central retinas, more contralateral projections of ganglion cells, and failure of foveal development. Melanin is a free-radical stabilizer and can bind many toxins. Some regard this feature as protective; others think that it contributes to tissue toxicity.

Retinal Adhesion

None of the aforementioned functions would be possible without another RPE function, namely maintenance of retinal adhesion. The subretinal space is never bridged by tissue, and yet the neural retina remains rather firmly attached to RPE throughout life. This adhesion is vital to the retina since a detached retina can no longer register focused images and detached photoreceptors eventually degenerate for lack of nutrients. Multiple systems keep the retina in place. These factors include passive hydrostatic forces, interdigitation of outer segments and RPE microvilli, active transport of subretinal fluid, and the complex structure and binding properties of the interphotoreceptor matrix. In situations of pathology, retinal adhesion can diminish and detachment of the retina occurs. Detachment does not occur simply because there is a hole in the retina or a leak in the RPE; there must be either positive traction pulling the neural retina or positive forces pushing fluid into the subretinal space.

The RPE in Disease

Clearly, the RPE is vital for normal visual function. Three retinal degenerations in humans appear to be due to defects unique to the RPE: Sorsby fundus dystrophy and two forms of autosomal recessive retinitis pigmentosa. There are two generalized retinal degenerations: Usher syndrome (type 1B) and sex-linked retinitis pigmentosa, in which the defective gene is expressed strongly in RPE and weakly in neural retina. It has been suggested that some forms of macular degeneration, such as vitelliform macular degeneration (Best disease) and malattia leventinese (dominant drusen), may be due to a primary defect in the RPE. Age-related macular degeneration and Stargardt disease appear to affect the RPE early in their course, though the genetic defect is in the rods. Finally, choroideremia and gyrate atrophy produce blindness by their early impact on the RPE. In certain pathologic situations (including proliferative vitreoretinopathy and subretinal neovascularization), RPE cells detach from the basement membrane and become migratory. Efforts are now underway to determine effective methods of RPE transplantation that may ameliorate the functional deficits in these diseases.

Gallemore RP, Hughes BA, Miller SS. Retinal pigment epithelial transport mechanisms and their contributions to the electroretinogram. *Prog Retin Eye Res.* 1997;16: 509–566.

Marmor MF, Wolfensberger TJ, eds. *The Retinal Pigment Epithelium: Function and Disease.* New York: Oxford; 1998:103–134.

CHAPTER XVIII

Free Radicals and Antioxidants

Adverse effects of reactive forms of oxygen have been repeatedly proposed as causal factors in many types of tissue pathology, including cataract and age-related macular degeneration. Lipid peroxides are formed when oxyradicals or singlet oxygen molecules react with unsaturated fatty acids, which are present in cells largely as glycerylesters in phospholipids or triglycerides. The oxidation of membrane phospholipids has been hypothesized to increase permeability of cell membranes and/or to inhibit membrane ion pumps. This loss of barrier function is thought to lead to edema, disturbances in electrolyte balance, and elevation of intracellular calcium, all of which contribute to cell malfunction.

Cellular Sources of Active Oxygen Species

Free radicals are molecules or atoms that possess an unpaired electron. This property makes them highly reactive toward other molecular species. For example, free-radical reactions with polyunsaturated fatty acids in the presence of oxygen lead to the rapid formation of fatty-acid hydroperoxides, which increase permeability of membranes, causing electrolyte disturbances. Some free-radical reactions are involved in normal cell functions; others are thought to be important mediators of tissue damage. Oxygen-derived free radicals and their metabolites are generated within aerobic organisms in several ways.

Oxygen necessary for normal metabolism usually undergoes tetravalent (four-electron) reduction by intracellular systems, such as cytochrome oxidase in mitochondria (Fig XVIII-1), and is finally discarded as water without leakage of reactive intermediates. However, a small percentage of the metabolized oxygen undergoes univalent reduction in four one-electron steps. Oxygen accepts an electron from a reducing agent in each of these steps, and several highly reactive intermediates are formed:

- The superoxide free radical (O_2^-)
- Hydrogen peroxide (H_2O_2)
- The hydroxyl radical (OH•)

Some of the reactive species leak out of their enzyme-binding sites and may damage other components of tissues such as proteins, membrane lipids, and DNA if not captured by detoxifying enzymes. Superoxide is not only produced in mitochondrial electron-transport systems but is also formed in some enzymatic reactions, such as the xanthine and xanthine oxidase system. Hydrogen peroxide is produced directly in peroxisomes, as well as by enzyme-catalyzed dismutation of superoxide. Hydroxyl radicals can be formed through the mechanism shown in Figure XVIII-1. In addition, any free iron (Fe^{2+}) present may catalyze formation of the hydroxyl radical from superoxide and hydrogen peroxide. Iron and other catalytic metals such as copper may also be involved in generation of these species by accelerating nonenzymatic oxidation of several molecules, including glutathione.

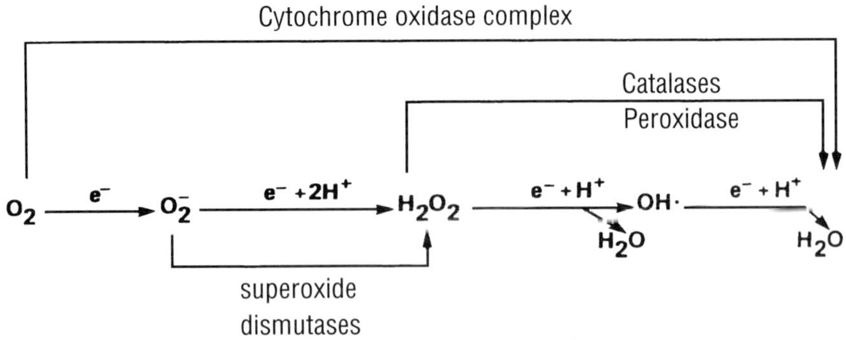

FIG XVIII-1—Enzymes involved in the metabolism of oxygen and in the detoxification of oxygen radicals generated by the univalent reduction of molecular oxygen. The univalent pathway involves a series of single electron transfers, producing the superoxide free radical (O_2^-), hydrogen peroxide (H_2O_2), water, and the hydroxyl radical (OH•). Superoxide dismutase catalyzes the conversion of superoxide to hydrogen peroxide without oxidizing other molecules. Catalase and peroxidase catalyze the reduction of hydrogen peroxide to water without formation of the toxic hydroxyl radical. These enzyme systems are capable of preventing the buildup of toxic species produced from the univalent reduction of oxygen. The cytochrome oxidase complex appears to catalyze the tetravalent reduction of oxygen to water without leakage of reactive intermediates. (Courtesy of F.J.G.M. van Kuijk, MD, PhD.)

Other sources of activated oxygen species include products from the enzymatic synthesis of prostaglandins, leukotrienes, and thromboxanes (see Chapter XII, Fig XII-1). The synthesis of these compounds starts with formation of arachidonic acid hydroperoxide by the enzyme lipoxygenase or cyclic peroxide by the enzyme cyclooxygenase. The NADPH oxidase system of phagocytes yields activated oxygen species, especially during inflammation reactions. Oxygen-radical production is also associated with ionizing radiation and metabolism of many chemicals and drugs, including carcinogenic compounds. Formation of reactive oxygen species—such as singlet oxygen by a light-mediated mechanism—is considered below.

Superoxide and hydrogen peroxide are relatively stable in biological systems, whereas the hydroxyl radical is extremely reactive and capable of producing broad, nonspecific oxidative damage. However, free radicals and other active oxygen species also are important in many biological reactions that maintain normal cell functions, such as mitochondrial and microsomal electron-transport systems.

Mechanisms of Lipid Peroxidation

The mechanism by which random oxidation of lipids takes place is called *auto-oxidation.* This oxidation is a free-radical chain reaction usually described as a series of three processes: initiation, propagation, and termination. The reaction sequence during the auto-oxidation of the fatty acid linoleate (18:2w6) is shown in Figure XVIII-2. During the initiation step, the fatty acid is converted to an intermediate radical after removal of an allylic hydrogen. Immediately, the propagation step follows, and the fatty-acid radical intermediate reacts with oxygen at either end to produce fatty-acid peroxy radicals, removing hydrogens from other fatty acids to form the 9

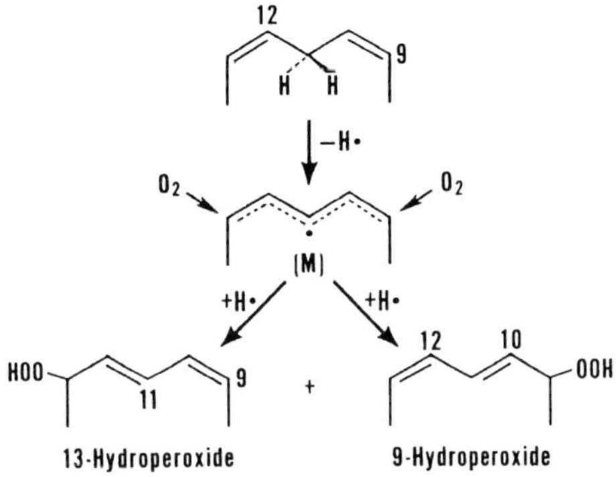

FIG XVIII-2—Mechanism of linoleate auto-oxidation. Formation of two conjugated isomers: 9-hydroperoxide and 13-hydroperoxide. (Courtesy of F.J.G.M. van Kuijk, MD, PhD.)

and 13 conjugated hydroperoxides. Thus, a new fatty-acid radical is formed, which again can react with oxygen. As long as oxygen is available, one free radical can lead to oxidation of thousands of fatty acids. A termination reaction, in which two radicals form a nonradical product, can interrupt the chain reaction. Auto-oxidation is also inhibited by free-radical scavengers such as vitamin E, which cause termination reactions. Auto-oxidation of arachidonic acid and docosahexaenoic acid yields 6 and 10 conjugated isomers, respectively.

Polyunsaturated fatty acids are susceptible to auto-oxidation because their allylic hydrogens are easily removed by several types of initiating radicals. The primary products of auto-oxidation formed during the propagation step are hydroperoxides (ROOH), which may decompose, especially in the presence of trace amounts of transition metal ions (eg, free iron or copper), to create peroxy radicals (ROO•), hydroxy radicals (HO•), and oxy radicals (RO•).

Photo-oxidation is a process by which oxygen is activated electronically by light to form singlet oxygen, which in turn reacts at a diffusion-controlled rate with unsaturated fatty acids or other cellular constituents. The mechanism of singlet-oxygen generation most often discussed involves exposure of a photosensitizer to light in the presence of normal triplet oxygen ($^3O_2$). A photosensitizer is excited by absorption of light energy to an excited singlet state, which rapidly relaxes to an excited triplet state. In this state, the sensitizer may react with triplet oxygen ($^3O_2$) to form singlet oxygen ($^1O_2$). Compared with auto-oxidation, for example, the photo-oxidation of linoleate yields not only the conjugated 9 and 13 hydroperoxides, but the nonconjugated 10 and 12 hydroperoxides as well (Fig XVIII-3). Therefore, the occurrence of the nonconjugated hydroperoxide isomers indicates singlet oxygen–induced damage to the polyunsaturated fatty acids. Photo-oxidation can be inhibited by singlet-oxygen quenchers such as carotenoids, which are discussed later in this chapter.

Lipid peroxidation not only causes direct damage to the cell membrane but also causes secondary damage in cells through its aldehydic breakdown products. Lipid hydroperoxides are unstable, and they break down to form many aldehydes such as malondialdehyde and 4-hydroxyalkenals. These aldehydes can react quickly with proteins, inhibiting their normal functions. For example, Na^+,K^+-ATPase is easily

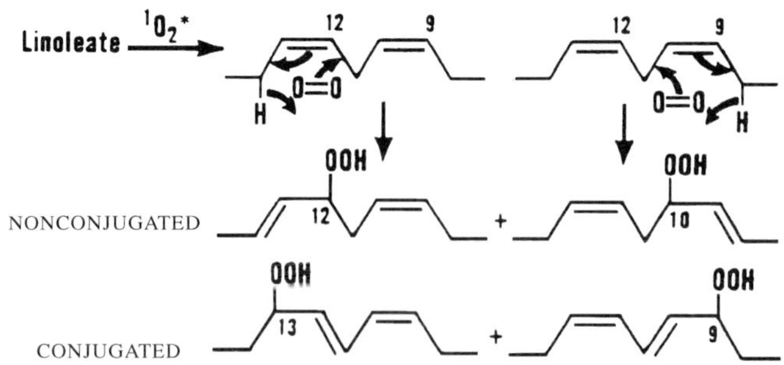

FIG XVIII-3—Mechanism of lineolated photo-oxidation. Formation of two conjugated and two nonconjugated isomers (10-hydroperoxide and 12-hydroperoxide). (Courtesy of F.J.G.M. van Kuijk, MD, PhD.)

inhibited by low concentrations of 4-hydroxyalkenals, causing severe electrolyte imbalances. Both the lens and the retina are susceptible to such oxidative damage.

Oxidative Damage to the Lens

The lens is susceptible to challenge by various active species of oxygen since it contains low levels of molecular oxygen and trace amounts of transition metals such as copper and iron. It is thought that metal-catalyzed auto-oxidation reactions of various reducing agents in the lens can lead to the production of potentially damaging oxidants such as oxidized glutathione (GSSG) and dehydroascorbic acid, as well as H_2O_2, which can go on to produce hydroxyl radicals via Fenton-type reactions. In addition, ultraviolet radiation entering the lens can generate oxidative species. Although most UVB radiation (<320 nm wavelength) striking the human eye is absorbed either by the cornea or by the high level of ascorbic acid in the aqueous humor, a certain proportion is able to reach the lens epithelium, where it can cause damage. UVA light (320–400 nm wavelength) is able to reach more deeply into the lens, where it can react with various chromophores to generate H_2O_2, superoxide anion, and singlet oxygen. Although repair or regeneration mechanisms are active in the lens epithelium and superficial cortex, this is not the case in the deep cortex and the nucleus, where any damage to lens proteins and membrane lipids is irreversible. One result of this damage can be cross-linking and insolubilization of proteins leading to loss of transparency (see Chapter XIV). Certain types of human cataracts appear to initiate at the site of the fiber cell plasma membrane; this may be associated with the fact that oxygen is five to seven times more soluble in membrane lipids than in the cytoplasm.

To defend against oxidative stress, the young, healthy lens possesses a variety of effective antioxidant systems. These include the enzymes glutathione peroxidase, catalase, and superoxide dismutase (Fig XVIII-4). By means of the glutathione redox cycle, GSSG is reconverted to glutathione (GSH) by glutathione reductase via the pyridine nucleotide NADPH provided by the hexose monophosphate shunt as the reducing pathway. Thus, GSH acts as a major scavenger of active oxygen species in the lens. The mammalian lens contains unusually high levels of protein sulfhydryl groups for reasons that are not well understood; however, it is clear that the groups

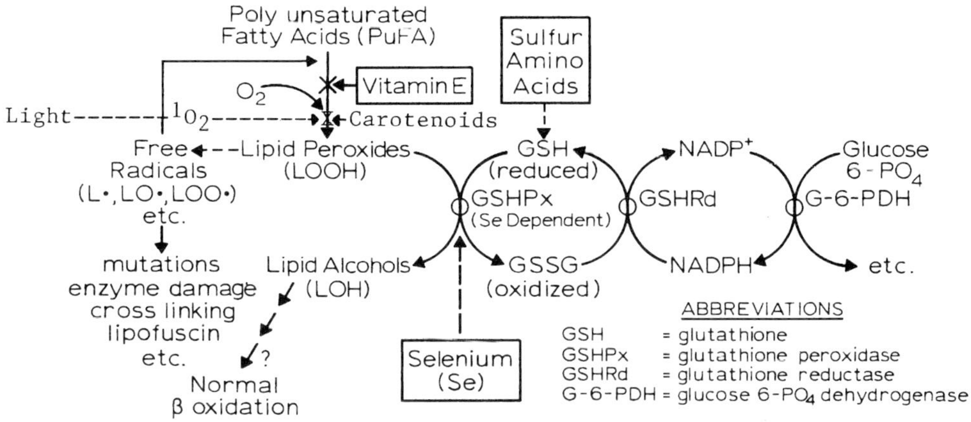

FIG XVIII-4—Mechanisms by which several antioxidants protect against oxidative damage. *Upper left,* Free radicals lead to the formation of lipid peroxides. Vitamin E inhibits this auto-oxidation process by scavenging free-radical intermediates. Carotenoids inhibit photo-oxidation by quenching singlet oxygen ($^1O_2$). *Center,* If lipid hydroperoxides are formed, they can be reduced by glutathione peroxidase (GSHPx), which requires selenium as a cofactor. If these protective enzymes are not fully active, more free radicals are formed by breakdown of lipid peroxides, which in turn leads to additional oxidation of polyunsaturated fatty acids. (Courtesy of F.J.G.M. van Kuijk, MD, PhD.)

must exist nearly completely in the reduced state for the tissue to remain transparent. The young human lens contains a high level of GSH that is first synthesized in the epithelium and that then migrates to the lens cortex and nucleus. With age, levels of GSH decline significantly in the human lens, particularly in the nucleus. Studies have indicated that a cortical/nuclear barrier may exist in the mature human lens, which inhibits the free flow of GSH to the nucleus. The result is that with age, the human lens nucleus becomes more susceptible to oxidative damage and cataract. Nuclear cataracts show high levels of oxidized cysteine and methionine in the lens proteins. Animals treated with hyperbaric oxygen, or humans treated therapeutically with oxygen, develop increased light scattering in the lens nucleus and nuclear cataracts.

The free-radical scavengers ascorbic acid and vitamin E are also present in the lens. These scavengers work in conjunction with GSH and the glutathione redox cycle to protect against oxidative damage. Carotenoids that can quench singlet oxygen also exist in the lens. Epidemiologic (observational) studies have shown that people with higher levels of plasma antioxidants, particularly vitamin E, have a reduced risk of cataract, particularly nuclear cataract. However, this observation does not automatically mean that antioxidant supplements would prevent cataract formation. The effectiveness of antioxidants remains to be established through double-blind, controlled trials.

Lyle BJ, Mares-Perlman JA, Klein BE, et al. Serum carotenoids and tocopherols and incidence of age-related nuclear cataract. *Am J Clin Nutr.* 1999;69:272–277.

Padgaonkar VA, Lin L-R, Leverenz VR, et al. Hyperbaric oxygen *in vivo* accelerates the loss of cytoskeletal proteins and MIP26 in guinea pig lens nucleus. *Exp Eye Res.* 1999;68:493–504.

Sweeney MH, Truscott RJ. An impediment to glutathione diffusion in older normal human lenses: a possible precondition for nuclear cataract. *Exp Eye Res.* 1998;67: 587–595.

Vulnerability of the Retina to Free Radicals

Experimental data have shown that retinal photoreceptors degenerate when they are exposed to oxidative challenges such as hyperbaric oxygen, iron overload, or injection of lipid hydroperoxides into the vitreous humor. The retina also degenerates when antioxidative defenses are reduced, which presumably elevates levels of lipid peroxidation in the absence of unusual oxidative stress. The retina is made vulnerable to damage from lipid peroxidation by several distinctive characteristics, four of which are considered here:

- Vertebrate retinal rod outer segments are susceptible to damage by oxygen because of their high content of polyunsaturated fatty acids. Their phospholipids typically contain about 50 mol% docosahexaenoic acid, the most highly polyunsaturated fatty acid that occurs in nature. It is well established that polyunsaturated fatty acids are sensitive to peroxidation in proportion to their number of double bonds.
- The rod inner segment is very rich in mitochondria, which may leak activated oxygen species.
- The excellent oxygen supply through the choroid and the retinal vessels elevates the risk of oxidative damage. Vertebrate retinas maintained in vitro showed at least a sevenfold higher rate of oxygen consumption per milligram of protein compared with all other tissues tested (except the adrenal gland). The oxygen tension is highest at the choroid and drops toward the inner segments because of the high metabolic demand of their mitochondria. Oxygen consumption has been reported to decrease when the retina is illuminated.
- Light exposure may trigger photo-oxidative processes mediated by singlet oxygen, and the RPE may play a key role.

The RPE is tightly packed with endoplasmic reticulum and appears to be rich in antioxidant enzymes in most species tested. The RPE of pigmented animals contains melanin granules, which function as a light trap. Although melanin is commonly assumed to be photoprotective, its role in prevention of light damage to ocular tissues is not clearly understood. Evidence suggests that the RPE is quite sensitive to dietary antioxidant deficiency, in which light-activated melanin may contribute to phototoxicity. If an RPE cell dies, then the numerous photoreceptors supported by that RPE cell may suffer severe damage or death.

Intense light at levels that may be encountered in daily life is phototoxic to the retina. Even though the cornea absorbs some UV radiation, the retinas of young people are exposed to a substantial amount of light in the range of 350–400 nm (young lenses transmit these wavelengths). The lens yellows with age, and the cutoff wavelength in the elderly moves up to about 430 nm. Because the adult lens absorbs nearly 100% of light below 400 nm, little or no UV light reaches the retina in older people. In addition to UV light, blue light (400–500 nm) can be harmful to the retina (blue-light hazard). Carotenoids present in the retina act as a blue-light filter, shielding the photoreceptors in the retina from this radiation.

Young RW. Solar radiation and age-related macular degeneration. *Surv Ophthalmol.* 1988;32:252–269.

Antioxidants in the Retina and RPE

Several antioxidant mechanisms have been established in biological systems, including free-radical scavenging, quenching of singlet oxygen, and enzymatic reduction of hydroperoxides. Antioxidants characterized in vertebrates include vitamin E, carotenoids, selenium, GSH, selenium-dependent glutathione peroxidase, and non–selenium-dependent glutathione peroxidase (glutathione-S-transferase). Antioxidants in vertebrates also include catalase and superoxide dismutase, and antioxidant roles for ascorbate and melanin have also been reported. The relation between some of these antioxidants and the protective mechanisms is shown in Figure XVIII-4.

Yu BP. Cellular defenses against damage from reactive oxygen species. *Physiol Rev.* 1994;74:139–162.

Selenium, Glutathione, Glutathione Peroxidase, and Glutathione-S-Transferase

A number of enzymes have been identified that can provide antioxidant protection by a peroxide-decomposing mechanism. For example, selenium-dependent glutathione peroxidase (GSH-Px) and several enzymes of the glutathione-S-transferase (GSH-S-Ts) group can reduce organic hydroperoxides. GSH-Px is also active with H_2O_2 as a substrate, although the GSH-S-Ts group cannot act on H_2O_2. All of these enzymes require GSH, which is converted to GSSG during the enzymatic reaction. The hexose monophosphate shunt enzymes produce NADPH, which is needed for reduction of GSSG by GSH-reductase. Both GSH-Px and GSH-S-Ts activities have been measured in human and animal retinas. It was found that the human retina also has GSH-S-Ts activities that specifically utilize 4-hydroxyalkenals as substrate and thus may constitute another defense mechanism. The highest concentration of selenium in the human eye is present in the RPE: 100–400 ng in RPE cells of a single human eye, up to 10 times more than in the retina (40 ng). The selenium level in the human retina is constant with age; in the human RPE, however, selenium increases with age. The two eyes of the same subject show no differences.

Superoxide Dismutase (SOD) and Catalase

Superoxide dismutase catalyzes the dismutation of superoxide to hydrogen peroxide, which is further reduced to water by catalase or peroxidase. Two types of SOD are usually isolated from mammalian tissues: *Cu-Zn SOD,* the cytoplasmic enzyme, which is inhibited by cyanide; and *Mn SOD,* the mitochondrial enzyme, which is not inhibited by cyanide.

Catalase catalyzes the reduction of hydrogen peroxide to water. Information on catalase activity in the retina is currently rather limited. Total retinal catalase activity was found to be very low but detectable in the rabbit. A protective role for catalase has been reported in rats with experimental allergic uveitis.

Vitamin E

Vitamin E acts by scavenging free radicals, thus terminating the propagation steps and leading to interruption of the auto-oxidation reaction. Reports on the vitamin E content of the retina of the adult rat raised on normal chow diets show values ranging from 215 to 325 ng. A detailed study on vitamin E content of microdissected parts of vertebrate eyes showed that the RPE is rich in vitamin E relative to the photoreceptors, and that photoreceptors are rich in vitamin E relative to most other tissues in the rat. An analysis of the effects of varying dietary vitamin E intake on the vitamin E content of the isolated rod outer segments and other retinal components showed that these tissues are depleted of vitamin E more slowly than most other tissues during dietary vitamin E deficiency.

Studies on vitamin E in postmortem human eyes found that vitamin E is higher in the RPE than in the retina. Furthermore, the vitamin E levels in human retinal tissues increase with age until the sixth decade of life and then decrease. This decrease begins at the age that also marks an increase in the incidence of age-related macular degeneration (ARMD). Epidemiologic studies on the relation between antioxidants and ARMD have yielded different results, and currently the Age-Related Eye Disease Study (AREDS) is under way to help understand the role of antioxidants in ARMD. A series of reviews on ARMD have been published on the World Wide Web in the online journal *Molecular Vision;* one example is listed below.

Friedrichson T, Kalbach HL, Buck P, et al. Vitamin E in macular and peripheral tissues of the human eye. *Curr Eye Res.* 1995;14:693–701.

Winkler BS, Boulton ME, Gottsch JD, et al. Oxidative damage and age-related macular degeneration. *Mol Vis.* 1999;5:32. *www.molvis.org/molvis/v5/p32/*

Ascorbate

Ascorbate (vitamin C) is thought to function synergistically with vitamin E to terminate free-radical reactions. It has been proposed that vitamin C can react with the vitamin E radicals formed when vitamin E scavenges free radicals. Vitamin E radicals are then regenerated to native vitamin E. The vitamin C radicals thus produced can be reduced by NADH reductase, with NADH as the electron acceptor. Ascorbic acid is found throughout the eye of many species in concentrations that are high relative to other tissues.

Delamere NA. Ascorbic acid and the eye. In: Harris JR, ed. *Subcellular Biochemistry, Volume 25: Ascorbic Acid: Biochemistry and Biomedical Cell Biology.* New York: Plenum Press; 1996:313–329.

Carotenoids

Various roles have been proposed for carotenoids (xanthophylls) in biological systems, including limiting chromatic aberration at the fovea of the retina and the quenching of singlet oxygen. β-Carotene is the precursor for vitamin A and can act as a free-radical trap at low oxygen tension. In postmortem human retinas, carotenoids have been shown to make up the yellow pigment in the macula. A mixture of the two carotenoids *lutein* and *zeaxanthin* (Fig XVIII-5) is present in the macula and located in the fibers of Henle. It has been demonstrated that in humans, zeaxanthin is concentrated primarily in the fovea, whereas lutein is dispersed throughout the retina. Interestingly, little β-carotene is present in the human eye.

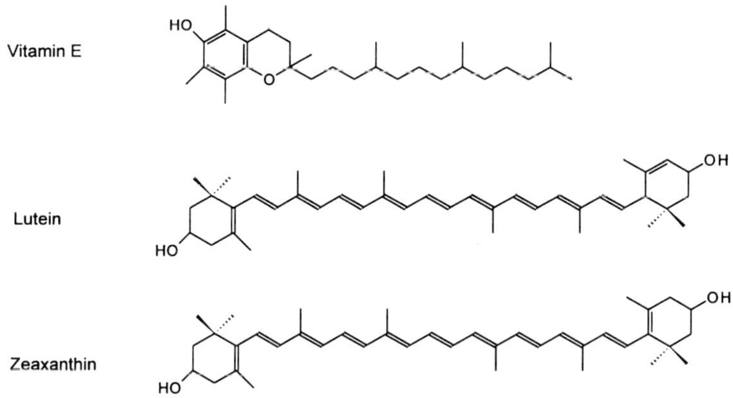

FIG XVIII-5—Structure of vitamin E and the carotenoids lutein and zeaxanthin. (Courtesy of F.J.G.M. van Kuijk, MD, PhD.)

Furthermore, carotenoids are present only in the retina and not at all in the RPE. In the peripheral retina, lutein and zeaxanthin are also concentrated in the photoreceptor outer segments and may act as antioxidants to protect the macula against short-wavelength visible light. It is not yet known whether carotenoids are also present in cone outer segments, because a good purification method for cones is not available (note that there is no yellow color in the cones in Figure XVIII-6B). Figure XVIII-5 shows the structure of vitamin E and the carotenoids. Figure XVIII-6A shows their localization in the human macula and peripheral retina, whereas Figure XVIII-6B shows their localization in a cross section of the peripheral retina.

Khachik F, Bernstein PS, Garland DL. Identification of lutein and zeaxanthin oxidation products in human and monkey retinas. *Invest Ophthalmol Vis Sci.* 1997;38: 1802–1811.

Mayne ST. Beta-carotene, carotenoids, and disease prevention in humans. *FASEB J.* 1996;10:690–701.

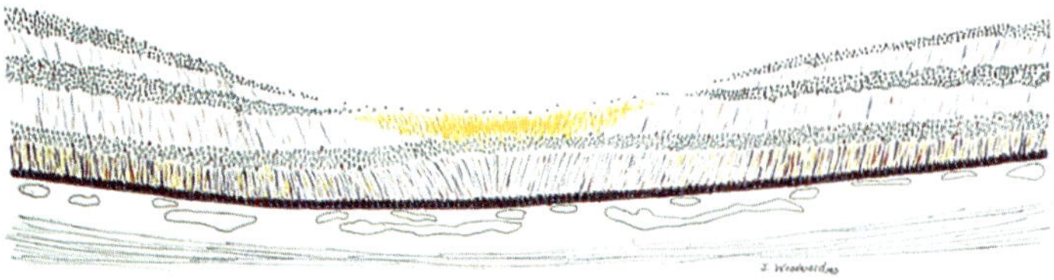

A

FIG XVIII-6—**A**, The localization of antioxidants in the human macula and peripheral retina. *Yellow* represents carotenoids, *blue* represents vitamin E, and *red* represents selenium. Vitamin E and selenium are primarily concentrated in the retinal pigmented epithelium *(RPE)*. In the macula, carotenoids are present in the fibers of Henle; in the peripheral retina, carotenoids are also present in the rods. **B**, The localization in a cross section of the peripheral retina. Vitamin E and selenium remain primarily concentrated in the RPE but are also enriched in the rod outer segments. Carotenoids have been found in rod outer segments in the peripheral retina, but it is not yet known whether they are also present in outer segments of cones. (Illustrations by J. Woodward, MD.) (Courtesy of F.J.G.M. van Kuijk, MD, PhD.)

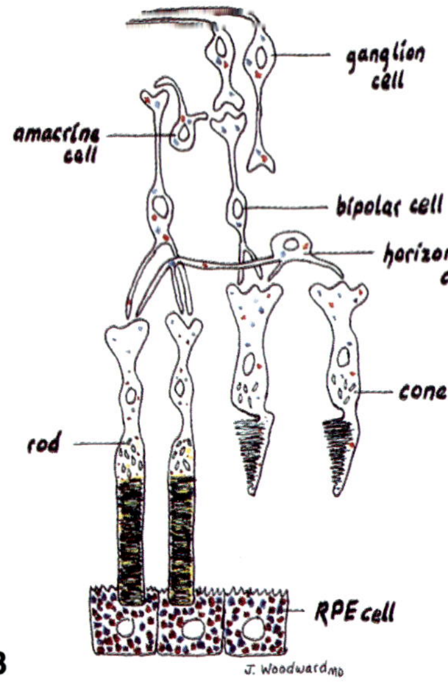

B

PART 5

OCULAR PHARMACOLOGY

CHAPTER XIX

Introduction

The majority of drugs used to treat ocular conditions either mimic or block the actions of neural or humoral transmitters. Pharmacologic interventions generally work by manipulation of physiologic processes, yet they differ from these processes in many ways. For example, the pharmacologic agents in dilating drops are applied in a location remote from their intended sites of action but in relatively enormous quantities that swamp homeostatic mechanisms; the neurotransmitters that control the pupil are released in precise locations in minute quantities and participate in a delicately balanced system.

Chronic pharmacologic interventions, as in the treatment of glaucoma, depend on a frequently unreliable mechanism—the patient—to deliver agents. These interventions use bolus administration and require empirical adjustment of doses. In contrast, physiologic processes in healthy patients are reliable mechanisms that continuously deliver appropriate quantities of agents via well-controlled avenues of feedback.

Pharmacologic agents often produce unwanted side effects by acting at sites other than the intended ones, often induce counteracting mechanisms (induction of drug-metabolizing enzymes, down-regulation of receptors) that alter their effectiveness, and occasionally induce allergic reactions. Physiologic agents are generally highly specific, work in concert with feedback mechanisms, and usually have immune privilege.

All of these differences between pharmacologic interventions and physiologic processes have practical implications for both the present and the future practice of ophthalmology. Even when underlying pathophysiologic lesions are unknown and current therapies merely create counteracting lesions, modeling therapies to mimic features of physiologic processes often creates an advantage. Attempts to deliver agents preferentially to the desired site of action and continuously from sustained-release devices or depots have been and will be rewarded by improved efficacy and compliance and reduction of side effects. Future advances in therapy might come from mimicking or harnessing physiologic processes, as in the following examples:

- Coupling drug release to feedback-control mechanisms (eg, glaucoma therapy regulated by a pressure-sensing strain gauge)
- Achieving higher specificity of action by designing drugs to complement the geometry of receptor sites
- Manipulating the genome to effect repair or regeneration through direction of physiologic processes

Contributors to this revised section include Moonyoung Chung, MD, and Arun Gulani, MD, from University of Florida.

CHAPTER XX

Pharmacologic Principles

The study of ocular pharmacology begins with a review of some general principles of pharmacology, with particular attention to special features of the eye that facilitate or impede certain therapeutic approaches.

Pharmacokinetics

To achieve a therapeutic effect, a drug must reach its site of action in sufficient concentration. The concentration at the site of action is a function of the following:

- Amount administered
- Extent and rate of absorption at the administration site
- Distribution and binding in tissues
- Movement by bulk flow in circulating fluids
- Transport between compartments
- Biotransformation
- Excretion

Eyedrops

Most ocular medications are administered as eyedrops. With this route of administration, adequate concentrations can be achieved in the anterior segment without incurring unwanted effects in other body systems, an advantage over systemic therapy for most drugs.

Some features of topical ocular therapy limit its effectiveness. Very little of an administered drop is retained by the eye. When a 50 μl eyedrop is delivered from the usual commercial dispenser, the volume of lacrimal fluid held by the eyelids and cul-de-sac rises from 7 μl to only 10 μl in the blinking eye of an upright patient. Thus, at most 20% of the administered drug is retained (10 μl/50 μl). A rapid turnover of fluid in the tear reservoir also occurs, 16% per minute in the undisturbed eye and much more if the drop elicits reflex tearing. Consequently, for slowly absorbed drugs, only 50% of the drug that reached the tear reservoir remains 4 minutes after instillation ($0.84^4 \cong 0.50$), and only 17% remains after 10 minutes.

Some simple measures have been shown to improve ocular absorption of materials that do not traverse the cornea rapidly. Patients taking more than one eyedrop medication should be instructed to wait 5 minutes between drops; otherwise, the second drop may simply wash out the first. Patients can also be instructed to compress the nasolacrimal duct with digital pressure at the medial canthus, both to prevent egress of tears by that route and to reduce systemic absorption through the nasal mucosa. The lacrimal pumping mechanism can also be halted by the simpler measure of instructing patients to keep their eyes closed for 5 minutes after taking drops.

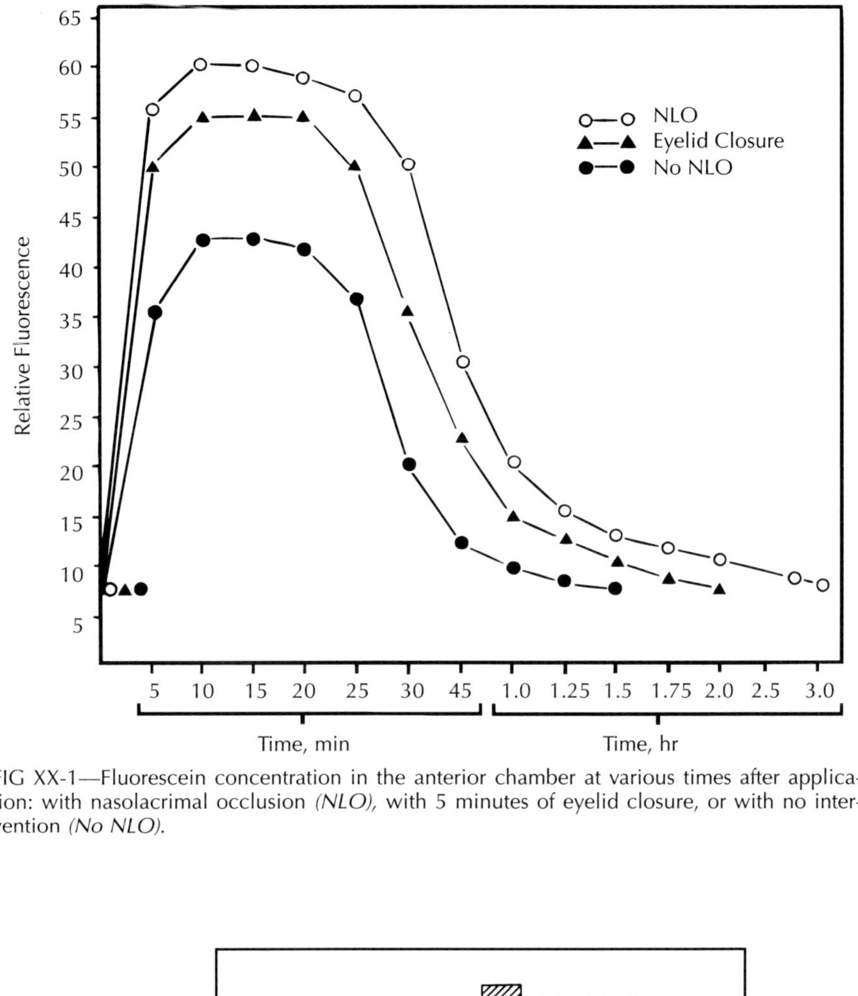

FIG XX-1—Fluorescein concentration in the anterior chamber at various times after application: with nasolacrimal occlusion *(NLO),* with 5 minutes of eyelid closure, or with no intervention *(No NLO).*

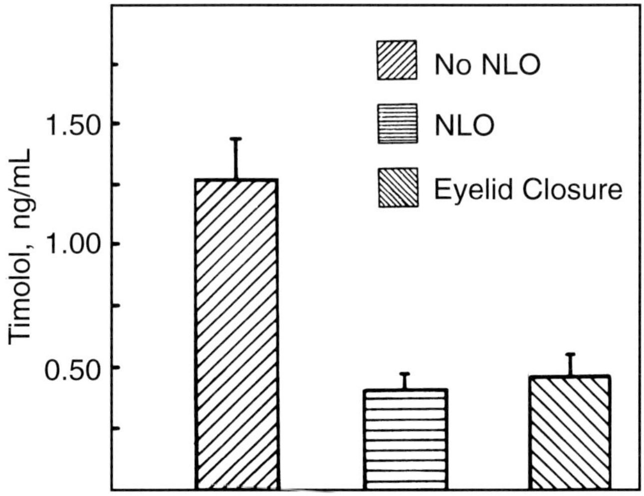

FIG XX-2—Systemic absorption of timolol at various times after application: with nasolacrimal obstruction *(NLO),* with 5 minutes of eyelid closure, or with no intervention *(No NLO).*

This step increases the ocular absorption of topically applied materials (Fig XX-1) and decreases systemic absorption (Fig XX-2).

Because the contact time of eyedrop medications is short, the rate of transfer from the tear fluid into the cornea is critical. The cornea lacks fenestrated barriers such as the vascular endothelium and mucosa of the stomach, which allow rapid diffusion of all but large molecules through extracellular passages. Instead, the corneal epithelium and endothelium have tight intercellular junctions that limit passage in the extracellular space; therefore, drugs must pass through cell membranes. Breakdown of these barriers in clinical situations such as corneal abrasion multiplies the rates of intraocular drug penetration.

Similar considerations apply to conjunctiva. There is growing evidence that the permeability of the conjunctiva to small water-soluble molecules is 20 times that of the cornea. Perilimbal conjunctiva offers an effective trans-scleral route for delivery of drugs to anterior segment structures.

Among the other factors that determine the amount of medication to penetrate the cornea are drug concentration and solubility, viscosity, lipid solubility, surfactants, and reflex tearing.

Drug concentration and solubility In order to get a sufficient amount of a drug through the corneal barriers, it is often necessary to load the small tear reservoir with concentrated solutions (eg, 1%–4% pilocarpine). A practical limit to exploiting these high concentrations is reached when the high tonicity of the resulting solutions elicits reflex tearing or when drugs that are poorly water-soluble reach their solubility limits.

Viscosity The addition of high-viscosity substances such as methylcellulose and polyvinyl alcohol increases drug penetration. However, because there is little correlation between effectiveness and solution viscosity, such substances may act by altering the barrier function of the corneal epithelium as well as by increasing drug contact time with the cornea.

Lipid solubility To traverse the cornea, a drug must pass in turn through the lipid-rich environment of the epithelial cell membranes, the water-rich environment of the stroma, and another lipid barrier at the endothelium. Studies of the permeability of isolated corneas to families of chemical compounds show that lipid solubility is more important than water solubility in promoting penetration.

First, the ratio of lipid solubility to water solubility is ascertained for each compound in the series. This ratio is determined by measuring the *phase separation* of a drug between two solvents (eg, octanol and water) and then calculating the ratio of concentrations in the two compartments *(partition coefficient).* The greater the relative lipid solubility, the higher the partition coefficient.

For substituted ethoxzolamides, the permeability coefficient is 70 times higher for compounds of high lipid solubility than for those of low lipid solubility (Fig XX-3). However, compounds with excessively high partition coefficients are often poorly soluble in tears. To develop the most effective drugs, studies of systematically substituted compounds need to account for the effects of substituents on potency and solubility as well as on the permeability coefficient.

Many eye medications are alkaloids, or weak bases. Such drugs as tropicamide, cyclopentolate, atropine, and epinephrine exist in both charged and uncharged forms at the slightly alkaline pH of tears. The partition coefficients of these drugs can

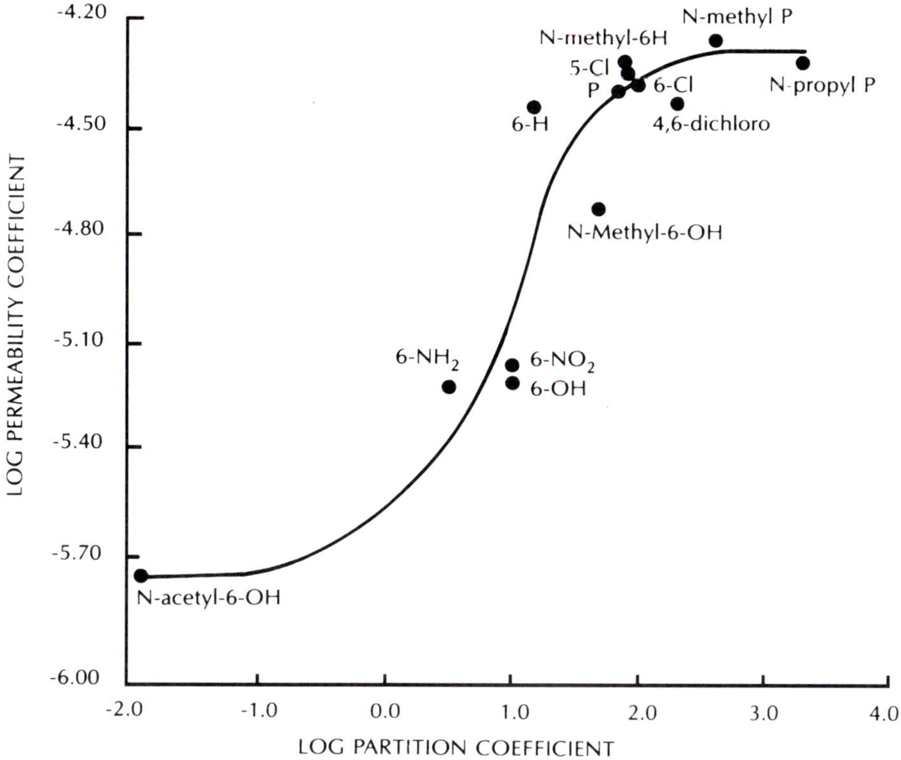

FIG XX-3—The log–log relationship of permeability coefficient (rabbit cornea) and partition coefficient (octanol/buffer). The more lipid-soluble compounds in the series (toward the right in the plot) are also the best able to penetrate the cornea (toward the top of the plot).

be increased by raising the pH of the water phase, thereby increasing the proportion of drug molecules in the more lipid-soluble, uncharged form.

Surfactants Many preservative agents used in eyedrops to prevent bacterial contamination are surface-active agents that alter cell membranes in the cornea as well as in bacteria. They reduce the barrier effect of the corneal epithelium and increase drug permeability. For example, a 0.1% carbachol solution containing 0.03% benzalkonium chloride can elicit the same miotic response as a 2% solution without it.

Reflex tearing Such tearing reduces the contact time of the drug with the cornea. It can occur when eyedrops have a pH value very different from the physiologic 7.4, are not isotonic, or contain irritants.

Ointments

Another strategy for increasing the contact time of ocular medications is the use of ointments. Commercial oil-based ointments usually consist of petrolatum and mineral oil. The mineral oil allows the ointment to melt at body temperature. Both ingredients are also effective lipid solvents. However, most water-soluble medications are insoluble in the ointment and are present as microcrystals. Only those microcrystals on the surface of the ointment dissolve in the tears; the rest are trapped until the ointment melts. Such protracted but slow release may prevent the drug from reaching a therapeutic level in the tears. Only if the drug has a high lipid solubility (which allows it to diffuse through the ointment) and some water solubility will it escape from the ointment into both the corneal epithelium and the tears. Fluorometholone, chloramphenicol, and tetracycline are examples of drugs that achieve higher aqueous levels when administered as ointment than as drops.

Periocular Injections

Injection of medication beneath the conjunctiva or Tenon's capsule allows drugs to bypass the conjunctival and corneal epithelial barriers. Injections into these sites and retrobulbar injections can allow medications to reach therapeutic levels behind the lens–iris diaphragm. This approach is especially useful for drugs with low lipid solubilities (such as penicillin), which do not penetrate the eye adequately if they are given topically.

Intraocular Injections

Intraocular injection of drugs delivers effective concentrations at the target site instantly. However, the dangers inherent in intraocular injections far outweigh possible benefits in almost all circumstances. The only *established* indication for this route is the administration of antibiotics in the treatment of endophthalmitis. Only minute amounts of antibiotic are tolerated within the eye. For example, the maximum safe anterior chamber dose of polymyxin B is 0.1 mg.

Sustained-Release Oral Preparations

The practical value of sustained-release preparations is significant. For example, a single dose of acetazolamide will reduce intraocular pressure for up to 10 hours, whereas a single dose of sustained-release acetazolamide will produce a comparable effect lasting 20 hours. Sustained-release medications offer a more steady blood level of the drug, avoid marked peaks and valleys, and reduce the frequency of administration.

Systemic Therapy

Just as the intercellular tight junctions of the corneal epithelium and endothelium limit anterior access to the interior of the eye, similar barriers limit access through vascular channels. The vascular endothelium of the retina, like that of the brain, is nonfenestrated and knitted together by tight junctions. Although both the choroid and the ciliary body have fenestrated vascular endothelia, the choroid is effectively bound by the retinal pigment epithelium and the ciliary body by its nonpigmented epithelium.

The blood–ocular barrier, like the cornea, is more readily penetrated by drugs with higher lipid solubilities. Thus, chloramphenicol, which is highly lipid-soluble, penetrates 20 times better than penicillin, which has poor lipid solubility.

The ability of systemically administered drugs to gain access to the eye is also influenced by the degree to which they are bound to plasma proteins. Only the unbound form can cross the blood–ocular barriers. Sulfonamides are lipid-soluble but penetrate poorly, because at therapeutic levels more than 90% of the medication is bound to plasma proteins. Similarly, the greater protein binding of oxacillin reduces its penetration (compared with methicillin). Bolus administration of a drug exceeds the binding capacity of plasma proteins and leads to higher intraocular drug levels than can be achieved by a slow intravenous drip. This approach is used for the administration of antibiotics, which must reach high peak intraocular levels.

Intravenous Injections

Continuous intravenous administration of an antibiotic is assumed to be an effective way of maintaining intraocular levels. Because of the barriers and possible reservoir effects of the eye, however, antibiotics such as ampicillin, chloramphenicol, and erythromycin penetrate the eye at higher initial levels and maintain at least comparable intraocular levels for 4 hours when given as a single intravenous bolus rather than by continuous infusion.

Inflammation may affect the barrier properties of the eye to permit better penetration of substances from the circulation. This effect is demonstrated by the appearance of fluorescein in the vitreous as a result of leakage from the retinal vessels in the presence of inflammation.

The distribution of ampicillin, tetracycline, and dexamethasone in rabbit eye tissues after intravenous administration has been studied. The highest levels were found in the sclera and conjunctiva, followed by the iris and ciliary body, cornea, aqueous humor, choroid, and retina. Very low levels appeared in the lens and vitreous. The drugs showed no marked differences in their distribution. The distribution pattern is determined by the vascularity of the tissue and the barriers that exist between the blood and the tissue.

Methods of Ocular Drug Design and Delivery

Ocular drugs are being designed with a focus on specificity and safety, and new delivery systems hold promise for improving convenience and compliance. Each of the approaches that are discussed below responds to a specific problem in ocular pharmacokinetics.

Prodrugs Prodrugs are compounds that are inactive until enzymatically activated. *Dipivefrin HCl* (DPE, Propine) is a prodrug of epinephrine. The presence of two pivalyl residues confers lipid solubility and increases 17-fold the parent compound's ability to penetrate the cornea (Fig XX-4). Thus, a 0.1% solution of DPE can be used in place of epinephrine 1%–2%. The pivalyl groups are cleaved by corneal esterases, releasing epinephrine into the anterior chamber. Because DPE has low intrinsic activity and is used in a lower concentration, it is virtually free of the systemic side effects of epinephrine. It is therefore a better-targeted drug.

Sustained-release devices and gels Drop therapy involves periodic delivery of relatively large quantities of a drug. Enough of the drug must reach receptors to achieve a therapeutic effect, while more remains in the surrounding tissues to act as a local reservoir between applications. This application of greater quantities of the drug than

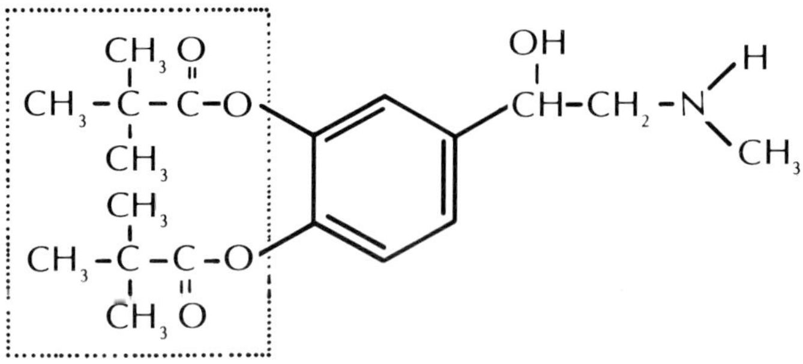

FIG XX-4—The structure of dipivefrin (DPE).

would be needed to achieve short-term effects often leads to unwanted side effects (eg, miosis and induced accommodation with the use of pilocarpine).

Devices have been developed that deliver an adequate supply of medication at a steady-state level, thus achieving beneficial effects with fewer side effects. The *Ocusert Delivery System* (Fig XX-5) delivers pilocarpine at a steady rate of 40 µg/hr, and its therapeutic effect in lowering intraocular pressure is equivalent to 2% pilocarpine used four times a day. Yet because the total daily dose of pilocarpine is only 960 µg (24 hours × 40 µg/hr) when delivered with the device as compared to 4000 µg (four doses × 2000 mg/100 ml × 0.05 ml/dose) with drops, miosis is less marked and the induced accommodation is reduced.

A newer form of timolol maleate, *Timoptic-XE,* is a novel example of sustained release. It contains a heteropolysaccharide that becomes a gel on contact with tear film and acts as a local reservoir, delivering therapeutic levels of the drug between applications.

Another example is the *ganciclovir sustained-release intraocular device (GIOD).* Ganciclovir, an antiviral agent, is incorporated into a polyvinyl alcohol coating and suspended from the sclera into the intravitreal cavity. An ethylvinyl acetate disc with polyvinyl alcohol coating serves as a reservoir. The thickness of the polyvinyl alcohol lid regulates the delivery of ganciclovir to target tissue. Typically, the implant is effective for 8 months. It was recently approved by the Food and Drug Administration for intraocular use.

Collagen cornea shields Porcine scleral tissue is extracted and molded into these contact lens–like shields, which are useful as a delivery system to prolong the contact between a drug and the cornea. Drugs can be incorporated into the collagen matrix during the manufacturing process, absorbed into the shield during rehydration, or applied topically over the shield in the eye. Because the shield dissolves in 12, 24, or 72 hours, depending on the manufacturing process for collagen cross-linking, the drug is released gradually into the tear film, and high concentrations are maintained on the corneal surface and conjunctival cul-de-sac.

The results of experimental investigations suggest that drug delivery by collagen shields may be more helpful in the early management of bacterial keratitis than other conventional modes. Clinical studies indicate that collagen-shield drug delivery may be useful in providing antibiotic prophylaxis against possible infection while promoting epithelial healing after ocular surgery, trauma, or spontaneous erosion.

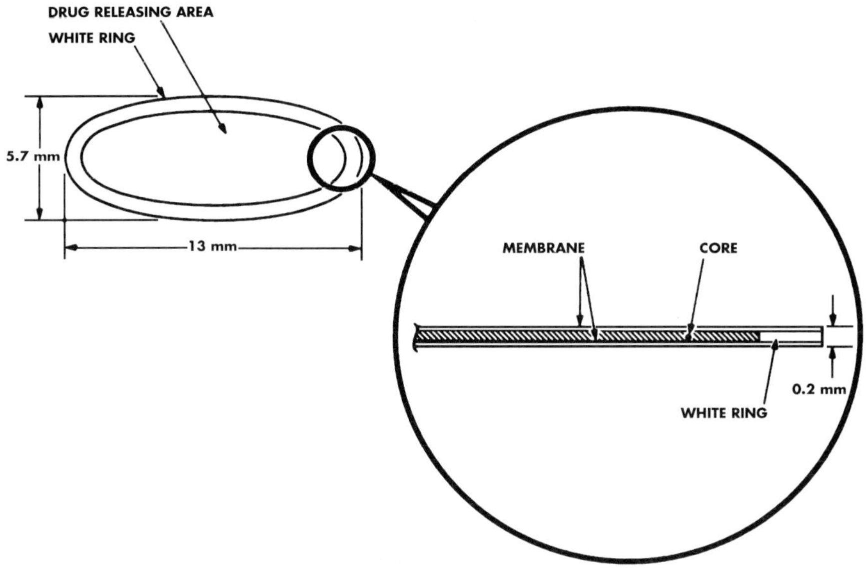

FIG XX-5—The Ocusert Delivery System.

Despite these therapeutic benefits, collagen shields are poorly tolerated because they are very uncomfortable.

Liposomes Liposomes are synthetic lipid microspheres that serve as multipurpose vehicles for the topical delivery of drugs, genetic material, and cosmetics. They are produced when certain phospholipid molecules interact to form a bilayer lipid membrane in an aqueous environment. The interior of the bilayer consists of the hydrophobic fatty-acid tails of the phospholipid molecule, whereas the outer layer is composed of hydrophilic polar-head groups of the molecule. The lipid membranes close up to produce a spherical envelope with the formation of either unilamellar or multilamellar liposomes.

Depending on their chemical properties, drugs can be incorporated into liposomes in several ways. A water-soluble drug is dissolved in the aqueous phase of the interior compartment, whereas a hydrophobic drug can be intercalated into the lipid bilayer itself. Liposome encapsulation of drugs for topical application and injection permits sustained levels of drug release over prolonged periods of time; it may also permit the use of high concentrations of drugs by substantially reducing toxicity to surrounding tissues. Gels that release pilocarpine and ocular lubricant slowly have also been developed.

Iontophoresis The physical process of moving charged molecules by an electrical current is called *iontophoresis.* This mechanism is based on the physical principle that ions are repelled by poles of the same charge and attracted by poles of opposite charge. A major advantage of using iontophoresis for drug delivery is the elimination of systemic toxicity. This procedure places a relatively high concentration of the drug locally, where it can achieve maximum benefit with little waste or systemic absorption. Animal studies have demonstrated that iontophoresis increases penetration of various antibiotics and antiviral drugs across ocular surfaces into the cornea and the interior of the eye. However, the disadvantages associated with the procedure (such as patient discomfort, ocular tissue damage, and necrosis) restrict the popularity of this mode of drug delivery.

Pharmacodynamics

Most drugs act by binding to and altering the function of regulatory macromolecules, usually neurotransmitter or hormone receptors or enzymes. Binding may be a reversible association mediated by electrostatic and/or van der Waals forces, or it may involve formation of a covalent intermediate. If the drug-receptor interaction stimulates the receptor's natural function, the drug is termed an *agonist.* Stimulation of an opposing effect characterizes an *antagonist.* Corresponding effectors of enzymes are termed *activators* and *inhibitors.* This terminology is crucial to understanding the next chapter.

The relationship between the initial drug-receptor interaction and the clinical drug's dose-response may be simple or complex. In some cases, the drug's clinical effect closely reflects the degree of receptor occupancy on a moment-to-moment basis. Such is usually the case for drugs that affect neural transmission or that are enzyme inhibitors. In contrast, some drug effects lag hours behind receptor occupancy or persist long after the drug is gone. Such is the case with many drugs acting on hormone receptors, because their effects are often mediated through a series of biochemical events.

In addition to differences in timing of receptor occupancy and drug effects, the degree of receptor occupancy can differ considerably from the corresponding drug effect. For example, because the amount of carbonic anhydrase present in the ciliary processes is 100 times that required to support aqueous secretion, more than 99% of the enzyme must be inhibited before secretion is reduced. On the other hand, some maximal hormone responses occur at concentrations well below that required for receptor saturation, indicating the presence of "inbound receptors."

Bochot A, Couvreur P, Fattal E: Intravitreal administration of antisense oligonucleotides: potential of liposomal delivery [Review]. *Prog Retin Eye Res.* 2000;19(2):131–147.

Eller MG, Schoenwald RD, Dixson JA, et al. Topical carbonic anhydrase inhibitors. III: Optimization model for corneal penetration of ethoxzolamide analogues. *J Pharm Sci.* 1985;74:155–160.

Kuwano M, Horibe Y, Kawashima Y. Effect of collagen cross-linking in collagen corneal shields on ocular drug delivery. *J Ocul Pharmacol Ther.* 1997;13(1):31–40.

Poland DE, Kaufman HE. Clinical uses of collagen shields. *J Cataract Refract Surg.* 1988;14:489–491.

Zimmerman TJ, Kooner KS, Kandarakis AS, et al. Improving the therapeutic index of topically applied ocular drugs. *Arch Ophthalmol.* 1984;102:551–553.

CHAPTER XXI

Ocular Pharmacotherapeutics

Cholinergic Agents

A number of commonly used ophthalmic medications affect the activity of acetylcholine receptors in synapses of the peripheral nervous system (Fig XXI-1). Such receptors are found in:

- The motor end plates of the extraocular muscles and levator palpebrae superioris (supplied by *somatic motor nerves*)
- The cells of the superior cervical ganglion (sympathetic) and the ciliary and sphenopalatine (parasympathetic) ganglia (supplied by *preganglionic autonomic nerves*)
- Parasympathetic effector sites in the iris sphincter and ciliary body and in the lacrimal, accessory lacrimal, and meibomian glands (supplied by *postganglionic parasympathetic nerves*)

Although all cholinergic receptors are by definition responsive to acetylcholine, they are not homogeneous in their response to other agents. Such agents fall into two categories:

- *Nicotinic agents* are supplied by somatic motor and preganglionic autonomic nerves and are responsive to nicotine.
- *Muscarinic agents* are supplied by postganglionic parasympathetic nerves and are responsive to muscarine.

Cholinergic agents are further divided into the following groups (Fig XXI-2):

- *Direct-acting agonists* act on the receptor to elicit an excitatory postsynaptic potential.
- *Indirect-acting agonists* inhibit the acetylcholinesterase of the synaptic cleft, preventing deactivation of endogenous acetylcholine.
- *Antagonists* block the action of acetylcholine on the receptor.

Muscarinic Drugs

Direct-acting agonists Topically applied direct-acting agonists have three actions. First, they cause contraction of the iris sphincter, which not only constricts the pupil (miosis) but also changes the anatomical relationship of the iris to the lens and the chamber angle. Second, they cause contraction of the circular fibers of the ciliary muscle, relaxing the zonular tension on the lens equator and allowing the lens to assume a more spherical shape (accommodation). The lens also undergoes a small forward displacement. Third, they cause contraction of the longitudinal fibers of the ciliary muscle, producing tension on the scleral spur (opening the trabecular meshwork) and increasing aqueous outflow facility. Contraction of the ciliary musculature also produces tension on the peripheral retina, occasionally resulting in a retinal tear or even rhegmatogenous detachment.

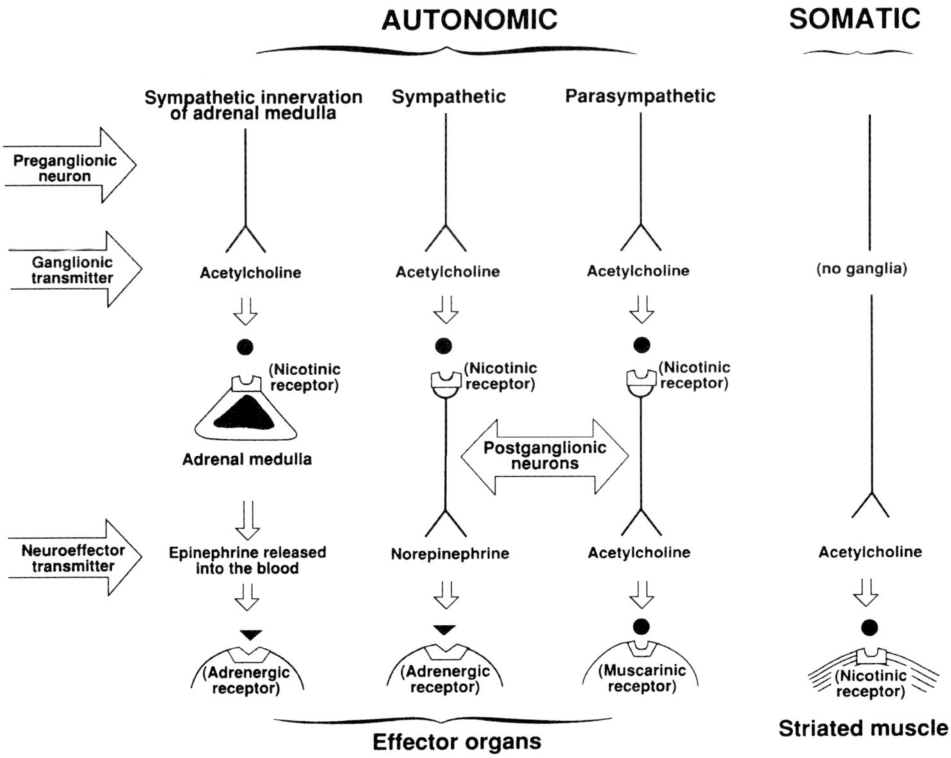

FIG XXI-1—Summary of the neurotransmitters released and the types of receptors found within the autonomic and somatic nervous systems. (Reproduced with permission from Mycek MJ, Harvey RA, Champe PC, eds. *Lippincott's Illustrated Reviews: Pharmacology.* 2nd ed. Philadelphia: Lippincott-Raven; 1997:32.)

Acetylcholine does not penetrate the corneal epithelium well, and it is rapidly degraded by acetylcholinesterase (Fig XXI-3). Thus, it is not used topically. Acetylcholine 1% (Miochol) and *carbachol* 0.01% (Miostat) are available for intracameral use in anterior segment surgery. These drugs produce prompt and marked miosis, which helps avoid iris capture by the optic of posterior chamber lenses and may prevent iris incarceration in surgical wounds.

Intracameral carbachol 0.01% is 100 times more effective and longer lasting than acetylcholine administered similarly. Maximal miosis is achieved within 5 minutes and lasts for 24 hours. In addition, carbachol 0.01% is an effective hypotensive agent and lowers intraocular pressure (IOP) during the critical 24-hour period after surgery. However, the drugs may also increase the risk of pupillary-block angle closure in the absence of a patent peripheral iridectomy.

Acetyl-β-methylcholine, also known as *methacholine* (Mecholyl) 2.5%, and *pilocarpine* 0.12% are used to confirm Adie tonic pupil, a condition in which the parasympathetic innervation of the iris sphincter and ciliary muscle is defective because of the loss of postganglionic fibers. Denervated muscarinic smooth muscle fibers in the affected segments of the iris exhibit supersensitivity and respond well to these weak miotics, whereas normal iris does not.

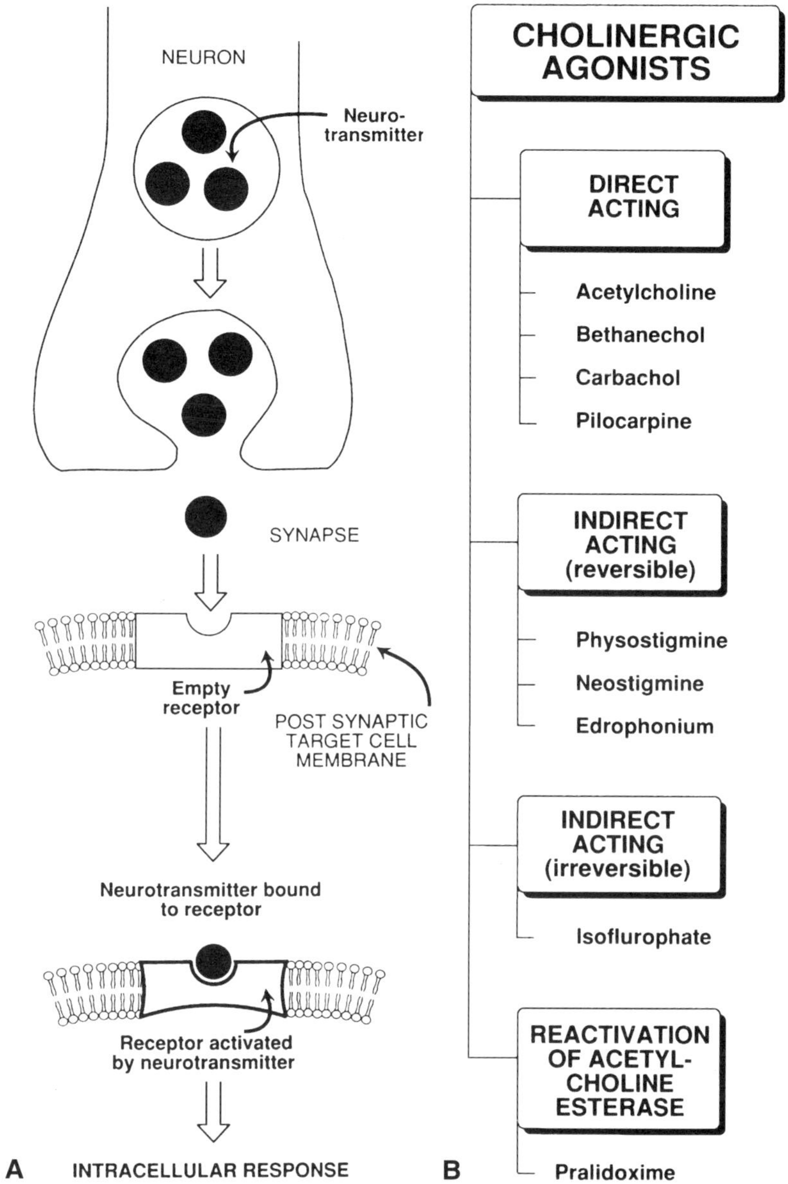

A **B**

FIG XXI-2—**A**, Neurotransmitter binding triggers an intracellular response. (Reproduced with permission from Harvey RA, Champe PC, eds. *Lippincott's Illustrated Reviews: Pharmacology*. Philadelphia: Lippincott; 1992:30.) **B**, Summary of cholinergic agonists. (Reproduced with permission from Mycek MJ, Harvey RA, Champe PC. *Lippincott's Illustrated Reviews: Pharmacology*. 2nd ed. Philadelphia: Lippincott-Raven; 1997:35.)

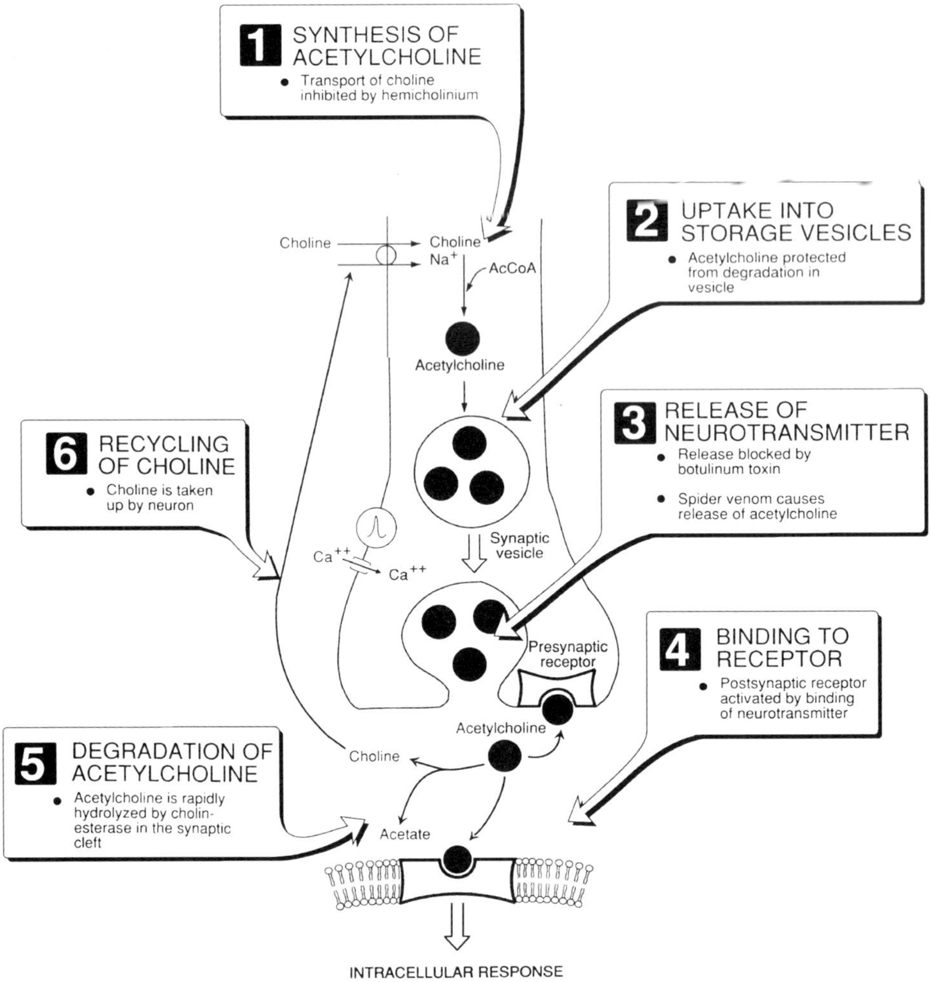

FIG XXI-3—Synthesis and release of acetylcholine from the cholinergic neuron. (Reproduced with permission from Mycek MJ, Harvey RA, Champe PC, eds. *Lippincott's Illustrated Reviews: Pharmacology*. 2nd ed. Philadelphia: Lippincott-Raven; 1997:37.)

Pilocarpine 1%–6% (QID) and carbachol 0.75%–3.00% (TID) are used in the treatment of primary open-angle glaucoma because they lower IOP by increasing outflow facility (Table XXI-1). Use of 4% pilocarpine is contraindicated in acute attacks, which may induce intense anterior movement of the lens–iris diaphragm, closing the angle completely. Miotic therapy is not an adequate substitute for laser iridotomy and should not be used for chronic control or prophylaxis of pupillary-block angle-closure glaucoma. (See also BCSC Section 10, *Glaucoma,* Chapter XI.)

Miosis, cataractogenesis, and induced myopia are generally unwelcome side effects of muscarinic therapy. Although the broad range of retinal dark adaptation

Table XXI-1

Miotics

GENERIC NAME	TRADE NAME	STRENGTHS (%)
Cholinergic Agents		
Carbachol	Isopto Carbachol	0.75%–3%
Pilocarpine hydrochloride	Akarpine	1%–4%
	Isopto Carpine	0.25%–10%
	Ocusert Pilo	20, 40 μg/hr
	Pilocar	0.5%–0.6%
	Pilopine HS gel	4%
	Piloptic	0.5%–0.6%
	Pilostat	1%–4%
	Available generically	0.5%–0.6%
Pilocarpine nitrate	Pilagan	1%–4%
Cholinesterase Inhibitors		
Physostigmine	Isopto Eserine	0.25%, 0.5%
	Available generically as Eserine Oph Oint	0.25%
Echothiophate iodide	Phospholine iodide	0.03%–0.25%
Pilocarpine polymer	Ocusert	20–40 μg
Pilopine gel	Pilopine gel HS	4.0%

usually compensates sufficiently for the effect of miosis on vision during daylight hours, patients may be visually incapacitated in dim illumination. In addition, miosis often compounds the effect of axial lenticular opacities; many cataract patients are unable to tolerate miotics. The myopia induced by ciliary muscle contraction may be disabling in persons younger than 50 years, who show substantial induced accommodation. Miotic cysts and increased incidence of retinal detachment are some of the other complications *when used at higher concentrations.* Systemic side effects after ocular use of pilocarpine are rare. They include salivation, diarrhea, vomiting, bronchial spasm, and diaphoresis (Fig XXI-4).

Ciliary muscle stimulation can be desirable in the management of accommodative esotropia. The near response is a synkinesis of accommodation, miosis, and convergence. Muscarinic agonists reduce the need to accommodate; thus, accommodative esotropia is also reduced because the patient not only experiences less accommodation but also less convergence.

Side effects may be reduced by using a device (Ocusert) that delivers the drug continuously at a low rate. Alternatively, a sufficient amount of the drug for a day's therapy can be delivered in a slowly dissolving gel (Pilopine HS gel) administered at bedtime so that the unwanted effects occur primarily during sleep. Both Ocusert and pilocarpine gel may be useful in some younger patients, in patients bothered by variable myopia or intense miosis, in older patients with lens opacities, and in patients who have difficulty complying with more frequent dosing regimens.

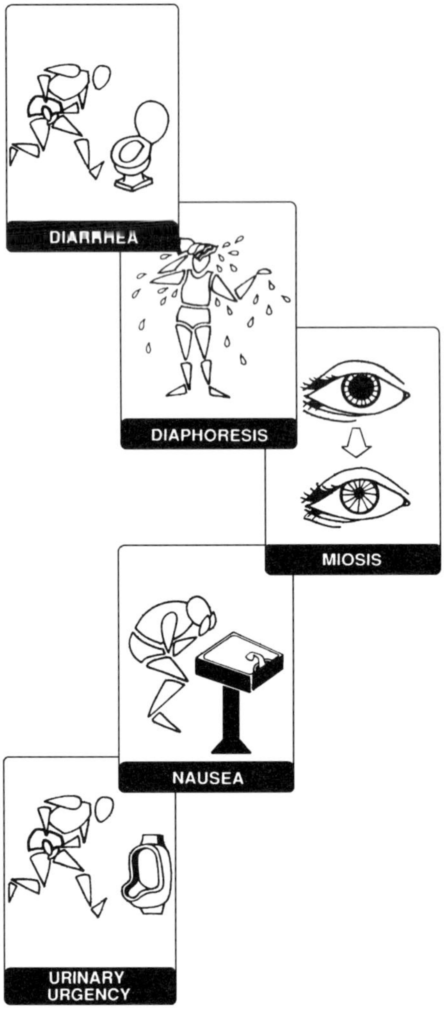

FIG XXI-4—Some adverse effects observed with cholinergic drugs. (Reproduced with permission from Mycek MJ, Harvey RA, Champe PC, eds. *Lippincott's Illustrated Reviews: Pharmacology*. 2nd ed. Philadelphia: Lippincott-Raven; 1997:40.)

Aceclidine is a synthetic cholinergic drug that acts directly on muscarinic end plates in a manner similar to pilocarpine. In human eyes with open angles, the pressure-lowering effect of 4% aceclidine was found to be comparable to that of 2% pilocarpine. Aceclidine produces far less accommodative spasm than does pilocarpine and could become the drug of choice for glaucoma patients young enough to retain considerable accommodation. In a study published in 1987, the induced myopia 45 minutes after administration of 2% aceclidine was 0.34 D, compared with 2.96 D from instillation of 2% pilocarpine. This drug is currently available only in Europe.

Abrams DA, Robin AL, Pollack IP, et al. The safety and efficacy of topical 1% ALO 2145 (p-aminoclonidine hydrochoride) in normal volunteers. *Arch Ophthalmol.* 1987;105: 1206–1207.

Indirect-acting agonists Topically applied indirect-acting muscarinic agonists (cholinesterase inhibitors) have the same actions as direct-acting muscarinic agonists, although they have a longer duration of action and are frequently more potent. Twice-daily treatment is sufficient. They work by reacting with the active serine hydroxyl site of cholinesterases, creating the formation of a slowly hydrolyzed intermediate (Fig XXI-5). Thus, they render the enzyme unavailable for hydrolyzing spontaneously released acetylcholine. There are two classes of cholinesterase inhibitors:

- *Reversible inhibitors,* such *as physostigmine* (Eserine), which carbamylate acetylcholinesterase
- *Irreversible inhibitors,* such as echothiophate (Phospholine Iodide) and *diisopropyl phosphorofluoridate* (DFP), which phosphorylate both the acetylcholinesterase of the synaptic cleft and the butyrylcholinesterase (pseudocholinesterase) of plasma

One carbamylating agent, *demecarium bromide* (Humorsol), is also irreversible; it contains two carbamyl groups and cross-links units of the enzyme.

Carbamylenzyme is regenerated by hydrolysis of the carbamyl-ester linkage over 3–4 hours. Regeneration of dialkylphosphorylated enzyme is so slow, however, that recovery of activity depends primarily on new enzyme synthesis.

The action of phosphorylating cholinesterase inhibitors can be reversed acutely by treatment with oxime-containing compounds that remove the dialkylphosphate moiety from the enzyme. This treatment must take place rapidly, before the spontaneous elimination of one of the alkyl residues ($T_{1/2}$ = 20 minutes for butyrylcholinesterase, 270 minutes for acetylcholinesterase), which makes the monoalkylphosphate intermediate no longer susceptible to regeneration by oxime. Thus, the oxime *pralidoxime* (2-PAM)—although useful in the treatment of acute organophosphate poisoning (eg, insecticide exposure)—is of little value in reversing the marked reduction of plasma butyrylcholinesterase activity that occurs with chronic irreversible cholinesterase-inhibitor therapy.

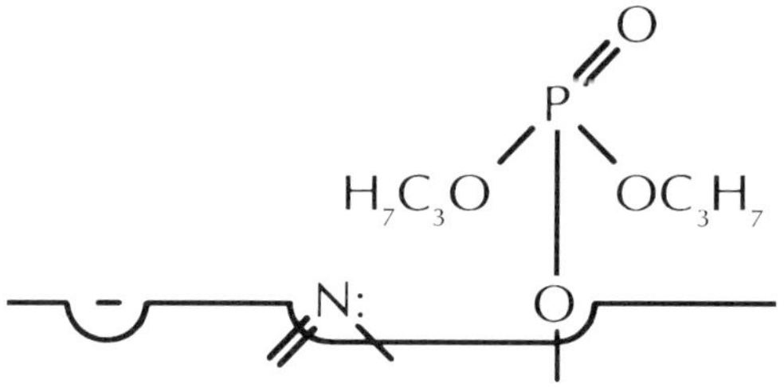

Phosphorylated Enzyme

FIG XXI-5—The enzyme-inhibitor intermediate.

Patients on such therapy may experience toxic reactions from systemic absorption of local anesthetics containing ester groups (eg, procaine) that are normally inactivated by plasma cholinesterase. Administration of the muscle relaxant succinylcholine during induction of general anesthesia is also hazardous in such patients, because the drug would not be metabolized and would result in prolonged respiratory paralysis.

Phosphorylating cholinesterase inhibitors also have local ocular toxicities. Children may develop cystlike proliferations of the iris pigment epithelium at the pupil margin that can block the pupil. For unknown reasons, cyst development can be minimized by concomitant use of phenylephrine (2.5%) drops. In adults, cataracts may develop or preexisting opacities may progress. Interestingly, children rarely if ever develop such cataracts, and adults rarely if ever develop significant epithelial cysts.

Therapy with cholinesterase inhibitors should not be combined with direct-acting cholinergic agonists, because the combination is less effective than either drug given alone.

Pilocarpine has less intrinsic muscarinic activity than acetylcholine. If cholinesterase inhibitors are given first, pilocarpine acts as a partial antagonist because it competes with acetylcholine. Thus, the miosis from physostigmine decreases slightly after administration of pilocarpine; a similar reduction in pressure-lowering effect probably occurs as well.

Because cholinesterase inhibitors are potent insecticides, they have been used in the treatment of lice infestations of the eyelashes. The adult form of the crab louse appears susceptible; the nits are more resistant and must be removed mechanically. Eserine destroys the louse but leaves the nits unaffected. Topical application must continue for 2 weeks so that the new crab lice are treated once they are hatched from the nit. *Demodex folliculorum*, or hair-follicle mite, is less sensitive and can survive 3 or more days of treatment with echothiophate or DFP.

Antagonists Topically applied muscarinic antagonists, such as *atropine*, react with postsynaptic muscarinic receptors and block the action of acetylcholine. The resultant paralysis of the iris sphincter, coupled with the unopposed action of the dilator muscle, causes pupillary dilation, or *mydriasis* (Table XXI-2). Mydriasis facilitates examination of the peripheral lens, ciliary body, and retina and is used therapeutically in the treatment of iritis, because it reduces contact between the posterior iris surface and the anterior lens capsule, thereby preventing the formation of iris–lens adhesions, or *posterior synechiae*.

Muscarinic antagonists also paralyze the ciliary muscles, which helps to relieve pain associated with iridocyclitis, inhibit accommodation for accurate refraction in children, and treat ciliary block (malignant) glaucoma. However, use of cycloplegic agents to dilate the pupils of patients with primary open-angle glaucoma frequently elevates the IOP dramatically, especially in patients requiring miotics for pressure control. It is advisable to use short-acting agents and to monitor the pressure in patients with severe optic nerve damage.

In situations requiring complete cycloplegia, such as the treatment of iridocyclitis or the full refractive correction of accommodative esotropia, the more potent agents *atropine* and *scopolamine* are preferred. Although some cycloplegic effect of a single drop of atropine lasts for days, two or three instillations a day may be required for maintenance of full cycloplegia to relieve pain in iritis. It may become necessary to change medications if atropine elicits a characteristic local irritation

TABLE XXI 2

MYDRIATRICS AND CYCLOPLEGICS

GENERIC NAME	TRADE NAME	CONCENTRATION (%)	ONSET	DURATION OF ACTION
Phenylephrine hydrochloride	AK-Dilate	Soln, 2.5%, 10%	30–60 min	3–5 h
	Mydfrin	Soln, 2.5%		
	Neo-Synephrine	Soln, 2.5%, 10%		
	Available generically	Soln, 2.5%, 10%		
Hydroxyamphetamine hydrobromide*	Paremyd	Soln, 1%	15–60 min	3–4 h
Atropine sulfate	Atropisol	Soln, 1%	45–120 min	7–14 days
	Atropine-Care	Soln, 1%		
	Isopto Atropine	Soln, 0.5%–1%		
	Available generically	Soln, 1% Ointment, 1%		
Cyclopentolate hydrochloride	Pentolair	Soln, 1%		
	AK–Pentolate	Soln, 1%	30–60 min	2 days
	Cyclogyl	Soln, 0.5%– 2%		
	Available generically	Soln, 1%		
Homatropine hydrobromide	Isopto Homatropine	Soln, 2%, 5%	30–60 min	3 days
	Available generically	Soln, 2%, 5%		
Scopolamine hydrobromide	Isopto Hyoscine	Soln, 0.25%	30–60 min	4–7 days
Tropicamide	Mydriacyl	Soln, 0.5%, 1%	20–40 min	4–6 h
	AK-Tropicacyl	Soln, 0.5%, 1%		
	Available generically	Soln, 0.5%, 1%		

*In combination with tropicamide

with swelling and maceration of the eyelids and conjunctival hyperemia. When mydriasis alone is necessary to facilitate examination or refraction, agents with shorter residual effect are preferred, because they allow quicker return of pupil response and reading ability.

Systemic absorption of topically administered muscarinic antagonists can produce dose-related toxicity, especially in children, whose dose is distributed in a smaller body mass. Flushing, fever, tachycardia, and even delirium can result from a combination of central and peripheral effects (Fig XXI-6). Mild cases may require only discontinuation of the drug, but severe cases can be treated with subcutaneous physostigmine, 0.25 mg every 15 minutes, until the symptoms subside. Physostigmine is used because it is a tertiary amine (uncharged) and can cross the blood–brain barrier.

Systemic administration of atropine blocks the oculocardiac reflex, a reflex bradycardia sometimes elicited during ocular surgery by manipulation of the conjunctiva, the globe, or the extraocular muscles. The reflex can also be prevented at the afferent end by retrobulbar anesthesia.

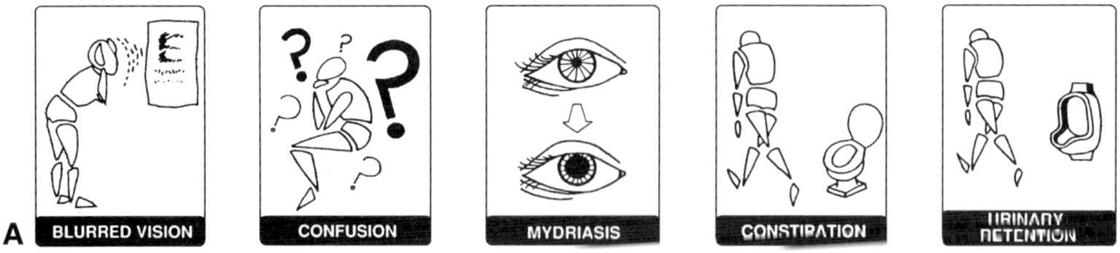

FIG XXI-6—**A**, Adverse effects commonly observed with cholinergic antagonists. **B**, Summary of cholinergic antagonists. (Reproduced with permission from Mycek MJ, Harvey RA, Champe PC, eds. *Lippincott's Illustrated Reviews: Pharmacology*. 2nd ed. Philadelphia: Lippincott-Raven; 1997: 45, 49.)

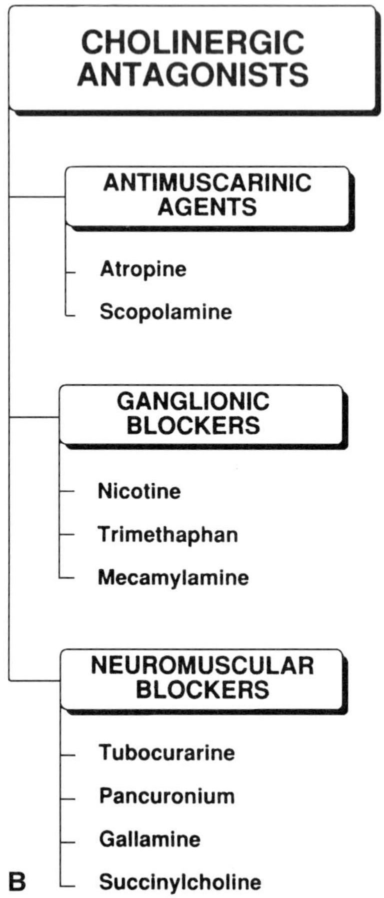

Nicotinic Drugs

Indirect-acting agonists The only cholinesterase inhibitor administered by ophthalmologists in a dose sufficient to allow it to act as an indirect-acting nicotinic agonist is *edrophonium* (Tensilon). It is a competitive inhibitor of acetylcholinesterase that binds to the enzyme's active site but does not form a covalent link with it.

Edrophonium is used in the diagnosis of myasthenia gravis, a neuromuscular disease characterized by muscle weakness and marked fatigability of skeletal muscles. Occasionally, this disease is manifested primarily by ptosis and diplopia. Myasthenia gravis is caused by an autoimmune mechanism in which antibodies deplete acetylcholine receptors in the neuromuscular junction with a resultant subsensitivity of synaptic transmission. The diagnosis is confirmed by a 2 mg dose rapidly injected intravenously, followed 60 seconds later by an additional 8 mg if the first dose has no effect; the patient is then examined for improvement in muscle function. BCSC Section 5, *Neuro-Ophthalmology,* discusses myasthenia gravis and the use of edrophonium in detail.

In myasthenic patients, the inhibition of acetylcholinesterase by edrophonium allows acetylcholine released into the synaptic cleft to accumulate to levels adequate to act through the reduced number of acetylcholine receptors. Because edrophonium also augments muscarinic transmission, muscarinic side effects (vomiting, diarrhea, urination, and bradycardia) may occur unless atropine, 0.4–0.6 mg, is coadministered intravenously.

Antagonists Nicotinic antagonists are administered as neuromuscular blocking agents to facilitate intubation for general anesthesia. They are of two types:

- *Nondepolarizing agents,* including curare-like drugs such as *gallamine* and *pancuronium,* which bind competitively to nicotinic receptors on striated muscle but do not cause contraction
- *Depolarizing* agents, such as *succinylcholine* and *decamethonium,* which bind competitively to nicotinic receptors and cause an initial receptor depolarization and muscle contraction

In singly innervated (en plaque) muscle fibers, this depolarization–contraction is followed by a prolonged unresponsiveness and flaccidity. However, these drugs produce sustained contractions of multiply innervated fibers, which make up one fifth of the muscle fibers of extraocular muscles. Such contractions of extraocular muscles (a nicotinic agonist action) exert force on the globe, an undesirable effect in cases in which the IOP is to be measured. The use of these agents in the induction of general anesthesia should be avoided in operations on lacerated eyes, because the force of the muscles on the globe could expel intraocular contents.

Adrenergic Agents

Several ophthalmic medications affect the activity of adrenergic receptors in synapses of the peripheral nervous system. Such receptors are found in the following locations:

- The cell membranes of the iris dilator muscle, the superior palpebral smooth muscle of Müller, the ciliary epithelium and processes, the trabecular meshwork, and the smooth muscle of ocular blood vessels (supplied by postganglionic autonomic fibers from the superior cervical ganglion)

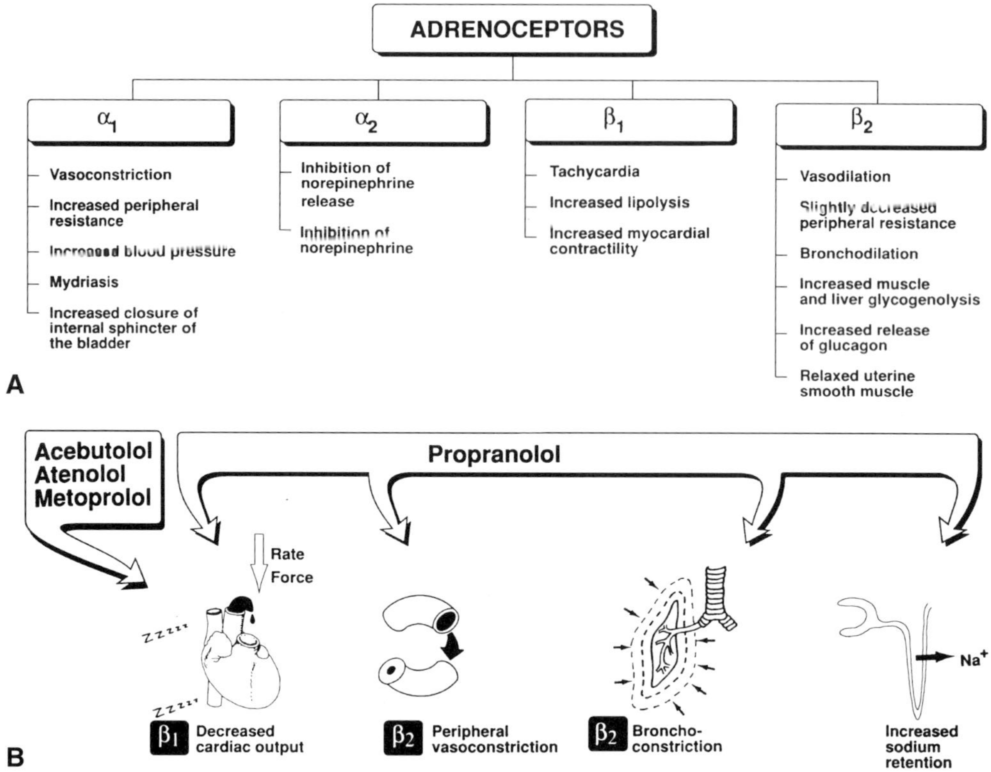

FIG XXI-7—**A**, Major effects mediated by α- and β-adrenoceptors. **B**, Actions of propranolol and β_1 blockers. (Reproduced with permission from Mycek MJ, Harvey RA, Champe PC, eds. *Lippincott's Illustrated Reviews: Pharmacology*. 2nd ed. Philadelphia: Lippincott-Raven; 1997:60, 75.)

- The presynaptic terminals of some sympathetic and parasympathetic nerves, where they have feedback inhibitory actions

Though adrenergic receptors were originally defined by their response to epinephrine (adrenaline), the transmitter of most sympathetic postganglionic fibers is actually norepinephrine. Adrenergic receptors have been subclassified into four categories, α_1, α_2, β_1, and β_2, based on their profile of responses to natural and synthetic catecholamines (Fig XXI-7). The *α_1 receptors* generally mediate smooth-muscle contraction, whereas *α_2 receptors* mediate feedback inhibition of presynaptic sympathetic (and sometimes parasympathetic) nerve terminals. The *β_1 receptors* are found predominantly in the heart, where they mediate stimulatory effects; *β_2 receptors* mediate relaxation of smooth muscle in most blood vessels and in the bronchi.

Systemic absorption of ocular adrenergic agents is frequently sufficient to cause systemic effects, which are manifested in the cardiovascular system, the bronchial airways, and the brain. Much remains to be determined regarding the ocular pharmacologic mechanisms of adrenergic agents currently in use, let alone those under investigation. Adrenergic agents may be direct-acting agonists, indirect-acting agonists, or antagonists at one or more of the four types of receptor.

α-Adrenergic Agents

Direct-acting α_1-adrenergic agonists The primary clinical use of direct-acting α_1-adrenergic agonists, such as *phenylephrine,* is stimulation of the iris dilator muscle to produce mydriasis. However, the parasympathetically innervated iris sphincter muscle is much stronger than the dilator muscle, and therefore dilation achieved with phenylephrine alone is largely overcome by the pupillary light reflex during ophthalmoscopy. Coadministration of a cycloplegic agent allows sustained dilation.

Systemic absorption of phenylephrine may elevate systemic blood pressure. This effect is of clinical significance only if the patient is an infant or has an abnormally increased sensitivity to α-agonists, which occurs with orthostatic hypotension and in association with the use of drugs that accentuate adrenergic effects (eg, reserpine, tricyclic antidepressants, cocaine, monoamine oxidase inhibitors—discussed below). Even with lower doses of phenylephrine (2.5%), infants may exhibit transient rise in blood pressure, because the dose received in an eyedrop is large for them on a per-weight basis.

Indirect-acting adrenergic agonists These agents are used to test for and localize defects in sympathetic innervation to the iris dilator muscle. Normally, nerve fibers from a hypothalamic nucleus for pupil response pass down the spinal cord to synapse with cells in the intermediolateral columns. In turn, preganglionic fibers exit the cord through the anterior spinal roots in the upper thorax to synapse in the superior cervical ganglion in the neck. Finally, postganglionic adrenergic fibers terminate in a neuroeffector junction with the iris dilator muscle. The norepinephrine released is inactivated primarily by reuptake into secretory granules in the nerve terminal (Fig XXI-8). Approximately 70% of released norepinephrine is recaptured.

If a defect occurs anywhere in the pathway, the baseline release of norepinephrine to the iris dilator will be lower on the side of the injury, resulting in a relatively miotic pupil. The presence of a lesion can be confirmed by applying 4% cocaine to each eye and comparing the pupil sizes at 1 hour. Cocaine blocks reuptake of norepinephrine into the presynaptic vesicles, causing it to accumulate and resulting in pupillary dilation. The injured side will have less accumulation and show less dilation.

The site of the lesion can be determined to be either preganglionic or postganglionic by a Paredrine test: a drop of 1% *hydroxyamphetamine* (Paredrine) is applied to each eye and the pupil responses are compared. Hydroxyamphetamine acts by penetrating the sympathetic nerve terminals adjacent to the dilator muscle and releasing stored norepinephrine, resulting in pupillary dilation. On the side of a postganglionic injury, fewer fibers will be intact, less norepinephrine will be released, and less dilation will ensue. In the case of a preganglionic lesion, however, the postganglionic fibers synthesize and store norepinephrine normally but are not neurally stimulated to release it. A Paredrine test will therefore dilate the pupil normally. If the dilator muscle has developed supersensitivity by a compensatory increase in receptor number or responsiveness, the dilation may be greater than that in the normal eye. (See also BCSC Section 5, *Neuro-Ophthalmology.*)

Apraclonidine hydrochloride (para-aminoclonidine) is an α_2-adrenergic agonist and a clonidine derivative that prevents release of norepinephrine at nerve terminals. It decreases aqueous production as well as episcleral venous pressure and improves

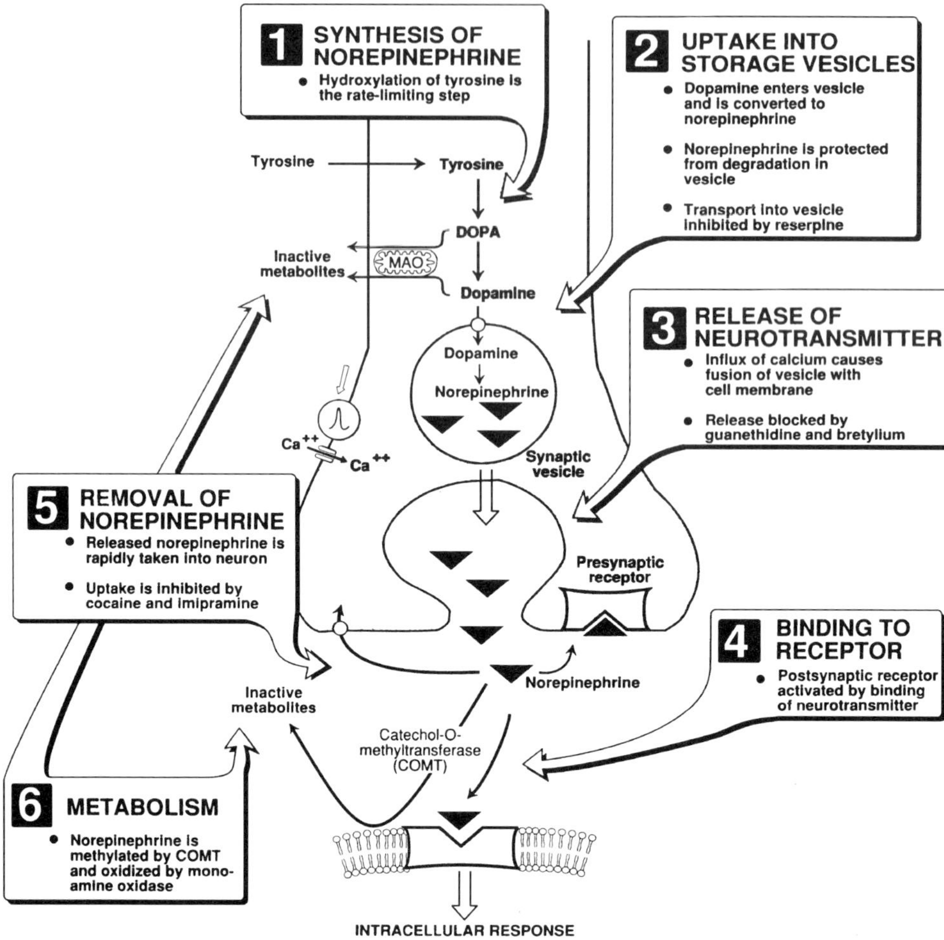

FIG XXI-8—Synthesis and release of norepinephrine from the adrenergic neuron. *MAO,* monoamine oxidase inhibitor. (Reproduced with permission from Mycek MJ, Harvey RA, Champe PC, eds. *Lippincott's Illustrated Reviews: Pharmacology.* 2nd ed. Philadelphia: Lippincott-Raven; 1997:57.)

trabecular outflow. However, its true ocular hypotensive mechanism is not fully understood. When administered pre- and postoperatively, the drug is effective in diminishing the acute IOP rise that follows argon laser iridectomy, argon laser trabeculoplasty, Nd:YAG laser capsulotomy, and cataract extraction. Apraclonidine hydrochloride may be effective for the short-term lowering of IOP, but development of topical sensitivity and tachyphylaxis often limits long-term use.

Brimonidine tartrate has less tachyphylaxis than apraclonidine in long-term use, and the rate of allergic reactions (such as follicular conjunctivitis and contact blepharitis–dermatitis) is also lower (up to 40% for apraclonidine and less than 15% for brimonidine). Cross-sensitivity to brimonidine in patients with known hypersensitivity to apraclonidine is minimal. Brimonidine's mechanism of lowering IOP is thought to involve both decreased aqueous production and increased uveoscleral outflow. Similar to the case with beta-blockers, a central mechanism may account for part of the IOP reduction from brimonidine 0.2%: A single eye treatment trial for 1 week caused a statistically significant reduction of 1.2 mm Hg in the fellow eye.

Brimonidine's peak IOP reduction is approximately 26% (2 hours post dose). At peak, it is comparable to a nonselective beta blocker and superior to the selective beta blocker betaxolol, although at trough (12 hours post dose) the reduction is only 14%–15%, or less effective than the nonselective beta blockers but comparable to betaxolol. Brimonidine may also have potential neuroprotective properties, as shown in animal models of optic nerve and retinal injury that are independent of IOP reduction. The proposed mechanism of neuroprotection is up-regulation of a neurotrophin, basic fibroblast growth factor, and cellular regulatory genes.

Caution is recommended when using apraclonidine or brimonidine in patients on a monoamine oxidase inhibitor or tricyclic antidepressant therapy and in patients with severe cardiovascular disease. Use of these drugs concomitantly with beta blockers, antihypertensives, and cardiac glycosides (ophthalmic and systemic) also requires prudence. Although effective in acutely lowering the IOP in angle-closure glaucoma, these drugs may also induce vasoconstriction that can prolong iris-sphincter ischemia and reduce the efficacy of concurrent miotics. Apraclonidine has much greater affinity for α_1-receptors than does brimonidine and is therefore more likely to produce vasoconstriction in the eye. Brimonidine has been shown to not induce vasoconstriction in the posterior segment or optic nerve.

Ligand binding to α_2-receptors in other systems has been shown to mediate inhibition of the enzyme adenylate cyclase. Adenylate cyclase is present in the ciliary epithelium and is thought to have a role in aqueous production.

Antagonists *Thymoxamine,* an α_1-adrenergic blocking agent, acts by competitive inhibition of norepinephrine at the receptor site. Because pupillary muscle, unlike ciliary muscle, has predominantly α-adrenergic receptors, use of thymoxamine inhibits α-adrenergic tone to the dilator muscle of the iris and results in pupil constriction. Thymoxamine does not have any significant activity on ciliary muscle contraction and therefore does not induce a significant change in anterior chamber depth, facility of outflow, IOP, or accommodation. It is useful for differentiating angle-closure glaucoma from POAG with closed angles and can also be used to reverse the pupil dilation caused by phenylephrine. Thymoxamine is not available commercially in the United States even though it has been used widely in Europe for several years.

Dapiprazole hydrochloride (Rev-Eyes), an α-adrenergic blocking agent, produces miosis through its effect on the dilator muscle of the iris. Like thymoxamine, it does not have significant activity on ciliary muscle contraction and therefore does not induce a significant change in anterior chamber depth, facility of outflow, or accommodation. Topical application of dapiprazole 0.5% solution reverses the mydriasis produced by phenylephrine and tropicamide in 30 minutes, but it is not effective in reversing the mydriasis produced by cycloplegics. Itching and stinging is often associated with administration of dapiprazole.

β-Adrenergic Agents

β_2-Adrenergic agonists These agents lower IOP by increasing uveoscleral outflow and perhaps also by increasing outflow through the trabecular meshwork. The beneficial effect on outflow more than compensates for a small increase in aqueous inflow as detected by fluorophotometry.

TABLE XXI-3

ADRENERGIC AGONISTS

GENERIC NAME	TRADE NAME	CONCENTRATION (%)
β_2-Adrenergic agonists		
Dipivefrin hydrochloride	AK Pro	0.1%
	Propine	0.1%
	Available generically	0.1%
Epinephrine borate	Epinal	0.5%, 1%
	Eppy/N	1%
Epinephrine hydrochloride	Epifrin	0.5%–2%
	Glaucon	1%, 2%
α_2-Selective agonists		
Apraclonidine hydrochloride	Iopidine	0.5%–1.0%
Brimonidine tartrate	Alphagan	0.2%

β_2-Receptors linked to adenylate cyclase are present in the ciliary epithelium and processes as well as in the trabecular meshwork. Treatment with *L-epinephrine*, an α- and β-agonist, increases intracellular levels of cyclic adenosine monophosphate (cAMP) in these tissues and in the aqueous humor. In other tissues, β-receptor–mediated generation of cAMP in turn activates cAMP-dependent enzymes that result in responses such as glycogenolysis and gluconeogenesis in the liver and lipolysis in adipose tissue. However, the biochemical mechanisms responsible for the lowering of IOP remain to be determined.

L-epinephrine is commercially available as a borate, a hydrochloride, and a bitartrate (Table XXI-3). The molecular weight of L-epinephrine bitartrate is approximately twice that of the borate and hydrochloride. Therefore, L-epinephrine bitartrate 2% contains as much free base as a 1% solution of either of the other forms. Borate form is more comfortable than other forms, whose low pH induces burning on application. Local allergic and irritative manifestations and systemic side effects (headache, palpitations, and cardiac arrhythmias) are common causes of intolerance of long-term epinephrine therapy. Oxidation products of epinephrine may produce black deposits in the conjunctiva. Although such deposits are harmless, they have been mistaken for foreign bodies or even melanomas. Epinephrine therapy has also been associated with a reversible cystoid maculopathy that occurs in approximately 25% of chronically treated aphakic eyes that lack a posterior lens capsule. Although epinephrine maculopathy does not occur in phakic eyes, it has not yet been established whether an intact posterior lens capsule, hyaloid membrane, and/or the presence of an intraocular lens retards the development of the condition.

The prodrug dipivefrin HCl (DPE, Propine) 0.1% is therapeutically equivalent to epinephrine compounds of 1%–2%, because the presence of two pivalyl residues (see Chapter XX, Fig XX-4) increases its lipid solubility and corneal penetration by a factor of 17. The drug has little adrenergic activity until the pivalyl groups are cleaved by corneal esterases. Together, the reductions in both intrinsic activity and concentration employed virtually eliminate the systemic side effects of epinephrine. Extraocular irritation is also reduced because of either the lower concentration or the

TABLE XXI-4

β-ADRENERGIC BLOCKING AGENTS (BETA BLOCKERS)

GENERIC NAME	TRADE NAME	CONCENTRATION (%)
Betaxolol hydrochloride	Betoptic-S Betoptic	0.25% 0.5%
Carteolol hydrochloride	Ocupress	1.0%
Levobunolol hydrochloride	AK Beta Betagan Available generically	0.25%, 0.5% 0.25%, 0.5% 0.25%, 0.5%
Metipranolol	OptiPranolol	0.3%
Timolol maleate	Timoptic Timoptic-XE (gel) Available generically	0.25%, 0.5% 0.25%, 0.5%
Timolol hemihydrate	Betimol	0.25%, 0.5%
Timolol maleate (preservative free)	Timoptic in OCUDOSE	0.25%, 0.5%

reduction in auto-oxidation of the phenolic hydroxyls. However, because epinephrine is liberated from the prodrug inside the eye, the risk of epinephrine maculopathy should be unchanged.

Chronic therapy with epinephrine has been shown in an animal model to result in a down-regulation of the number of β-receptors. This phenomenon may underlie the loss of some of its therapeutic effectiveness over time (tachyphylaxis).

β-Adrenergic antagonists (Table XXI-4) These agents, also known as *beta blockers,* lower IOP by reducing aqueous humor production as much as 50%. Six agents are approved for use in the treatment of glaucoma: timolol maleate, timolol hemihydrate, metipranolol, carteolol, levobunolol, and betaxolol. Although it is likely that the site of action resides in the ciliary body, it is not known whether the vasculature of the ciliary processes or the pumping mechanism of the ciliary epithelium is primarily affected. One possible mechanism may be an effect on the β-adrenergic receptor–coupled adenylate cyclase of the ciliary epithelium. Although systemic administration of beta blockers has been reported to elevate blood lipid, such elevation has not been demonstrated with topical beta blockers such as timolol.

Timolol maleate 0.25%–0.50% (Timoptic) and *levobunolol* (Betagan) 0.5% are mixed β_1/β_2-antagonists. Tests of more specific beta blockers suggest that β_2-antagonists have a greater effect on aqueous secretion than β_1-antagonists. For example, comparative studies have shown that the specific β_1-antagonist *betaxolol* 0.5% (Betoptic) is about 85% as effective in lowering IOP as timolol. However, preliminary evidence indicates that some or all of the deficit in betaxolol's ability to lower IOP can be made up by its greater additive effect in combined therapy with epinephrine or dipivefrin. Mixed β_1- and β_2-agonists show little or no additive effect on the lowering of IOP with the use of an α-adrenergic agonist.

Metipranolol hydrochloride (OptiPranolol) is a nonselective β_1- and β_2-adrenergic receptor–blocking agent. As a 0.3% topical solution, it is similar in effect to other topical nonselective beta blockers and is efficacious in reducing IOP.

Carteolol (Ocupress) demonstrates intrinsic sympathomimetic activity, which means that, while acting as a competitive antagonist, it also causes a slight to moderate activation of receptors. Thus, even though carteolol produces beta-blocking effects, these may be tempered, reducing the effect on cardiovascular and respiratory systems. Carteolol may also be less likely to adversely affect the systemic lipid profile compared with other beta blockers.

Betaxolol is a selective β_1-antagonist that is significantly safer than the nonselective beta blockers when pulmonary, cardiac, central nervous system, or other systemic conditions are considered. Betaxolol may be useful in patients with a history of bronchospastic disorders, although other therapies should be tried in lieu of betaxolol (β-selectivity is only relative and not absolute, and some β_2 effect can therefore remain). In general, the IOP-lowering effect of betaxolol is less than the nonselective beta-adrenergic antagonists.

Prodrugs of nonselective beta blockers are being developed, and they may offer the benefit of the higher potency of β_1/β_2-blocking agents while reducing their potential systemic side effects.

It is curious that both β-agonist and β-antagonist drugs can lower IOP. This paradox is compounded by the observation that β-agonist and β-antagonist drugs have slightly additive effects in lowering IOP.

Carbonic Anhydrase Inhibitors (CAIs)

Aqueous humor is secreted into the posterior chamber by the nonpigmented epithelium of the ciliary processes. Although the physiologic mechanisms of secretion are not fully understood, secretion is known to depend largely on active transport of sodium by Na^+,K^+-ATPase on the surface of nonpigmented epithelial cells. Inhibition of that enzyme by *ouabain* injected into the vitreous cavity of experimental animals reduces secretion markedly. Unfortunately, ouabain and other cardiac glycosides cannot be used clinically to treat glaucoma because effective doses for the eye would require systemic doses toxic to the heart.

However, Na^+ transport and coupled aqueous secretion can be inhibited indirectly. Na^+ transport and fluid flow seem to be partially linked to HCO_3^- formation in the ciliary epithelium, and HCO_3^- formation can be substantially reduced by inhibition of the enzyme carbonic anhydrase (Table XXI-5). The linkage between HCO_3^- and Na^+ transport is demonstrated by the fact that the effect on aqueous flow is no greater when ouabain and a carbonic anhydrase inhibitor (CAI) are given together than when each drug is given alone.

Carbonic anhydrase catalyzes the hydration of dissolved CO_2 to H_2CO_3, which ionizes into HCO_3^- and H^+. The HCO_3^- is then available to accompany secreted Na^+. When the enzyme is inhibited, the transport of radiolabeled HCO_3^- into the posterior chamber is reduced by up to 60%. That portion (60%) of Na^+ transport accompanied by Cl^- is unaffected by inhibition of carbonic anhydrase and, curiously, is unaffected by coadministration of ouabain. The mechanism(s) of the residual secretion and means of inhibiting it are yet to be determined. CAIs such as acetazolamide are also effective in treating certain cases of cystoid macular edema.

The concentration of carbonic anhydrase in the ciliary epithelium of rabbits is 0.3 µm, and it is 10 times more concentrated in the kidney and in the choroid plexus of the brain. Carbonic anhydrase is present in considerable excess of what is needed to supply the amount of HCO_3^- transported. Calculations based on the K_{cat} (catalysis constant) and K_m (apparent affinity constant) of the enzyme and the concentrations of substrates and product indicate that 100 times as much enzyme is present

TABLE XXI-5

CARBONIC ANHYDRASE INHIBITORS

GENERIC NAME	TRADE NAME	DOSAGE	ONSET	DURATION OF ACTION
Systemic				
Acetazolamide	Diamox Available generically	125, 250, 500 mg (time-released)	2 h	4–6 h
Acetazolamide sodium	Diamox Parenteral	500 mg, 5–10 mg/kg[3]	5–10 min	2 h
Dichlorphenamide	Daranide	50 mg	30 min	6 h
Methazolamide	Glauctabs MZM Neptazane Available generically	2 h 20, 50 mg	4–6 h	
Topical				
Dorzolamide HCl	Trusopt	2% solution	5 min	8 h
Brinzolamide	Azopt	1% suspension		

as is needed in the ciliary body. Correspondingly, in clinical use the enzyme must in fact be more than 99% inhibited in order to significantly reduce aqueous flow. The enzyme in the kidney, present in 1000-fold excess, must be more than 99.9% inhibited to affect the usual pathway for HCO_3^- reabsorption. With the inhibitor *methazolamide,* the difference in concentration of carbonic anhydrase in the ciliary body and the kidney can be exploited to lower IOP without incurring renal HCO_3^- loss, and an unpleasant metabolic acidosis can be avoided. Even though renal stone formation has been reported with use of methazolamide, it is significantly less than with other agents because of this specific property. In contrast, *acetazolamide* is actively secreted into the renal tubules, and renal effects are unavoidable. Topical forms—dorzolamide and brinzolamide—are agents that are also available for chronic treatment of glaucoma.

The necessity to inhibit more than 99% of the enzyme has made it difficult to achieve a clinical effect with topically applied agents until recently. *Dorzolamide* (Trusopt) and *brinzolamide* (Azopt) are currently approved for use by the FDA. They are water-soluble, penetrate the cornea easily, and are specially formulated for topical ophthalmic use. When administered as solution three times a day, they effectively inhibit carbonic anhydrase II and avoid the systemic side effects of oral administration. Both agents are equally effective and reduce IOP by 14%–17%. The concomitant administration of dorzolamide and oral CAIs is not recommended because of potentially additive side effects.

The four compounds other than topical agents that are used clinically are all administered orally and/or parenterally. The longer half-life of methazolamide allows it to be used twice daily, and acetazolamide is also available in a 500 mg sustained-release form used twice a day. The others must be used four times a day. None of the compounds has an ideal combination of features:

- High potency (low K_i)
- Good ocular penetration (high percentage in the non-ionized form and high lipid solubility to facilitate passage through the blood–ocular barrier)
- High proportion of the drug present in the blood in the unbound form
- Long plasma half-life

In addition to lowering IOP by inhibiting ciliary body carbonic anhydrase, each of the agents at high doses further lowers IOP by causing renal metabolic acidosis. The mechanism by which acidosis lowers secretion is uncertain, but it plausibly involves reduction in HCO_3^- formation.

At the onset of acidosis, the renal effects cause alkaline diuresis, with loss of Na^+, K^+, and HCO_3^-. Severe hypokalemia can result in patients concurrently receiving diuretics, steroids, or adrenocorticotropic hormone (ACTH). This situation may be dangerous for patients using digitalis, in whom hypokalemia may elicit arrhythmias. Patients on chronic CAI therapy should have potassium levels checked at regular intervals, preferably by their primary care physician.

The acidosis prompts a renal mechanism for HCO_3^- reabsorption unrelated to carbonic anhydrase; this mechanism limits the degree of acidosis and halts both the diuresis and K^+ loss after the first few days of treatment. However, *dichlorphenamide* also acts as a chloruretic agent and may cause continued K^+ loss.

CAI therapy may interact unfavorably with certain systemic conditions. The alkalinization of the urine present during initial CAI treatment prevents excretion of NH_4^+, a factor to consider in patients with cirrhosis of the liver. Metabolic acidosis may exacerbate diabetic ketoacidosis. In patients with severe chronic obstructive pulmonary disease, respiratory acidosis may be caused by impairment of CO_2 transfer from the pulmonary vasculature to the alveoli.

The use of acetazolamide has been linked to the formation of stones in the urinary tract. In a retrospective case-control series, the incidence of stones was 11 times higher in patients using this drug. The increased risk occurred primarily during the first year of therapy. Continued use after occurrence of a stone was associated with a high risk of recurrent stone formation. However, a history of spontaneous stone formation more than 5 years prior to acetazolamide therapy did not appear to be associated with a special risk. The mechanisms responsible for such stone formation may be related to metabolic acidosis and the associated pH changes as well as decreased excretion of citrate.

Nearly 50% of patients are intolerant of CAIs because of distressing central nervous system and gastrointestinal side effects, including:

- Numbness and tingling of the hands, feet, and lips
- Malaise
- Metallic taste of carbonated beverages
- Anorexia and weight loss
- Nausea
- Somnolence

- Impotence and loss of libido
- Depression

Whenever the clinical situation allows, it is wise to begin therapy at low doses (eg, 25–50 mg methazolamide BID or 125 mg acetazolamide QID), because side effects will be less severe and weaning may actually reduce their incidence. Patients should be informed of the potential side effects of these agents; otherwise, many may fail to associate their systemic symptoms with the use of a medicine given by the ophthalmologist.

Rare side effects from this class of drugs include those common to other members of the sulfonamide family, such as transient myopia, hypersensitive nephropathy, skin rash, thrombocytopenia, and aplastic anemia. The latter effect is idiosyncratic and dose-related; white blood cell counts do not detect susceptible patients. CAIs have been associated with teratogenic effects (forelimb deformity) in rodents, and their use is not advised during pregnancy.

Prostaglandin Analogues

Prostaglandin (PG) analogues are a relatively new class of ocular hypotensive agents. Currently, two PG analogues have been approved for clinical use, one of which is available in the United States: *latanoprost* (Xalatan). The other, isopropyl unoprosterone (Rescula), is undergoing clinical trials. Both are analogues of $PGF_{2\alpha}$. Other new PG analogues are currently under investigation.

Latanoprost is a prodrug that penetrates the cornea and becomes biologically active after being hydrolyzed by corneal tissue esterase. It appears to lower IOP by enhancing uveoscleral outflow and may reduce the pressure by 6–9 mm Hg (25%–35%). One potential advantage of this agent is once-daily dosing. Other advantages include the lack of cardiopulmonary effects and the additivity to other antiglaucoma medications except, perhaps, higher-concentration miotic agents.

An ocular side effect unusual to this class of drugs is the darkening of the iris and periocular skin as a result of increased numbers of melanosomes (increased melanin content—melanogenesis) within the melanocytes. The risk of iris pigmentation correlates with baseline iris pigmentation. Light-colored irides may experience increased pigmentation in 10%–20% of eyes in the initial 18–24 months of therapy, whereas nearly 60% of eyes that are light brown or two-toned may experience increased pigmentation over the same time period. The long-term sequelae of this side effect are unknown. Other side effects reported in association with the use of a topical PG analogue include hypertrichosis of the eyelashes, cystoid macular edema, and uveitis. The two latter side effects are more common in eyes with preexisting risk factors for either macular edema or uveitis. It is recommended that topical application be performed at night to mitigate the conjunctival injection. Reported systemic reactions include flulike symptoms, skin rash, and possible uterine bleeding in postmenopausal women. Reactivation of herpetic keratitis has been reported with use of latanoprost.

Combined Medications

Medications that are combined and placed in a single bottle have the potential benefits of improved efficacy, convenience, and compliance, as well as reduced cost. Adrenergic agonists and parasympathomimetic agents (epinephrine and pilocarpine) have been available for many years as a combined agent. They are weakly additive in IOP-lowering effect, satisfying FDA guidelines that the fixed combination be more efficacious than either agent given alone. Epinephrine combined with a beta-blocker agent (Probeta = levobunolol + dipivefrin 0.1%) is currently available in Canada. Beta blocker (timolol 0.5%) combined with miotic pilocarpine (either 2% or 4% concentration) is available for twice-daily therapy. Betanorm (beta-blocker metipranolol and pilocarpine) is available in many countries outside the United States.

Cosopt, the combination of a beta blocker (timolol 0.5%) and topical CAI (dorzolamide 2%), has demonstrated similar efficacy compared with the two agents given separately with twice-daily administration. The advantage of this combined therapy may be the convenience and lessened confusion of one bottle rather than two, which may increase the potential for greater compliance. However, the twice-daily dosing may create greater exposure to the potential systemic side effects of a beta blocker, as beta blockers are generally equally effective when given only once daily. The ocular side effects are the same as for both drugs individually. The indications for this combined medication may be as a substitute for both a beta blocker and a topical CAI. If Cosopt is used as monotherapy, a monocular trial of timolol should be tried first. If timolol is effective in significantly lowering the IOP, a monocular trial of dorzolamide should be used with timolol. An alternative trial could involve Cosopt in one eye twice daily and timolol in the opposite eye. It is important to prove that both the timolol component and the dorzolamide component each have an effect on IOP before using the combined medication.

Osmotic Agents

Actions and Uses

Increased serum osmolarity reduces IOP and vitreous volume by drawing fluid out of the eye across vascular barriers. The osmotic activity of an agent depends on the number of particles in solution and the maintenance of an osmotic gradient between the plasma and the intraocular fluids. It is independent of the molecular weight. Low-molecular-weight agents such as urea that penetrate the blood–ocular barriers produce a small rebound in IOP after an initial lowering because of a reversal of the osmotic gradient when the kidneys clear the blood of excess urea. Osmotic agents are used in the short-term management of acute glaucoma and in reducing vitreous volume prior to cataract surgery.

Agents

Osmotic agents should be used with care in patients in whom cardiovascular overload might occur with moderate vascular volume expansion, such as patients with a history of congestive heart failure, angina, systemic hypertension, or recent myocardial infarct. The osmotic agents that have been used clinically are mannitol, urea, glycerol (glycerin), sodium ascorbate, and isosorbide (Table XXI-6).

Intravenous agents *Mannitol* must be given intravenously because it is not absorbed from the gastrointestinal tract. *Urea* is unpalatable and thus is used intravenously. Use of urea is out of favor because of the rebound mentioned above and its tendency to cause tissue necrosis if it extravasates during administration. IV administration produces a rapid onset of action, which is usually desirable, but both mannitol and urea have been associated with subarachnoid hemorrhage attributed to rapid volume overload of the blood vessels and/or rapid shrinkage of the brain with traction of the subarachnoid vessels.

Table XXI-6

Hyperosmotic Agents

GENERIC NAME	TRADE NAME	PREPARATION	DOSE	ROUTE
Glycerin	Osmoglyn	50%	1–1.5 g/kg	Oral
Isosorbide	Ismotic	45%	1.5 g/kg	Oral
Mannitol	Osmitrol	5%–20%	0.5–2 g/kg	IV
Urea	Ureaphil	Powder or 30% soln	0.5–2 g/kg	IV

These agents are cleared by the kidneys and produce a marked osmotic diuresis that may be troublesome in the operating room. The conscious patient should void shortly before surgery, and a urinal should be available. If general anesthesia is employed, an indwelling urethral catheter may be required to prevent bladder distension.

Oral agents *Glycerol* (Osmoglyn) is the most frequently used oral osmotic agent. Its nauseating sweet taste can be minimized by giving it with cracked ice. In diabetic patients, the nonmetabolized sugar *isosorbide* (Ismotic) is often preferred.

Sodium ascorbate is metabolized, and thus its osmotic diuretic effects are less marked. *Ethanol* has the disadvantages of producing inebriation and suppressing antidiuretic hormone release, causing additional diuresis. *Ascorbate* is unstable in solution and crosses the blood–ocular barriers.

Abrams DA, Robin AL, Pollack IP, et al. The safety and efficacy of topical 1% ALO 2145 (p-aminoclonidine hydrochloride) in normal volunteers. *Arch Ophthalmol.* 1987; 105:1206–1207.

Allen RC, Epstein DL. Additive effect of betaxolol and epinephrine in primary open-angle glaucoma. *Arch Ophthalmol.* 1986;104:1178–1184.

Berry DP Jr, Van Buskirk EM, Shields MB. Betaxolol and timolol. A comparison of efficacy and side effects. *Arch Ophthalmol.* 1984;102:42–45.

Chrisp P, Sorkin EM. Ocular carteolol: a review of its pharmacological properties, and therapeutic use in glaucoma and ocular hypertension. *Drugs Aging.* 1992;2:58–77.

Clineschmidt CM, Williams RD, Snyder E, et al. A randomized trial in patients inadequately controlled with timolol alone comparing the dorzolamide-timolol combination to monotherapy with timolol or dorzolamide. *Ophthalmology.* 1999;106 (Suppl 12):17–24.

Friedland BR, Maren TH. Carbonic anhydrase: the pharmacology of inhibitors related to the treatment of glaucoma. In: Chader G, Sears ML, eds. *Pharmacology of the Eye: Handbook of Experimental Pharmacology.* New York: Springer-Verlag; 1983: 279–309.

Ingram CJ, Brubaker RF. Effect of brinzolamide and dorzolamide on aqueous humor flow in human eyes. *Am J Ophthalmol.* 1999;128(3):292–296.

Kolker AE, Becker B. Epinephrine maculopathy. *Arch Ophthalmol.* 1968;79:552–562.

Moroi SE, Gottfredsdottir MS, Schteingart MT, et al. Cystoid macular edema associated with latanoprost therapy in a case series of patients with glaucoma and ocular hypertension. *Ophthalmology.* 1999;106(5):1024–1029.

Netland PA, Allen RC. Glaucoma Medical Therapy: Principles and Management. *American Academy of Ophthalmology Monograph Series No. 13.* 1999:1–278.

Schoene RB, Abuan T, Ward RL, et al. Effects of topical betaxolol, timolol and placebo on pulmonary function in asthmatic bronchitis. *Am J Ophthalmol.* 1984;97:86–92.

Shaw BR, Lewis RA. Intraocular pressure elevation after pupillary dilation in open-angle glaucoma. *Arch Ophthalmol.* 1986;104:1185–1188.

Silver LH. Clinical efficacy and safety of brinzolamide (Azopt), a new topical carbonic anhydrase inhibitor for primary open-angle glaucoma and ocular hypertension. Brinzolamide Primary Therapy Study Group. *Am J Ophthalmol.* 1998;126(3):400–408.

Silver LH, the Brinzolamide Dose-Response Study Group. Dose-response evaluation of the ocular hypotensive effect of brinzolamide ophthalmic suspension (Azopt). *Surv Ophthalmol.* 2000;44(4 Suppl 2):147–153.

Smith SL, Sine CS, Pruitt CA, et al. The use of latanoprost 0.005% once daily and its effect on intraocular pressure as primary or adjunctive therapy. *J Ocul Pharmacol Ther.* 1999;15(1):29–39.

Stewart RH, Kimbrough RL, Ward RL. Betaxolol vs timolol. A six-month double-blind comparison. *Arch Ophthalmol.* 1986;104:46–48.

Anti-Inflammatory Agents

Ocular inflammation can be treated with drugs administered topically, by local injection, or systemically. These drugs may be classified as glucocorticoids (a form of corticosteroid), nonsteroidal anti-inflammatory agents (NSAIDs), antihistamines, histamine release blockers, or antifibrotics.

Glucocorticoids

Steroids are applied topically to prevent or suppress corneal graft rejection, anterior chamber reaction after anterior segment surgery, filtering bleb scarring, and immune or traumatic iritis and uveitis (Table XXI-7). Subconjunctival and retrobulbar injections of steroids are used to treat severe ocular inflammations. Systemic steroid therapy is used to treat giant-cell arteritis and severe ocular inflammations. Treatment of acute inflammatory ischemic optic neuritis with steroids is a subject of controversy and ongoing research. BCSC Section 5, *Neuro-Ophthalmology,* discusses these issues in depth.

The ocular anti-inflammatory action of glucocorticoids is achieved by separate cell-specific effects on lymphocytes, macrophages, polymorphonuclear leukocytes, vascular endothelial cells, fibroblasts, and other cells. In each of these types of cells, glucocorticoids must:

- Penetrate the cell membrane
- Bind to soluble receptors in the cytosol
- Bring about a conformational shift that allows translocation of the glucocorticoid receptor complex to nuclear binding sites
- Induce or suppress the transcription of specific messenger ribonucleic acids

Table XXI-7

Topical Anti-inflammatory Agents

NAME	DOSAGE FORM	TRADE NAME
Steroids		
Dexamethasone	Ophthalmic suspension 0.1%	Maxidex
Dexamethasone sodium phosphate	Ophthalmic ointment 0.05%	AK-Dex Decadron Available generically
Dexamethasone sodium phosphate	Ophthalmic solution 0.1%	AK-Dex Decadron Available generically
Fluorometholone	Ophthalmic ointment 0.1% Ophthalmic suspension 0.1% Ophthalmic suspension 0.1% Ophthalmic suspension 0.25%	FML S.O.P. Fluor-Op FML FML Forte Available generically (0.1%)
Fluorometholone acetate	Ophthalmic suspension 0.1%	Flarex Eflone
Medrysone	Ophthalmic suspension 1%	HMS
Prednisolone acetate	Ophthalmic suspension 0.12% Ophthalmic suspension 0.125% Ophthalmic suspension 1% Ophthalmic suspension 1%	Pred Mild Econopred Econopred Plus Pred Forte Available generically (1%)
Prednisolone sodium phosphate	Ophthalmic solution 0.125% Ophthalmic solution 0.125% Ophthalmic solution 0.125% Ophthalmic solution 1% Ophthalmic solution 1%	AK-Pred Inflamase Mild Available generically AK-Pred Inflamase Forte Available generically (1%)
Rimexolone	Ophthalmic solution 1%	Vexol
Nonsteroidal Anti-Inflammatory Drugs*		
Diclofenac	Ophthalmic solution 0.1%	Voltaren
Flurbiprofen	Ophthalmic solution† 0.03%	Ocufen Available generically
Ketorolac	Ophthalmic solution 0.5%	Acular
Suprofen	Ophthalmic solution† 1%	Profenal

* Recently, concerns have been raised regarding the possible association between postsurgical corneal complications and the use of NSAIDs following cataract or refractive surgery. Investigations were under way as this book went to press.

† Indicated for intraoperative miosis only.

The proteins produced in the eye under the control of these mRNAs are not known, and only resultant effects have been described.

At the tissue level, glucocorticoids prevent or suppress the local hyperthermia, vascular congestion, edema, and pain of initial inflammatory responses, whether the cause is traumatic (radiant, mechanical, or chemical), infectious, or immunologic. They also suppress the late inflammatory responses of capillary proliferation, fibroblast proliferation, collagen deposition, and scarring.

At the biochemical level, the most important effect of anti-inflammatories may be the inhibition of arachidonic acid release from phospholipids (see Part 4, Biochemistry). Liberated arachidonic acid is otherwise converted into PGs, PG endoperoxides, leukotriene, and thromboxane, which are potent mediators of inflammation. Glucocorticoids also suppress the liberation of lytic enzymes from lysozymes.

The effects of glucocorticoids on immune-mediated inflammation are complicated. Glucocorticoids do not affect the titers of either IgE, which mediates allergic mechanisms, or IgG, which mediates autoimmune mechanisms. Nor do glucocorticoids appear to interfere with the normal processes in the afferent limb of cell-mediated immunity, as in graft rejection. Apparently, they interfere instead with the efferent limb of the immune response. For example, glucocorticoids prevent macrophages from being attracted to sites of inflammation by interfering with their response to lymphocyte-released migration-inhibiting factor. Systemically administered glucocorticoids cause sequestration of lymphocytes, especially the T lymphocytes that mediate cellular immunity. However, the post-transcriptional molecular mechanisms of these responses are as yet unknown. BCSC Section 9, *Intraocular Inflammation and Uveitis*, discusses immune responses in detail.

Adverse effects Glucocorticoids may cause a number of adverse effects in the eye and elsewhere in the body. Complications in the eye include the following:

- Glaucoma
- Posterior subcapsular cataracts
- Exacerbation of bacterial and viral (especially herpetic) infections through suppression of protective immune mechanisms
- Ptosis
- Mydriasis
- Scleral melting
- Eyelid skin atrophy

In the body, oral doses can cause the following:

- Suppression of the pituitary–adrenal axis
- Gluconeogenesis resulting from hyperglycemia, muscle wasting, osteoporosis
- Redistribution of fat from the periphery to the trunk
- Central nervous system effects such as euphoria
- Insomnia
- Aseptic necrosis of hip
- Peptic ulcer
- Diabetes
- Occasionally, psychosis

The systemic side effects of steroids and the benefits and limitations of alternate-day therapy are discussed in BCSC Section 1, *Update on General Medicine.*

Steroid-induced elevation in IOP may occur with topical, periocular, nasal, and systemic glucocorticoid therapy. Individuals differ in their responsiveness: approximately 4% develop pressures higher than 31 mm Hg after 6 weeks of therapy with topical dexamethasone. High levels of response are generally reproducible. The mechanism by which steroids decrease the facility of aqueous outflow through the trabecular meshwork remains unknown. In vitro studies have shown an endoreplication of DNA in the trabecular cells, as well as the production of aberrant sialated polypeptides.

Individual response to steroids is highly dependent on the duration, strength, and frequency of therapy and the potency of the agent used. Steroid-induced IOP elevation almost never occurs in less than 5 days and rarely in less than 2 weeks. It is not generally appreciated that late responses to therapy are common and that failure of the IOP to rise after 6 weeks of therapy does not ensure that the patient will not develop a marked elevation in pressure after several months of therapy. For this reason, *IOP monitoring is required at periodic intervals during the entire course of chronic steroid therapy* to prevent iatrogenic glaucomatous nerve damage.

Steroid-induced IOP rises are usually reversible by discontinuance of therapy if the drug has not been used for more than 1 year, but permanent elevations of pressure are common if therapy has continued for 18 months or more. Pressure usually returns to baseline within 2 weeks in reversible cases.

The anti-inflammatory and pressure-elevating potencies of six steroids used in ophthalmic therapy are given in Table XXI-8. The anti-inflammatory potency was determined by an in vitro assay of inhibition of lymphocyte transformation, and the IOP effects were determined by testing in persons already known to be highly responsive to topical dexamethasone. Some dissociation of effects was observed. However, until all these agents are compared in a model of ocular inflammation relevant to human disease, no conclusion can be reached about the dissociation of effects. The lower-than-expected effect on pressure of some of these agents may be explained by more rapid metabolism of fluorometholone in the eye compared with dexamethasone and by the relatively poor penetration of medrysone. The efficacy of these agents for intraocular inflammation may be similarly reduced.

TABLE XXI-8

COMPARISON OF ANTI-INFLAMMATORY AND IOP-ELEVATING POTENCIES

GLUCOCORTICOID	RELATIVE POTENCY	RISE IN IOP (mm Hg)
Dexamethasone 0.1%	24	22
Fluorometholone 0.1%	21	6
Prednisolone 1%	2.3	10
Medrysone 1%	1.7	1
Tetrahydrotriamcinolone 0.25%	1.4	2
Hydrocortisone 0.5%	1.0	3

When a steroid-induced pressure rise is suspected but continued steroid therapy is warranted, the physician faces the following choices:

- Continue the same treatment and closely monitor the status of the optic nerve
- Attempt to offset the pressure rise with other agents
- Reduce the potency, concentration, or frequency of the steroid used, while monitoring both pressure and inflammation

When alternative classes of anti-inflammatory agents could be employed, a change may be advisable.

Agents and regimens Choice of available corticosteroid agents and dosage regimens remains somewhat empirical. Steroids can be used topically (iritis), intravenously (optic neuritis), intravitreally (endophthalmitis), or in a periocular fashion (uveitis) (Table XXI-9).

Rimexolone (Vexol) is a newly designed steroid for topical use. It is available as multidose topical ophthalmic suspension (1%), and its mechanism of action is very similar to that of other steroids. Clinical studies have demonstrated that rimexolone 1% ophthalmic suspension is efficacious for treatment of anterior chamber inflammation following cataract surgery. After five to seven doses given hourly, the mean serum concentrations were approximately 130 pg/ml with a half-life of approximately 1–2 hours.

Rimexolone is indicated for the treatment of postoperative inflammation following ocular surgery and of anterior uveitis. The main advantage of rimexolone is that it offers a low steroid response similar to that of fluorinated steroids while still exerting full anti-inflammatory activity equal to that of glucocorticoids. Common side effects include visual field defects and posterior subcapsular cataracts. Elevated IOP has been reported, even though it is rare. Systemic side effects including headache, hypotension, rhinitis, pharyngitis, and taste perversion occur in fewer than 2% of patients.

TABLE XXI-9

USUAL ROUTE OF STEROID ADMINISTRATION IN OCULAR INFLAMMATION

CONDITION	ROUTE
Blepharitis	Topical
Conjunctivitis	Topical
Episcleritis	Topical
Scleritis	Topical and/or systemic
Keratitis	Topical
Anterior uveitis	Topical and/or periocular
Posterior uveitis	Systemic and/or periocular, intravitreal
Endophthalmitis	Systemic/periocular, intravitreal
Optic neuritis	Systemic or periocular
Cranial arteritis	Systemic
Sympathetic ophthalmia	Systemic and topical

Application of 1–2 drops of rimexolone into the conjunctival sac of the affected eye every hour during waking hours effectively controls anterior uveitis initially. Dosage can be modified according to the intensity of uveitis later.

Loteprednol etabonate 0.5% (Lotemax), a new ester-based steroid, targets inflammation with a unique, site-active mechanism of action, penetrating through the cornea directly to the site of inflammation similar to other steroids. It is structurally similar to other steroids but lacks a ketone group at position 20.

In clinical studies, loteprednol postoperatively resolved or controlled inflammation in 84% of patients following cataract surgery (versus placebo). It is also effective for giant papillary conjunctivitis. Moreover, in corticosteroid responders, studies show that patients treated with loteprednol demonstrate a low incidence of clinically significant, increased IOP.

As with other ophthalmic corticosteroids, loteprednol is contraindicated in most viral diseases of the corneas and conjunctiva and in mycobacterial and fungal diseases of the eye. Prolonged use may result in secondary glaucoma, cataract formation, and secondary ocular infections following suppression of the host response and/or perforation of the globe. The most common adverse events in patients treated with loteprednol were abnormal vision/blurring, burning, chemosis, discharge, and dry eyes.

Nonsteroidal Anti-Inflammatory Drugs (NSAIDs)

Derivation Derivatives of the 20-carbon essential fatty acid arachidonic acid have been shown to be mediators of a wide variety of biological functions, including regulation of smooth muscle tone (in blood vessels, bronchi, uterus, and gut), platelet aggregation, hormone release (growth hormone, ACTH, insulin, renin, and progesterone), and inflammation.

The synthetic cascade that results in the production of a wide variety of derivatives (depending on the stimulus and tissue) begins with stimulation of phospholipase A_2, the enzyme that liberates arachidonic acid from phospholipids of the cell membrane. (Phospholipase A_2 is inhibited by corticosteroids.) Arachidonic acid is then converted into either cyclic endoperoxides by *cyclo-oxygenase* (PG synthase) or into hydroperoxides by *lipoxygenase*. Among the subsequent products of the endoperoxides are the PGs, which mediate inflammation and other responses; prostacyclin, a vasodilator and platelet antiaggregant; and thromboxane, a vasoconstrictor and platelet aggregant. The hydroperoxides form a chemotactic agent and the leukotrienes C_4, D_4, and E_4, previously known as the slow-reacting substance of anaphylaxis.

Classification The currently available NSAIDs inhibit the production and, thus, the inflammation-inducing effects of PGs through the cyclo-oxygenase pathway. On the basis of chemical structures, NSAIDs can be classified as:

- *Salicylates*: aspirin, etofenamic acid, flufenamic acid, meclofenamate, mefenamic acid, tolfenamic acid
- *Indoles*: indomethacin, sulindac, tolmetin
- *Phenylalkanoic acids*: diclofenac, fenoprofen, flurbiprofen, ibuprofen, ketoprofen, ketorolac, naproxen, piroxicam, sutoprofen
- *Pyrazolones*: oxyphenbutazone, phenylbutazone

Table XXI-10

Nonsteroidal Anti-inflammatory Drugs

DRUG	STARTING DOSE
Aspirin	650 mg qid
Indomethacin (Indocin)	25 mg tid
Fenoprofen (Nalfon)	600 mg qid
Ibuprofen (Motrin)	400 mg qid
Ketoprofen (Orudis)	75 mg tid
Naproxen (Naprosyn)	250 mg bid
Piroxicam (Feldene)	20 mg qid
Sulindac (Clinoril)	150 mg bid
Tolmetin (Tolectin)	400 mg tid

Agents Table XXI-10 lists a number of these agents with their starting dosages. *Aspirin* and a number of other compounds have been found to inactivate cyclo-oxygenase in circulating platelets and megakaryocytes. They inhibit a host of biological functions, including the local signs of inflammation (local heat, vasodilation, edema, swelling) as well as pain and fever, and they have complex effects on clotting. At low doses (300 mg every other day), aspirin inhibits the cyclo-oxygenase in platelets that is essential for conversion of arachidonic acid to PGG_2 and thromboxane. Inhibition of thromboxane production prevents coagulation. Platelets have a life span of 7–10 days and, as anucleate platelets cannot produce new cyclo-oxygenase in their lifetime, this anticoagulant effect lasts for 7–10 days even after the discontinuation of aspirin administration.

In contrast, production of prostacyclin by the vascular endothelium is only briefly suppressed, because cyclo-oxygenase can be replenished by these nucleated cells. As a consequence, aspirin therapy for postoperative pain or for pain associated with traumatic hyphema may increase the risk of hemorrhage. However, low-dose aspirin therapy may benefit patients having platelet emboli (as in some cases of amaurosis fugax). Diversion of arachidonic acid to the lipoxygenase pathway by inhibition of cyclo-oxygenase may explain why aspirin use can be associated with asthma attacks and hypersensitivity reactions (mediated by leukotrienes C_4, D_4, and E_4) in susceptible persons.

High doses of aspirin, such as those employed in the treatment of arthritis, may occasionally have toxic effects such as headache, dizziness, tinnitus, dimmed vision, mental confusion, drowsiness, hyperventilation, nausea, and vomiting. These effects may be potentiated by the concomitant use of carbonic anhydrase inhibitors at doses sufficient to cause systemic acidosis. During metabolic acidosis, a higher proportion of aspirin molecules are shifted into the more lipid-soluble un-ionized form, which more readily penetrates the blood–brain barrier.

Systemic therapy with aspirin and other cyclo-oxygenase inhibitors is occasionally effective in the treatment of scleritis and uveitis but is generally disappointing. These agents do not appear to be as effective as steroids.

Flurbiprofen (Ocufen) was the first commercially available topical ocular NSAID. When applied preoperatively, it helps to retard the PG-mediated pupillary constriction that can otherwise interfere with extracapsular cataract surgery.

Diclofenac (Voltaren) is another topical NSAID. It has an FDA-approved indication for the postoperative prophylaxis and treatment of ocular inflammation and has also been used successfully to prevent and treat cystoid macular edema (CME).

Suprofen (Profenal), available in 1% ophthalmic solution, is indicated for inhibition of intraoperative miosis only.

The role of topical NSAIDs in the treatment of ocular inflammation and in the prevention of aphakic CME is under investigation. NSAIDs such as *indomethacin* can be effective in treating orbital inflammatory diseases with minimal side effects. The prophylactic use of indomethacin in cataract patients has been reported to reduce the incidence of angiographically detected CME, but its effect on visually significant CME has yet to be reported.

Ketorolac (Acular) is an NSAID that blocks the metabolism of arachidonic acid by cyclo-oxygenase. Arachidonic acid metabolites are present in higher quantities in the tears of ocular allergic disease patients. Two recent double-masked studies revealed that patients with ocular allergies who were treated with ketorolac tromethamine had significantly less conjunctival inflammation, ocular itching, and tearing compared with placebo. Even though ketorolac is effective in allergic disease, it does not have a decongestant effect and does not relieve redness. The recommended dose of ketorolac is 1 drop (0.25 mg) four times per day. The most common side effects are stinging and burning on instillation (40%) and occasional corneal melts.

Mast-Cell Stabilizers and Antihistamines

Allergic conjunctivitis is an immediate hypersensitivity reaction in which triggering antigens couple to reaginic antibodies (IgE) on the cell surface of mast cells and basophils, leading to release of histamine from secretory granules. The released histamine causes capillary dilatation and increased permeability and thus conjunctival injection and swelling. It also stimulates nerve endings, causing pain and itching. Table XXI-11 lists agents for the relief of allergic conjunctivitis by class.

Short-term relief for mild allergic symptoms may be achieved with topical antihistamines such as *antazoline*. More recalcitrant cases often require short-term use of topical steroids.

Hydroxymethyl progesterone is well suited for this use, since it has a low potential for elevating IOP. Unfortunately, more potent steroids are often used and abused for this relief, because of either limited response to other agents or chronic disease.

Cromolyn sodium 4% (Crolom), a blocker of histamine release, is currently the therapy of choice for severe vernal and atopic conjunctivitis. However, it has no direct antihistamine effect and must be used prophylactically for several weeks to be effective. It has few, if any, side effects and is thus much preferred to potent steroids.

Lodoxamide (Alomide) is a mast-cell stabilizer that has been used for the treatment of allergic conjunctivitis. The human eye has approximately 50 million mast cells. Each contains several hundred granules that in turn contain preformed chemical mediators. Chronic exposure to antigen results in an antigen–IgE antibody bound to the mast-cell membrane. The release of a cascade of mediators such as histamine, PG, leukotrienes, and chemotactic factors follows. These mediators cause the itching and hyperemia associated with allergic conjunctivitis. Mast-cell stabilizers have traditionally been viewed as preventing calcium influx across mast-cell membranes, thereby preventing mast-cell degranulation and mediator release. Recent studies, however, demonstrate that cromolyn sodium inhibits neutrophil, eosinophil, and monocyte activation in vitro.

TABLE XXI-11

AGENTS FOR RELIEF OF ALLERGIC CONJUNCTIVITIS

GENERIC NAME	TRADE NAME	CLASS
Cromolyn sodium	Crolom	Mast-cell inhibitor
Emedastine	Emadine	H_1-antagonist
Ketorolac tromethamine	Acular	NSAID
Levocabastine	Livostin	H_1-antagonist
Lodoxamide	Alomide	Mast-cell inhibitor
Loteprednol	Alrex	Corticosteroid
Naphazoline/antazoline	Vasocon-A	Antihistamine/ decongestant
Naphazoline/pheniramine	Naphcon-A Opcon-A	Antihistamine/ decongestant
Olopatadine	Patanol	H_1-antagonist/ mast-cell inhibitor

Lodoxamide has beneficial effects in both allergic conjunctivitis and vernal keratoconjunctivitis. It has been shown to produce stabilization of the mast-cell membrane 2500 times greater than cromolyn sodium. Clinically, lodoxamide onset is quicker with less stinging than cromolyn sodium in treating allergic conjunctivitis. One recent multicenter, double-masked study showed that lodoxamide was superior to cromolyn sodium in treating vernal keratoconjunctivitis. However, as with all mast-cell stabilizers, it takes several weeks to become clinically effective. For patients who are very uncomfortable, it may be necessary to use topical steroids concurrently with mast-cell stabilizers for the first weeks until these agents are fully effective.

The usual dose of lodoxamide 0.1% for adults and children older than 2 years is 1–2 drops in the affected eye four times daily for up to 3 months. The most frequently reported adverse reactions are burning, stinging, and discomfort upon instillation (15%).

Pemirolast potassium 0.1% (Alamast) is a mast-cell stabilizer that received FDA approval in September 1999 for the prevention of itchy eyes due to allergic conjunctivitis. In clinical studies, the most common side effects were headache, rhinitis, and cold/flu symptoms, which were generally mild.

Nedocromil sodium (Abrevia) is a mast-cell stabilizer now in phase III clinical trials. Studies have demonstrated that this agent is more potent than sodium cromoglycate. It also has a more convenient twice-daily dosing regimen.

Olopatadine hydrochloride 0.1% (Patanol) has both H_1 antihistamine and mast-cell stabilizing properties. This agent has a rapid onset and at least an 8-hour duration of action. Recommended dosing is 1–2 drops in the affected eye two times a day at an interval of 6–8 hours. Adverse reactions included headache at an incidence of 7%. The following adverse reactions were reported at an incidence of less than 5%: ocular burning or stinging, dry eye, foreign body sensation, hyperemia, keratitis, lid edema, pruritus, asthenia, cold syndrome, pharyngitis, rhinitis, sinusitis, and taste perversion.

Ketotifen fumarate 0.025% (Zaditor) received FDA approval in July 1999. This agent is both a combined mast-cell stabilizer and H_1-receptor antihistamine. Recommended dosing is 1 drop every 8–12 hours. Onset of action is within minutes after administration. Side effects of conjunctival injection, headaches, and rhinitis were reported at an incidence of 10%–25%.

Levocabastine (Livostin) is a newly synthesized H_1-receptor antagonist. It has been shown to be significantly effective in reducing the signs and symptoms of allergen-induced allergic conjunctivitis. Its onset of action occurs within minutes and lasts for at least 4 hours. Levocabastine has been shown to be more effective than placebo or vasoconstrictor–antihistamine combinations. It appears to be as effective as cromolyn sodium in some studies, and it has the advantage of a quicker onset than mast-cell stabilizers.

The usual dose of levocabastine 0.05% is 1 drop four times per day for up to 2 weeks. The most frequent side effects reported were mild, transient stinging and burning (15%) and headache (5%).

Emedastine difumarate 0.05% (Emadine) is a relatively selective H_1-receptor antagonist indicated for the temporary relief of the signs and symptoms of allergic conjunctivitis. Recommended dosing is 1 drop up to four times per day. The most common side effect was headache (11%). Bad taste, blurred vision, burning or stinging, corneal infiltrates, dry eye, rhinitis, and sinusitis are other noted side effects.

Azelastine hydrochloride is an H_1 antihistamine in phase III clinical trials for use in eyedrop form for the treatment of allergic conjunctivitis. In one study, azelastine eye drops were effective and well tolerated at a dose of 0.05% for the treatment of seasonal allergic conjunctivitis. This drug is currently available as a nasal spray for the treatment of allergic rhinitis.

Antagonists of leukotriene, platelet-activating factor, bradykinin, and nitric oxide synthase are some of the other promising drugs currently under investigation. Manipulation of T-cell activity with cyclosporine or immunosuppressives such as azathioprine may also prove to be useful. Topical cyclosporine has also been demonstrated to be an effective and safe steroid-sparing agent for the treatment of atopic keratoconjunctivitis.

Bito LZ. Prostaglandins: old concepts and new perspectives. *Arch Ophthalmol.* 1987; 105:1036–1039.

Flach AJ. Nonsteroidal anti-inflammatory drugs. In: Zimmerman TJ, Kooner KS, eds. *Ophthalmol Clin North Am.* 1989:151–161.

Foster CS, Forstot SL, Wilson LA. Mortality rate in rheumatoid arthritis patients developing necrotizing scleritis or peripheral ulcerative keratitis. Effects of systemic immunosuppression. *Ophthalmology.* 1984;91:1253–1263.

Giede-Tuch C, Westhoff M, Zarth A. Azelastine eye-drops in seasonal allergic conjunctivitis or rhinoconjunctivitis. A double-blind, randomized, placebo-controlled study. *Allergy.* 1998;53(9):857–862.

Hingorani M, Moodaley L, Calder VL, et al. A randomized, placebo-controlled trial of topical cyclosporin A in steroid-dependent atopic keratoconjunctivitis. *Ophthalmology.* 1998;105(9):1715–1720.

Leibowitz HM, Bartlett JD, Rich R, et al. Intraocular pressure-raising potential of 1.0% rimexolone in patients responding to corticosteroids. *Arch Ophthalmol.* 1996;114: 933–937.

Mishima HK, Masuda K, Kitazawa Y, et al. A comparison of latanoprost and timolol in primary open-angle glaucoma and ocular hypertension. *Arch Ophthalmol.* 1996; 114:929–932.

Noble AG, Tripathi RC, Levine RA. Indomethacin for the treatment of idiopathic orbital myositis. *Am J Ophthalmol.* 1989;108:336–338.

Tripathi BJ, Millard CB, Tripathi RC. Corticosteroids induce a sialated glycoprotein (Cort-GP) in trabecular cells in vitro. *Exp Eye Res.* 1990;51:735–737.

Vorin P. Treating severe eye allergy. *Clin Exp Allergy.* 1998;28(Suppl 6):44–48.

Antifibrotic Agents

Antiproliferative agents, also known as *antimetabolites,* are occasionally required in the treatment of severe ocular inflammatory diseases, such as Behçet syndrome and sympathetic ophthalmia, or when the ocular disease is part of a systemic vasculitis. Systemic therapy with such agents is best carried out in consultation with a chemotherapist. The uses and side effects of these agents are discussed in BCSC Section 9, *Intraocular Inflammation and Uveitis.*

Fluorouracil is a fluorinated pyrimidine nucleoside analogue that blocks production of thymidylate and interrupts normal cellular DNA and RNA synthesis. Its primary action may be to cause cellular thymine deficiency and resultant cell death. The effect of fluorouracil is most pronounced on rapidly growing cells, and its use as an antiviral agent is primarily related to destruction of infected cells (warts) by topical application. Intravitreal injection of *5-fluorouracil* has been reported to be beneficial in preventing recurrent proliferative vitreoretinopathy after surgery for complex retinal detachments in an experimental model, although a suitable delivery system for use in patients has yet to be developed.

Subconjunctival injection improves the success of filtering surgery in difficult glaucoma cases. Studies are also underway to evaluate the beneficial effect of intraoperative topical use during filtering surgery. The drug is thought to inhibit the cellular proliferation that could otherwise occur in response to inflammation. In high-risk patients, including young glaucoma patients (≤40 years), the initial trabeculectomy with adjunctive 5-fluorouracil had a higher success rate than surgery without the adjunct.

Mitomycin-C is a compound isolated from the fungus *Streptomyces caespitosus.* The parent compound becomes a bifunctional alkylating agent after enzymatic alteration within the cell; it then inhibits DNA synthesis and cross-links DNA. Mitomycin's immunosuppressive properties are fairly weak; however, it is a potent inhibitor of fibroblast proliferation.

Topical mitomycin-C has also been used in filtering surgery. It has the advantage of functioning with a single intraoperative application and not requiring the repeated postoperative injections of 5-fluorouracil. Randomized comparative studies of mitomycin-C with 5-fluorouracil in high-risk patients show lower average pressures with fewer corneal-surface and hypotony-related complications in the groups treated with mitomycin-C. It is used as a single topical application during glaucoma filtering operations to impede scarring and prevent surgical failure. Complications of therapy are wound leakage, hypotony, and localized scleral melting. Severe toxicity has been reported in an animal model with intraocular instillation of mitomycin-C, resulting in irreversible progressive bullous keratopathy in three of four rabbits.

Topical mitomycin-C has also been recommended for use in the prevention of recurrence after pterygium excision. Recommended dosage is 0.02%–0.04% four times daily for 1–2 weeks after surgery. The recurrence rate with such therapy has been reported as low as 0%–11%. Unfortunately, several side effects—such as corneal edema, corneal perforation, corectopia, iritis, cataract, and intractable pain—have been reported. A primary conjunctival graft after pterygium removal may offer similar low recurrence rates without these serious complications. It has also been used to reduce haze in photorefractive keratectomy (PRK) patients.

Whiteside-Michel J, Liebmann J, Ritch R. Initial 5-fluorouracil trabeculectomy in young patients. *Ophthalmology*. 1992;99:7–13.

Antibiotics (Table XXI-12)

Penicillins and Cephalosporins

The penicillins and cephalosporins are β-lactam–containing antibacterial agents that react with and inactivate a particular bacterial transpeptidase that is essential for bacterial cell-wall synthesis (Fig XXI-9). The amide bond of the β-lactam group is surrounded by structural features in the antibiotic molecule that resemble the portion of the natural substrate with which the transpeptidase reacts. The peptidase reacts with the antibiotic, forming an inactive acyl intermediate.

Some bacteria are resistant to the action of penicillins and cephalosporins. The lipopolysaccharide outer coat of many gram-negative bacteria may prevent certain hydrophilic antibiotics from reaching their cytoplasmic membrane site of action. Furthermore, some bacteria produce β-*lactamases* (penicillinase), enzymes capable of cleaving the critical amide bond within these antibiotics. The different penicillins and cephalosporins vary in susceptibility to the β-lactamases produced by different bacterial species.

The penicillins and cephalosporins penetrate the blood–ocular and blood–brain barriers poorly and are actively transported out of the eye by the organic-acid transport system of the ciliary body. However, their penetration into the eye increases with inflammation and with coadministration of probenecid.

Serious and occasionally fatal hypersensitivity (anaphylactoid) reactions can occur in association with penicillin and cephalosporin therapy. Although anaphylaxis is more frequent following parenteral administration, it can occur with oral therapy. Such reactions are more likely to occur in persons with a history of sensitivity to multiple allergens. A history of immediate allergic response (anaphylaxis or rapid onset of hives) to any penicillin is a strong contraindication to the use of any other penicillin. Approximately 10% of persons allergic to a penicillin will have cross-reactivity to cephalosporins.

Penicillins There are five classes of penicillins, which differ in their spectrum of antibiotic activity and in their resistance to penicillinase:

- *Penicillin G, penicillin V,* and *phenethicillin* are highly effective against most gram-positive and gram-negative cocci, anaerobes, and *Actinomyces, Leptospira,* and *Treponema* organisms. However, most strains of *Staphylococcus aureus* and many strains of *Staphylococcus epidermidis* and *Neisseria gonorrhoeae* are now resistant, often through production of penicillinase. Penicillin V and phenethicillin are absorbed well orally, whereas penicillin G is better given intravenously because it is inactivated by stomach acid. These penicillins are excreted rapidly by the kidneys and have short half-lives unless given in depot forms (ie, procaine penicillin G) or administered with probenecid, which competitively inhibits excretion by the kidneys.

TABLE XXI-12

PRINCIPAL ANTIBIOTIC AGENTS

DRUG NAME	TOPICAL	SUBCONJUNCTIVAL	INTRAVITREAL	INTRAVENOUS (ADULT)
Amikacin sulfate	10 mg/ml	25 mg	400 μg	15 mg/kg daily in 2–3 doses
Ampicillin sodium	50 mg/ml	50–150 mg	500 μg	4–12 g daily in 4 doses
Bacitracin zinc	10,000 units/ml	5,000 units		
Carbenicillin disodium	4–6 mg/ml	100 mg	250–2000 μg	8–24 g daily in 4–6 doses
Cefazolin sodium	50 mg/ml	100 mg	2250 μg	2–4 g daily in 3–4 doses
Ceftazidime		200 mg	2200 μg	1 g daily in 2–3 doses
Clindamycin	50 mg/ml	15–50 mg	1000 μg	900–1800 mg daily in 2 doses
Colistimethate sodium	10 mg/ml	15–25 mg	100 μg	2.5–5.0 mg/kg daily in 2–4 doses
Erythromycin	50 mg/ml	100 mg	500 μg	
Gentamicin sulfate	8–15 mg/ml	10–20 mg	100–200 μg	3–5 mg/kg daily in 2–3 doses
Imipenem/Cilastatin sodium	5 mg/ml			2 g daily in 3–4 doses
Kanamycin sulfate	30–50 mg/ml	30 mg		
Methicillin sodium	50 mg/ml	50–100 mg	1000–2000 μg	6–10 g daily in 4 doses
Neomycin sulfate	5–8 mg/ml	125–250 mg		
Penicillin G	100,000 units/ml	0.5–1.0 million units		12–24 million units daily in 4 doses
Polymyxin B sulfate	10,000 units/ml	100,000 units		
Ticarcillin disodium	6 mg/ml	100 mg		200–300 mg/kg daily
Tobramycin sulfate	8–15 mg/ml	10–20 mg	100–200 μg	3–5 mg/kg daily in 2–3 doses
Vancomycin hydrochloride	20–50 mg/ml	25 mg	1000 μg	15–30 mg/kg daily in 1–2 doses

- The *penicillinase-resistant penicillins* include *methicillin, nafcillin, oxacillin, cloxacillin, dicloxacillin,* and *floxacillin.* They are less potent than penicillin G against susceptible organisms but are the drugs of choice for infections caused by penicillinase-producing *S. aureus.* Methicillin and nafcillin are acid-labile and are therefore given either parenterally or by subconjunctival injection. The other

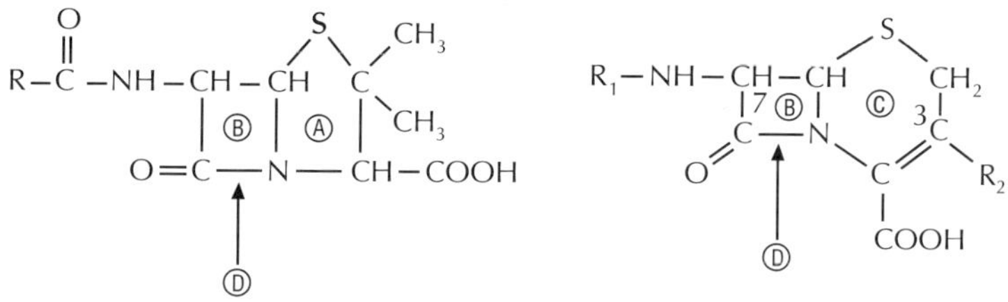

FIG XXI-9—Structure of the penicillins and cephalosporins. *A,* thiazole ring; *B,* β-lactam ring; *C,* dihydrothizine ring; *D,* site of action of beta-lactam. (From Mandell GL, Sande MA. Penicillins and cephalosporins. In: Gilman AG, Goodman LS, Gilman A, et al, eds. *The Pharmacological Basis of Therapeutics.* 6th ed. New York: Macmillan; 1980:1126–1161.)

agents in this group have reasonable oral absorption. When they are given systemically, coadministration of probenecid reduces renal excretion and outward transport from the eye.

- The broad-spectrum penicillins such as *ampicillin* and *amoxicillin* have antibacterial activity that extends to such gram-negative organisms as *Haemophilus influenzae* and *Proteus mirabilis.* Resistant strains of *H. influenzae* are becoming more common. These drugs are stable in acid and may be given orally. They are not resistant to penicillinase.
- *Carbenicillin, ticarcillin,* and *azlocillin* have antimicrobial activity extended to include *Pseudomonas, Enterobacter,* and indole-positive strains of *Proteus.* These drugs are given parenterally or subconjunctivally, though the indanyl ester of carbenicillin may be given orally. They are not resistant to penicillinase.
- *Piperacillin* and *mezlocillin* are particularly potent against *Pseudomonas.* They are administered parenterally or subconjunctivally, and they are not resistant to penicillinase.

Cephalosporins Bacterial susceptibility patterns and resistance to β-lactamases have determined the classification of the cephalosporins as first-, second-, or third-generation:

- *First generation: Cephalothin, cefazolin, cephalexin,* and *cephradine* have antimicrobial activity against gram-positive and gram-negative organisms, especially group-A *Streptococcus pyogenes, viridans,* and *pneumoniae; S. aureus* and *epidermidis; Clostridium perfringens; Bacillus subtilis;* and *Corynebacterium diphtheriae.* They are also active against *P. mirabilis, Salmonella, Shigella, Klebsiella,* and *Escherichia coli* but only half of the isolates of *H. influenzae.* They are not active against *Enterobacter,* other *Proteus* species, *Pseudomonas aeruginosa, Bacteroides, Serratia,* and enterococci. Cephalothin is the most resistant of these agents to staphylococcal β-lactamase. Cefazolin has somewhat greater activity against *Klebsiella.* Cephalexin and cephradine are stable in acid and available in oral forms.
- *Second-generation: Cefamandole* and *cefoxitin* display greater activity against three additional gram-negative organisms: *H. influenzae, Enterobacter aerogenes,* and *Neisseria* species. *Cefuroxime* has a similar spectrum of activity. Cefamandole has increased activity against *Enterobacter* species, indole-positive *Proteus,*

H. influenzae, and *Bacteroides.* Cefoxitin is active against indole-positive *Proteus* and *Serratia* organisms. Cefuroxime is valuable in the treatment of penicillinase-producing *N. gonorrhoeae* and ampicillin-resistant *H. influenzae,* and its penetration of the blood–brain barrier is adequate for initial treatment of suspected pneumococcal, meningococcal, or *H. influenzae* meningitis.

- *Third-generation: Cefotaxime* and *moxalactam* have enhanced activity against gram-negative bacilli, but they are inferior to first-generation cephalosporins with regard to their activity against gram-positive cocci. *Cefoperazone* and *ceftizoxime* have a similar spectrum of activity against gram-positive and -negative organisms, anaerobes, *Serratia, Proteus,* and some *Pseudomonas* isolates. Cefoperazone is particularly effective against *Pseudomonas.* Cefotaxime is able to penetrate the blood–brain barrier better than the other cephalosporins, and it presumably also penetrates the blood–ocular barrier.

Other Antibacterial Agents (Table XXI-13)

Sulfonamides Sulfonamides are derivatives of para-aminobenzenesulfonamide. They are structural analogues and competitive antagonists of para-aminobenzoic acid for the bacterial synthesis of folic acid. Only bacteria that must synthesize their own folic acid are affected by these drugs. Mammalian cells are not affected, because they are unable to synthesize folic acid. Sulfonamides are only bacteriostatic. They are more effective when administered with *trimethoprim,* a potent inhibitor of bacterial dihydrofolate reductase; together, they block successive steps in the synthesis of tetrahydrofolic acid.

Sulfacetamide ophthalmic solution (10%–30%) penetrates the cornea well. Susceptible organisms include *S. pneumoniae, C. diphtheriae, H. influenzae, Actinomyces,* and *C. trachomatis.* However, a 3-week course of systemic sulfonamide therapy is required to eradicate *Chlamydia.* Local irritation, itching, periorbital edema, and transient stinging are some of the common side effects from topical administration. As with all sulfonamide preparations, severe sensitivity reactions such as toxic epidermal necrolysis and Stevens-Johnson syndrome have been reported.

Tetracyclines Tetracyclines enter bacteria by an active transport across the cytoplasmic membrane. They inhibit protein synthesis by binding to the 30 S ribosomal subunit, thus preventing access of aminoacyl tRNA to the acceptor site on the mRNA–ribosome complex. Host cells are less affected because they lack an active-transport system.

Tetracyclines are broad-spectrum bacteriostatic antibiotics, active against many gram-positive and -negative bacteria and also against *Rickettsiae, Mycoplasma pneumoniae,* and *Chlamydiae.* However, many strains of *Klebsiella* and *H. influenzae* and nearly all strains of *Proteus vulgaris* and *P. aeruginosa* are resistant. Tetracycline is poorly water-soluble but is soluble in eyedrops containing mineral oil; it penetrates the corneal epithelium readily.

Systemic therapy with tetracycline is used for chlamydial infections; because the drug is excreted into oil glands, it is also used for staphylococcal infections of the meibomian glands. It chelates to calcium in milk and antacids and is best taken on an empty stomach. Tetracycline should not be given to children or pregnant women, because it can be deposited in growing teeth, discoloring them. Tetracycline depresses plasma prothrombin activity and thereby potentiates coumadin.

Table XXI-13

Ophthalmic Antibacterial Agents

GENERIC NAME	TRADE NAME	CONCENTRATION OPHTHALMIC SOLUTION (%)
Individual Agents		
Bacitracin zinc	AK-Tracin	Not available, ointment 500 units/g
Chloramphenicol	AK-Chlor	0.5%
	Chloromycetin	0.16%–0.5%, ointment (1%)
	Chloroptic	0.5%, ointment (1%)
	Available generically	0.5%
Ciprofloxacin hydrochloride	Ciloxan	0.3%, ointment (0.3%)
Erythromycin	Ilotycin	Not available; ointment (0.5%)
	Available generically	Not available; ointment (0.5%)
Gentamicin sulfate	Garamycin	0.3%, ointment (0.3%)
	Genoptic	0.3%, ointment (0.3%)
	Gentacidin	0.3%, ointment (0.3%)
	Gentak	0.3%, ointment (0.3%)
	Available generically	0.3%, ointment (0.3%)
Norfloxacin	Chibroxin	0.3%
Ofloxacin	Ocuflox	0.3%
Sulfacetamide sodium	AK-Sulf	10%, ointment (10%)
	Bleph-10	10%, ointment (10%)
	Cetamide	Not available; ointment (10%)
	Isopto Cetamide	15%
	Ophthacet	10%
	Sulamyd sodium	10%, 30%, ointment (10%)
	Sulf-10	10%
	Available generically	10%–30%, ointment (10%)
Tobramycin sulfate	Tobralcon	0.3%, ointment (0.3%)
	Tobrex	0.3%, ointment (0.3%)
	Available generically	0.3%
Mixtures		
Polymyxin B/Bacitracin zinc	AK-Poly-Bac	Not available; ointment
	Polysporin	(10,000 units - 500 units/g)
Polymyxin B/Neomycin/Bacitracin	AK-Spore	Not available; ointment
	Neosporin	(10,000 units - 3.5–400 units/g)
	Available generically	
Polymyxin B/Neomycin/Gramicidin	AK-Spore	10,000 units
	Neosporin	1.75 mg
	Available generically	0.025 mg/ml
Polymyxin B/Oxytetracycline	Terramycin	Not available; ointment
	Terak	(10,000 units - 5 mg/g)
Polymyxin B/Trimethoprim	Polytrim	10,000 units 1 mg/ml
	Available generically	

Chloramphenicol This broad-spectrum bacteriostatic agent inhibits bacterial protein synthesis by binding reversibly to the 50 S ribosomal subunit, preventing aminoacyl tRNA from binding to the ribosome. Chloramphenicol is effective against most *H. influenzae* and *Neisseria meningitidis* and *gonorrhoeae* and all anaerobic bacteria. It has some activity against *S. pneumoniae, S. aureus, Klebsiella pneumoniae, Enterobacter, Serratia,* and *P. mirabilis. P. aeruginosa* is resistant.

Chloramphenicol penetrates the corneal epithelium well during topical therapy and penetrates the blood–ocular barriers readily when given systemically. However, the use of this agent is limited because it has been implicated in an idiosyncratic and potentially lethal aplastic anemia. Although most cases of this anemia have occurred after oral administration, some have been associated with parenteral and even topical ocular therapy.

Aminoglycosides The aminoglycosides consist of amino sugars in glycosidic linkage. They are bacteriocidal agents that are transported across the cell membrane into bacteria and bind to the 30 S and 50 S ribosomal subunits, interfering with initiation of protein synthesis. The antibacterial spectrum of these agents is determined primarily by the efficiency of their transport into bacterial cells. Such transport is energy-dependent and may be reduced in the anaerobic environment of an abscess. Resistance to aminoglycosides may be caused by failure of transport, low affinity for the ribosome, or plasmid-transmitted ability to enzymatically inactivate the drug. The coadministration of drugs such as penicillin that alter bacterial cell-wall structure can markedly increase aminoglycoside penetration, resulting in a synergism of antibiotic activity against gram-positive cocci, especially enterococci. Amikacin is remarkably resistant to enzymatic inactivation.

Gentamicin, tobramycin, kanamycin, and *amikacin* have antibacterial activity against aerobic, gram-negative bacilli such as *P. mirabilis, P. aeruginosa, Klebsiella, Enterobacter,* and *Serratia.* Gentamicin and tobramycin are active against *S. aureus* and *epidermidis.* Kanamycin is generally less effective than the others against gram-negative bacilli. Resistance to gentamicin and tobramycin has gradually increased as a result of a plasmid-transmitted ability to synthesize inactivating enzymes. Thus, amikacin, which is generally impervious to these enzymes, is particularly valuable in treating such resistant organisms.

Aminoglycosides are not absorbed well orally but are given systemically by intramuscular or intravenous routes. They do not readily penetrate the blood–ocular barrier but may be administered as eyedrops, ointments, or periocular injections. Gentamicin and carbenicillin should not be mixed for IV administration because the carbenicillin inactivates the gentamicin over several hours. Similar incompatibilities exist in vitro between gentamicin and other penicillins and cephalosporins.

Use of *streptomycin* is now limited to *Streptococcus viridans* bacterial endocarditis, tularemia, plague, and brucellosis. *Neomycin* is a broad-spectrum antibiotic, effective against *Enterobacter, K. pneumoniae, H. influenzae, N. meningitidis, C. diphtheriae,* and *S. aureus.* It is given topically in ophthalmology and orally as a bowel preparation for surgery. Neomycin is too toxic to be used intravenously but can be given orally because it is not absorbed from the gut. Topical allergy to ocular use of neomycin occurs in about 8% of cases. It can cause punctate epitheliopathy and retard re-epithelialization of abrasions.

All aminoglycosides can cause dose-related vestibular and auditory dysfunction and nephrotoxicity when they are given systemically. Systemic use of aminoglycosides should be limited to serious infections, and the plasma concentration of drug and blood urea nitrogen should be monitored to avoid overdosing.

Fluoroquinolones These fluorinated derivatives of nalidixic acid are available in a variety of chemical structures that include *norfloxacin, ofloxacin, pefloxacin, levofloxacin, ciprofloxacin, enoxacin, lomefloxacin, temafloxacin, fleroxacin,* and *tosufloxacin.* These agents are highly effective broad-spectrum antimicrobials with potent activity against common gram-positive and -negative ocular pathogens. Their mechanism of action targets bacterial DNA supercoiling through the inhibition of DNA gyrase, which is one of the enzymes responsible for replication, genetic recombination, and DNA repair.

Studies in vitro have demonstrated that the fluoroquinolones, especially ciprofloxacin and temafloxacin, inhibit 90% of common bacterial corneal pathogens and have a lower minimum inhibitory concentration than gentamicin, tobramycin, and cefazolin. They are also less toxic to the corneal epithelium than are the aminoglycosides.

Three fluoroquinolones currently available are ofloxacin ophthalmic solution 0.3% (Ocuflox), ciprofloxacin HCl 0.3% (Ciloxan), and norfloxacin 0.3% (Chibroxin). They are indicated for the treatment of corneal ulcers caused by susceptible strains of *S. aureus, S. epidermidis, S. pneumoniae, P. aeruginosa, Serratia marcescens* (efficacy studied in fewer than 10 infections), and *Propionibacterium acnes.* They are also indicated for bacterial conjunctivitis due to susceptible strains of *S. aureus, S. epidermidis, S. pneumoniae, Enterobacter cloacae, H. influenzae, P. mirabilis,* and *P. aeruginosa.*

These fluoroquinolones have a high rate of penetration into ocular tissue. Their sustained tear concentration levels exceed the minimum inhibitory concentrations of key ocular pathogens for up to 12 hours or more after one dose. They also deliver excellent susceptibility kill rates, with a recent in vitro study confirming eradication of 87% to 100% of indicated pathogenic bacteria including *P. aeruginosa.* In addition, ofloxacin provides patient comfort and prevents precipitates from forming. Ofloxacin has a high intrinsic solubility that enables it to be formulated at a near-neutral 6.4 pH. Ciprofloxacin is formulated at a pH of 4.5. The combination of high efficacy and safety has made fluoroquinolones the most prescribed anti-infectives for the treatment of ocular pathogens.

The most frequently reported drug-related adverse reaction is transient ocular burning or discomfort. Other reported reactions include stinging, redness, itching, chemical conjunctivitis/keratitis, periocular/facial edema, foreign body sensation, photophobia, blurred vision, tearing, dryness, and eye pain. Rare reports of dizziness have been received.

Miscellaneous antibiotics *Vancomycin* is a tricyclic glycopeptide derived from cultures of *Nocardia orientalis.* It is bactericidal for most gram-positive organisms through the inhibition of glycopeptide polymerization in the cell wall. It is useful in the treatment of staphylococcal infections in patients who are allergic to or have not responded to the penicillins and cephalosporins as well as in the treatment of methicillin-resistant streptococci. It can also be used in combination with aminoglycosides to treat *S. viridans* or *bovis* endocarditis. Vancomycin is especially indicated in the treatment of pseudomembranous colitis caused by *Clostridium difficile.*

Vancomycin may be used topically or intraocularly to treat sight-threatening infections of the eye, including infectious keratitis and endophthalmitis caused by methicillin-resistant staphylococci or streptococci. It is a preferred substitute for a cephalosporin used in combination with an aminoglycoside in the empirical treatment of endophthalmitis. See BCSC Sections 8 *(External Disease and Cornea)* and 9 *(Intraocular Inflammation and Uveitis)* for further discussion.

The intravenous dosage of vancomycin in adults with normal renal function is 500 mg every 6 hours or 1 g every 12 hours. Dosing must be adjusted in subjects who have renal impairment. Topical vancomycin may be given in a concentration of 50 mg/ml in the treatment of infectious keratitis. Lower concentrations (5 mg/ml) have been used successfully in the treatment of susceptible staphylococcal blepharoconjunctivitis. Intravitreal vancomycin with an aminoglycoside is recommended for initial empirical therapy for exogenous bacterial endophthalmitis. A dose of 1 mg in 0.1 ml establishes intraocular levels significantly higher than the minimum inhibitory concentration for most gram-positive organisms.

Unlike systemic treatment, topical and intraocular vancomycin have not been associated with ototoxicity or nephrotoxicity. Hourly use of 50 mg of vancomycin per milliliter delivers a dose of 36 mg/day, which is well below the recommended systemic dose.

Erythromycin is a macrolide (many-membered lactone ring attached to deoxy sugars) antibiotic that binds to the 50 S subunit of bacterial ribosomes and interferes with protein synthesis. It is effective against gram-positive cocci such as *S. pyogenes* and *S. pneumoniae*, gram-positive bacilli such as *C. diphtheriae* and *Listeria monocytogenes,* and a few gram-negative organisms such as *N. gonorrhoeae.* It is the treatment of choice for *Legionella pneumophila,* the agent of legionnaires' disease, as well as for *M. pneumoniae.* Erythromycin is administered orally as enteric-coated tablets or in esterified forms to avoid inactivation by stomach acid. It can also be administered parenterally or topically as an ophthalmic ointment. The drug penetrates the blood–ocular and blood–brain barriers poorly.

Polymyxin B is a mixture of basic peptides that function as cationic detergents to dissolve phospholipids of bacterial cell membranes, thus disrupting the cells. It is used topically or by local injection to treat corneal ulcers. Gram-negative bacteria are susceptible, including *Enterobacter, Klebsiella,* and *P. aeruginosa.*

Bacitracin is a mixture of polypeptides that inhibit bacterial cell-wall synthesis. It is active *against Neisseria, H. influenzae, Actinomyces,* and most gram-positive bacilli and cocci. It is available as an ophthalmic ointment either alone or in combination with polymyxin and neomycin.

Antifungal Agents (Table XXI-14)

Polyenes The polyene antibiotics are named for a component sequence of four to seven conjugated double bonds. That lipophilic region allows them to bind to sterols in the cell membrane of susceptible fungi, an interaction that results in damage to the membrane and leakage of essential nutrients. Other antifungals (such as flucytosine and the imidazoles) and even other antibiotics (such as tetracycline and rifampin) can enter through the damaged membrane, yielding synergistic effects.

Natamycin is available as a 5% suspension for topical ophthalmic use (once an hour). Local hypersensitivity reactions of the conjunctiva and eyelid and corneal epithelial toxicity may occur. *Amphotericin B* may be reconstituted at 0.25%–0.50% in sterile water (with deoxycholate to improve solubility) for topical use (every 30 minutes). It may also be administered systemically for disseminated disease, although careful monitoring for renal and other toxicities is required. These agents penetrate the cornea poorly. They are active topically against a variety of filamentous fungi, including *Aspergillus, Cephalosporium, Curvularia, Fusarium,* and *Penicillium,* and the yeast *Candida albicans.* Systemic amphotericin B is used in the treatment of systemic *Blastomyces, Coccidioides, Histoplasma, Cryptococcus, Candida, Mucorales,* and *Aspergillus* infections.

Table XXI-14

Antifungal Agents

GENERIC (TRADE) NAME	ROUTE	DOSAGE	SPECTRUM
Amphotericin B (Fungizone)	Topical	0.1–0.5% solution; dilute with water for injection or dextrose 5% in water	*Blastomyces* *Candida* *Coccidioides* *Histoplasma*
	Subconjunctival	0.8–1.0 mg	
	Intravitreal	5 µg	
	Intravenous	Because of side effects and toxicity, dose needs to be cautiously adjusted	
Fluconazole (Diflucan)	Oral	800 mg on day 1, then 400 mg daily in divided doses	*Candida*
Flucytosine (Ancobon)	Oral	50–150 mg/kg daily in 4 divided doses	*Candida* *Cryptococcus*
	Topical	1% solution	
Natamycin (Natacyn)	Topical	5% suspension	*Candida* *Aspergillus* *Cephalosporium* *Fusarium* *Penicillium*
Miconazole nitrate (Monistat)	Topical	1% solution	*Candida*
	Subconjunctival	5–10 mg	*Cryptococcus*
	Intravitreal	10 µg	*Aspergillus*
Ketoconazole (Nizoral)	Oral	200–400 mg daily	*Candida* *Cryptococcus* *Histoplasma*

Imidazoles The imidazole-derived antifungal agents also increase fungal cell-membrane permeability. *Miconazole* is available in a 1% solution that may be injected subconjunctivally (5 mg in 0.5 ml, once or twice daily) or applied topically. Miconazole penetrates the cornea poorly. *Ketoconazole* is available in 200 mg tablets for oral therapy (every 6–8 hours). Ketoconazole normally penetrates the blood–brain barrier and, presumably, the blood–ocular barrier poorly, but therapeutic levels can be achieved in inflamed eyes. The imidazole antifungals act against various species of *Aspergillus, Coccidioides, Cryptococcus,* and *Candida.*

Flucytosine Flucytosine (5-fluorocytosine) is converted by some species of fungal cells to 5-fluorouracil by cytosine deaminase, and then to 5-fluorodeoxyuridylate. The latter compound inhibits thymidylate synthase, an important enzyme in DNA synthesis. Host cells lack cytosine deaminase activity and are less affected. Only fungi that have both a permease to facilitate flucytosine penetration and cytosine deaminase are sensitive to flucytosine. It is taken orally at 50–150 mg/kg daily, divided into four doses (one every 6 hours). Although the drug is well absorbed and penetrates the blood–ocular barrier well, the majority of *Aspergillus* and half of *Candida* isolates are resistant. Flucytosine is used primarily as an adjunct to systemic amphotericin B therapy.

Leibowitz HM. Antibacterial effectiveness of ciprofloxacin 0.3% ophthalmic solution in the treatment of bacterial conjunctivitis. *Am J Ophthalmol.* 1991;112:29S–33S.

Leibowitz HM. Clinical evaluation of ciprofloxacin 0.3% ophthalmic solution for treatment of bacterial keratitis. *Am J Ophthalmol.* 1991;112:34S–47S.

Mandell GL, Sande MA. Antimicrobial agents: sulfonamides, trimethoprim-sulfamethoxazole, and urinary tract antiseptics. In: Gilman AG, Goodman LS, Gilman A, et al, eds. *The Pharmacological Basis of Therapeutics.* 6th ed. New York: Macmillan; 1980: 1106–1125.

Mandell GL, Sande MA. Antimicrobial agents: the aminoglycosides. In: Gilman AG, Goodman LS, Gilman A, et al, eds. *The Pharmacological Basis of Therapeutics.* 6th ed. New York: Macmillan; 1980:1162–1180.

Antiviral Agents (Table XXI-15)

Topical Antiviral Agents Three agents that compete with natural nucleotides for incorporation into viral and mammalian DNA are available for treatment of herpes simplex virus (HSV) keratitis. *Idoxuridine* (5-iodo-2′-deoxyuridine) and *trifluridine* (Viroptic) are structural analogues of thymidine, and *vidarabine* (adenine arabinoside, ara-A) is an analogue of adenine.

Trifluridine (1% drops, every 2–4 hours) has the advantage of being more soluble than the other agents. As a result, it can be used in drop form and can penetrate diseased corneas to treat herpetic iritis. It also has a greater efficacy (95% vs. 75% of cases healed in 2 weeks) than idoxuridine (0.5% ointment, every 4 hours) and equivalent or greater efficacy than vidarabine (3% ointment, every 3 hours). Cross-resistance does not seem to occur among the three agents.

Synthetic oligonucleotide analogues complementary to specific mRNAs of HSV are emerging as promising agents in the treatment of HSV keratitis.

Systemic Antiviral Agents Figure XXI-10 summarizes antiviral drugs.

Acyclovir is a synthetic guanosine analogue that requires phosphorylation to become active. It undergoes monophosphorylation by viral thymidine kinase. Because the viral thymidine kinase in HSV types 1 and 2 has many times more affinity to acyclovir than does host thymidine kinase, high concentrations of acyclovir monophosphate accumulate in infected cells. Acyclovir monophosphate is then further phosphorylated to the active compound acyclovir triphosphate (Fig XXI-11). The triphosphate cannot cross cell membranes and accumulates further. This increased concentration of acyclovir triphosphate is 50–100 times greater in infected cells than in uninfected cells.

Acyclovir triphosphate inhibits virus growth in three ways:

- It can function as a competitive inhibitor of DNA polymerases, with viral DNA polymerases being significantly more susceptible to acyclovir triphosphate than human DNA polymerases.
- It can be a DNA chain terminator.
- It can produce irreversible binding between viral DNA polymerase and the interrupted chain, causing permanent inactivation.

Table XXI-15

Antiviral Agents

GENERIC NAME	TRADE NAME	TOPICAL CONCENTRATION/ OPHTHALMIC SOLUTION (%)	SYSTEMIC DOSAGE
Trifluridine	Viroptic	1.0%	
Vidarabine monohydrate	Vira-A	3.0% (ophth ointment)	
Acyclovir sodium	Zovirax		Oral: Herpes simplex keratitis 200 mg 5 times daily for 7–10 days Oral: Herpes zoster ophthalmicus 600–800 mg 5 times daily for 10 days IV if patient is immunocompromised
Cidofovir	Vistide		IV induction: 5 mg/kg constant infusion over 1 hour once weekly for 2 consecutive weeks Maintenance: 5 mg/kg constant infusion over 1 hour administered every 2 weeks
Foscarnet sodium	Foscavir		IV: by controlled infusion only, either by central vein or by peripheral vein induction: 60 mg/kg (adjusted for renal function) given over 1 h every 8 h for 14–21 days Maintenance: 90–120 mg/kg given over 2 h once daily
Ganciclovir	Vitrasert		Intravitreal: 4.5 mg Sterile intravitreal insert designed to release the drug over a 5–8 mo period
Ganciclovir sodium	Cytovene		IV induction: 5 mg/kg every 12 h for 14–21 days Maintenance: 5 mg/kg daily for 7 days

Because of potential side and toxic effects with systemic dosage, the possible dosage adjustments and warnings should be followed properly.

The result is a several-hundred-fold inhibition of HSV growth with minimal toxicity to uninfected cells.

Acyclovir-resistant thymidine kinase HSVs have evolved. They occur primarily in patients receiving multiple courses of therapy or in patients with AIDS. Thymidine kinase mutants are susceptible to vidarabine and foscarnet. Changes in viral DNA polymerase structures can also mediate resistance to acyclovir.

Acyclovir can be used topically, orally, or intravenously. Usual oral dosage is 800 mg 5 times a day. Oral acyclovir is only 15%–30% bioavailable, and food does not affect absorption. For unknown reasons, bioavailability is lower in patients with

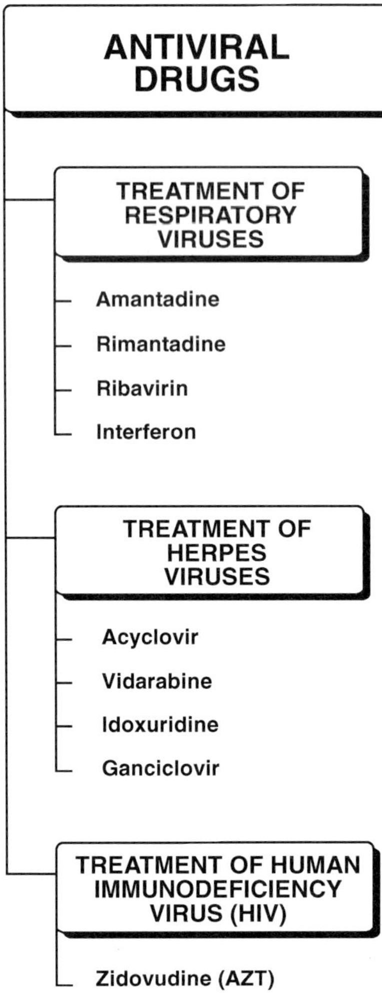

FIG XXI-10—Summary of antiviral drugs. (Reproduced with permission from Mycek MJ, Harvey RA, Champe PC, eds. *Lippincott's Illustrated Reviews: Pharmacology*. 2nd ed. Philadelphia: Lippincott-Raven; 1997:329.)

transplants. Acyclovir is minimally protein-bound (10%–30%), and drug interactions through binding displacement have not been reported. The drug is well distributed, with cerebrospinal fluid (CSF) and brain concentrations equaling approximately 50% of serum values. Concentrations of acyclovir in zoster vesicle fluid are equivalent to those in plasma. Aqueous humor concentrations are 35% and salivary concentrations 15% that of plasma. Vaginal concentrations are equivalent to those of plasma, and breast-milk concentrations exceed them.

The percutaneous absorption of topical acyclovir is low and occurs primarily when large areas are treated. Plasma concentrations of 0.3 μg/ml were noted in

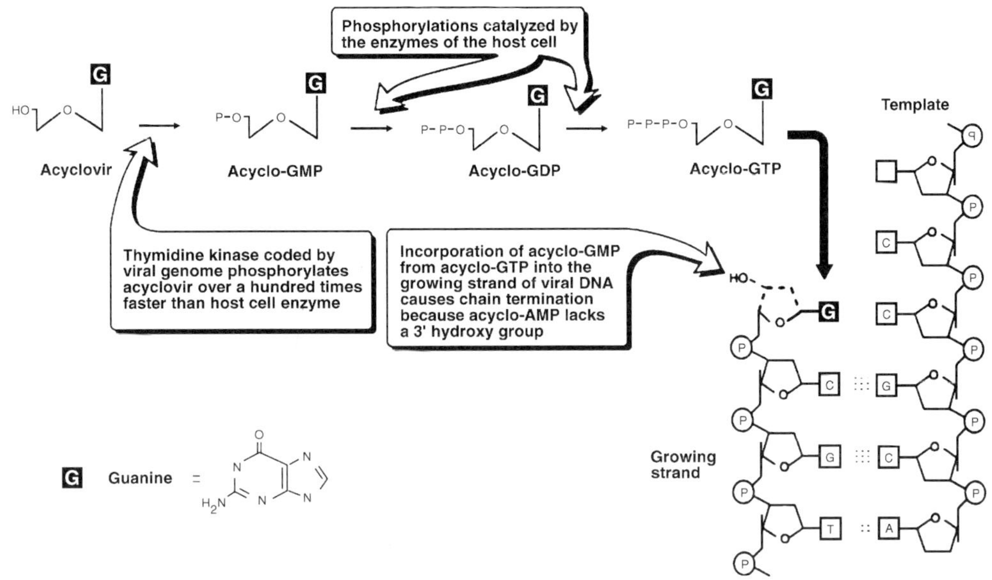

FIG XXI-11—Conversion of acyclovir to acyclovir triphosphate and the subsequent incorporation into viral DNA, causing chain termination. (Reproduced with permission from Mycek MJ, Harvey RA, Champe PC, eds. *Lippincott's Illustrated Reviews: Pharmacology*. 2nd ed. Philadelphia: Lippincott-Raven; 1997.)

patients treated topically with this drug for herpes zoster. Peak serum concentrations after oral ingestion of acyclovir average 0.6 mg/ml and occur 90 minutes after dosing, but peak serum concentrations after intravenous administration reach approximately 10 mg/ml.

The plasma half-life for normal adults and neonates is 3.3 and 3.8 hours, respectively. It increases to 20 hours in anuric patients. In the urine, 60%–90% of acyclovir is excreted unchanged through both glomerular filtration and tubular secretion. As a result, acyclovir may interfere with the renal excretion of drugs such as methotrexate that are eliminated through the renal tubules; probenecid significantly decreases the renal excretion of acyclovir. A major metabolite of acyclovir, carboxymethoxymethylguanine, accounts for 10%–20% of the total administered dose and is excreted in urine. Acyclovir is effectively removed by hemodialysis (60%) but only minimally removed by peritoneal dialysis. A commonly used intravenous dosage for acyclovir is 1500 mg/m^2/day.

Acyclovir has proven to be effective in preventing the recurrence of HSV keratitis in oral doses of 400 mg twice a day in a recently concluded study. However, it was of no added benefit when used with topical steroids and trifluridine in the treatment of HSV stromal keratitis. The addition of oral acyclovir to a regimen of topical antiviral agent may be considered in patients with HSV iridocyclitis even though benefit has not proven to be statistically significant.

Valacyclovir is currently approved for management of herpes zoster infections in immunocompetent persons. It is an amino acid ester prodrug of acyclovir. The bioavailability is much higher than with acyclovir (54% vs. 20%). The recommended dosage is 1 g three times a day for 7–14 days.

Famciclovir is currently indicated for the management of uncomplicated acute herpes zoster. It has demonstrated efficacy in relieving acute zoster signs and symptoms and reducing the duration of postherpetic neuralgia when administered during acute zoster. The recommended dosage for the management of acute herpes zoster is 500 mg three times a day for 7 days. The efficacy of famciclovir in herpes zoster ophthalmicus as well as HSV keratitis is currently being evaluated in an ongoing multicenter multinational study.

Ganciclovir (9-2-hydroxypropoxymethylguanine) is a synthetic guanosine analogue active against many herpes viruses. As with acyclovir, it must be phosphorylated to become active. Infection-induced kinases, viral thymidine kinase, or deoxyguanosine kinase of various herpes viruses can catalyze this reaction. After monophosphorylation, cellular enzymes convert ganciclovir to the triphosphorylated form, and the triphosphate inhibits viral DNA polymerase rather than cellular DNA polymerase. Ganciclovir triphosphate competitively inhibits the incorporation of guanosine triphosphate into DNA. Because of its toxicity and the availability of acyclovir for treatment of many herpes virus infections, its use is currently restricted to treatment of cytomegalovirus (CMV) retinitis.

Ganciclovir is used only intravenously, since less than 5% of an oral dose is absorbed. The intravenous induction dose is 5 mg/kg every 12 hours for 14–21 days. Once the infection is under control, a daily dose of 6 mg/kg is required for maintenance of the virus-free state. CSF concentrations are approximately 50% those of plasma, with peak plasma concentrations reaching 4–6 sLg/ml. The plasma half-life is 3–4 hours in persons with normal renal function, increasing to over 24 hours in patients with severe renal insufficiency. Over 90% of systemic ganciclovir is eliminated unchanged in urine, and dose modifications are necessary for persons with compromised renal function. Ganciclovir is approximately 50% removed by hemodialysis. It can be administered intravitreally or as a sustained-release intraocular device.

Foscarnet (phosphonoformic acid) inhibits DNA polymerases, RNA polymerases, and reverse transcriptases. In vitro, it is active against herpes viruses, influenza virus, and HIV. Foscarnet is used primarily in the treatment of AIDS patients with CMV retinitis. It acts by blocking the pyrophosphate receptor site of CMV DNA polymerase. Viral resistance is attributable to structural alterations in this enzyme. Foscarnet inhibits herpes viruses and cytomegaloviruses that are resistant to acyclovir and ganciclovir. It is administered intravenously in doses of 20 mg/kg as bolus followed by 0.16 mg/kg/min infusion for maintenance.

Foscarnet bioavailability is approximately 20%. Because it can bind with calcium and other divalent cations, foscarnet becomes deposited in bone and may be detectable for many months. Distribution follows a three-compartment model and produces peak serum concentrations of approximately 30 pg/ml. It is eliminated by both glomerular filtration and tubular secretion, with 80%–90% of the administered dose appearing unchanged in the urine. Dosage adjustment is required in persons with impaired renal function.

Zidovudine (azidodeoxythymidine [AZT]) is a thymidine nucleoside analogue with activity against HIV. Zidovudine becomes phosphorylated to monophosphate, diphosphate, and triphosphate forms by cellular kinases in infected and uninfected cells. It has two primary methods of action:

- The triphosphate acts as a competitive inhibitor of viral reverse transcriptase.
- The azido group prevents further chain elongation and acts as a DNA chain terminator.

Zidovudine inhibits HIV reverse transcriptase at much lower concentrations than those needed to inhibit cellular DNA polymerases. Zidovudine is currently indicated as treatment for some stages of HIV infection.

Zidovudine is administered orally in doses of 1500 mg/day and is about 60% bioavailable with 40% metabolized by first pass. Peak concentrations occur within 30–90 minutes to give steady-state peak concentrations of 0.05–1.5 fg/ml. Zidovudine is about 40% protein-bound. CSF concentrations vary widely and range from 25% to 100% of serum values. Zidovudine enters the brain, phagocytic cells, liver, muscle, and placenta. Plasma half-life is about 1 hour. IV dosage is 1–2 mg/kg four times a day.

Cidofovir (HPMPC) is another effective antiviral drug that was shown to inhibit CMV replication when administered intravitreally in a small series of patients. Dosage administered was 20 µg/0.1 ml. Long-lasting suppression of CMV retinitis was noted, with an average time to progression of 55 days. Ocular side effects included mild uveitis and a small drop in IOP.

Fluorouracil is a fluorinated pyrimidine nucleoside analogue that blocks production of thymidylate and interrupts normal cellular DNA and RNA synthesis. Its primary action may be to cause cellular thymine deficiency and resultant cell death. The effect of fluorouracil is most pronounced on rapidly growing cells, and its use as an antiviral agent is primarily related to destruction of infected cells (warts) by topical application.

Medications for Acanthamoeba *Infections*

Acanthamoeba is a ubiquitous, free-living amoeba that inhabits soil, water, and air. Its appearance as a corneal pathogen is no longer infrequent because of a number of factors, including the increased use of contact lenses. The species responsible for corneal infections, which include *A. polyphaga, A. castellanii, A. hatchetti,* and *A. culbertsoni,* exist as both trophozoites and double-walled cysts. Because of the variations among species of *Acanthamoeba,* no single drug is effective in treating all *Acanthamoeba* keratitis. Polyhexamethylene biguanide (0.02% solution) is the first-line agent with lowest minimal amoebicidal concentration. Other medications that are effective include chlorhexidine, neomycin, neomycin-polymyxin B-gramicidin mixtures, natamycin 5% topical suspension, imidazoles such as miconazole 1% topical solution, propamidine isethionate 0.1% drops (Brolene), and topical dibromopropamidine 0.15% ointment.

Metcalf JF, Cosper CS, Rich LS. A synthetic antisense oligonucleotide analogue prevents herpetic stromal keratitis in CF-1 mice. *Invest Ophthalmol Vis Sci.* 1991;32:806.

Sudesh S, Laibson PR. The impact of the herpetic eye disease studies on the management of herpes simplex virus ocular infections [Review]. *Curr Opin Ophthalmol.* 1999;10(4):230–233.

Local Anesthetics

Local anesthetic agents are used extensively in ophthalmology. Topical preparations yield corneal and conjunctival anesthesia for comfortable performance of examination techniques such as tonometry, gonioscopy, removal of superficial foreign bodies, corneal scraping for bacteriologic studies, and paracentesis and for use of contact lenses associated with fundus examination and laser procedures. Local retrobulbar and eyelid blocks yield excellent anesthesia and akinesia for intraocular and orbital surgery (Tables XXI-16, XXI-17).

The local anesthetic agents used in ophthalmology are tertiary amines linked by either *ester* or *amide* bonds to an aromatic residue. Because the protonated form is far more soluble and these compounds undergo hydrolysis more slowly in acidic solutions, they are supplied in the form of their hydrochloride salts. When exposed to tissue fluids at pH 7.4, about 5%–20% of the anesthetic agent molecules will be in the unprotonated form, as determined by the pKa (8.0–9.0) of the individual agent. The more lipid-soluble unprotonated form penetrates the lipid-rich myelin sheath and cell membrane of axons. Once inside, most of the molecules are again protonated. The protonated form gains access to and blocks the sodium channels on the inner wall of the cell membrane and increases the threshold for electrical excitability. As increasing numbers of sodium channels are blocked, nerve conduction is impeded and finally blocked.

After administration of a local anesthetic, nerve fibers that are small or unmyelinated are blocked most quickly because their higher discharge rates open sodium channel gates more frequently and because conduction can be prevented by the disruption of a shorter length of axon. The action potential in unmyelinated fibers spreads continuously along the axon; in myelinated fibers, the action potential spreads by saltation. Therefore, only a short length of an unmyelinated fiber need be functionally interrupted, whereas one or more nodes must be blocked in a myelinated fiber. In larger myelinated fibers, the nodes are farther apart.

Clinically, local anesthetics first block the poorly myelinated and narrow parasympathetic fibers (as evidenced by pupil dilation) and sympathetic fibers (vasodilation), followed by sensory fibers (pain and temperature), and finally the larger and more myelinated motor fibers (akinesia). The optic nerve, enclosed in a meningeal lining, is often not blocked by retrobulbar injections.

Amide local anesthetics are preferred to ester agents for retrobulbar blocks because the amides have a longer duration of action and less systemic toxicity. However, this duration of action is limited by diffusion from the site of injection because amide agents are not metabolized locally but are metabolized and inactivated in the liver, primarily by dealkylation.

Ester agents are susceptible to hydrolysis by serum cholinesterases in ocular vessels as well as by metabolism in the liver. Toxicity of ester anesthetics may occur at lower doses when serum cholinesterase levels are low because of echothiophate eyedrop treatment or a hereditary serum cholinesterase deficiency.

The toxic manifestations of local anesthetics are generally related to dose. However, patients with severe hepatic insufficiency may have symptoms of toxicity with either amide or ester local anesthetics even at lower doses. These manifestations include restlessness and tremor that may proceed to convulsions, and respiratory and myocardial depression. Central nervous system stimulation can be counteracted by IV diazepam; respiratory depression calls for ventilatory support.

Because local anesthetics block sympathetic vascular tone and dilate vessels, a 1:200,000 concentration of epinephrine is frequently added to shorter-acting agents

Table XXI-16

Regional Anesthetics

GENERIC	CONCENTRATION (%) /MAXIMUM DOSE	ONSET OF ACTION	DURATION OF ACTION	MAJOR ADVANTAGES/ DISADVANTAGES
Procaine*	1%–4%/500 mg	7–8 min	30–45 min 60 min (with epinephrine)	Short duration; poor absorption from mucous membranes
Tetracaine*	0.25%	5–9 min	120–140 min (with epinephrine)	
Bupivacaine†	0.25%–0.75%	5–11 min	480–720 min (with epinephrine)	Long duration of action
Lidocaine†	1%–2%/500 mg	4–6 min	40–60 min 120 min (with epinephrine)	Spreads readily without hyaluronidase
Mepivacaine†	1%–2%/500 mg	3–5 min	120 min	Duration of action greater without epinephrine
Prilocaine†	1%–2%/600 mg	3–4 min	90–120 min (with epinephrine)	As effective as lidocaine
Etidocaine†	1%	3 min	300–600 min	

*Ester type compound
†Amide type compound

Table XXI-17

Topical Anesthetic Agents

GENERIC	TRADE NAME	CONCENTRATION (%)
Cocaine hydrochloride		1%–4%
Proparacaine hydrochloride	AK-Tain Alcain Ophthetic	0.5% 0.5% 0.5%
Tetracaine hydrochloride	AK-T-Caine Pontocaine hydrochloride	0.5% 0.5%

to retard vascular absorption. Such use of epinephrine raises circulating catecholamine levels and may result in systemic hypertension and cardiac arrhythmias.

Topically applied anesthetics disrupt intercellular tight junctions, resulting in increased corneal epithelial permeability to subsequently administered agents (ie, dilating drops). They also interfere with corneal epithelial metabolism and repair and thus cannot be used for chronic pain relief.

Lidocaine (Xylocaine) is an amide local anesthetic used in strengths of 1%–4% (with or without epinephrine) for injection and 2%–4% for topical mucosal anesthesia. It yields a rapid (5-minute) retrobulbar or eyelid block that lasts 1–2 hours. The topical solution, applied to the conjunctiva with a cotton swab for 1–2 minutes, reduces the discomfort of subconjunctival injections. Topical lidocaine is preferable to cocaine or proparacaine as anesthesia for conjunctival biopsy because it has less effect on epithelial morphology. Lidocaine is also extremely useful for suppressing cough during ocular surgery. The maximum safe dose of the 2% solution for local injection is 15 ml in adults. A common side effect is drowsiness.

Mepivacaine (Carbocaine) is an amide agent used in strengths of 1%–3% (with or without a vasoconstrictor). It has a rapid onset and lasts about 2–3 hours. The maximum safe dose is 25 ml of a 2% solution.

Bupivacaine (Marcaine) is an amide agent that has a slower onset of action than lidocaine. It may yield relatively poor akinesia but has the advantage of long duration of action, up to 8 hours. It is available in 0.25%–0.75% solutions (with or without epinephrine) and is frequently administered in a mixture with lidocaine or mepivacaine to achieve a rapid, complete, and long-lasting effect. The maximum safe dose is 25 ml of a 0.75% solution.

Etidocaine (Duranest) is a rapid-onset (5 minutes), long-acting (over 5 hours) derivative of lidocaine, used in 0.5%–1.5% solutions (with or without epinephrine). It usually blocks motor fibers before sensory fibers. The maximum safe dose is 30 ml of a 1% solution.

The following agents are commonly used for topical anesthesia. They have higher lipid solubilities and therefore more rapid onset than other agents. Thus, the period of initial discomfort caused by the drops is shortened.

Proparacaine (eg, Ophthaine, Ophthetic) is an ester topical anesthetic available as a 0.5% solution. It is the least irritating topical anesthetic, has a rapid onset of about 15 seconds, and lasts about 20 minutes. It has been reported that proparacaine used without a preservative does not inhibit the growth of *Staphylococcus, Candida,* or *Pseudomonas* and thus might be preferred to other agents for corneal anesthesia prior to obtaining a scraping for culture from a corneal ulcer. Its structure is different enough from the other local anesthetics that cross-sensitization apparently does not occur.

Benoxinate is an ester topical anesthetic available in a 0.4% solution with fluorescein (Fluress) for use in tonometry. It has an onset and duration similar to proparacaine.

Tetracaine is an ester topical anesthetic available in 0.5% solution. It has a longer onset of action and duration of action than proparacaine and causes more extensive corneal epithelial toxicity.

Topical Anesthesia in Anterior Segment Surgery

The first modern use of topical anesthesia was Koller's use of cocaine in 1884. Since then, synthetic drugs have become available; cocaine is no longer used because of the potential risk of side effects and drug abuse. Benoxinate 0.4% (oxybuprocaine; 2-diethylaminoethyl-4-amino-3-butoxybenzoate), an ester anesthetic, is currently the most frequently used because of its high degree of safety. Other commonly used agents are tetracaine 0.5% or 1% (amethocaine) and proparacaine 0.5%; both are short-acting (20 minutes) and are the least toxic to the corneal epithelium. Lidocaine 4% (Lignocaine) and bupivacaine 0.5% and 0.75% have a longer duration of action but an increased associated corneal toxicity.

Technique The aim is to block the nerves that supply the superficial cornea and conjunctiva—namely, the long and short ciliary, nasociliary, and lacrimal nerves. The patient should be warned that application of the drops onto the surface of the cornea stings (except proxymetacaine).

Preparation of the unblocked eyelid requires the patient to keep the eye closed, but the eye is kept open when the plastic drape is applied in order to secure the lid and lashes. Since visual perception is not lost, the patient is asked to focus on the source of the light, the intensity of which is reduced subsequently. The subconjunctival injection of antibiotics can be painful, but this can be avoided by including the antibiotics in the infusion bottle.

Topical anesthesia may be combined with subconjunctival anesthesia. This allows subconjunctival and scleral manipulations to be carried out and is well tolerated by patients.

Intraocular Lidocaine

Recently, intraocular lidocaine has been used to provide analgesia during surgery. The solution used is 0.3 ml of 1% isotonic, nonpreserved lidocaine administered intracamerally.

No side effects have been reported, except for possible transient retinal toxicity if lidocaine is injected posteriorly in the absence of a posterior capsule. Lidocaine obviates the need for intravenous and regional anesthetic supplementation in most patients. Adequate anesthesia is obtained in about 10 seconds. As with topical techniques, the ability of the patient to cooperate during surgery is desirable.

Crandall DC. Pharmacology of ocular anesthetics. In: Tasman W, Jaeger EA, eds. *Duane's Foundations of Clinical Ophthalmology.* Vol 3. Philadelphia: Lippincott; 1999:1–22.

Purified Neurotoxin Complex

Botulinum toxin type A (Botox, formerly called Oculinum) is produced from cultures of the Hall strain of *Clostridium botulinum.* It blocks neuromuscular conduction by binding to receptor sites on motor nerve terminals, entering the nerve terminals and inhibiting the release of acetylcholine. Botulinum toxin type A injections provide effective relief of the excessive, abnormal contractions associated with blepharospasm. Typically, 5 units are placed subcutaneously at each of the two sites over the brow, on the upper eyelid and on the lower eyelid of one side, for a total of six injections per side. Botulinum can also be used to treat strabismus, possibly by inducing an atrophic lengthening of the injected muscle and a corresponding shortening of the muscle's antagonist.

Jordan DR, Anderson RL. Essential blepharospasm. In: *Focal Points: Clinical Modules for Ophthalmologists.* San Francisco: American Academy of Ophthalmology; 1988:6.

Scott AB. Botulinum toxin treatment of strabismus. In: *Focal Points: Clinical Modules for Ophthalmologists.* San Francisco: American Academy of Ophthalmology; 1989:12.

Medications for the Dry Eye

Artificial tear preparations (demulcents) and emollients form an occlusive film over the corneal surface to lubricate and protect the eye from drying. The active ingredients in demulcent preparations are polyvinyl alcohol, cellulose, and methylcellulose and their derivatives: hydroxypropyl cellulose, hydroxyethylcellulose, hydroxypropyl methylcellulose, and carboxymethylcellulose. Multidose preparations contain preservatives, including benzalkonium chloride, chlorobutanol, thimerosal, and sorbic acid. However, the new generation of ophthalmic preservatives is virtually nontoxic. Unit-dose preparations are unpreserved drops that eliminate the cytotoxic effects of preservatives.

Ocular emollients are ointments prepared with sterile petrolatum, liquid lanolin, mineral oil, methylparaben, and polyparaben. Future development of medications for dry eye should be aimed at providing nourishment for the keratoconjunctival surface as well as revitalizing the tear-secreting system. (See also BCSC Section 8, *External Disease and Cornea.*)

Ocular Decongestants

Common agents such as naphazoline, tetrahydrozoline, and phenylephrine hydrochloride are used as topical drops to cause temporary vasoconstriction of conjunctival vessels. Side effects include rebound vasodilation and conjunctival hyperemia.

Irrigating Solutions

Sterile isotonic solutions are for general ophthalmic use. Sterile, physiologically balanced salt solutions that are isotonic to eye tissues are used for intraocular irrigation during surgical procedures. Irrigating solutions contain sodium chloride, potassium chloride, calcium chloride, magnesium chloride, sodium acetate, dextrose, glutathione disulfidehydrochloric acid, and sodium hydroxide. Intraocular irrigating solutions contain no preservatives.

Diagnostic Agents

Fluorescein 2%, lissamine green 1%, and rose bengal 1% are some of the most commonly used solutions in the examination and diagnosis of external ocular diseases. The first two stains outline the defects of the conjunctival and corneal epithelium, whereas the rose bengal indicates abnormal devitalized epithelial cells. For the study of retinal and choroidal circulation as well as abnormal changes of retinal pigment epithelium, sodium fluorescein solution in concentrations of 5%, 10%, or 25% is injected intravenously. Fundus fluorescein angiography is helpful in diagnosing various vascular diseases and neoplastic disorders. Indocyanine green, a tricarbocyanine type of dye, is used in selected cases to study choroidal vasculature in a variety of retinal disorders. Typically, 25 mg of dye is injected as intravenous solution. Indocyanine green angiography is particularly helpful in identification and delineation of poorly defined choroidal neovascular membranes in age-related macular degeneration.

Indocyanine green and trypan blue dye are useful in delineating the anterior capsule during phacoemulsification of mature cataracts.

Viscoelastic Agents

Viscoelastic agents possess certain chemical and physical properties that include the capacity to resist flow and deformation. Viscoelastics for ophthalmic use must also be inert, isosmotic, sterile, nonpyrogenic, nonantigenic, and optically clear. In addition, they must be sufficiently hydrophilic to allow easy dilution and irrigation from the eye. Naturally occurring and synthetic compounds include *sodium hyaluronate, chondroitin sulfate, hydroxypropyl methylcellulose,* and *polyacrylamide*. Combined chondroitin sulfate/sodium hyaluronate materials are also available. Viscoelastic agents protect ocular tissues such as the corneal endothelium and epithelium, help to maintain intraocular space, and facilitate tissue manipulation; thus, they are indispensable tools in cataract or glaucoma surgery, penetrating keratoplasty, anterior segment reconstruction surgery, and retinal surgery. (See also the discussions of hyaluronic acid and vitreous collagen cross-linking in Chapter XV and BCSC Section 11, *Lens and Cataract*.)

Fibrinolytic Agents

Tissue plasminogen activator (tPA), urokinase, and *streptokinase* are all fibrinolytic agents. tPA is a naturally occurring serine protease with a molecular mass of 68 kD. Because tPA is normally present at a higher concentration in the aqueous humor of the human eye than it is in blood, it is less toxic to ocular tissues and is specific for dissolution of fibrin clots. tPA has been used successfully in the resolution of fibrin clots after vitrectomy, keratoplasty, and glaucoma filtering procedures. These drugs are not approved by the FDA for ocular use.

Thrombin

Thrombin, a sterile protein substance, is a useful adjunct for maintaining hemostasis during complicated intraocular surgery. Intravitreal thrombin may be used to control intraocular hemorrhage during vitrectomy. The addition of thrombin (100 units/ml) to the vitrectomy infusate significantly shortens intraocular bleeding time. Thrombin produced by DNA recombinant techniques minimizes the degree of postoperative inflammation. Thrombin causes significant ultrastructural corneal endothelial changes when human corneas are perfused with 1000 units/ml.

Antifibrinolytic Agents

Antifibrinolytic agents, such as *ε-aminocaproic acid* and *tranexamic acid,* inhibit the activation of plasminogen. The action of plasminogen activators and plasmin is inhibited. These agents may be used systemically to treat cases of hemorrhage secondary to excessive fibrinolysis and to prevent recurrent hyphema. They are contraindicated in the presence of active intravascular clotting. Recurrent hyphema most commonly occurs 2–6 days after the original hemorrhage.

ε-Aminocaproic acid is used in a dosage of 50–100 mg/kg every 4 hours, up to 30 g daily. Adverse reactions include nausea, vomiting, muscle cramps, conjunctival suffusion, nasal stuffiness, headache, rash, pruritus, dyspnea, tonic toxic confusional states, cardiac arrhythmias, and systemic hypotension. Gastrointestinal side effects are similar with doses of either 50 or 100 mg/kg. The drug should be continued for a full 5–6 days to achieve maximal clinical effectiveness. Topical ε-aminocaproic acid may be an attractive alternative to systemic delivery in the treatment of traumatic hyphema. Optimal topical concentration to maximize aqueous levels and minimize corneal epithelial toxicity is 30% ε-aminocaproic acid in 2% carboxypolymethylene.

Tranexamic acid is another antifibrinolytic agent that reduces the incidence of rebleeding after traumatic hyphema. It is 10 times more potent in vitro than ε-aminocaproic acid. Usual dosage is 25 mg of tranexamic acid/kg three times daily for 3–5 days. Gastrointestinal side effects are rare.

Corneal Storage Medium

Corneal storage medium helps to prolong the viability of donor corneas to be used for transplantation. The main components of the various kinds of media include a bicarbonate-buffered minimum essential medium (MEM) or a hybrid medium of MEM/TC 199, chondroitin sulfate, and dextran (to retard proteoglycan loss during storage and reduce intraoperative and postoperative rebound swelling), as well as gentamicin sulfate or other antibiotics as prophylactic agents. Corneal tissue storage can be prolonged with the addition of recombinant growth factors such as epidermal growth factor, antioxidants, insulin, adenosine triphosphate precursors, anticollagenases, and antiproteases.

Drugs on the Horizon

Interferon

A naturally occurring species-specific defense against viruses, interferon is synthesized intracellularly and increases resistance to virus infection. Synthetic analogues such as polyinosinic acid–polycytidylic acid have been used to induce patients to form their own interferon.

Topically administered interferon has been found to be ineffective in the treatment of epidemic keratoconjunctivitis caused by adenovirus. Interferon used in conjunction with acyclovir showed significantly quicker healing time in herpes simplex keratitis patients than treatment with acyclovir alone (5.8 vs. 9.0 days). Interferon has also been found to speed healing of an epithelial defect when used in combination with trifluridine. The dosage of interferon used with 30 million IU/ml was 2 drops per day for the first 3 days of treatment. Interferon alone has little effect on the treatment of herpes simplex keratitis. In combination, it seems to act as a topical adjuvant to traditional antiviral therapy in resistant herpes simplex keratitis.

Interferon has also been shown to inhibit vascular endothelial cell proliferation and differentiation. It is particularly effective in the treatment of juvenile pulmonary hemangiomatosis, which used to be a fatal condition before the invention of interferon. Intralesional administration of interferon has been reported to be especially effective in ocular Kaposi sarcoma.

Growth Factors

Growth factors are a diverse group of proteins that act at autocrine and paracrine levels to affect various cellular processes, including:

- Metabolic regulation
- Tissue differentiation
- Cell growth and proliferation
- Maintenance of viability
- Changes in cell morphology

They are synthesized in a variety of cells and have a spectrum of target cells and tissues. The presence of various growth factors in retina, vitreous humor, aqueous humor, and corneal tissues has been demonstrated. These include:

- Epidermal growth factor
- Fibroblast growth factors
- Transforming growth factors
- Vascular endothelial growth factor
- Insulin-like growth factors

These growth factors are capable of diverse, synergistic, and sometimes antagonistic biological activities.

Under normal physiologic conditions, the complex and delicate coordination of the effects of and the interactions among growth factors maintains the homeostasis of intraocular tissues. The net effect of a growth factor depends on its bioavailability, which is determined by its concentration, its binding to carrier proteins, the level of its receptor in the target tissue, and the presence of other complementary or antagonistic regulatory factors.

Pathologically, the breakdown of blood–ocular barriers disrupts the balance among growth factors in the ocular media and tissues and may result in various abnormalities. The disruption in the balance among isoforms of transforming growth factor β, basic fibroblast growth factor, vascular endothelial growth factor, and insulin-like growth factors is suspected to cause ocular neovascularization. Transforming growth factor βs and platelet-derived growth factor are implicated in the pathogenesis of proliferative vitreoretinopathy and excessive proliferation of Tenon's capsule fibroblasts that can result in the scarring of the glaucoma filtration bleb. Increased concentrations of insulin-like growth factors in plasmoid aqueous humor may be responsible for the abnormal hyperplastic response of the lens epithelium and corneal endothelium seen in inflammatory conditions and in traumatic insults to the eye.

Identifying growth factors and understanding their mechanisms of action in the eye offers great potential for providing the ophthalmologist with new methods for manipulation of and intervention in ocular disorders. Epidermal growth factor and fibroblast growth factor can accelerate corneal wound repair after surgery, chemical burns, or ulcers and can increase the number of corneal endothelial cells. Fibroblast growth factor has also been shown to delay the process of retinal dystrophy in Royal College of Surgeons rats.

Vascular endothelial growth factor (VEGF), also known as *vasculotropin,* deserves special mention. It is a dimeric, heparin-binding polypeptide mitogen and has four isoforms that are generated from alternative splicing of mRNA. VEGF gene is widely expressed in actively proliferating vascular tissue and is implicated in the pathogenesis of various neovascular retinopathies such as diabetes and age-related choroidal neovascularization. Currently, the role of monoclonal antibodies specifically targeted against VEGF as a therapeutic modality in angiogenic eye disease is being evaluated, and this approach holds great promise.

The development of a "cocktail" of growth factors and the use of growth factor antagonists will provide an effective therapeutic modality for the treatment of various ocular disorders. With the advance of genetic engineering and innovations in drug delivery systems, growth factors will probably make up the next generation of ophthalmic pharmaceuticals for the physiologic regulation of tissue maintenance, restoration, and repair.

Calcium-Channel Blockers

Calcium-channel blockers like *verapamil* have several ocular effects that may be beneficial in glaucoma patients. These drugs generally lower IOP, in particular following topical administration. The mechanism of action appears to be a result of increased outflow facility. In addition, the use of calcium-channel antagonists is associated with slowed rate of progression of low-tension glaucoma, possibly by increasing optic nerve blood flow, reducing vasospasm, or creating a neuroprotective effect. Blood pressure should be monitored in patients treated with these drugs because of their hypotensive effect. Systemic calcium-channel blockers are probably the antihypertensive drug of choice in patients with hypertension and low-tension glaucoma.

Neuroprotective Agents

Glutamate neurotoxicity has been speculated as one of the factors that accelerate the loss of ganglion cells in glaucoma. Glutamate concentration was noted to be elevated in the vitreous of open-angle glaucoma patients undergoing cataract as well as vitreous surgery. Memantine, a N-methyl-D-aspartate (NMDA) receptor blocker, is effective in preventing the toxicity of glutamate on ganglion cell in a rat model. This compound, in contrast with magnesium, blocks the NMDA receptor in the presence of sustained release of low glutamate concentrations, prevents the influx of calcium, and protects the ganglion cells from neuronal excitotoxicity and subsequent death. Memantine is currently in phase III trials for treatment of glaucoma.

Compounds for Lowering Intraocular Pressure

In addition to $PGF_{2\alpha}$ and its derivatives (Travopost) and topically active carbonic anhydrase inhibitor, the following compounds are under further investigation for their ocular hypotensive effects: AGN 191151; AGN 191024; selective α_2-adrenergic agonists BHT-920, BHT-933, and UK-14303 (topical application reduces IOP in rabbits, monkeys, and normal volunteers). These α_2-adrenergic agonists appear to act by reducing formation of aqueous humor. Topical application of vanadate (a Na^+, K^+–ATPase inhibitor) 1% drops in normotensive and laser-induced glaucomatous monkey eyes reduces IOP by decreasing the rate of aqueous humor flow. However, clinical studies have failed to demonstrate a therapeutic effect in treating ocular hypertension. *Forskolin,* a novel compound that stimulates the cAMP secondary messenger system at the site of the catalytic subunit, has received conflicting reports regarding its IOP-lowering effects. Aqua-Flow, a collagen glaucoma drainage device, is being evaluated and is in phase I FDA trial.

Bioadhesives

The use of bioadhesives is experimental. *Cyanoacrylates* are used to seal penetrating corneal wounds and leaking filtering blebs, with or without bandage contact lens, as well as to seal leaks during lamellar or penetrating keratoplasty so the globe remains firm enough for the surgical maneuver. BCSC Section 8, *External Disease and Cornea,* discusses these procedures in greater depth. Other adhesives such as *collagen* and *fibrin* are under investigation. The use of bioadhesives for the precise delivery of growth factors is also being explored.

Protein Kinase C Inhibitors

LY333531, a protein kinase C inhibitor, is in phase II trial as treatment for diabetic retinopathy.

Gene Therapy

Instead of delivering a drug to the circulation aimed to affect a specific condition, gene therapy involves the treatment of disease by the delivery of a human gene to a target organ. The subsequent production of a *gene product,* or protein, then acts as the "drug." The phases of gene delivery, expression, and action of the gene product are analogous to conventional drug therapy, and each gene therapy system has its own specific pharmacologic characteristics.

Gene therapy currently follows two basic approaches. The more obvious is *gene replacement therapy* to correct hereditary conditions in which defective production of an essential enzyme or protein is attributable to the gene. Here, the goal is to insert the normal gene into a cell so that a sufficient amount of product becomes available. The second approach, called *gene addition therapy,* attempts to treat acquired diseases. An example is the use of *cytokine gene transfer* into tumor cells to stimulate the immune response against these cells.

Genes that are transferred to cells must contain all components necessary to direct the production of RNA that will result in functional protein production. The transferred DNA must therefore contain promoter elements necessary for transcription and the entire protein coding region of the gene. Other elements to enhance RNA stability, such as a polyadenylation sequence, are often included. (See Part 3, Genetics, for illustrations and more detail.)

Several methods can be used to introduce DNA into mammalian cells. DNA can be introduced in vitro, into cells by microinjection, by coprecipitation with calcium phosphate, or by increasing cell permeability by electrical shock (electroporation). These techniques allow insertion of DNA into only a small number of cells and therefore have limited use for therapeutic application. Instead, more efficient vehicles have been developed for gene transfer that permit in vivo gene delivery.

Vehicles for DNA delivery include a variety of plasmid- and virus-based vectors (Table XXI-18). Examples include recombinant plasmids; plasmid mixed with lipid micelles called *liposomes;* and genetically engineered, recombinant DNA or RNA viruses that can carry exogenous genes, infect mammalian cells, and transfer the functional genetic material to that cell. The most widely investigated vehicles for gene transfer are genetically engineered retrovirus vectors derived from the Moloney murine leukemia retrovirus. Additionally, gene-transfer vectors have been developed from adenovirus and herpes simplex virus. Virus-based vectors are engineered to maximize the safety of gene transfer by modification of the viral genome so that normal virus replication does not occur.

Regardless of the vector used, the introduction of foreign genes into cells begins with the identification and cloning of a gene. That gene then can be inserted into a plasmid or grown in bacteria. Then, alone or combined with one of the many vehicles for gene transfer, it can be delivered to target cells. The vehicle carrying the DNA enters the cell by passing through the host cell membrane or by active transport through a specific receptor site. The DNA is then taken up into the nucleus, where it can function.

If a retrovirus vector is used, DNA can be permanently, though randomly, integrated into the host cell genome. In most systems, the host cell machinery supplies

Table XXI-18

Gene Therapy Vectors

PLASMID-BASED VECTORS	VIRUS-BASED VECTORS
Plasmid	Retrovirus
Plasmid with liposome	Adenovirus
Plasmid linked to ligand	Adeno-associated virus
	Herpes simplex virus

enzymes necessary for the transcription of the DNA into messenger RNA, which can then be translated by the host cell into protein. This protein can function intracellularly or extracellularly either to replace a hereditary deficient or defective protein or to provide an additional therapeutic function. It can be one of the following:

- A protein that functions intracellularly, such as the enzyme adenosine deaminase used to correct the mutation in lymphocytes responsible for the severe combined immunodeficiency syndrome
- A cell membrane protein, such as the chloride-channel cystic fibrosis transmembrane conductance regulator that is mutant in cystic fibrosis
- A secreted protein, such as the cytokine interleukin-2 used for anticancer therapy

Two basic experimental approaches have been used for gene therapy. The first is to remove the target cells from the affected person, add the normal gene to the cells in vitro, and return these modified cells to the donor. If a target cell expressing a mutation can be removed from an organ, modified in vitro, and then returned to that organ to function correctly, the disease may be ameliorated. This strategy has been widely used in animal models for the transfer of new genes to bone marrow cells and was used in the first approved human gene therapy protocol for the correction of adenosine deaminase deficiency. In these studies, T lymphocytes (which are known to be the critical functional site of the deficient enzyme) were removed from the blood and infected in vitro with a retrovirus vector containing the normal adenosine deaminase gene. The genetically modified lymphocytes were reinfused, where they functioned normally and improved the clinical status of the treated individuals.

This strategy cannot be applied to all diseases because it is virtually impossible to remove and then replant a large number of cells from organs such as the heart, lung, or brain. Therefore, a second approach is needed: Genes must be delivered to the target organ or organs in vivo in a highly efficient manner. This method, called *in vivo gene transfer,* has been successful in several experimental models, such as the transfer of the chloride-channel transmembrane regulator gene to lung cells by aerosolization into the airway to treat cystic fibrosis, and the injection of a "healthy" dystrophin gene for the treatment of muscular dystrophy into skeletal muscle.

Although gene therapy is in its infancy, several clinical trials involving a variety of creative approaches for the treatment of hereditary and acquired diseases are under way. For example, gene therapy is being used for the treatment of hereditary hypercholesterolemia caused by low-density lipoprotein receptor deficiency: the normal gene for the low-density lipoprotein receptor is added into resected defective hepatocytes, which are then injected into the liver. In addition, cancer therapies that attempt to enhance the immune response to tumors are in progress. These therapies involve the use of a retrovirus for the transfer of genes for interleukin-2 and tumor necrosis factor to tumor cells, or the use of liposomes for the transfer of human leukocyte-associated antigens by direct injection into tumors. Using analysis identical to that employed in the assessment of other kinds of pharmacotherapy, a major research effort is under way to determine the utility of gene therapy for the treatment of many other diseases, including AIDS, leukemia, and hemophilia.

Fraunfelder FT. *Drug-Induced Ocular Side Effects and Drug Interactions.* 4th ed. Baltimore: Williams & Wilkins; 1996.

Lee PP, Yang JC. The nonapproved use of medications. *Ophthalmology.* 1991;99: 1071–1074.

Netland PA, Erickson KA. Calcium channel blockers in glaucoma management. *Ophthalmol Clin North Am.* 1995;8:327–334.

Samples JR. Benign essential blepharospasm. In: Fraunfelder FT, Roy FH, eds. *Current Ocular Therapy.* 4th ed. Philadelphia: Saunders; 1995.

Tripathi RC, Borisuth NSC, Tripathi BJ. Growth factors in the aqueous humor and their therapeutic implications in glaucoma and anterior segment disorders of the human eye. *Drug Dev Res.* 1991;22:1–23.

Tripathi RC, Tripathi BJ, Park JK, et al. Intracameral tissue plasminogen activator for resolution of fibrin clots after glaucoma filtering procedures. *Am J Ophthalmol.* 1991; 111:247–248.

CHAPTER XXII

Legal Aspects of Medical Therapy

The United States Food and Drug Administration (FDA) has statutory authority both to approve the marketing of prescription drugs and to specify the uses of these drugs. The FDA has created a three-step process regulating human testing of new drugs before they are approved for marketing. After animal and in vitro studies, *phase I* testing begins. This testing varies with the drug but generally involves 10–80 persons and collects toxicology and pharmacokinetic data concerning dosage range, absorption, metabolism, and toxicity. *Phase II* testing involves randomized, controlled clinical trials on a minimum of 50–100 affected persons to determine safety and effectiveness. A drug then enters *phase III* testing, which involves expanded controlled and uncontrolled trials, to evaluate the overall risk-benefit relationship and to provide an adequate basis for physician labeling. The data gathered from these tests are then submitted as part of a new drug application for marketing. The FDA's approval of each drug and its specific uses is based on documentation submitted by manufacturers that supports the safety and efficacy of specific drug applications.

Once approved for any use, a drug may be prescribed by individual physicians for any indication in all age groups without violating federal law. However, physicians remain liable to malpractice actions. In particular, a nonapproved use that does not adhere to an applicable standard of care places a practitioner in a difficult legal position. If a respectable minority of similarly situated physicians prescribe in the same manner, a standard of care could be met in most jurisdictions. In addition, informed consent in equivocal cases would be helpful.

The FDA has established clear guidelines on investigational drugs, and their use must meet specified commercial and investigative requirements.

BASIC TEXTS

Anatomy

Beard C, Quickert MH. *Anatomy of the Orbit: A Dissection Manual.* 3rd ed. Birmingham, AL: Aesculapius; 1988.

Bron AJ, Tripathi RC, Tripathi BJ, eds. *Wolff's Anatomy of the Eye and Orbit.* 8th ed. London: Chapman and Hall; 1997.

Duke-Elder S, ed. *System of Ophthalmology.* Vol II, *The Anatomy of the Visual System.* St Louis: Mosby; 1961.

Dutton JJ. *Atlas of Clinical and Surgical Orbital Anatomy.* Philadelphia: Saunders; 1994.

Fine BS, Yanoff M. *Ocular Histology: A Text and Atlas.* 2nd ed. Hagerstown, MD: Harper & Row; 1979.

Hogan MJ, Alvarado JA, Weddell JE. *Histology of the Human Eye.* Philadelphia: Saunders; 1971.

Mausolf FA. *The Anatomy of the Ocular Adnexa.* Springfield, IL: Thomas; 1975.

Miller NR, Newman NJ, eds. *Walsh and Hoyt's Clinical Neuro-Ophthalmology.* 5th ed. Baltimore: Williams & Wilkins; 1997.

Reeh MJ, Wobig JL, Wirtschafter JD. *Ophthalmic Anatomy.* San Francisco: American Academy of Ophthalmology; 1981.

Snell RS, Lemp MA. *Clinical Anatomy of the Eye.* Boston: Blackwell; 1997.

Tasman W, Jaeger EA, eds. *Duane's Clinical Ophthalmology.* Philadelphia: Lippincott; 2000.

Zide BM, Jelks GW, eds. Surgical Anatomy of the Orbit. New York: Raven Press; 1985.

Embryology

Jakobiec FA, ed. *Ocular Anatomy, Embryology, and Teratology.* Philadelphia: Harper & Row; 1982.

Genetics

Emery AEH, Rimoin DL, Connor JM, et al, eds. *Principles and Practice of Medical Genetics.* 3rd ed. New York: Churchill Livingstone; 1996.

McKusick VA. *Mendelian Inheritance in Man: A Catalog of Human Genes and Genetic Disorders.* 12th ed. Baltimore: Johns Hopkins University Press; 1998.

Renie WA, ed. *Goldberg's Genetic and Metabolic Eye Disease.* 2nd ed. Boston: Little, Brown; 1986.

Thompson MW, McInnes RR, Willard HF. *Genetics in Medicine.* 5th ed. Philadelphia: Saunders; 1991.

Traboulsi EI, ed. *Genetic Diseases of the Eye.* New York: Oxford University Press; 1998.

Biochemistry and Metabolism

Anderson RE, Hollyfield JG, LaVail MM, eds. *Retinal Degenerations.* Boca Raton, FL: CRC Press; 1991.

Berman ER. *Biochemistry of the Eye.* New York: Plenum Press; 1991.

LaVail MM, Anderson RE, Hollyfield JG, eds. *Inherited and Environmentally Induced Retinal Degenerations.* New York: Liss; 1989.

Shichi H. *Biochemistry of Vision.* New York: Academic Press; 1983.

Sies H, ed. Antioxidants in Disease Mechanisms and Therapy. Orlando, FL: Academic Press; 1996.

Ocular Pharmacology

Gilman AG, Goodman LS, Gilman A, eds. *Goodman and Gilman's The Pharmacological Basis of Therapeutics.* 8th ed. New York: Pergamon Press; 1990.

Sears ML, ed. *Handbook of Experimental Pharmacology: Pharmacology of the Eye.* New York: Springer-Verlag; 1984.

RELATED ACADEMY MATERIALS

Focal Points: Clinical Modules for Ophthalmologists

Brazis PW, Lee AG. Neuro-ophthalmic problems caused by medications (Module 11, 1998).

Hodge WG, Hwang DG. Antibiotic use in corneal and external eye infections (Module 10, 1997).

Jampol LM. Nonsteroidal anti-inflammatory drugs: 1997 update (Module 6, 1997).

Mieler WF. Systemic therapeutic agents and retinal toxicity (Module 12, 1997).

Serle JB, Podos SM. New therapeutic options for the treatment of glaucoma (Module 5, 1999).

Tessler HH, Goldstein DA. Update on systemic immunosuppressive agents (Module 11, 2000).

Publications

Bradford CA, ed. *Basic Ophthalmology for Medical Students and Primary Care Residents* (1999).

Jordan DR, Anderson RA. *Surgical Anatomy of the Ocular Adnexa: A Clinical Approach* (Ophthalmology Monograph 9, 1996).

Mannis MJ, Smith ME. *Case Studies in Ophthalmology for Medical Students.* 2nd ed (1993).

Reinecke RD, Farrell TA. *Fundamentals of Ophthalmology: A Programmed Text* (1987).

Wilson FM II, ed. *Practical Ophthalmology: A Manual for Beginning Residents.* 4th ed (1996).

Slide-Script

Tang R. *Ocular Manifestations of Systemic Disease* (Eye Care Skills for the Primary Care Physician Series, 1996).

Preferred Practice Patterns

Preferred Practice Patterns Committee. *Quality of Ophthalmic Care* (1988).

Ethics

AAO Ethics Committee. *The AAO Code of Ethics and You* (1987).

AAO Ethics Committee. *Clinical Trials and Investigative Procedures* (1992).

AAO Ethics Committee. *Code of Ethics* (reviewed annually).

AAO Ethics Committee. *The Ethical Ophthalmologist: A Primer* (1993).

AAO Ethics Committee. *Ethics in Ophthalmology: A Practical Guide* (reviewed annually).

To order any of these materials, please call the Academy's Customer Service number at (415) 561-8540.

CREDIT REPORTING FORM

BASIC AND CLINICAL SCIENCE COURSE
Section 2
2001–2002

CME Accreditation

The American Academy of Ophthalmology is accredited by the Accreditation Council for Continuing Medical Education to provide continuing medical education for physicians.

The American Academy of Ophthalmology designates this CME activity for up to 40 hours in category 1 credit toward the AMA Physician's Recognition Award. Each physician should claim only those hours of credit that he/she has actually spent in the educational activity.

If you wish to claim continuing medical education credit for your study of this section, you must send this page and the following 3 pages (by mail or FAX) to the Academy office. Please make sure to:

1. Fill in and sign the statement below.
2. Write your answers to the questions on the back of this form.
3. Complete the study questions and mark your answers on the Section Completion Form.
4. Complete the Section Evaluation.

Important: These completed forms must be received at the Academy within 3 years of purchase.

I hereby certify that I have spent _____ (up to 40) hours of study on the curriculum of this section and that I have completed the study questions. (The Academy, *upon request,* will send you a verification of your Academy credits earned within the last 3 years.)

☐ *Please send credit verification now.*

Signature ____
Date

Name: ____

Address: ____

City and State: ______________________________ Zip: ______________

Telephone: (_______) ________________ *Academy Member ID# ______________
area code

* *Your ID number is located following your name on any Academy mailing label and on your Monthly Statement of Account.*

Please return completed form to: **American Academy of Ophthalmology**
P.O. Box 7424
San Francisco, CA 94120-7424
ATTN: Clinical Education Division

READER'S RESPONSES

1. Please list several ways in which your study of this section will affect your practice.

2. How can we improve this section to better meet your continuing educational needs?

3. (OPTIONAL) Please list any topics you would like to see covered in other Academy clinical education products.

2001–2002

SECTION COMPLETION FORM

BASIC AND CLINICAL SCIENCE COURSE

ANSWER SHEET FOR SECTION 2

Question	Answer	Question	Answer	Question	Answer
1	a b c d	19	a b c d	36	a b c d
2	a b c d	20	a b c d	37	a b c d
3	a b c d	21	a b c d	38	a b c d
4	a b c d	22	a b c d	39	a b c d
5	a b c d	23	a b c d	40	a b c d
6	a b c d	24	a b c d	41	a b c d
7	a b c d	25	a b c d	42	a b c d
8	a b c d	26	a b c d	43	a b c d
9	a b c d	27	a b c d	44	a b c d
10	a b c d	28	a b c d	45	a b c d
11	a b c d	29	a b c d	46	a b c d
12	a b c d	30	a b c d	47	a b c d
13	a b c d	31	a b c d	48	a b c d
14	a b c d	32	a b c d	49	a b c d
15	a b c d	33	a b c d	50	a b c d
16	a b c d	34	a b c d	51	a b c d
17	a b c d	35	a b c d	52	a b c d
18	a b c d			53	a b c d

Please complete the Section Evaluation on the back of this page.

SECTION EVALUATION

Please indicate your response to the statements listed below by placing the appropriate number to the left of each statement.

1 = agree strongly
2 = agree
3 = no opinion
4 – disagree
5 = strongly disagree

_____ This section meets its stated objectives.

_____ This section helped me keep current on this topic.

_____ I will apply knowledge gained from this section to my practice.

_____ This section covers topics in sufficient depth and detail.

_____ This section's illustrations are of sufficient number and quality.

_____ The references included in the text provide an appropriate amount of additional reading.

_____ The study questions at the end of the book are useful.

STUDY QUESTIONS

STUDY QUESTIONS

The following multiple-choice questions are designed to be used after your course of study with this book. Record your responses on the answer sheet (the Section Completion Form) by circling the appropriate letter. For the most effective use of this exercise, *complete the entire test* before consulting the answers.

Although a concerted effort has been made to avoid ambiguity and redundancy in these questions, the authors recognize that differences of opinion may occur regarding the "best" answer. The discussions are provided to demonstrate the rationale used to derive the answer. They may also be helpful in confirming that your approach to the problem was correct or, if necessary, in fixing the principle in your memory.

1. Which one of the following extraocular muscles does *not* originate in the orbital apex?

   a. Superior rectus
   b. Inferior rectus
   c. Superior oblique
   d. Inferior oblique

2. Which one of the following nerve fibers synapses in the ciliary ganglion?

   a. Sympathetic fibers of the carotid plexus
   b. Parasympathetic fibers of cranial nerve III (oculomotor)
   c. Motor fibers of cranial nerve III to the inferior oblique muscle
   d. Sensory fibers of cranial nerve V (trigeminal)

3. Which one of the following rectus muscle tendons inserts closest to the corneal limbus?

   a. Medial rectus
   b. Lateral rectus
   c. Superior rectus
   d. Inferior rectus

4. The posterior lamella of the upper eyelid contain all *except* which one of the following?

   a. Levator aponeurosis
   b. Müller's muscle
   c. Orbicularis muscle
   d. Conjunctival lining

5. The mucin-producing tissue or cells of the eye are which one of the following?

   a. Goblet cells
   b. Glands of Krause and Wolfring
   c. Glands of Zeis and Moll
   d. Meibomian glands

6. Which one of the following has the correct sequence of angle structures in the anterior segment of the eye, proceeding from anterior to posterior?

   a. Corneal endothelium, scleral spur, trabecular meshwork, Schwalbe's line, ciliary recess
   b. Ciliary recess, trabecular meshwork, Schwalbe's line, scleral spur, corneal endothelium
   c. Corneal endothelium, Schwalbe's line, trabecular meshwork, scleral spur, ciliary recess
   d. Ciliary recess, Schwalbe's line, trabecular meshwork, scleral spur, corneal endothelium

7. Retinal pigment epithelial functions include all *except* which one of the following?

   a. Vitamin A metabolism
   b. Formation of the basal lamina
   c. Heat exchange
   d. Conversion of 11-*cis*-retinaldehyde to all-*trans*-retinaldehyde

8. Transmission of neuronal impulses in the retina from photoreceptor rods and cones to the optic nerve follows which one of the following sequences?

   a. Outer nuclear layer, inner nuclear layer, ganglion cell layer, ganglion cell nerve fiber layer
   b. Ganglion cell nerve fiber layer, ganglion cell layer, inner nuclear layer, outer nuclear layer
   c. Inner nuclear layer, outer nuclear layer, ganglion cell layer, ganglion cell nerve fiber layer
   d. Outer nuclear layer, inner nuclear layer, ganglion cell nerve fiber layer, ganglion cell layer

9. Which one of the following cranial nerves is *not* a true nerve?

   a. Cranial nerve II (optic)
   b. Cranial nerve III (oculomotor)
   c. Cranial nerve IV (trochlear)
   d. Cranial nerve V (trigeminal)

10. Which one of the following cranial nerves exits from the dorsal aspect of the midbrain?

   a. Cranial nerve III (oculomotor)
   b. Cranial nerve IV (trochlear)
   c. Cranial nerve V (trigeminal)
   d. Cranial nerve VI (abducens)

11. A lesion in the lower, posterior aspect of the left occipital lobe is most likely to affect which quadrant of visual field in the right eye?

a. Upper nasal
b. Lower nasal
c. Upper temporal
d. Lower temporal

12. Aneurysms that affect the oculomotor nerve (cranial nerve III) commonly occur at the junction of which two arteries?

a. Internal and external carotid arteries
b. Anterior cerebral and anterior communicating arteries
c. Basilar and posterior cerebral arteries
d. Posterior communicating and internal carotid arteries

13. Postganglionic parasympathetic fibers for reflex lacrimation are carried in which one of the following cranial nerves?

a. Cranial nerve III (oculomotor)
b. Cranial nerve IV (trochlear)
c. Cranial nerve V (trigeminal)
d. Cranial nerve VI (abducens)

14. An expanding lesion in the cavernous sinus is *least* likely to affect which one of the following nerves?

a. Ophthalmic division of cranial nerve V (trigeminal)
b. Maxillary division of cranial nerve V
c. Mandibular division of cranial nerve V
d. Cranial nerve VI (abducens)

15. Reiger syndrome includes all *except* which one of the following?

a. Oligodontia
b. Maxillary hypoplasia
c. Posterior keratoconus
d. Posterior embryotoxon

16. Which one of the following statements about colobomas is *not* correct?

a. Colobomas are inherited as an autosomal recessive trait.
b. Colobomas may involve the retina.
c. Colobomas may involve the iris.
d. Colobomas can create an absolute scotoma in a visual field.

17. Clinical findings in congenital rubella, acquired in the first trimester, may include all *except* which one of the following?

a. Deafness
b. Pigmentary retinopathy
c. Cataract
d. Facial maldevelopment

18. Bilateral optic nerve hypoplasia may occur in conjunction with which one of the following clinical findings?

    a. Absence of the septum pellucidum
    b. Glaucoma
    c. Cataract
    d. Pigmentary retinopathy

19. Persistent hyperplastic primary vitreous is *not* associated with which one of the following?

    a. Cataract
    b. Peripheral retinal neovascularization
    c. Microphthalmos
    d. Retinal detachment

20. Which one of the following statements is *not* correct about oculodermal melanocytosis (nevus of Ota)?

    a. It is occasionally associated with glaucoma.
    b. Hyperpigmentation may involve the uveal tract.
    c. It is usually unilateral.
    d. There is no risk of malignant melanoma.

21. Which one of the following gene groups is a homeotic gene group that appears to regulate activity or expression of genes guiding embryonic development of cells into body parts?

    a. *PAX*
    b. *RET*
    c. *GLC*
    d. *BAX*

22. Which one of the following statements about the anti-oncogene p53 is correct?

    a. It codes for a protein with tumor antiangiogenesis effect.
    b. It prevents cells from proliferating if their DNA is irreparably damaged.
    c. It codes for a protein with direct anti-tumor effect.
    d. It stimulates tumorigenesis.

23. Which one of the following disorders is *not* associated with a defect in a mitochondrial gene?

    a. Leber hereditary optic neuropathy
    b. Chronic progressive external ophthalmoplegia
    c. Neuropathy, ataxia, and retinitis pigmentosa (NARP)
    d. Retinoblastoma

24. Which one of the following is an indication for chromosomal analysis?

   a. Clinical diagnosis in newborns affected by multisystem malformations
   b. Multiple miscarriages
   c. Studies of malignancies
   d. All of the above

25. Which one of the following is *not* correct about retinoblastoma?

   a. It may be associated with chromosome 13 long arm (13q14) deletion syndrome.
   b. It affects between 1 and 15,000–34,000 live births in the United States.
   c. About 50% of patients with hereditary retinoblastoma have a family history of the disease.
   d. The hereditary pattern of familial retinoblastoma is autosomal dominant.

26. Tay-Sachs disease is associated with all *except* which one of the following?

   a. Autosomal dominant inheritance
   b. Macular cherry-red spot
   c. Optic atrophy
   d. A defect in the enzyme hexosaminidase

27. Homocystinuria is associated with which one of the following ocular findings?

   a. Glaucoma
   b. Cataract
   c. Dislocated lens
   d. Retinitis pigmentosa

28. Which one of the following statements is *not* correct about autosomal recessive inheritance?

   a. The mutant gene usually causes a milder form of the disease in the heterozygote.
   b. The ratio of normal to affected in siblings is 3:1.
   c. Both genders are equally affected.
   d. Affected individuals may have children who are asymptomatic carriers of the gene.

29. A certain ocular disorder is found to occur more often in males than females, but affected males do not transmit the disorder. However, virtually every son and daughter of an affected female inherits the trait. What is the most likely mode of inheritance of this disorder?

   a. Autosomal dominant
   b. Autosomal recessive
   c. X-linked recessive
   d. Mitochondrial

30. Which is associated with a missing chromosome?

a. Trisomy 13 (Patau) syndrome
b. Trisomy 21 (Down) syndrome
c. Turner syndrome
d. Klinefelter syndrome

31. Which one of the following can be a mutagen?

a. Chemicals
b. Radiation
c. Viruses
d. All of the above

32. Which one of the following immunoglobulins is found in the tear film?

a. IgA
b. IgM
c. IgE
d. IgD

33. Which one of the following is found in a significantly higher concentration in the aqueous humor than in blood?

a. Sodium (Na+)
b. Potassium (K+)
c. Bicarbonate (HCO_3^-)
d. Ascorbic acid (vitamin C)

34. Which one of the following is an α_1-adrenergic agonist?

a. Phenylephrine (Neo-Synephrine)
b. Apraclonidine (Iopidine)
c. Betaxolol (Betoptic)
d. Thymoxamine

35. The blood–aqueous barrier has the greatest effect on aqueous concentration compared with blood concentration of which one of the following?

a. Protein
b. Ascorbic acid (vitamin C)
c. Bicarbonate
d. Glucose

36. Which sugar has been implicated in cataractogenesis in diabetics?

a. Galactose
b. Glucose
c. Sorbitol
d. Mannitol

37. Which one of the following is the largest constituent of vitreous?

a. Collagen
b. Hyaluronic acid
c. Sodium
d. Water

38. Photoreceptor cones have one of three visual pigments with absorptive maxima of 570 nm, 540 nm, and 440 nm. Which one of these would absorb blue light the most?

a. 570 nm
b. 540 nm
c. 440 nm
d. None of the above

39. Cone density is greatest in which area of the retina?

a. Macula
b. Peripapillary region
c. Arcuate regions
d. Peripheral retina

40. Which vitamin is most critical for photoreceptor response to light?

a. A
b. B
c. C
d. D

41. Functions of the retinal pigment epithelium (RPE) include all *except* which one of the following?

a. Maintenance of retinal adhesion
b. Development of photoreceptors during embryogenesis
c. Transport of metabolites to and from the retina
d. Transmission of light-generated neuronal impulses from photoreceptors to the ganglion cells

42. Methods of increasing drug penetration through the normal cornea include all except which one of the following?

a. Increasing drug concentration
b. Reducing drug lipid solubility
c. Adding a surfactant (benzalkonium chloride)
d. Adding a high-viscosity agent (methylcellulose, polyvinyl alcohol)

43. Clinical effects of direct-acting muscarinic agents (miotics) include all *except* which one of the following?

   a. Increased myopia
   b. Decreased range of accommodation
   c. Central anterior chamber shallowing
   d. Reduced night vision

44. Which one of the following is a direct-acting cholinergic agent?

   a. Pilocarpine (Isoptocarpine, Pilocar, Pilostat)
   b. Echothiophate iodide (Phospholine iodide)
   c. Physostigmine (Escrine)
   d. Demecarium bromide (Humorsol)

45. Which one of the following statements is correct about cocaine?

   a. It directly stimulates adrenergic receptors.
   b. It directly stimulates muscarinic receptors.
   c. It blocks re-uptake of norepinephrine.
   d. It blocks adrenergic receptors.

46. Dapiprazole hydrochloride (Rev-Eyes) most effectively reverses the effect of which one of the following agents?

   a. Tropicamide (Mydriacyl)
   b. Cyclopentolate (Cyclogyl)
   c. Scopolamine (Hyoscine)
   d. Phenylephrine (Neo-Synephrine)

47. Which one of the following antiglaucoma agents is preferred in moderate to severe congestive heart failure?

   a. Timolol (Timoptic, Betimol)
   b. Betaxolol (Betoptic)
   c. Carteolol (Ocupress)
   d. None of the above

48. Which one of the following antiglaucoma agents is preferred in mild asthma?

   a. Timolol (Timoptic, Betimol)
   b. Betaxolol (Betoptic)
   c. Carteolol (Ocupress)
   d. None of the above

49. Systemic side effects of oral carbonic anhydrase inhibitors include all *except* which one of the following?

   a. Paresthesias
   b. Weight loss
   c. Hyperkalemia
   d. Metabolic alkalosis

50. Which one of the following is *not* a mast cell stabilizer or antihistamine?

a. Lodoxamide (Alomide)
b. Cromolyn sodium (Crolom)
c. Olopatadine hydrochloride (Patanol)
d. Ketorolac tromethamine (Acular)

51. Compared with first-generation cephalosporins, third-generation cephalosporins have enhanced activity against all but which of the following organisms?

a. Gram-positive cocci
b. Gram-negative bacilli
c. *Haemophilus influenzae*
d. *Pseudomonas aeruginosa*

52. Antiviral agents include all *except* which one of the following?

a. Trifluridine (Viroptic)
b. Amphotericin-B
c. Vidarabine (ara-A)
d. Idoxuridine (Stoxil)

53. Which one of the following local/regional anesthetic agents has the longest duration of effect?

a. Lidocaine
b. Tetracaine
c. Bupivacaine
d. Mepivicaine

ANSWERS

1. Answer—d. The inferior oblique muscle does not originate in the orbital apex.

2. Answer—b. Each of the four nerves passes through the ciliary ganglion, but only parasympathetic fibers of cranial nerve III (oculomotor) to the iris sphincter and ciliary muscle synapse in the ciliary ganglion.

3. Answer—a. The tendon of the medial rectus muscle inserts closest to the limbus. After that, in order, are the inferior, lateral, and superior rectus muscles.

4. Answer—c. The posterior lamella of the upper eyelid include the conjunctiva, levator aponeurosis, and Müller's muscle. The skin, vascular structures, and orbicularis muscle are in the anterior lamella.

5. Answer—a. Goblet cells produce mucin. The glands of Krause and Wolfring produce tears, and the glands of Zeiss and Moll and the meibomian glands produce oil.

6. Answer—c.

7. Answer—d. The conversion of 11-*cis*-retinaldehyde to all-*trans*-retinaldehyde occurs in the photoreceptor outer segments and not the retinal pigment epithelium.

8. Answer—a. After photoreceptor stimulation, visual system impulses travel from the outer nuclear layer, to the inner nuclear layer, to the ganglion cell layer, and finally to the ganglion cell nerve fibers (axons).

9. Answer—a. The optic nerve (cranial nerve II) is a tract of the brain and, unlike the other options, is not a true peripheral nerve.

10. Answer—b. Cranial nerve IV (trochlear) exits from the dorsal aspect of the midbrain. The other three exit ventrally.

11. Answer—c. A lesion in the posterior aspect of the left occipital lobe is most likely to produce a congruous right upper quadrant anoptic visual field defect. This would be in the upper temporal quadrant of the visual field of the right eye and the upper nasal quadrant of the visual field of the left eye.

12. Answer—d. Aneurysms at the junction of the posterior communicating and internal carotid arteries may affect cranial nerve III (oculomotor).

13. Answer—c. The lacrimal nerve, of the first (ophthalmic) division of cranial nerve V, carries parasympathetic nerves to the lacrimal gland.

14. Answer—c. The third (mandibular) division of cranial nerve V (trigeminal) does not pass through the cavernous sinus.

15. Answer—c. Posterior keratoconus is not seen in Reiger syndrome. The other conditions are all seen in Reiger syndrome.

16. Answer—a. Colobomas may involve the iris, choroid, retina, and optic nerve, and they may cause a large, dense scotoma. Colobomas are inherited as an autosomal dominant trait with incomplete penetrance and expressivity.

17. Answer—d. Cataract, deafness, and cardiac defects are the classic triad of congenital rubella contracted in the first trimester of pregnancy. Pigmentary retinopathy may also be seen. Facial maldevelopment is not a finding.

18. Answer—a. Bilateral optic nerve hypoplasia and absent septum pellucidum is de Morsier syndrome. Bergmeister's papilla and a Mittendorf dot may be seen, as may a serous detachment of the retina.

19. Answer—b. Persistent hyperplastic primary vitreous (PHPV) is not associated with peripheral retinal neovascularization. Peripheral retinal neovascularization may be seen on retinopathy of prematurity and sickle cell retinopathy. Cataract, microphthalmia, retinal detachment, and glaucoma are seen in PHPV.

20. Answer—d. Malignant melanomas can arise from nevi later in life in patients with oculodermal melanocytosis.

21. Answer—a. The *PAX* genes are homeotic genes. *PAX6* is seen in aniridia and Peters anomaly, and *PAX2* is seen in coloboma of the optic nerve. *GLC* is associated with open-angle glaucoma. *RET* is associated with thyroid cancer and multiple endocrine neoplasia types IIA and IIB. *BAX* gene codes for a protein that promotes apoptosis.

22. Answer—b. Anti-oncogene p53 prevents cells from proliferating if their DNA is irreparably damaged. Mutations of p53 predispose to tumorigenesis.

23. Answer—d. The hereditary pattern in familial retinoblastoma is autosomal dominant. The other three conditions appear to be a result of a defect in mitochondrial genes.

24. Answer—d. All three listed are potential indications for chromosomal analyses.

25. Answer—c. Only about 10% of patients with hereditary retinoblastoma have a family history of the disease. The remaining 90% appear to have a new mutation in their germ cells.

26. Answer—a. Tay-Sachs disease is an autosomal recessive genetic disorder with a defect in the enzyme hexosaminidase. Heterozygotes are asymptomatic.

27. Answer—c. Homocystinuria is associated with dislocated lens.

28. Answer—a. In most autosomal recessive disorders, the heterozygote is asymptomatic.

29. Answer—a. Mitochondrial gene disorders are transmitted only via maternal lines.

30. Answer—c. Of the four options, only Turner syndrome (XO) involves a chromosomal deletion. The others all involve a trisomy, which in Klinefelter syndrome is XXY.

31. Answer—d. Each of the three choices can cause mutagenesis.

32. Answer—a. IgA and IgG are found in the tear film.

33. Answer—d. Ascorbic acid (vitamin C) is secreted into aqueous by active transport and is found in a much higher concentration in the aqueous than in blood.

34. Answer—a. Phenylephrine (Neo-Synephrine) is an α_1-adrenergic agonist, and thymoxamine (not available in the United States) is an α_1-adrenergic antagonist. Apraclonidine (Iopidine) is an α_2-adrenergic antagonist, and betaxolol (Betoptic) is a selective β_1-adrenergic antagonist.

35. Answer—a. The blood–aqueous barrier serves to block entry of serum protein into the anterior chamber except when the eye is inflamed.

36. Answer—c. Sorbitol accumulates in the lens, creating an osmotic load, and contributes to cataractogenesis in the diabetic.

37. Answer—d. The vitreous is 99% water with small amounts of collagen, hyaluronic acid, and solutes.

38. Answer—c. Blue light is predominantly absorbed by cones absorbing the shorter wavelengths (ie, 440 nm).

39. Answer—a. Cone photoreceptor density is greatest in the macular region of the retina.

40. Answer—a. Vitamin A, a metabolic precursor of 11-*cis*-retinaldehyde, is most necessary for the light-induced photoreceptor response.

41. Answer—d. Retinal pigment epithelium is not directly involved in the transmission of light-generated responses from the photoreceptors to the ganglion cells.

42. Answer—b. Increasing drug lipid solubility increases drug penetration through the normal intact cornea.

43. Answer—b. Miotic agents constrict the pupillary sphincter and the ciliary muscle. Increasing myopia and decreased central anterior chamber are a result of ciliary muscle contraction. Pupillary constriction causes decreased night vision but *increases* the range of accommodation (pinhole effect).

44. Answer—a. In this list, pilocarpine (Isoptocarpine, Pilocar, Pilostat) is the only direct-acting cholinergic agent.

45. Answer—c. Cocaine blocks re-uptake of norepinephrine at adrenergic receptor terminals. This increases the adrenergic response but does not directly stimulate or block receptor response.

46. Answer—d. Dapiprazole (Rev-Eyes) is an α_1-adrenergic antagonist and, as such, has its greatest effect in reversing the effect of phenylephrine (Neo-Synephrine). It has little effect against tropicamide (Mydriacyl) and the cycloplegic agents.

47. Answer—d. In moderate to severe congestive heart failure, both selective (betaxolol) and nonselective (carteolol, levobunolol, metipranolol, timolol) beta-adrenergic antagonists are contraindicated. An alternative glaucoma medication is preferred.

48. Answer—b. In mild reactive airway disease (asthma), the selective β_1-adrenergic antagonist, betaxolol, is preferred over the other four currently available FDA-approved nonselective beta-adrenergic antagonists. In moderate to severe asthma, both selective and nonselective beta-adrenergic agents should be avoided.

49. Answer—d. Oral carbonic anhydrase inhibitors can have many systemic side effects. One of them is a systemic metabolic acidosis, not alkalosis.

50. Answer—d. Ketorolac tromethamine (Acular) is a nonsteroidal anti-inflammatory agent. Each of the others is a mast cell stabilizer, is an antihistamine, or combines the effects of both.

51. Answer—a. First-generation cephalosporins (cephalothin, cephalexin, cefazolin, and cephradine) have greater activity against gram-positive cocci than third-generation cephalosporins (cefotaxime, moxalactam, cefoperazone, ceftizoxime). However, third-generation cephalosporins have greater gram-negative activity.

52. Answer—b. Amphotericin-B is an antifungal agent.

53. Answer—c. Bupivacaine (Marcaine) has the longest duration of effect (8–12 hours) of the local/regional agents listed.

INDEX

ILLUSTRATIONS

The authors submitted the following figures for this revision. (Illustrations that were reproduced from other sources or submitted by contributors not on the committee are credited in the captions.)

Ata A. Abdel-Latif, PhD: Figs XII-1, XII-2

Gerhard W. Cibis, MD: Fig VI-16

Ramesh C. Tripathi, MD, PhD: Figs III-9, III-11